Directed Algebraic Topology

This is the first authored book to be dedicated to the new field of directed algebraic topology that arose in the 1990s, in homotopy theory and in the theory of concurrent processes. Its general aim can be stated as 'modelling non-reversible phenomena' and its domain should be distinguished from that of classical algebraic topology by the principle that directed spaces have privileged directions and directed paths therein need not be reversible. Its homotopical tools (corresponding in the classical case to ordinary homotopies, fundamental group and fundamental groupoid) should be similarly 'non-reversible': directed homotopies, fundamental monoid and fundamental category. Homotopy constructions occur here in a directed version, which gives rise to new 'shapes', like directed cones and directed spheres. Applications deal with domains where privileged directions appear, including rewrite systems, traffic networks and biological systems. The most developed examples can be found in the area of concurrency.

MARCO GRANDIS is Professor in the Department of Mathematics at the University of Genoa, Italy.

NEW MATHEMATICAL MONOGRAPHS

All the titles listed below can be obtained from good booksellers or from Cambridge University Press. For a complete series listing visit http://www.cambridge.org/uk/series/sSeries.asp?code=NMM

Directed Algebraic Topology

Models of Non-Reversible Worlds

MARCO GRANDIS

Università degli Studi di Genova, Italy

CAMBRIDGE UNIVERSITY PRESS
Cambridge, New York, Melbourne, Madrid, Cape Town, Singapore, São Paulo, Delhi

Cambridge University Press
The Edinburgh Building, Cambridge CB2 8RU, UK

Published in the United States of America by Cambridge University Press, New York

www.cambridge.org
Information on this title: www.cambridge.org/9780521760362

© M. Grandis 2009

First published 2009

Printed in the United Kingdom at the University Press, Cambridge

A catalogue record for this publication is available from the British Library

ISBN 978-0-521-76036-2 Hardback

To

Maria Teresa
and
Marina

Contents

Introduction

1 Aims and applications

Directed algebraic topology is a recent subject which arose in the 1990s, on the one hand in abstract settings for homotopy theory, like [G1], and, on the other hand, in investigations in the theory of concurrent processes, like [FGR1, FGR2]. Its general aim should be stated as 'modelling non-reversible phenomena'. The subject has a deep relationship with category theory.

The domain of directed algebraic topology should be distinguished from the domain of classical algebraic topology by the principle that *directed spaces have privileged directions and directed paths therein need not be reversible*. While the classical domain of topology and algebraic topology is a reversible world, where a path in a space can always be travelled backwards, the study of non-reversible phenomena requires broader worlds, where a directed space can have non-reversible paths.

The homotopical tools of directed algebraic topology, corresponding in the classical case to ordinary homotopies, the fundamental group and fundamental n-groupoids, should be similarly 'non-reversible': *directed homotopies, the fundamental monoid* and *fundamental n-categories*. Similarly, its homological theories will take values in 'directed' algebraic structures, like *preordered* abelian groups or abelian *monoids*. Homotopy constructions like mapping cone, cone and suspension, occur here in a directed version; this gives rise to new 'shapes', like (lower and upper) directed cones and directed spheres, whose elegance is strengthened by the fact that such constructions are determined by universal properties.

Applications will deal with domains where privileged directions appear, such as concurrent processes, rewrite systems, traffic networks, space-time models, biological systems, etc. At the time of writing, the

1

most developed ones are concerned with concurrency: see [FGR1, FGR2, FRGH, Ga1, GG, GH, Go, Ra1, Ra2].

A recent issue of the journal *Applied Categorical Structures*, guest-edited by the author, has been devoted to Directed Algebraic Topology and Category Theory (Vol. 15, no. 4, 2007).

2 Some examples

As an elementary example of the notions and applications we are going to treat, consider the following (partial) order relation in the cartesian plane

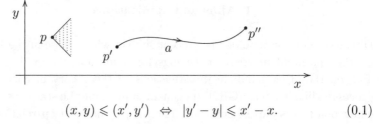

$$(x, y) \leqslant (x', y') \quad \Leftrightarrow \quad |y' - y| \leqslant x' - x. \tag{0.1}$$

The picture shows the 'cone of the future' at a point p (i.e. the set of points which follow it) and a *directed path* from p' to p'', i.e. a continuous mapping $a \colon [0, 1] \to \mathbf{R}^2$ which is (weakly) *increasing*, with respect to the natural order of the standard interval and the previous order of the plane: if $t \leqslant t'$ in $[0, 1]$, then $a(t) \leqslant a(t')$ in the plane.

Take now the following (compact) subspaces X, Y of the plane, with the induced order (the cross-marked open rectangles are taken out). A directed path in X or Y satisfies the same conditions as above

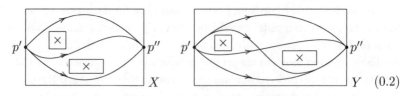
$$\tag{0.2}$$

We shall see that – as displayed in the illustrations above – there are, respectively, three or four 'homotopy classes' of directed paths from the point p' to the point p'', in the fundamental categories $\uparrow\Pi_1(X)$, $\uparrow\Pi_1(Y)$; in both cases there are none from p'' to p', and every loop is constant. (The prefixes $\uparrow$ and d- are used to distinguish a directed notion from the corresponding 'reversible' one.)

First, we can view each of these 'directed spaces' as a stream with two islands, and the induced order as an upper bound for the relative velocity feasible in the stream. Secondly, one can interpret the horizontal coordinate as (a measure of) time, the vertical coordinate as position in a one-dimensional physical medium, and the order as the possibility of going from (x, y) to (x', y') with velocity $\leqslant 1$ (with respect to a 'rest frame' of the medium). The two forbidden rectangles are now linear obstacles in the medium, with a bounded duration in time. Thirdly, our illustrations can be viewed as execution paths of concurrent automata subject to some conflict of resources, as in [FGR2], fig. 14.

In all these cases, the fundamental category distinguishes between obstructions (islands, temporary obstacles, conflict of resources) that intervene essentially together (in the earlier diagram on the left) or one after the other (on the right). On the other hand, the underlying topological spaces are homeomorphic, and topology, or algebraic topology, cannot distinguish these two situations. Notice also that, here, all the fundamental monoids $\uparrow\pi_1(X, x_0)$ are trivial: as a striking difference with the classical case, the fundamental monoids often carry a very minor part of the information of the fundamental category $\uparrow\Pi_1(X)$.

The study of the fundamental category of a directed space, via minimal models up to directed homotopy of categories, will be developed in Chapter 3.

3 Directed spaces and other directed structures

The framework of ordered topological spaces is a simple starting point but is too poor to develop directed homotopy theory.

We want a 'world' sufficiently rich to contain a 'directed circle' $\uparrow\mathbf{S}^1$ and higher directed spheres $\uparrow\mathbf{S}^n$ – all of them arising from the discrete two-point space under directed suspension (of pointed objects). In $\uparrow\mathbf{S}^1$, directed paths will move in a particular direction, with fundamental monoids $\uparrow\pi_1(\uparrow\mathbf{S}^1, x_0) \cong \mathbf{N}$; its directed homology will give $\uparrow H_1(\uparrow\mathbf{S}^1) \cong \uparrow\mathbf{Z}$, i.e. the group of integers *equipped with the natural order*, where the positive homology classes are generated by cycles which are directed paths (or, more generally, positive linear combinations of directed paths).

Our main structure, to fulfil this goal, will be a topological space X *equipped* with a set dX of *directed paths* $[0, 1] \to X$, closed under: constant paths, partial increasing reparametrisation and concatenation

(Section 1.4). Such objects are called *d-spaces* or *spaces with distinguished paths*, and a morphism of d-spaces $X \to Y$ is a continuous mapping which preserves directed paths. All this forms a category d**Top** where limits and colimits exist and are easily computed – as topological limits or colimits, equipped with the adequate d-structure.

Furthermore, the standard *directed interval* $\uparrow\mathbf{I} = \uparrow[0,1]$, i.e. the real interval $[0,1]$ with the natural order and the associated d-structure, is an *exponentiable* object: in other words, the (directed) cylinder $I(X) = X \times \uparrow\mathbf{I}$ determines an object of (directed) paths $P(Y) = Y^{\uparrow\mathbf{I}}$ (providing the functor right adjoint to I), so that a directed homotopy can equivalently be defined as a map of d-spaces $IX \to Y$ or $X \to PY$. The underlying set of the d-space $P(Y)$ is the set of distinguished paths dY.

Various d-spaces of interest arise from an ordinary space equipped with an order relation, as in the case of $\uparrow\mathbf{I}$, the directed line $\uparrow\mathbf{R}$ and their powers; or, more generally, from a space equipped with a *local preorder* (Sections 1.9.2 and 1.9.3), as for the directed circle $\uparrow\mathbf{S}^1$. But other d-spaces of interest, which are able to build a bridge with non-commutative geometry, cannot be defined in this way: for instance, the quotient d-space of the directed line $\uparrow\mathbf{R}$ modulo the action of a dense subgroup (see Section 6 of this introduction).

The category **Cub** of cubical sets is also an important framework where directed homotopy can be developed. It actually has some advantages on d**Top**: in a cubical set K, after observing that an element of K_1 need not have any counterpart with reversed vertices, we can also note that an element of K_n need not have any counterpart with faces permuted (for $n \geqslant 2$). Thus, a cubical set has 'privileged directions', in any dimension. In other words, **Cub** allows us to break *both* basic symmetries of topological spaces, the reversion of paths and the transposition of variables in two-dimensional paths, parametrised on $[0,1]^2$, while d**Top** is essentially based on one-dimensional information and only allows us to break the symmetry of reversion. As a consequence, pointed directed homology of cubical sets is much better behaved than that of d-spaces, and yields a *perfect* directed homology theory (Section 2.6.3).

On the other hand, **Cub** presents various drawbacks, beginning with the fact that elementary paths and homotopies, based on the obvious interval, cannot be concatenated; however, higher homotopy properties of **Cub** can be studied with the geometric realisation functor **Cub** $\to$ d**Top** and the notion of *relative equivalence* that it provides (Section 5.8.6).

The *breaking of symmetries* is an essential feature that distinguishes directed algebraic topology from the classical one; a discussion of these aspects can be found in Section 1.1.5.

Directed homotopies have been studied in various structures, either because of general interest in homotopy theory, or with the purpose of modelling concurrent systems, or in both perspectives. Such structures comprise: differential graded algebras [G3], ordered or locally ordered topological spaces [FGR2, GG, Go, Kr], simplicial, precubical and cubical sets [FGR2, GG, G1, G12], inequilogical spaces [G11], small categories [G8], flows [Ga2], etc. Our main structure, d-spaces, was introduced in [G8]; it has also been studied by other authors, e.g. in [FhR, FjR, Ra2].

4 Formal foundations for directed algebraic topology

We will use settings based on an abstract *cylinder functor* $I(X)$ and natural transformations between its powers, like faces, degeneracy, connections, and so on. Or, dually, on a *cocylinder functor* $P(Y)$, representing the object of (directed) paths of an object Y. Or also, on an adjunction $I \dashv P$ that allows one to see directed homotopies as morphisms $I(X) \to Y$ or equivalently $X \to P(Y)$, as mentioned above for d-spaces.

As a crucial aspect, such a formal structure is based on endofunctors and 'operations' on them (natural transformations between their powers). In other words, it is 'categorically algebraic', in much the same way as the theory of monads, a classical tool of category theory (Section A4, in the Appendix). This is why such structures can generally be lifted from a ground category to categorical constructions on the latter, like categories of diagrams, or sheaves, or algebras for a monad (Chapter 5).

After a basic version in Chapter 1, which covers all the frameworks we are interested in, we develop stronger settings in Chapter 4. *Relative settings*, in Section 5.8, deal with a basic world, satisfying the basic axioms of Chapter 1, which is equipped with a forgetful functor with values in a strong framework; such a situation has already been mentioned above, for the category **Cub** of cubical sets and the (directed) geometric realisation functor **Cub** → d**Top**.

A peculiar fact of all 'directed worlds' (categories of 'directed objects') is the presence of an involutive covariant endofunctor R, called *reversor*, which turns a directed object into the *opposite* one, $R(X) = X^{op}$; its action on preordered spaces, d-spaces and (small) categories is obvious; for cubical sets, one interchanges lower and upper faces. Then, the ordinary reversion of paths is replaced with a *reflection* in the opposite directed object. Notice that the classical *reversible* case is a *particular instance* of the directed one, where R is the identity functor.

In the classical case, settings based on the cylinder (or path) endofunctor go back to Kan's well-known series on 'abstract homotopy', and in particular to [Ka2] (1956); the book [KP], by Kamps and Porter, is a general reference for such settings. In the directed case, the first occurrence of such a system, containing a reversor, is probably a 1993 paper of the present author [G1].

Quillen model structures [Qn] seem to be less suited to formalise directed homotopy. But, *in the reversible case*, we prove (in Theorem 4.9.6) that our strong setting based on the cylinder determines a structure of 'cofibration category', a non-self-dual version of Quillen's model categories introduced by Baues [Ba].

5 Interactions with category theory

On the one hand, category theory intervenes in directed algebraic topology through the fundamental category of a directed space, viewed as a sort of algebraic model of the space itself. On the other hand, directed algebraic topology can be of help in providing a sort of geometric intuition for category theory, in a sharper way than classical algebraic topology – the latter can rather provide intuition for the theory of groupoids, a reversible version of categories.

The interested reader can see, in Section 1.8.9, how the pasting of comma squares of categories only works up to convenient notions of 'directed homotopy equivalence' of categories – in the same way as, in **Top**, the pasting of homotopy pullbacks leads to homotopy equivalent spaces.

The relationship of directed algebraic topology and category theory is even stronger in 'higher dimensions'. It consists of higher fundamental categories for directed spaces, on the one hand, and geometric intuition for the – very complex – theory of higher dimensional categories, on the other. Such aspects are still under research and will not be treated in this book. The interested reader is referred to [G15, G16, G17] and references therein.

Finally, we should note that category theory has also been of help in fixing the structures that we explore here, according to general principles discussed in the Appendix, Section A1.6.

6 Interactions with non-commutative geometry

While studying the directed homology of cubical sets, in Chapter 2, we also show that cubical sets (and d-spaces) can express topological facts missed by ordinary topology and already investigated within non-commutative geometry. In this sense, they provide a sort of 'non-commutative topology', without the metric information of C*-algebras

This happens, for instance, in the study of group actions or foliations, where a topologically trivial quotient (the orbit set or the set of leaves) can be enriched with a natural cubical structure (or a d-structure) whose directed homology agrees with Connes' analysis in non-commutative geometry.

Let us only recall here that, if ϑ is an irrational number, $G_\vartheta = \mathbf{Z} + \vartheta\mathbf{Z}$ is a dense subgroup of the additive group $\mathbf{R}$, and the topological quotient $\mathbf{R}/G_\vartheta$ is trivial (has the indiscrete topology). Non-commutative geometry 'replaces' this quotient with the well-known irrational rotation C*-algebra A_ϑ (Section 2.5.1). Here we replace it with the cubical set $C_\vartheta = (\Box\!\uparrow\!\mathbf{R})/G_\vartheta$, a quotient of the singular cubical set of the directed line (or the quotient d-space $D_\vartheta = \uparrow\!\mathbf{R}/G_\vartheta$, cf. Section 2.5.2). Computing its directed homology, we prove that the (pre)ordered group $\uparrow\!H_1(C_\vartheta)$ is isomorphic to the totally ordered group $\uparrow\!G_\vartheta \subset \mathbf{R}$. It follows that the classification up to isomorphism of the family C_ϑ (or D_ϑ) coincides with the classification of the family A_ϑ up to strong Morita equivalence. Notice that, *algebraically* (i.e. forgetting order), we only get $H_1(C_\vartheta) \cong \mathbf{Z}^2$, which gives no information on ϑ: here, the information content provided by the ordering is much finer than that provided by the algebraic structure.

7 From directed to weighted algebraic topology

In Chapter 6 we end this study by investigating 'spaces' where paths have a 'weight', or 'cost', expressing length or duration, price, energy, etc. The general aim is now: measuring the cost of (possibly non-reversible) phenomena.

The weight function takes values in $[0, \infty]$ and is *not* assumed to be invariant up to path-reversion. Thus, 'weighted algebraic topology' can be developed as an enriched version of directed algebraic topology, where illicit paths are penalised with an infinite cost, and the licit ones are measured. Its algebraic counterpart will be 'weighted algebraic structures', equipped with a sort of directed seminorm.

A generalised metric space in the sense of Lawvere [Lw1] yields a prime structure for this purpose. For such a space we define a *fundamental weighted category*, by providing each homotopy class of paths with a weight, or seminorm, which is subadditive with respect to composition.

We also study a more general framework, *w-spaces* or *spaces with weighted paths* (a natural enrichment of d-spaces), *whose relationship with non-commutative geometry also takes into account the metric aspects* – in contrast with cubical sets and d-spaces. Here, the irrational rotation C*-algebra A_ϑ corresponds to the w-space $W_\vartheta = \mathrm{w}\mathbf{R}/G_\vartheta$, a quotient of the standard weighted line, whose classification up to isometric isomorphism (resp. Lipschitz isomorphism) is the same as the classification of A_ϑ up to isomorphism (resp. strong Morita equivalence).

8 Terminology and notation

The reader is assumed to be acquainted with the basic notions of topology, algebraic topology and category theory. However, most of the notions and results of category theory that are used here are recalled in the Appendix.

In a category $\mathbf{A}$, the set of morphisms (or maps, or arrows) $X \to Y$, between two given objects, is written as $\mathbf{A}(X, Y)$. A natural transformation between the functors $F, G\colon \mathbf{A} \to \mathbf{B}$ is written as $\varphi\colon F \to G\colon \mathbf{A} \to \mathbf{B}$, or $\varphi\colon F \to G$.

Top denotes the category of topological spaces and continuous mappings. A homotopy φ between maps $f, g\colon X \to Y$ is written as $\varphi\colon f \to g\colon X \to Y$, or $\varphi\colon f \to g$. $\mathbf{R}$ is the euclidean line and $\mathbf{I} = [0, 1]$ is the standard euclidean interval. The concatenation of paths and homotopies is written in additive notation: $a + b$ and $\varphi + \psi$; trivial paths and homotopies are written as $0_x, 0_f$. **Gp** (resp. **Ab**) denotes the category of groups (resp. abelian groups) and their homomorphisms.

Cat denotes the 2-category of small categories, functors and natural transformations. In a small category, the composition of two consecutive arrows $a\colon x \to x'$, $b\colon x' \to x''$ is either written in the usual notation ba or in additive notation $a + b$. In the first case, the identity of the object x is written as $\mathrm{id}\, x$ or 1_x, in the second as 0_x. Loosely speaking, we tend to use additive notation in the fundamental category of some directed object, or in a small category that is itself 'viewed' as a directed object; on the other hand, we follow the usual notation when we are applying the standard techniques of category theory, which would look unfamiliar in additive notation.

A *preorder* relation, generally written as $x \prec y$, is assumed to be reflexive and transitive; an *order*, often written as $x \leqslant y$, is also assumed to be antisymmetric (and need *not* be total). A mapping that preserves preorders is said to be *increasing* (always used in the weak sense). As usual, a preordered set X will be *identified* with the (small) category whose objects are the elements of X, with precisely one arrow $x \to x'$ when $x \prec x'$ and none otherwise. We shall distinguish between the ordered real line $\mathbf{r}$ and the ordered topological space $\uparrow\mathbf{R}$ (the euclidean line with the natural order), whose fundamental category is $\mathbf{r}$. $\uparrow\mathbf{Z}$ is the *ordered* group of integers, while $\mathbf{z}$ is the underlying ordered set.

The index α takes values $0, 1$; these are often written as $-, +$, e.g. in superscripts.

9 Acknowledgements

I would like to thank the many colleagues with whom I had helpful discussions about parts of the subject matter of this book, including J. Baez, L. Fajstrup, E. Goubault, E. Haucourt, W.F. Lawvere, R. Paré, T. Porter, and especially M. Raussen for his useful suggestions.

While preparing this text, I have much appreciated the open-minded policy of Cambridge University Press and the kind, effective, informal assistance of the editor, Roger Astley. The technical services at CUP have also been of much help.

I am grateful to the dear memory of Gabriele Darbo, who aroused my interest in algebraic topology and category theory.

This work was supported by MIUR Research Projects and by a research grant of Università di Genova.

Part I
First-order directed homotopy and homology

1

Directed structures and first-order homotopy properties

We begin by studying basic homotopy properties, which will be sufficient to introduce directed homology in the next chapter.

Section 1.1 explores the transition from classical to directed homotopy, comparing topological spaces with the simplest topological structure where privileged directions appear: the category p**Top** of preordered topological spaces.

In Sections 1.2 and 1.3 we begin a formal study of directed homotopy in a *dI1-category*, i.e. a category equipped with an abstract cylinder endofunctor I endowed with a basic structure. Dually, we have a *dP1-category*, with a cocylinder (or path) functor P, while a *dIP1-category* has both endofunctors, under an adjunction $I \dashv P$. The higher order structure, developed in Chapter 4, will make substantial use of the 'second-order' functors, I^2 or P^2.

Sections 1.4 and 1.5 introduce our main directed world, the category d**Top** of *spaces with distinguished paths*, or *d-spaces*, which – with respect to preordered spaces – also contains objects with non-reversible loops, like the directed circle $\uparrow S^1$. Then, in Section 1.6, we explore the category **Cub** of cubical sets and their left or right directed homotopy structures.

Coming back to the general theory, in Section 1.7, we deal with *dI1-homotopical categories*, i.e. dI1-categories which have a terminal object and all homotopy pushouts, and therefore also mapping cones and suspensions. This leads to the (lower or upper) cofibre sequence of a map, whose classical counterpart for topological spaces is the well-known Puppe sequence [Pu].

These results are dualised in Section 1.8, which is concerned with *dP1-homotopical categories*, homotopy pullbacks and the fibre sequence of a map. Pointed *dIP1-homotopical categories* combine both aspects and cover pointed preordered spaces and pointed d-spaces.

We end in Section 1.9 by discussing other topological settings for directed algebraic topology: inequilogical spaces, c-sets, generalised metric spaces, bitopological spaces, locally preordered spaces.

Note. The index α takes values $0, 1$; these are often written as $-, +$ (e.g. in superscripts).

1.1 From classical homotopy to the directed case

We explore here the transition from classical to directed homotopy, comparing topological spaces with the simpler topological structure where privileged directions appear: a preordered topological space (Section 1.1.3).

Small categories can also be interpreted as directed structures, viewing an arrow as a path and a natural transformation as a directed homotopy from one functor to another (Section 1.1.6).

1.1.0 The structure of the classical interval

In the category **Top** of topological spaces and continuous mappings, a path in the space X is a map $a \colon \mathbf{I} \to X$ defined on the standard interval $\mathbf{I} = [0, 1]$, with euclidean topology.

The basic, 'first-order' structure of $\mathbf{I}$ consists of four maps, linking it to its 0-th cartesian power, the singleton $\mathbf{I}^0 = \{*\}$

$$
\begin{aligned}
&\partial^\alpha \colon \{*\} \rightrightarrows \mathbf{I}, &&\partial^-(*) = 0, \;\; \partial^+(*) = 1 &&(\textit{faces}), \\
&e \colon \mathbf{I} \to \{*\}, &&e(t) = * &&(\textit{degeneracy}), \quad (1.1) \\
&r \colon \mathbf{I} \to \mathbf{I}, &&r(t) = 1 - t &&(\textit{reversion}).
\end{aligned}
$$

Identifying a point x of the space X with the corresponding map $x \colon \{*\} \to X$, this basic structure determines:

(a) the endpoints of a path $a \colon \mathbf{I} \to X$, $a\partial^- = a(0)$ and $a\partial^+ = a(1)$;
(b) the trivial path at the point x, which will be written as $0_x = xe$;
(c) the reversed path of a, written as $-a = ar$.

Two consecutive paths $a, b \colon \mathbf{I} \to X$ ($a\partial^+ = b\partial^-$, i.e. $a(1) = b(0)$) have a concatenated path $a + b$

$$
(a + b)(t) = \begin{cases} a(2t) & \text{if } 0 \leqslant t \leqslant 1/2, \\ a(2t - 1) & \text{if } 1/2 \leqslant t \leqslant 1. \end{cases} \quad (1.2)
$$

Formally, this can be expressed by saying that the *standard concatenation pushout* – pasting two copies of the interval, one after the other – is homeomorphic to $\mathbf{I}$ and can be realised as $\mathbf{I}$ itself

$$
\begin{array}{ccc}
\{*\} & \xrightarrow{\partial^+} & \mathbf{I} \\
\partial^- \downarrow & \overset{c^-}{\dashv} & \downarrow c^- \\
\mathbf{I} & \xrightarrow{c^+} & \mathbf{I}
\end{array}
\qquad
\begin{array}{l}
c^-(t) = t/2, \\[12pt]
c^+(t) = (t+1)/2.
\end{array}
\qquad (1.3)
$$

Indeed, the concatenated path $a + b\colon \mathbf{I} \to X$ comes from the universal property of the pushout, and is characterised by the conditions

$$(a+b).c^- = a, \qquad (a+b).c^+ = b. \qquad (1.4)$$

Finally, there is a 'second-order' structure which involves the standard square $\mathbf{I}^2 = [0,1] \times [0,1]$ and is used to construct homotopies of paths

$$
\begin{array}{llll}
g^-\colon \mathbf{I}^2 \to \mathbf{I}, & g^-(t,t') = \max(t,t') & (lower\ connection), \\
g^+\colon \mathbf{I}^2 \to \mathbf{I}, & g^+(t,t') = \min(t,t') & (upper\ connection), & (1.5) \\
s\colon \mathbf{I}^2 \to \mathbf{I}^2, & s(t,t') = (t',t) & (transposition).
\end{array}
$$

These maps, together with (1.1), complete the structure of $\mathbf{I}$ as an *involutive lattice* in **Top** (or, better, a 'dioid' with symmetries, see Section 1.1.7).

The choice of the superscripts of g^-, g^+ comes from the fact that the unit of g^α is $\partial^\alpha(*)$. Within homotopy theory, the importance of these binary operations has been highlighted by R. Brown and P.J. Higgins [BH1, BH3], who introduced the term *connection*, or *higher degeneracy*. Algebraically and categorically, the 'soundness' of introducing these operations is made evident by the notions of 'dioid' and 'diad', which correspond, respectively, to monoid and monad (Sections 1.1.7–1.1.9).

1.1.1 The cylinder

Given two continuous mappings $f, g\colon X \to Y$ (in **Top**), a homotopy $\varphi\colon f \to g$ 'is' a map $\varphi\colon X \times \mathbf{I} \to Y$ defined on the cylinder $I(X) = X \times \mathbf{I}$, which coincides with f on the lower basis of the cylinder and with g on the upper one

$$\varphi(x,0) = f(x), \qquad \varphi(x,1) = g(x) \qquad \text{(for all } x \in X). \quad (1.6)$$

This map will be written as $\hat{\varphi}\colon X \times \mathbf{I} \to Y$ when we want to distinguish it from the homotopy $\varphi\colon f \to g$ which it represents. All this is based on the *cylinder* endofunctor:

$$I: \mathbf{Top} \to \mathbf{Top}, \qquad\qquad I(X) = X \times \mathbf{I}, \qquad (1.7)$$

and on the four natural transformations which it inherits from the basic structural maps of the standard interval (written as the latter)

$$\partial^\alpha: X \to IX, \quad \partial^-(x) = (x,0), \ \partial^+(x) = (x,1) \qquad\quad (faces),$$
$$e: IX \to X, \quad e(x,t) = x \qquad\qquad\qquad\quad (degeneracy), \quad (1.8)$$
$$r: IX \to IX, \quad r(x,t) = (x,1-t) \qquad\qquad\quad (reversion).$$

Notice that we often define a functor by its action on objects, as in (1.7), provided its extension to morphisms is evident: $I(f) = f \times \mathrm{id}\mathbf{I}$ in the present case (also written $f \times \mathbf{I}$). As a general fact of notation, a component $\varphi X: FX \to GX$ of the natural transformation $\varphi: F \to G$ is often written as $\varphi: FX \to GX$, as above.

Identifying $I\{*\} = \{*\} \times \mathbf{I} = \mathbf{I}$, the structural maps of the standard interval coincide with the components of the transformations (1.8) on the singleton.

The transformations ∂^α, e, r give rise to the *faces* $\varphi \partial^\alpha$ of a homotopy, the *trivial homotopy* $0_f = fe$ of a map and the *reversed homotopy* $-\varphi = \varphi r$. (More precisely, the homotopy $-\varphi: g \to f$ is represented by the map $(-\varphi)\hat{} = \hat{\varphi} r: IX \to Y$.)

A path $a: \mathbf{I} \to X$ is the same as a homotopy $a: x \to x'$ between its endpoints (always by identifying $I\{*\} = \mathbf{I}$).

Two consecutive homotopies $\varphi: f \to g$, $\psi: g \to h$ can be concatenated, extending the procedure for paths, in (1.2). This can be formally expressed by noting that the *concatenation pushout of the cylinder* – pasting two copies of a cylinder IX, 'one on top of the other' – can be realised as the cylinder itself

$$
\begin{array}{ccc}
X & \xrightarrow{\ \partial^+\ } & IX \\
{\scriptstyle \partial^-}\downarrow & \quad\downarrow{\scriptstyle c^-} & \\
IX & \xrightarrow[\ c^+\ ]{} & IX
\end{array}
\qquad
\begin{array}{l}
c^-(x,t) = (x,t/2), \\[2ex]
\\
c^+(x,t) = (x,(t+1)/2).
\end{array}
\qquad (1.9)
$$

In fact, the subspaces $c^-(IX) = X \times [0,1/2]$ and $c^+(IX) = X \times [1/2,1]$ form a finite closed cover of IX, so that a mapping defined on IX is continuous if and only if its restrictions to such subspaces are. The concatenated homotopy $\varphi + \psi: f \to h$ is represented by the map $(\varphi + \psi)\hat{}: IX \to Y$ which reduces to φ on c^-, and to ψ on c^+.

Note that the fact that 'pasting two copies of the cylinder gives back the cylinder' is rather *peculiar of spaces*; e.g., it does not hold for chain complexes, where the concatenation of homotopies is based on a more general procedure, dealt with in Section 4.2; nor does it hold for the homotopy structure of **Cat**, cf. Sections 1.1.6 and 4.3.2.

Here also, there is a 'second-order' structure, with three natural transformations which involve the second-order cylinder $I^2(X) = I(I(X)) = X \times \mathbf{I}^2$ and will be used to construct higher homotopies (i.e. homotopies of homotopies)

$$
\begin{aligned}
&g^\alpha : I^2 X \to IX, \quad g^\alpha(x,t,t') = (x, g^\alpha(t,t')) \quad (connections), \\
&s : I^2 X \to I^2 X, \quad s(x,t,t') = (x,t',t) \quad (transposition).
\end{aligned}
\tag{1.10}
$$

For instance, it is important to note that, given a homotopy $\varphi \colon f^- \to f^+ \colon X \to Y$, we need the transposition to construct a homotopy $I\varphi$, by modifying the mapping $I(\hat\varphi)$, *which does not have the correct faces*

$$
\begin{aligned}
&I\varphi \colon If^- \to If^+, \qquad (I\varphi)\hat{} = I(\hat\varphi).sX \colon I^2 X \to IY, \\
&I(\hat\varphi).sX.\partial^\alpha(IX) = I(\hat\varphi).I(\partial^\alpha X) = If^\alpha.
\end{aligned}
\tag{1.11}
$$

1.1.2 The cocylinder

We conclude this brief review of the formal bases of classical homotopy recalling that a homotopy $\varphi \colon f \to g$, described by a map $\hat\varphi \colon IX \to Y$, also has a dual description as a map $\check\varphi \colon X \to PY$ with values in the *path-space* $PY = Y^{\mathbf{I}}$.

In fact, it is well known (and rather easy to verify) that every *locally compact Hausdorff space A* is exponentiable in **Top**, which means that the functor $- \times A \colon \mathbf{Top} \to \mathbf{Top}$ has a right adjoint, written $(-)^A \colon \mathbf{Top} \to \mathbf{Top}$ (cf. Sections A4.2 and A4.3, in the Appendix). Concretely, the space Y^A is the set of maps $\mathbf{Top}(A, Y)$ equipped with the compact-open topology. The adjunction consists of the natural bijection

$$
\mathbf{Top}(X \times A, Y) \to \mathbf{Top}(X, Y^A), \quad f \mapsto f', \quad f'(x)(a) = f(x,a), \tag{1.12}
$$

which is called the *exponential law*, as it gives a bijection $Y^{X \times A} \to (Y^A)^X$.

In particular, the cylinder functor $I = - \times \mathbf{I}$ has a right adjoint

$$
P \colon \mathbf{Top} \to \mathbf{Top}, \qquad P(Y) = Y^{\mathbf{I}}, \tag{1.13}
$$

called the *cocylinder* or *path functor*: $P(Y)$ is the space of paths $\mathbf{I} \to Y$, with the compact-open topology (also called the topology of uniform convergence on $\mathbf{I}$). The counit of the adjunction

$$\mathrm{ev}\colon P(Y) \times \mathbf{I} \to Y, \qquad (a, t) \mapsto a(t), \qquad (1.14)$$

is also called *evaluation* (of paths).

The functor P inherits from the interval $\mathbf{I}$, contravariantly, a *dual* structure, which we write with the same symbols. It consists of a basic, first-order part:

$$\begin{aligned}
&\partial^\alpha \colon PY \to Y, \quad \partial^-(a) = a(0),\ \partial^+(a) = a(1) &&(\textit{faces}), \\
&e\colon Y \to PY, \quad e(y)(t) = y &&(\textit{degeneracy}), \quad (1.15)\\
&r\colon PY \to PY, \quad r(a)(t) = a(1-t) &&(\textit{reversion}),
\end{aligned}$$

and a second-order structure:

$$\begin{aligned}
&g^\alpha \colon PY \to P^2Y, \quad g^\alpha(a)(t, t') = a(g^\alpha(t, t')) &&(\textit{connections}), \\
&s\colon P^2Y \to P^2Y, \quad s(a)(t, t') = a(t', t) &&(\textit{transposition}).
\end{aligned} \quad (1.16)$$

In this description, the faces of a homotopy $\check{\varphi}\colon X \to PY$ are defined as $\partial^\alpha \check{\varphi}\colon X \to Y$.

Concatenation of homotopies can now be performed with the *concatenation pullback* QY (which can be realised as *the object of pairs of consecutive paths*) and the concatenation map c

$$
\begin{array}{ccc}
QY & \xrightarrow{\ c^+\ } & PY \\
{\scriptstyle c^-}\downarrow & \quad\vert\quad & \downarrow{\scriptstyle \partial^-} \\
PY & \xrightarrow[\ \partial^+\ ]{} & Y
\end{array}
$$

$$\begin{aligned}
&QY = \{(a, b) \in PY \times PY \mid \partial^+(a) = \partial^-(b)\}, \\
&c\colon QY \to PY, \quad c(a, b) = a + b \quad (\textit{concatenation map}).
\end{aligned} \quad (1.17)$$

Again, as a peculiar property of topological spaces, the natural transformation c is invertible (splitting a path into its two halves), and we can also realise QY as PY.

1.1.3 Preordered topological spaces

The simplest topological setting where one can study directed paths and directed homotopies is likely the category p**Top** of *preordered topological spaces* and *preorder-preserving continuous mappings*; the latter will be simply called *morphisms* or *maps*, when it is understood we are in this category. (Recall that a preorder relation, generally written $\prec$, is only assumed to be reflexive and transitive; it is an order if it is also anti-symmetric.)

Here, the *standard directed interval* $\uparrow\mathbf{I} = \uparrow[0,1]$ has the euclidean topology and the natural order. A (directed) *path* in a preordered space X is, by definition, a map $a\colon \uparrow\mathbf{I} \to X$ (continuous and preorder-preserving).

The category p**Top** has all limits and colimits (see Section A2), constructed as for topological spaces and equipped with the initial or final preorder for their structural maps; for instance, in a product $X = \Pi X_j$, we have the product preorder: $(x_j) \prec_X (x_j')$ if and only if, for each index j, $x_j \prec x_j'$ in X_j.

The forgetful functor $U\colon \text{p}\mathbf{Top} \to \mathbf{Top}$ has both a left and a right adjoint, $D \dashv U \dashv D'$ where DS (resp. $D'S$) is the space S equipped with the *discrete* order (resp. the *chaotic*, or *indiscrete*, *pre*order). The standard embedding of **Top** in p**Top** will be the one given by the *indiscrete preorder*, so that all (ordinary) paths in S are directed in $D'S$.

Note that the category of ordered *spaces does not allow for such an embedding*, and would not allow us to view classical algebraic topology within the directed one; furthermore, its colimits are 'different' from the topological ones.

Our category is not cartesian closed (Section A4.3), of course; but it is easy to transfer here the classical result for topological spaces, recalled above (Section 1.1.2). Thus, every preordered space A having a *locally compact Hausdorff topology* and an *arbitrary preorder* is exponentiable in p**Top**, with Y^A consisting of the set $\text{p}\mathbf{Top}(A,Y) \subset \mathbf{Top}(UA,UY)$ of *preorder-preserving continuous mappings*, equipped with the (induced) compact-open topology and the *pointwise preorder*

$$f \prec g \quad \text{if} \quad (\forall x \in A)(f(x) \prec_Y g(x)). \tag{1.18}$$

The natural bijection $\text{p}\mathbf{Top}(X{\times}A, Y) \to \text{p}\mathbf{Top}(X, Y^A)$ is a restriction of the classical one (1.12) to *preoreder-preserving* continuous mappings, and is described by the same formula.

A richer setting, the category d**Top** of d-spaces, or spaces with distinguished paths (mentioned in Section 3 of the Introduction), will be studied starting from Section 1.4.

1.1.4 The basic structure of directed homotopies

Let us examine the first-order structure of the standard directed interval $\uparrow\mathbf{I} = \uparrow[0,1]$, in the category p**Top** of preordered topological spaces.

Faces and *degeneracy* are as in the classical case (1.1)

$$\partial^\alpha : \{*\} \rightrightarrows \uparrow\mathbf{I} : e, \qquad \partial^-(*) = 0, \ \partial^+(*) = 1, \ e(t) = *, \qquad (1.19)$$

where $\uparrow\mathbf{I}^0 = \{*\}$ is now an ordered space, with the unique order relation on the singleton.

On the other hand, the classical reversion $r(t) = 1 - t$ is not an *endomap* of $\uparrow\mathbf{I}$, but becomes a map which we prefer to call *reflection*

$$r : \uparrow\mathbf{I} \to \uparrow\mathbf{I}^{\mathrm{op}}, \quad r(t) = 1 - t \qquad (reflection), \qquad (1.20)$$

as it takes values in the *opposite* preordered space $\uparrow\mathbf{I}^{\mathrm{op}}$ (with the opposite preorder).

Here also, the *standard concatenation pushout* can be realised as $\uparrow\mathbf{I}$ itself

$$
\begin{array}{ccc}
\{*\} & \xrightarrow{\ \partial^+\ } & \uparrow\mathbf{I} \\
\partial^- \downarrow & {}^{\text{--}\;\text{--}} \searrow {}_{c^-} & \downarrow c^- \\
\uparrow\mathbf{I} & \xrightarrow[\ c^+\ ]{} & \uparrow\mathbf{I}
\end{array}
\qquad
\begin{array}{l}
c^-(t) = t/2, \\[2em]
c^+(t) = (t+1)/2,
\end{array}
\qquad (1.21)
$$

since a mapping $a : \uparrow\mathbf{I} \to X$ with values in a preordered space is a *map* (continuous and preorder-preserving) if and only if its two restrictions ac^α (to the first or second half of the interval) are maps.

For every preordered topological space X, we have the (directed) cylinder $I(X) = X \times \uparrow\mathbf{I}$, with the product topology and the product preorder:

$$(x, t) \prec (x', t') \qquad \text{if} \qquad (x \prec x' \text{ in } X) \quad \text{and} \quad (t \leqslant t' \text{ in } \uparrow\mathbf{I}). \qquad (1.22)$$

The cylinder functor has a first-order structure, formed of four natural transformations. Faces and degeneracy are as in the classical case (1.8), but the reflection r, again, has to be expressed via the *reversor*, i.e. the involutive endofunctor R which reverses the preorder relation

$$
\begin{array}{ll}
\partial^\alpha : 1 \rightrightarrows I : e, & (faces, \ degeneracy), \\
R : \text{p}\mathbf{Top} \to \text{p}\mathbf{Top}, \ R(X) = X^{\mathrm{op}} & (reversor), \\
r : IR \to RI, \ r : I(X^{\mathrm{op}}) \to (IX)^{\mathrm{op}}, & \\
r(x, t) = (x, 1 - t) & (reflection).
\end{array}
\qquad (1.23)
$$

Since $\uparrow\mathbf{I}$ is exponentiable in $\mathrm{p}\mathbf{Top}$ (by Section 1.1.3), we also have a path functor, right adjoint to the cylinder functor, where the preordered space $P(Y)$ has the compact-open topology and the pointwise preorder:

$$P\colon \mathrm{p}\mathbf{Top} \to \mathrm{p}\mathbf{Top}, \qquad\qquad P(Y) = Y^{\uparrow\mathbf{I}},$$
$$a \prec b \ \text{ if } \ (\forall t \in [0,1]) \, (a(t) \prec_Y b(t)) \ \text{ (for } a, b \colon \uparrow\mathbf{I} \to Y). \tag{1.24}$$

The path functor is equipped with a dual (first-order) structure, formed of four natural transformations (with the same reversor R as above)

$$\partial^\alpha \colon P \to 1, \qquad e \colon 1 \to P, \qquad r \colon RP \to PR. \tag{1.25}$$

A (directed) *homotopy* $\varphi \colon f^- \to f^+ \colon X \to Y$ is defined by a map $\hat{\varphi} \colon X \times \uparrow\mathbf{I} \to Y$ with $\hat{\varphi}\partial^\alpha = f^\alpha$; or, equivalently, by a map $\check{\varphi} \colon X \to Y^{\uparrow\mathbf{I}}$ with $\partial^\alpha\check{\varphi} = f^\alpha$. It yields a *reflected* homotopy between the opposite spaces:

$$\varphi^{\mathrm{op}} \colon Rf^+ \to Rf^- \colon X^{\mathrm{op}} \to Y^{\mathrm{op}},$$
$$(\varphi^{\mathrm{op}})\hat{} = R(\hat{\varphi}).rX \colon IRX \to RIX \to RY. \tag{1.26}$$

1.1.5 Breaking the symmetries of classical algebraic topology

A topological space X has *intrinsic symmetries*, which act on its singular n-cubes $a \colon \mathbf{I}^n \to X$.

We have already recalled the standard reversion r and the standard transposition s (Section 1.1.0)

$$r \colon \mathbf{I} \to \mathbf{I}, \ \ r(t) = 1 - t; \qquad s \colon \mathbf{I}^2 \to \mathbf{I}^2, \ \ s(t, t') = (t', t). \tag{1.27}$$

Their n-dimensional versions

$$r_i = \mathbf{I}^{i-1} \times r \times \mathbf{I}^{n-i} \colon \mathbf{I}^n \to \mathbf{I}^n \qquad\qquad (i = 1, ..., n),$$
$$s_i = \mathbf{I}^{i-1} \times s \times \mathbf{I}^{n-i-1} \colon \mathbf{I}^n \to \mathbf{I}^n \qquad\quad (i = 1, ..., n-1), \tag{1.28}$$

generate the group of symmetries of the n-cube, i.e. the hyperoctahedral group $(\mathbf{Z}/2)^n \rtimes S_n$ (a semidirect product): the reversions r_i commute with each other and generate the first factor, while the transpositions s_i generate the symmetric group S_n. Plainly, the hyperoctahedral group acts on the set of n-cubes $a \colon \mathbf{I}^n \to X$.

Generally speaking, we will call 'reversible' a framework which has reversions, and 'symmetric' (or also 'permutable') a framework which has transpositions. Topological spaces have both kinds of symmetries, while, in directed algebraic topology, the first kind *must* be broken and the second *can*.

(a) *Reversion versus reflection.* We have already recalled that the prime effect of the reversion $r\colon \mathbf{I} \to \mathbf{I}$ is reversing the paths, in any topological space. This map also gives the reversion of homotopies, by means of the reversion of the cylinder functor (Section 1.1.1); or, equivalently, by the reversion of the path functor (Section 1.1.2).

Preordered topological spaces, studied above, lack a reversion. But we now have a sort of 'external reversion', i.e. a *reflection pair* (R, r) consisting of an involutive endofunctor $R\colon \mathrm{p}\mathbf{Top} \to \mathrm{p}\mathbf{Top}$ (which reverses the preorder, and will be called *reversor*) and a *reflection* $r\colon IR \to RI$ for the cylinder functor; or, equivalently, $r\colon RP \to PR$, for the path functor (Section 1.1.4). This behaviour will be shared by all the structures for directed homotopy which we will consider. Notice that the reversible case is a particular instance, when R is the identity.

(b) *Transposition.* Coming back to topological spaces, the transposition $s(t, t') = (t', t)$ of the standard square $\mathbf{I}^2$ yields the transposition symmetry of the iterated cylinder functor $I^2(X) = X \times \mathbf{I}^2$ (Section 1.1.1) and of the iterated path functor $P^2(Y) = Y^{\mathbf{I}^2}$ (Section 1.1.2). It follows, as we have seen in (1.11), that the cylinder functor is homotopy invariant (and the same holds for P).

This *second-order* symmetry (acting on I^2 and P^2) exists in various directed structures, for instance in $\mathrm{p}\mathbf{Top}$ and $\mathrm{d}\mathbf{Top}$, but does not exist in other cases, e.g. for cubical sets (studied in Section 1.6). *Its role, within directed algebraic topology, is double-edged.* On the one hand, its presence yields the important consequence recalled above: the homotopy invariance of the cylinder and path functors. On the other hand, it restricts the interest of directed homology (and prevents a good relation of the latter with suspension): we will see that, for a cubical set X having this sort of symmetry, the directed homology group $\uparrow H_n(X)$ has a trivial preorder, for $n \geqslant 2$ (Proposition 2.2.6).

The presence of the transposition symmetry, for preordered spaces and d-spaces, reveals that the directed character of these structures *does not go beyond the one-dimensional level*: after distinguishing some paths $\uparrow\mathbf{I} \to X$ and forbidding others, no higher choice is needed: namely, a continuous mapping $a\colon \uparrow\mathbf{I}^n \to X$ (in $\mathrm{p}\mathbf{Top}$ or $\mathrm{d}\mathbf{Top}$) is a map of the category if and only if, for every increasing map $f\colon \uparrow\mathbf{I} \to \uparrow\mathbf{I}^n$, the path $af\colon \uparrow\mathbf{I} \to X$ is a map.

On the other hand, in an abstract cubical set K, after observing that an element of K_1 need not have any counterpart with reversed vertices, we can also notice that an element of K_n need not have any counterpart with faces permuted (for $n \geqslant 2$). Thus, a cubical set has a real choice of 'privileged directions', in any dimension.

It is interesting to note that, for cubical sets (and other directed structures lacking a transposition), there is again a sort of (weak) 'surrogate', consisting of an *external transposition pair* (S, s), where $S \colon \mathbf{Cub} \to \mathbf{Cub}$ is the involutive endofunctor which reverses the order of faces (cf. (1.134) and (1.155)).

1.1.6 Categories and directed homotopy

We give now a brief description of *directed homotopy in* **Cat**, the cartesian closed category of small categories. Directed homotopy equivalence in **Cat** will be studied in Chapter 3, as a crucial tool to classify the fundamental categories of directed spaces, and thus analyse such spaces.

The *reversor* functor R takes a small category to the opposite one

$$R \colon \mathbf{Cat} \to \mathbf{Cat}, \qquad R(X) = X^{\mathrm{op}}. \qquad (1.29)$$

Let us make precise that X^{op} has precisely the same objects as X, with $X^{\mathrm{op}}(x, y) = X(y, x)$ and the opposite composition, so that R is (strictly) involutive.

The role of the standard point is played by the terminal category $\mathbf{1} = \{*\}$, while the *directed interval* $\uparrow\mathbf{i} = \mathbf{2} = \{0 \to 1\}$ is an order category on two objects. It has obvious *faces* $\partial^\pm \colon \mathbf{1} \to \mathbf{2}$ and the (unique) isomorphism $r \colon \mathbf{2} \to \mathbf{2}^{\mathrm{op}}$ as *reflection*.

A *point* $x \colon \mathbf{1} \to X$ of the small category X is an object of the latter; we will also write $x \in X$. A (directed) *path* $a \colon \mathbf{2} \to X$ from x to x' is an arrow $a \colon x \to x'$ of X, their concatenation is the composition, strictly associative and unitary. This is a motivation of our frequent use of the additive notation for composition, as above for topological paths (see Section 8, in the Introduction). Notice that the standard concatenation pushout (1.3) gives here the ordinal category $\mathbf{3}$ (with non-trivial arrows $0 \to 1 \to 2$), which is not isomorphic to the interval $\mathbf{2}$ (in contrast with the behaviour of the topological interval); this aspect will be dealt with in Section 4.3.2.

The (directed) *cylinder* functor $IX = X \times \mathbf{2}$ has a right adjoint, $PY = Y^{\mathbf{2}}$ (the category of morphisms of Y and its commutative squares, see Section A1.7). According to this adjunction, a (directed) *homotopy*

$\varphi\colon f \to g\colon X \to Y$, represented by a functor $X\times\mathbf{2} \to Y$ or, equivalently, $X \to Y^{\mathbf{2}}$, is the same as a natural transformation $f \to g$.

Operations of homotopies coincide with the 2-categorical structure of **Cat** (Section A5.2). In particular, homotopy concatenation is the vertical composition of natural transformations, which is strictly associative and unitary. This operation will be written in the same notation we are using for composition in small categories: either the ordinary one or the additive one (below, x is an object of the domain category of the natural transformations $\varphi\colon f \to g$ and $\psi\colon g \to h$)

$$(\psi\varphi)(x) = \psi x.\varphi x, \qquad \mathrm{id}(f)(x) = \mathrm{id}(fx) \qquad (\textit{usual notation}),$$
$$(\varphi + \psi)(x) = \varphi x + \psi x, \quad 0_f(x) = 0_{f(x)} \qquad (\textit{additive notation}).$$

We have already said that directed homotopy equivalence in **Cat** will be studied later. Ordinary equivalence of categories is a stricter, far simpler notion. It is based on the *standard reversible interval*, the groupoid on two objects linked by an isomorphism and its inverse:

$$\mathbf{i} = \{0 \rightleftarrows 1\}, \qquad r\colon \mathbf{i} \to \mathbf{i}, \qquad r(u) = u^{-1}, \qquad (1.30)$$

with the obvious reversion r, defined above. This gives rise to a *reversible cylinder* functor $X \times \mathbf{i}$, with right adjoint $Y^{\mathbf{i}}$ (the full subcategory of $Y^{\mathbf{2}}$ whose objects are the isomorphisms of Y); thus, a *reversible homotopy* $\varphi\colon f \to g\colon X \to Y$ is the same as a natural isomorphism of functors. This *reversible* homotopy structure will be written as **Cat$_\mathbf{i}$**.

Its restriction to the full subcategory **Gpd** of small groupoids is also of interest. (Notice that both structures on **Cat** reduce to the same homotopies for groupoids, represented by the restriction of the (co)cylinder of **Cat$_\mathbf{i}$**.)

1.1.7 Dioids

Let us say something more about the higher structure of the standard interval **I** of topological spaces. We have already remarked (in Section 1.1.0) that it is an involutive lattice, with respect to the binary operations $g^-(t, t') = \max(t, t')$, $g^+(t, t') = \min(t, t')$.

However, the idempotence of these operations is of little interest for homotopy (and does not hold for other structures of interest on **I**, see Section 1.1.9). What is really relevant is the fact that **I** is a *dioid* [G1], i.e. a set equipped with two monoid operations, such that the unit element of each of them is an absorbent element for the other. (In [G1], this structure is also called a 'cubical monoid', because a dioid is to a monoid what a cubical set is to an augmented simplicial set; but this term might suggest a different structure, namely a cubical object in the category of monoids, and will not be used here.)

In general, in a monoidal category $\mathbf{A} = (\mathbf{A}, \otimes, E)$ (see Section A4.1), a *dioid* is an object $\mathbf{I}$ equipped with maps (still called *faces* or *units*, *degeneracy*, and *connections* or *main operations*)

$$E \mathrel{\mathop{\rightrightarrows}^{\partial^\alpha}_{e}} \mathbf{I} \mathrel{\mathop{\longleftarrow}^{g^\alpha}} \mathbf{I} \otimes \mathbf{I} \qquad (\alpha = -, +) \qquad (1.31)$$

which satisfy the following axioms, also displayed in the commutative diagrams below (for $\alpha \neq \beta$):

$$
\begin{aligned}
e\partial^\alpha &= 1, \quad eg^\alpha = e.\mathbf{I} \otimes e = e.e \otimes \mathbf{I} && (degeneracy), \\
g^\alpha.\mathbf{I} \otimes g^\alpha &= g^\alpha.g^\alpha \otimes \mathbf{I} && (associativity), \\
g^\alpha.\mathbf{I} \otimes \partial^\alpha &= 1 = g^\alpha.\partial^\alpha \otimes \mathbf{I} && (unit), \\
g^\beta.\mathbf{I} \otimes \partial^\alpha &= \partial^\alpha e = g^\beta.\partial^\alpha \otimes \mathbf{I} && (absorbency),
\end{aligned}
\qquad (1.32)
$$

In the cartesian case (i.e. if the tensor is the cartesian product), E is the terminal object of the category and the degeneracy axiom is automatically satisfied.

An *involutive* dioid is further equipped with an involutive *reversion* $r\colon \mathbf{I} \to \mathbf{I}$ exchanging the lower and upper structure, and can be equivalently defined as a monoid equipped with an involutive mapping (of sets) which takes the unit of multiplication to an absorbent element. On the other hand, a *symmetric* dioid has an involutive *transposition* $s\colon \mathbf{I} \otimes \mathbf{I} \to \mathbf{I} \otimes \mathbf{I}$ exchanging $\mathbf{I} \otimes \partial^\alpha$ with $\partial^\alpha \otimes \mathbf{I}$ and leaving the connections invariant. These notions will be adapted to the directed case in Section 4.2.8.

1.1.8 Diads

The 'categorical version' of a monoid is a *monad* (see Section A4.4), which is the basis of 'algebricity' in category theory. Similarly, as a 'categorical version' of a dioid, we define a *diad* on the category **A** [G1] as a collection $(I, \partial^-, \partial^+, e, g^-, g^+)$ consisting of an endofunctor $I \colon \mathbf{A} \to \mathbf{A}$ (the *cylinder* endofunctor) and natural transformations (with the usual names of *faces*, *degeneracy* and *connections*)

$$\mathrm{id}\mathbf{A} = 1 = I^0 \underset{e}{\overset{\partial^\alpha}{\rightrightarrows}} I \overset{g^\alpha}{\Leftarrow} I^2 \qquad (\alpha = -, +) \qquad (1.33)$$

making the following diagrams commute (for $\alpha \neq \beta$)

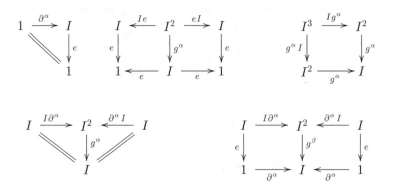

(If **A** is a small category, this is a dioid in the non-symmetric strict monoidal category of endofunctors of the category **A**, with tensor product the composition of endofunctors. This interpretation can be extended to arbitrary categories, in a set-theoretical setting which allows one to consider general categories of endofunctors.)

An *involutive* diad is further equipped with a *reversion* $r \colon I \to I$ exchanging the lower and upper structure, while a *symmetric* diad has a *transposition* $s \colon I^2 \to I^2$ exchanging $I\partial^\alpha$ with $\partial^\alpha I$ and leaving the connections invariant. (Again, we will not use here the term 'cubical monad' which was used in [G1] as a synonym of diad.)

1.1.9 A digression on dioids

This subsection is about dioids which are not lattices, their use in homotopy and some physical interpretations; it will not be used in the sequel.

Every unital ring R has a structure of involutive dioid, with respect to ring-multiplication and the operation $x * y$ given by the following set-theoretical involution r (which takes the unit 1 of multiplication to the absorbent element 0)

$$r(x) = 1 - x, \qquad x * y = r(rx.ry) = x + y - x.y. \qquad (1.34)$$

On the real field $\mathbf{R}$ this (non-idempotent) structure is *smooth* (while lattice operations are not) and can be used in *smooth homotopy theory*; one can restrict the structure to the involutive subdioid $[0, 1]$, or not.

Another interesting non-idempotent example, on the compact real interval, is the *dioidal half-line*, i.e. the Alexandroff compactification $H^1 = [0, \infty]$ of the additive monoid $[0, \infty[$, with the extended sum and the involution $r(x) = x^{-1}$ (taking 0 to the absorbent element ∞). The resulting operation

$$x * y = (x^{-1} + y^{-1})^{-1}, \qquad (1.35)$$

can be called the *inverse sum*, or also *harmonic sum*, since $(1/n) * 1 = 1/(n + 1)$.

The dioid H^1 has a physical interpretation, referring to electric circuits of pure resistors. Interpret its elements as *resistances*, '0' as the *perfect conductor*, '∞' as the *perfect insulator*, '+' as *series combination* and '$*$' as *parallel combination* (where conductances, the inverses of resistances, are added). The operation $*$ of H^1 is also of use in geometric optics, where $p * q = f$ is the well-known formula relating corresponding points with respect to a lens of focal length f. H^1 is a topological dioid, since the sum is proper over $[0, \infty[$.

$\mathbf{P}^1\mathbf{R} = \mathbf{R} \cup \{\infty\}$ and $\mathbf{P}^1\mathbf{C} = \mathbf{C} \cup \{\infty\}$ have a similar structure of involutive dioid, given by the extended sum and the involution $r(x) = x^{-1}$. They are not topological dioids. The dioid $\mathbf{P}^1\mathbf{C}$ formalises the calculus of *impedances* (and their inverses, *admittances*) for networks of resistors, inductors and capacitors in a steady sinusoidal state.

Extending H^1, one can also consider the *dioidal n-orthant* $H^n = \mathbf{R}_+^n \cup \{\infty\}$ as the one-point compactification of the additive monoid $\mathbf{R}_+^n = [0, \infty[^n$, with the extended sum and the involution, $r(x) = x/||x||^2$. It is a topological dioid, since the sum is proper over $\mathbf{R}_+^n$. Notice that H^n is not a cartesian power of H^1, and H^2 is not a subdioid of $\mathbf{P}^1\mathbf{C}$ (whose inverse sum is given by the involution $r(z) = z^{-1} = \bar{z}/|z|^2$).

As a final remark, if M is a topological monoid, equipped with a continuous involution r over the subspace $M \setminus \{0\}$, one gets a topological involutive dioid $X = M \cup \{\infty\}$ by one-point compactification, provided that the operation $M \times M \to M$ is a proper map and that $r(x) \to \infty$ for $x \to 0$.

1.2 The basic structure of the directed cylinder and cocylinder

The phenomena which we want to study, hinted at in the previous section, make sense in a category $\mathbf{A}$ equipped with homotopies which, generally, cannot be reversed, but are *reflected* in *opposite* objects.

These homotopies can be produced by a formal cylinder endofunctor I (Section 1.2.1), or – dually – by a cocylinder functor P (Section 1.2.2); or by both, under an adjunction $I \dashv P$ (Section 1.2.2). At the end of the section, we briefly consider a more general self-dual setting based on formal directed homotopies, which, however, would give no real advantage (Section 1.2.9).

Most of this material has been introduced in [G1, G3, G4, G21]. The starting point, in the classical (reversible) case, goes back to Kan's abstract cylinder approach [Ka2].

1.2.1 The basic setting, I

Our first basic setting is based on a formal cylinder endofunctor equipped with natural transformations, as in (1.23) for preordered spaces.

More precisely, a *dI1-category* $\mathbf{A}$ comes equipped with:

(a) a *reversor* $R \colon \mathbf{A} \to \mathbf{A}$, i.e. an involutive (covariant) automorphism (also written $R(X) = X^{\mathrm{op}}$, $R(f) = f^{\mathrm{op}}$);

(b) a *cylinder* endofunctor $I \colon \mathbf{A} \to \mathbf{A}$, with four natural transformations: two *faces* (∂^{α}), a *degeneracy* (e) and a *reflection* (r)

$$\partial^{\alpha} \colon 1 \rightrightarrows I \colon e, \qquad r \colon IR \to RI \qquad (\alpha = \pm), \qquad (1.36)$$

satisfying the equations

$$\begin{aligned} e\partial^{\alpha} &= 1 \colon \mathrm{id}\mathbf{A} \to \mathrm{id}\mathbf{A}, & RrR.r &= 1 \colon IR \to IR, \\ Re.r &= eR \colon IR \to R, & r.\partial^{-}R &= R\partial^{+} \colon R \to RI. \end{aligned} \qquad (1.37)$$

Since $RR = 1$, the transformation r is invertible with $r^{-1} = RrR \colon RI \to IR$. It is easy to verify that $r.\partial^{+}R = R\partial^{-}$.

A *homotopy* $\varphi \colon f^{-} \to f^{+} \colon X \to Y$ is defined as a map $\varphi \colon IX \to Y$ with $\varphi.\partial^{\alpha}X = f^{\alpha}$. When we want to distinguish the homotopy from the *map* which represents it, we write the latter as $\hat{\varphi}$. Each map

$f\colon X \to Y$ has a *trivial* endohomotopy, $0_f\colon f \to f$, represented by $f.eX = eY.If\colon IX \to Y$.

Every homotopy $\varphi\colon f \to g\colon X \to Y$ has a *reflected homotopy*

$$\varphi^{\mathrm{op}}\colon g^{\mathrm{op}} \to f^{\mathrm{op}}\colon X^{\mathrm{op}} \to Y^{\mathrm{op}}, \quad (\varphi^{\mathrm{op}})\hat{} = R(\hat{\varphi}).r\colon IRX \to RY, \quad (1.38)$$

and $(\varphi^{\mathrm{op}})^{\mathrm{op}} = \varphi$, $(0_f)^{\mathrm{op}} = 0_{f^{\mathrm{op}}}$.

An object X is said to be *reversive* or *self-dual* if it is isomorphic to X^{op}. A dI1-category will be said to be *reversible* when R is the identity; then, every homotopy has a *reversed* homotopy $-\varphi = \varphi^{\mathrm{op}}\colon g \to f$. Reversible structures have no 'privileged directions' and will only be considered here in as much as they can be of help in studying non-reversible situations.

In a dI1-category, the pair (R, r) will also be called the *reflection pair* of $\mathbf{A}$. The reader is warned that the following two bodies of notions should not be confused: on the one hand we have the *reversor* R, the *reflection* r and *reversive* objects; on the other hand, *reflective* subcategories and their *reflectors* (a standard notion of category theory), which will be frequently used in Section 3.3.

A *dI1-subcategory* of a dI1-category $(\mathbf{A}, R, I, \partial^\alpha, e, r)$ is a subcategory $\mathbf{A}' \subset \mathbf{A}$ which is closed with respect to the whole structure, and inherits therefore a dI1-structure.

1.2.2 The basic setting, II

Dually, a *dP1-category* has a *reversor* $R\colon \mathbf{A} \to \mathbf{A}$ (as above) and a *cocylinder*, or *path endofunctor* $P\colon \mathbf{A} \to \mathbf{A}$, with natural transformations in opposite directions, satisfying the dual equations (with $\alpha = \pm$)

$$\partial^\alpha\colon P \rightrightarrows 1\colon e, \qquad r\colon RP \to PR \qquad (R^2 = \mathrm{id}),$$
$$\partial^\alpha e = 1\colon \mathrm{id}\mathbf{A} \to \mathrm{id}\mathbf{A}, \qquad RrR.r = 1\colon RP \to RP, \qquad (1.39)$$
$$r.Re = eR\colon R \to RP, \qquad \partial^- R.r = R\partial^+\colon RP \to R.$$

Again, (R, r) is the *reflection pair* of $\mathbf{A}$. A homotopy $\varphi\colon f^- \to f^+\colon X \to Y$ is now defined as a map $\varphi\colon X \to PY$ with $\partial^\alpha Y.\varphi = f^\alpha$, which will be written as $\check{\varphi}$ when useful.

Given a dI1-structure on the category $\mathbf{A}$, if the cylinder endofunctor I has a right adjoint P, the latter *automatically* inherits a dP1-structure $(R, \partial'^\alpha, e', r')$, with the same reversor R and natural transformations which are *mates* to the transformations of I under the given adjunction (cf. A5.3). Explicitly, writing $\eta\colon 1 \to PI$ the unit and $\varepsilon\colon IP \to 1$ the counit of the adjunction, we have:

$$e' = (Pe.\eta \colon 1_\mathbf{A} \to PI \to P),$$
$$\partial'^\alpha = (\varepsilon.\partial^\alpha P \colon P \to IP \to 1_\mathbf{A}), \tag{1.40}$$
$$r' = (PR\varepsilon.PrP.\eta RP \colon RP \to P(IR)P \to P(RI)P \to PR).$$

We say then that $\mathbf{A}$ is a *dIP1-category*. Such a structure, of course, is formed of the adjunction $I \dashv P$ together with the reversor R and two triples of natural transformations, (∂^α, e, r) and $(\partial'^\alpha, e', r')$, each triple determining the other, as in (1.40) or dually. Then, both endofunctors give rise to the same homotopies, represented equivalently by maps $\hat{\varphi} \colon IX \to Y$ or $\check{\varphi} \colon X \to PY$. The counit of the adjunction $I \dashv P$ will also be called *path-evaluation* (cf. (1.14), for spaces) and written

$$\mathrm{ev} \colon IP \to 1. \tag{1.41}$$

Here also, a dP1 or dIP1-category is said to be *reversible* when R is the identity, so that every homotopy has a *reversed* homotopy $-\varphi = \varphi^{\mathrm{op}} \colon g \to f$. The notions of *dP1-subcategory* and *dIP1-subcategory* are defined as in the dI1-case.

For instance, **Top** is a reversible dIP1-category, with the obvious structure based on the standard interval (recalled in Section 1.1). The category p**Top** of preordered topological spaces is a non-reversible dIP1-category, with the structure described in Section 1.1.4, based on the reversor $R(X) = X^{\mathrm{op}}$ (which reverses preorder) and the cylinder functor $I(X) = X \times {\uparrow}\mathbf{I}$. **Cat** has a non-reversible dIP1-structure, where homotopies are natural transformations (Section 1.1.6), $R(X) = X^{\mathrm{op}}$ is the opposite category and the cylinder is $I(X) = X \times \mathbf{2}$. **Cat** has also a reversible dIP1-structure, written **Cat$_\mathfrak{i}$**, based on the self-dual category $\mathfrak{i}$, where homotopies are natural isomorphisms, and **Gpd** is a dIP1-subcategory of the latter (Section 1.1.6). Chain complexes have a canonical reversible structure, studied in Section 4.4. But we will see in Section 5.7 that they also admit a *weakly reversible* structure, where R is coherently isomorphic to the identity; the latter structure is also of interest, because it can be lifted to chain algebras (losing any reversibility property), while the first cannot.

As in these examples, a dIP1-structure is often generated by a standard directed interval, by cartesian (or tensor) product and internal hom, respectively (see Section 1.2.5).

Concatenation and the second-order structure will be dealt with in Part II.

1.2.3 The basic structure of homotopies

Let **A** be a dI1-category. One can define a *whisker composition* for maps and homotopies

$$X' \xrightarrow{h} X \underset{g}{\overset{f}{\rightrightarrows}} {\downarrow \varphi} \; Y \xrightarrow{k} Y'$$

$$k \circ \varphi \circ h \colon kfh \to kgh \colon X' \to Y',$$
$$(k \circ \varphi \circ h)\hat{} = k.\hat{\varphi}.Ih \colon IX' \to Y', \tag{1.42}$$

which will also be written as $k\varphi h$, *even though one should recall that the computational formula is* $k.\hat{\varphi}.Ih$.

This ternary operation satisfies:

$$k' \circ (k \circ \varphi \circ h) \circ h' = (k'k) \circ \varphi \circ (hh') \qquad (associativity),$$
$$1_Y \circ \varphi \circ 1_X = \varphi, \qquad k \circ 0_f \circ h = 0_{kfh} \qquad (identities), \quad (1.43)$$
$$(k \circ \varphi \circ h)^{\mathrm{op}} = k^{\mathrm{op}} \circ \varphi^{\mathrm{op}} \circ h^{\mathrm{op}} \colon (kgh)^{\mathrm{op}} \to (kfh)^{\mathrm{op}} \quad (reflection).$$

(These properties will be abstracted in a self-dual setting based on *assigning* homotopies, see Section 1.2.9.)

Actually, we can define a richer *cubical* structure on **A** (which will be important starting with Chapter 3): a *p-dimensional homotopy* $\varphi \colon X \to_p Y$ is a map $\hat{\varphi} \colon I^p X \to Y$. The composition with $\psi \colon Y \to_q Z$ is *n*-dimensional (with $n = p + q$) and defined as:

$$\psi \circ \varphi \colon X \to_n Z, \qquad (\psi \circ \varphi)\hat{} = \hat{\psi}.I^q \hat{\varphi} \colon I^n X \to I^q Y \to Z. \tag{1.44}$$

As we have already seen for topological spaces (cf. (1.11)), to extend the endofunctor I to homotopies requires a transposition $s \colon I^2 \to I^2$. This will be done in Section 4.1.5.

1.2.4 Concrete structures

A *concrete* dI1-category will be a dI1-category **A** equipped with a reversive object E, called the *standard point*, or *free point*, of **A**, and with a specified isomorphism

$$E \to E^{\mathrm{op}} \tag{1.45}$$

This is essentially equivalent to assigning a representable *forgetful* functor (Section A1.7), invariant up to isomorphism under the reversor R

$$U = |-| = \mathbf{A}(E, -) \colon \mathbf{A} \to \mathbf{Set}, \qquad UR \cong U. \tag{1.46}$$

Whenever possible, we will identify $E = E^{\mathrm{op}}$ and $UR = R$. The functor U is faithful if and only if E is a *separator* (or *generator*) in $\mathbf{A}$: given two maps $f \neq g\colon X \to Y$, there exists some $x\colon E \to X$ such that $fx \neq gx$. (Notice that a *concrete category* is generally assumed to be equipped with a *faithful* forgetful functor with values in **Set**, which here need not be true.)

In the examples below, we start by choosing the relevant forgetful functor, as an 'underlying set' in some sense, and then define the standard point as its representative object. Notice that the free point need *not* be terminal or initial (typically, it is not so in a pointed dI1-category); furthermore, the set $U(E) = \mathbf{A}(E, E)$ may have more than one element, as in case (d).

(a) If $\mathbf{A}$ is **Top**, p**Top** (or d**Top**), the forgetful functor is given by the underlying set, and the standard point is the singleton $\{*\}$, with its unique $\mathbf{A}$-structure.

(b) For pointed preordered spaces (see Section 1.5.5) the forgetful functor is given by the underlying set, and the free point is the *two-point* set $\mathbf{S}^0 = \{-1, 1\}$, with the discrete topology and the discrete order (i.e. the equality relation), pointed at 1 (for instance). The isomorphism $|X| = \mathbf{A}(\mathbf{S}^0, X)$ comes precisely from the fact that $\mathbf{S}^0$ has one 'free point', which can be mapped to any point of the pointed preordered space X.

(c) For **Cat**, the (non-faithful) forgetful functor is given by the set of objects, and the standard point is the singleton category $\mathbf{1} = \{*\}$.

(d) For chain complexes, the (non-faithful) forgetful functor is given by the underlying set of the 0-component, and the free point is the abelian group $\mathbf{Z}$ (in degree 0). For *directed chain complexes* (see Section 2.1.1), we will take the set of (weakly) positive elements of the 0-component; the free point will be the abelian group $\uparrow\mathbf{Z}$ with natural order (in degree 0): it is a reversive object but cannot be identified with $\uparrow\mathbf{Z}^{\mathrm{op}}$ (which has the opposite order). Similarly, the sets $U(A)$ and $UR(A)$, of (weakly) positive or negative elements of A_0, are in canonical bijection but cannot be identified.

Now, let $\mathbf{A}$ be concrete dI1-category; for the sake of simplicity, we assume that $E = E^{\mathrm{op}}$ (which simply allows us to omit the specified isomorphism $E \to E^{\mathrm{op}}$). The object $\mathbf{I} = I(E)$ inherits the structure of a *dI1-interval* in $\mathbf{A}$: by this we mean that it is equipped with four maps: two *faces* (∂^α), a *degeneracy* (e) and a *reflection* (r)

$$\partial^\alpha\colon E \rightrightarrows \mathbf{I}\colon e, \qquad r\colon \mathbf{I} \to \mathbf{I}^{\mathrm{op}} \qquad (\alpha = \pm), \qquad (1.47)$$

satisfying the equations

$$e\partial^\alpha = 1 \colon E \to E, \qquad r^{\mathrm{op}}.r = 1 \colon \mathbf{I} \to \mathbf{I},$$
$$e^{\mathrm{op}}.r = e \colon \mathbf{I} \to E, \qquad r.\partial^- = (\partial^+)^{\mathrm{op}} \colon E \to \mathbf{I}^{\mathrm{op}}. \tag{1.48}$$

A *point* of X is an element $x \in |X|$, i.e. a map $x \colon E \to X$, while a (directed) *path* in X is a map $a \colon \mathbf{I} \to X$, defined on $\mathbf{I} = I(E)$; thus, a path is a homotopy on the standard point E, between its endpoints

$$a \colon x^- \to x^+ \colon E \to X, \qquad x^\alpha = \partial^\alpha(a) = a\partial^\alpha. \tag{1.49}$$

Every point $x \colon E \to \mathbf{A}$ has a *trivial path*, $0_x = xe \colon x \to x$ (or degenerate path). Every path $a \colon x \to y$ has a reflected path $r(a) \colon y^{\mathrm{op}} \to x^{\mathrm{op}}$ in X^{op}

$$r(a) = a^{\mathrm{op}}.r \colon \mathbf{I} \to X^{\mathrm{op}}, \quad \partial^- r(a) = a^{\mathrm{op}}.r\partial^- = a^{\mathrm{op}}.\partial^{+\,\mathrm{op}} = y^{\mathrm{op}}. \tag{1.50}$$

Furthermore, given two paths $a^-, a^+ \colon x^- \to x^+$ with the same endpoints, a *2-homotopy* $h \colon a^- \to a^+$, or *homotopy with fixed endpoints*, is a map defined on the standard square $\mathbf{I}^{(2)} = I^2(E)$, with two parallel faces a^α and the other two degenerate at x^α

$$h \colon I^2(E) \to X, \qquad h.(I\partial^\alpha) = a^\alpha, \qquad h.(\partial^\alpha I) = x^\alpha e. \tag{1.51}$$

The *fundamental graph* $\uparrow\!\Gamma_1(X)$ has, for arrows, the classes of paths up to the equivalence relation *generated* by homotopy with fixed endpoints. In the same setting, extending the notions above, we will define the cubical set of singular cubes of X (Section 2.6.5). In a richer setting, allowing for concatenation of homotopies (and paths), $\uparrow\!\Gamma_1(X)$ will become the fundamental category $\uparrow\!\Pi_1(X)$ of X (cf. Chapter 3 and Section 4.5).

On the other hand, a *concrete dP1-category* $\mathbf{A}$ comes, by definition, with a *forgetful* functor $U = | - | \colon \mathbf{A} \to \mathbf{Set}$ such that $UR \cong U$. *This notion is* not *dual to the previous one* (necessarily, since the general notion of a category equipped with a functor to $\mathbf{Set}$ is not self-dual). Notice that we are not assuming that U be representable, nor faithful.

Here, a point in X will be an element of $|X|$, and a path in X will be an element $a \in |PX|$. The faces of a are obtained by applying the mappings $U(\partial^\alpha) \colon |PX| \to |X|$. By arguments parallel to the previous ones, one constructs, also here, the *fundamental graph* $\uparrow\!\Gamma_1(X)$.

Finally, a *concrete dIP1-category* $\mathbf{A}$ is a dIP1-category with a representable forgetful functor $U = \mathbf{A}(E, -) \colon \mathbf{A} \to \mathbf{Set}$, and $E = R(E)$.

Then the previous constructions coincide, since, because of the adjunction $I \dashv P$, we have a canonical bijection:

$$|PX| = \mathbf{A}(E, PX) = \mathbf{A}(I(E), X) = \mathbf{A}(\mathbf{I}, X). \tag{1.52}$$

1.2.5 Monoidal and cartesian structures

Let $\mathbf{A} = (\mathbf{A}, \otimes, E, s)$ be a symmetric monoidal category (see [M3, EK, Ke2]; or Section A4.1 for a brief review of definitions). We shall always omit the isomorphisms pertaining to the monoidal structure, except for the symmetry isomorphism

$$s(X, Y) \colon X \otimes Y \to Y \otimes X. \tag{1.53}$$

Let us assume that $\mathbf{A}$ is equipped with a reversor $R \colon \mathbf{A} \to \mathbf{A}$. Here, this will mean an involutive (covariant) automorphism $R(X) = X^{\mathrm{op}}$, which is strictly monoidal and consistent with the symmetry:

$$E^{\mathrm{op}} = E, \qquad\qquad (X \otimes Y)^{\mathrm{op}} = X^{\mathrm{op}} \otimes Y^{\mathrm{op}},$$
$$(s(X, Y))^{\mathrm{op}} = s(X^{\mathrm{op}}, Y^{\mathrm{op}}). \tag{1.54}$$

Thus, $\mathbf{A}$ has a forgetful functor $U = \mathbf{A}(E, -) \colon \mathbf{A} \to \mathbf{Set}$. (All the examples of Section 1.2.4 can be obtained in this way, for a suitable monoidal structure on $\mathbf{A}$.)

A *dI1-interval* $\mathbf{I}$ in $\mathbf{A}$ has, by definition, the structure described above (within a concrete dI1-category, see (1.47) and (1.48)). Then, the tensor product $- \otimes \mathbf{I}$ yields a dI1-structure:

$$I(X) = X \otimes \mathbf{I}, \qquad\qquad \partial^\alpha X = X \otimes \partial^\alpha \colon X \to IX,$$
$$eX = X \otimes e \colon IX \to X, \qquad rX = X^{\mathrm{op}} \otimes r \colon IRX \to RIX, \tag{1.55}$$

called the *symmetric monoidal dI1-structure* defined by the interval $\mathbf{I}$; this structure is concrete (1.2.4), with standard point E and standard interval $I(E) = E \otimes \mathbf{I} = \mathbf{I}$. We will see, in Section 4.1.4, that this structure is always a *symmetric* dI1-category, in the sense that the transposition $s \colon I^2 \to I^2$ given by the symmetry $s(\mathbf{I}, \mathbf{I})$ of the tensor product satisfies the relevant axioms of consistency with the rest of the structure.

Now, if the dI1-interval $\mathbf{I}$ is an exponentiable object (Section A4.2) in $\mathbf{A}$, we have a *symmetric monoidal dIP1-structure*, with

$$P(Y) = Y^{\mathbf{I}}, \qquad\qquad \partial'^\alpha Y = Y^{\partial^\alpha} \colon PY \to Y,$$
$$e'Y = Y^e \colon Y \to PY, \quad r'Y = (Y^{\mathrm{op}})^r \colon RPY \to PRY. \tag{1.56}$$

A *cartesian dI1-category* (or dIP1-category) is a symmetric monoidal dI1-category (or dIP1-category) where the tensor product is the cartesian one (equipped with the canonical symmetry isomorphism). For instance, p**Top**, **Top** and **Cat** are cartesian dIP1-categories; the first two are not cartesian closed.

On the other hand, a dI1-structure on a symmetric monoidal category **A** with reversor gives a dI1-interval $\mathbf{I} = I(E)$, with $\partial^\alpha = \partial^\alpha E \colon E \to \mathbf{I}$, and so on. But notice that this dI1-interval may not give back the original structure, when the latter is not monoidal. (For instance, the dI1-structure of *pointed* d-spaces is monoidal with respect to the smash product, but is not so with respect to the cartesian product; see Section 1.5.5.)

Occasionally, we will also consider a formal interval in a *non-symmetric* monoidal category with reversor. This situation requires more care, as it originates a left cylinder $\mathbf{I} \otimes X$, a *right cylinder* $X \otimes \mathbf{I}$ and – as a consequence – left and right homotopies; for instance, this is the case of cubical sets (Section 1.6).

1.2.6 Functors and preservation of homotopies

Let **A** and **X** be dI1-categories; we shall write their structures with the same letters $(R, I, \partial^\alpha, e, r)$.

We say that a functor $H \colon \mathbf{A} \to \mathbf{X}$ *preserves homotopies* if for every homotopy $f \to g$ in **A** there exists some homotopy $Hf \to Hg$ in **X**.

Replacing this existence condition with structure, a *lax dI1-functor* $H = (H, i, h) \colon \mathbf{A} \to \mathbf{X}$ will be a functor H equipped with two natural transformations i, h which satisfy the following conditions (implying that i is invertible)

$$i \colon RH \to HR, \quad h \colon IH \to HI \quad (comparisons),$$
$$(RiR).i = 1_{RH},$$
(1.57)

$$
\begin{array}{ccccc}
H & \xrightarrow{\partial^\alpha H} & IH & \xrightarrow{eH} & H \\
& H\partial^\alpha \searrow & \downarrow h & \nearrow He & \\
& & HI & &
\end{array}
\qquad
\begin{array}{ccccc}
IRH & \xrightarrow{Ii} & IHR & \xrightarrow{hR} & HIR \\
rH \downarrow & & & & \downarrow Hr \\
RIH & \xrightarrow{Rh} & RHI & \xrightarrow{iI} & HRI
\end{array}
$$

It follows that every homotopy $\varphi \colon f^- \to f^+ \colon A \to B$ in **A** (represented by a map $\hat{\varphi} \colon IA \to B$ with $\varphi \partial^\alpha = f^\alpha$) gives a homotopy in $\mathbf{X}$

$$H\varphi\colon Hf^- \to Hf^+ \colon HA \to HB,$$
$$(H\varphi)\hat{\ } = H(\hat{\varphi}).hA \colon IHA \to HIA \to HB, \qquad (1.58)$$
$$H(\hat{\varphi}).hA.\partial^\alpha HA = H(\hat{\varphi}.\partial^\alpha A) = Hf^\alpha,$$

consistently with faces (as checked above), trivial homotopies and reflection: $H(0_f) = 0_{H(f)}$, $H(\varphi^{\mathrm{op}}) = (H\varphi)^{\mathrm{op}}$.

The dual notion is a *lax dP1-functor* $K = (K, i, k) \colon \mathbf{A} \to \mathbf{X}$ between dP1-categories, with natural transformations $i \colon KR \to RK$ and $k \colon KP \to PK$ satisfying dual conditions

$$i \colon KR \to RK, \quad k \colon KP \to PK \quad (\textit{comparisons}),$$
$$(RiR).i = 1_{KR}, \qquad\qquad\qquad\qquad\qquad\qquad (1.59)$$

$$
\begin{array}{ccccc}
K \xrightarrow{Ke} KP \xrightarrow{K\partial^\alpha} K & & KRP \xrightarrow{iP} RKP \xrightarrow{Rk} RPK \\
\end{array}
$$

Now, a lax dI1-functor $H \colon \mathbf{A} \to \mathbf{X}$ between *dIP1-categories* becomes automatically a lax dP1-functor, with the inverse natural isomorphism $i^{-1} \colon HR \to RH$ and the comparison k mates to h under the adjunction $I \dashv P$ (Section A5.3)

$$k \colon HP \to PH, \quad kA = (HPA \to PIHPA \to PHIPA \to PHA). \quad (1.60)$$

The latter necessarily satisfies the coherence conditions (1.59). We say that H, equipped with the natural isomorphism i and the natural transformations $h \colon IH \to HI$, $k \colon KP \to PK$ (mates under the adjunction $I \dashv P$), is a *lax dIP1-functor*.

A *strong dI1-functor* $U \colon \mathbf{A} \to \mathbf{X}$ between dI1-categories is a lax dI1-functor whose comparisons are invertible. More particularly, U is *strict* if all comparisons are identities; then, the functor U commutes strictly with the structures

$$RU = UR, \quad\quad IU = UI, \quad\quad \partial^\alpha U = U\partial^\alpha \colon U \to IU = UI,$$
$$eU = Ue \colon IU = UI \to U, \quad\quad rU = Ur \colon IRU \to IRU. \qquad (1.61)$$

In this case, a functor $H \colon \mathbf{X} \to \mathbf{A}$ right adjoint to U inherits comparisons $i \colon RH \to HR$ and $h \colon IH \to HI$ making it a lax dI1-functor

(generally non-strong)

$$iX = (RHX \to HURHX = HRUHX \to HRX),$$
$$hX = (IHX \to HUIHX = HIUHX \to HIX). \qquad (1.62)$$

(The involutions R of **A** and **X** give $UR \dashv RH$ and $RU \dashv HR$; since $UR = RU$, the uniqueness of the right adjoint gives a canonical isomorphism $RH \to HR$, which can be computed combining units and counits as above.)

On the other hand, a left adjoint $D \dashv U$ inherits a transformation $DI \to ID$, *which is of no utility for transforming homotopies*. Various examples will be seen below (starting with Section 1.2.8).

1.2.7 Dualities

First, *categorical duality* turns a dI1-category **A** into the dual category $\mathbf{A}^*$ with a dP1-category structure: the reversor R^*, the path functor $P = I^*$, the natural transformations $(\partial^\alpha)^* \colon P \to 1$, etc. This duality *reverses the direction of arrows* and takes a homotopy $\varphi \colon f \to g \colon X \to Y$ in **A** (defined by a map $\varphi \colon IX \to Y$) to a homotopy $\varphi^* \colon f^* \to g^* \colon Y \to X$ of $\mathbf{A}^*$ (defined by the map $\varphi^* \colon Y \to PX$).

Second, we call *R-duality*, or *reflection-duality*, the fact of turning a dI1-category **A** into the dI1-category $\mathbf{A}^R$ with reversed faces $(\partial^R)^- = \partial^+$, $(\partial^R)^+ = \partial^-$ (all the rest being unchanged). This *reverses the direction of homotopies* (including paths), as it takes a homotopy $\varphi \colon f \to g \colon X \to Y$ in **A** to the homotopy $\varphi^R \colon g \to f \colon X \to Y$ in $\mathbf{A}^R$. The functor R can be viewed as a dI1-isomorphism $\mathbf{A}^R \to \mathbf{A}$, and one can replace φ^R with the reflected homotopy, in **A** (see (1.26))

$$R(\varphi^R) = \varphi^{\mathrm{op}} \colon g^{\mathrm{op}} \to f^{\mathrm{op}} \colon X^{\mathrm{op}} \to Y^{\mathrm{op}}, \quad (\varphi^{\mathrm{op}})\hat{} = R(\hat{\varphi}).rX. \qquad (1.63)$$

In other words, R-duality amounts to applying the reversor R to objects and maps, and reflecting homotopies with the procedure $\varphi \mapsto \varphi^{\mathrm{op}}$.

Notice that we are writing $\mathbf{A}^*$ the dual of a (generally large) dI1-category, while *within the directed structure of* **Cat** (Section 1.1.6), the dual of a small category is written as $RX = X^{\mathrm{op}}$. Small categories are viewed here as directed spaces, in their own right, and also as directed algebraic structures used to study other directed spaces via their fundamental categories.

1.2.8 Some forgetful functors and their adjoints

The categories **Top** (of topological spaces) and p**Top** (of preordered topological spaces) have a canonical dIP1-structure, described in the previous section (the former is reversible).

The forgetful functor $U:$ p**Top** $\to$ **Top** is a strict dI1-functor (Section 1.2.6) and – as a consequence – a *lax* dP1-functor; note that the comparison

$$UP \to PU, \qquad UP(Y) = U(Y^{\uparrow \mathbf{I}}) \subset (UY)^{\mathbf{I}} = PU(Y), \qquad (1.64)$$

is indeed *not* invertible.

We also know (by Section 1.2.6, again) that the right adjoint $D':$ **Top** $\to$ p**Top** (which equips a topological space with the chaotic preorder, see Section 1.1.3) inherits the structure of a *lax* dI1-functor, with $RD' = D'R$. In fact, the (non-invertible) comparison

$$ID' \to D'I, \qquad ID'(S) = D'(S) \times {\uparrow}\mathbf{I} \to D'(S \times \mathbf{I}) = D'I(S), \qquad (1.65)$$

is the identity mapping between two preordered spaces having the same underlying topological space and comparable preorders: the second is coarser, and strictly so (unless S is empty).

Always because of general reasons (Section 1.2.6), the left adjoint $D:$ **Top** $\to$ p**Top** (providing the discrete order) inherits a natural transformation

$$DI \to ID, \qquad DI(S) = D(S \times \mathbf{I}) \to D(S) \times {\uparrow}\mathbf{I} = ID(S), \qquad (1.66)$$

which is not invertible, and of no utility for homotopies and paths; and indeed, it is obvious that the functor $D:$ **Top** $\to$ p**Top** cannot take paths to directed paths.

We have also seen, in Section 1.1.6, that **Cat** has a non-reversible dIP1-structure, based on the directed interval **2**, and a reversible one, based on the groupoid **i** and written **Cat**$_\text{i}$. The category **Gpd** of small groupoids is a dIP1-subcategory of the latter (Section 1.2.2), and the embedding $U:$ **Gpd** $\to$ **Cat** is a lax dI1-functor, with comparisons:

$$\begin{aligned} &i: RU \to UR, \qquad iX: X^{\text{op}} \to X, \quad a \mapsto a^{-1}, \\ &h: IU \to UI, \qquad hX: X \times \mathbf{2} \subset X \times \mathbf{i}. \end{aligned} \qquad (1.67)$$

1.2.9 A setting based on directed homotopies

As foreshadowed in Section 1.2.3, it may be interesting to establish a *self-dual* setting based on directed homotopies, without having to choose between cylinder or path representations. (This setting is related to the notion of h-category recalled in Section A5.1, which is here enriched with a reversor.)

Say that a *dh1-structure* on the category $\mathbf{A}$ consists of the following data:

(a) a *reversor* $R \colon \mathbf{A} \to \mathbf{A}$ (again, an involutive covariant automorphism, written as $R(X) = X^{\mathrm{op}}$);

(b) for each pair of parallel morphisms $f, g \colon X \to Y$, a set of (directed) *homotopies* $\mathbf{A}_2(f, g)$, whose elements are written as $\varphi \colon f \to g \colon X \to Y$ (or $\varphi \colon f \to g$), so that each map f has a *trivial* (or *degenerate*, or *identity*) *endohomotopy* $0_f \colon f \to f$;

(c) an involutive action of R on homotopies, which turns $\varphi \colon f \to g$ into $\varphi^{\mathrm{op}} \colon g^{\mathrm{op}} \to f^{\mathrm{op}}$;

(d) a *whisker composition* for maps and homotopies

$$X' \xrightarrow{h} X \underset{g}{\overset{f}{\rightrightarrows}} {\downarrow \varphi}\; Y \xrightarrow{k} Y' \qquad k \circ \varphi \circ h \colon kfh \to kgh. \qquad (1.68)$$

These data must satisfy the following axioms:

$$k' \circ (k \circ \varphi \circ h) \circ h' = (k'k) \circ \varphi \circ (hh') \qquad (associativity),$$

$$1_Y \circ \varphi \circ 1_X = \varphi, \qquad k \circ 0_f \circ h = 0_{kfh} \qquad (identities), \qquad (1.69)$$

$$(k \circ \varphi \circ h)^{\mathrm{op}} = k^{\mathrm{op}} \circ \varphi^{\mathrm{op}} \circ h^{\mathrm{op}}, \qquad (0_f)^{\mathrm{op}} = 0_{f^{\mathrm{op}}} \qquad (reflection).$$

(More formally, this structure can be viewed as a category with reversor, enriched over the category of reflexive graphs; the latter is endowed with a suitable symmetric monoidal closed structure, see Section 4.3.3.)

Implicitly, we have already seen in Section 1.2.3 that every dI1-category has a dh1-structure. Dually, the same is true of dP1-categories.

However, developing the theory of dh1-categories, *the advantage of having a self-dual setting would soon disappear.* We would soon be obliged to assume the existence of homotopy pushouts (see Sections 1.3 and 1.7), in order to construct the cofibre sequence of a map; or, dually, the existence of homotopy pullbacks, to construct the fibre sequence of a map (Section 1.8). This would reintroduce the cylinder, as the homotopy pushout of two identities (Section 1.3.5); or, dually, the cocylinder as the homotopy pullback of two identities. More precisely, it is easy to prove that a dh1-category with all homotopy pushouts is the same as a dI1-category with all homotopy pushouts.

This is why we prefer to work from the beginning with the cylinder or the path endofunctor.

1.3 First-order homotopy theory by the cylinder functor, I

We study here various notions of directed homotopy equivalence and the homotopy pushout of two maps, in an arbitrary dI1-category **A** (Section 1.2.1).

Higher order properties of homotopy pushouts need a richer setting, and are deferred to Chapter 4. The present matter comes mostly from [G3, G7, G8].

1.3.1 Future and past homotopy equivalence

Let **A** be a dI1-category (or a dP1-category). A *future homotopy equivalence* $(f, g; \varphi, \psi)$ between the objects X, Y (or *homotopy equivalence in the future*) consists of a pair of maps and a pair of homotopies, the *units*

$$f\colon X \rightleftarrows Y \colon g, \qquad \varphi\colon 1_X \to gf, \quad \psi\colon 1_Y \to fg, \qquad (1.70)$$

which go *from* the identities of X, Y *to* the composed maps.

Future homotopy equivalences compose: given a second future homotopy equivalence

$$h\colon Y \rightleftarrows Z \colon k, \qquad \vartheta\colon 1_Y \to kh, \quad \zeta\colon 1_Z \to hk, \qquad (1.71)$$

their *composite* will be:

$$hf\colon X \rightleftarrows Z \colon gk, \quad g\vartheta f.\varphi\colon 1_X \to gk.hf, \quad h\psi k.\zeta\colon 1_Z \to hf.gk. \quad (1.72)$$

Thus, being future homotopy equivalent objects is an equivalence relation.

By R-duality (Section 1.2.7), a *past homotopy equivalence*, or *homotopy equivalence in the past*, has homotopies in the opposite direction, called *counits*, *from* the composed maps (gf, fg) *to* the identities

$$f\colon X \rightleftarrows Y \colon g, \qquad \varphi\colon gf \to 1_X, \quad \psi\colon fg \to 1_Y. \qquad (1.73)$$

More particularly, given a pair of morphisms $i\colon X_0 \rightleftarrows X \colon p$ such that $pi = \mathrm{id}X_0$ and there is a homotopy $\varphi\colon \mathrm{id}X \to ip$ (resp. $\varphi\colon ip \to \mathrm{id}X$), we say that X_0 is a *future* (resp. *past*) *deformation retract* of X, that i is the *embedding of a future* (resp. *past*) *deformation retract* and that p is a *future* (resp. *past*) *deformation retraction*. We add the adjective *strong* to each of these terms if one can choose the homotopy φ so that $\varphi i = 0_i$. In all these cases i is a split monomorphism (i.e. it admits a retraction p with $pi = \mathrm{id}$) and X_0 is a regular subobject of X.

(Recall that, in an arbitrary category, a *regular subobject* of an object X is an equaliser $i\colon X_0 \to X$ of some pair of maps $X \rightrightarrows Y$, or more precisely the class of all equalisers of such a pair; see Section A1.3). In **Top**, a regular subobject of X amounts to a subspace $X_0 \subset X$; in p**Top**, it amounts to a subspace equipped with the induced preorder.)

A lax dI1-functor $H = (H, i, h)\colon \mathbf{A} \to \mathbf{X}$ acts on homotopies (see (1.58)), whence it also preserves future and past homotopy equivalences, as well as future and past deformation retracts. The same holds for a lax dP1-functor. More generally, it also holds for a functor $H\colon \mathbf{A} \to \mathbf{X}$ which preserves homotopies, in the sense of Section 1.2.6, even though now this preservation is less constructive: H is not provided with a precise way of operating on homotopies.

If the structure of $\mathbf{A}$ is reversible, future and past homotopy equivalences coincide, and are simply called *homotopy equivalences*; the same holds for deformation retracts.

1.3.2 Contractible and co-contractible objects

Let $\mathbf{A}$ be always a dI1-category (or a dP1-category). If $\mathbf{A}$ has a terminal object $\top$, an object X is said to be *future contractible* if it is future homotopy equivalent to $\top$. Equivalently, X admits the terminal object $\top$ as a future deformation retract, or also, there is a homotopy $\varphi\colon 1_X \to f$ from the identity to a *constant* map $f\colon X \to X$ (a morphism which factorises through the terminal object).

Categorical duality turns the terminal object $\top$ of $\mathbf{A}$ into the initial object $\bot$ of $\mathbf{A}^*$, and future contractible objects of $\mathbf{A}$ into *future co-contractible objects* of $\mathbf{A}^*$ (i.e. objects which are future homotopy equivalent to the initial object). On the other hand, R-duality preserves the terminal and turns future contractible objects of $\mathbf{A}$ into *past contractible* objects of $\mathbf{A}^R$ (Section 1.2.7).

Future (or past) contractibility is adequate for categories of an 'extensive character', like (directed) topological spaces and cubical sets. Future (or past) *co*-contractibility is adequate for categories of an 'intensive character', like unital differential graded algebras (or the opposite categories of the previous ones). Obviously, the two notions coincide for pointed categories, like pointed spaces or 'general' differential graded algebras (without unit assumption).

Let us also remark that, when the initial object is *absolute* (i.e. every map $X \to \perp$ is invertible), as happens with 'spaces' and cubical sets, then the only co-contractible object is the initial one. Dual facts hold for unital differential algebras: the terminal object (the null algebra) is absolute and the unique contractible object.

One should also be aware that, while *contractibility* and *constant maps* are always based on the terminal object (and the dual notions on the initial one), the abstract notion of a *point*, already considered in Section 1.2.4, behaves differently. Thus, for pointed topological spaces, the free point is the unit $\mathbf{S}^0$ of the smash product, while (co)contractibility is based on the zero object.

Examples in p**Top** will be considered below (Section 1.3.4). The terminal object will be re-examined in Section 1.7.0.

1.3.3 Coarse d-homotopy equivalence

Let **A** be always a dI1-category (or a dP1-category). We consider here some coarse equivalence relations produced by the homotopies of **A**. But we will see, in Chapter 3, that **Cat** (with the directed homotopies considered above, in Section 1.1.6) has *finer* equivalence relations, of a deeper interest for the study of the fundamental category of a directed space.

First, the existence of a homotopy $\varphi \colon f \to g$ simply defines a reflexive relation, closed under composition with maps (Section 1.2.3). We shall denote as $f \sim_1 g$ the equivalence relation generated by the latter, which amounts to the existence of a finite sequence of homotopies, forward or backward

$$f = f_0 \to f_1 \leftarrow f_2 \ldots \to f_n = g. \tag{1.74}$$

This relation is a congruence of categories, because the original relation is closed under whisker composition, and will be called the *homotopy congruence* in **A**. One can define the *homotopy category* of **A** as the quotient

$$\mathrm{Ho}_1(\mathbf{A}) = \mathbf{A}/\sim_1, \tag{1.75}$$

but we will make little use of this construction, which gives a very poor model of **A**.

Second, the equivalence relation between *objects* of **A** generated by future and past homotopy equivalence will be called *coarse d-homotopy* *equivalence.* This can be controlled, *step by step*: we shall speak of an *n-step coarse d-homotopy equivalence* for a sequence of n homotopy equivalences in the future or in the past

$$X = X_0 \rightleftarrows X_1 \rightleftarrows \ldots \rightleftarrows X_n = Y. \tag{1.76}$$

The composed map $f: X \to Y$ will also be called a *coarse d-homotopy equivalence.* Note that f and the other composite $g: Y \to X$ in (1.76) give $gf \sim_1 \operatorname{id} X$, $fg \sim_1 \operatorname{id} Y$, so that X and Y are isomorphic in the homotopy category $\operatorname{Ho}_1(\mathbf{A})$.

Finally, a subobject $u: X_0 \to X$ will be said to be the embedding of a *coarse directed deformation retract in n steps* if it is the composite of a finite sequence

$$X_0 \to X_1 \to \ldots \to X_n = X, \tag{1.77}$$

where each morphism is the embedding of a future or past deformation retract; and in *precisely n steps* if a shorter similar chain does not exist. Then X_0 is coarse d-homotopy equivalent to X in n steps.

If $\mathbf{A}$ has a terminal object $\top$, we say that the object X is *coarsely d-contractible in n steps*, if there is a map $\top \to X$ which is a coarse directed deformation retract in n steps (note that this condition is stronger than saying coarsely d-homotopy equivalent to the terminal, in n steps).

If the structure of $\mathbf{A}$ is reversible and homotopies can be concatenated, the existence of a homotopy $\varphi: f \to g$ amounts to the equivalence relation $f \sim_1 g$. Homotopy equivalences coincide with the maps of $\mathbf{A}$ which become isomorphisms in $\operatorname{Ho}_1(\mathbf{A})$, and thus satisfy the 'two out of three' property: namely, if in a composite $h = gf$ two maps out of f, g, h are homotopy equivalences, so is the third.

1.3.4 Examples

Let us consider these notions in p**Top**, where the terminal object is the singleton $\{*\}$ (as a preordered space).

A preordered subspace $X_0 \subset X$ is (the embedding of) a future deformation retract of X if and only if there is a map φ (of preordered spaces) such that

$$\varphi: X \times \uparrow\mathbf{I} \to X, \qquad \varphi(x, 0) = x, \quad \varphi(x, 1) \in X_0 \qquad (\text{for } x \in X). \tag{1.78}$$

This implies that X is *upper bounded* by X_0 for the path preorder (each point of X has some upper bound in X_0). Thus, a preordered space X

which is future contractible has *a* maximum for its preorder: the point $i(*)$, where $i\colon \{*\} \to X$ is the embedding of a future deformation retract.

For instance, the cylinder $X \times {\uparrow}\mathbf{I}$ has a strong future deformation retract at its upper basis $\partial^+ (X) \subset IX$, by the lower connection (reaching the upper basis at time $t' = 1$)

$$g^-\colon (X \times {\uparrow}\mathbf{I}) \times {\uparrow}\mathbf{I} \to X \times {\uparrow}\mathbf{I}, \qquad g^-(x, t, t') = (x, \max(t, t')). \quad (1.79)$$

Symmetrically, the lower basis $\partial^-(X)$ is a strong past deformation retract of IX.

The left half-line ${\uparrow}]-\infty, 0] \subset {\uparrow}\mathbf{R}$ is future contractible to 0 (which is reached by the d-homotopy $\varphi(x, t) = -x(t - 1)$, at time $t = 1$), but not past contractible. The d-line ${\uparrow}\mathbf{R}$ is coarsely d-contractible in two steps, as ${\uparrow}]-\infty, 0]$ is a past deformation retract of ${\uparrow}\mathbf{R}$ (reached by the homotopy $\psi(x, t) = \min(x, tx)$ at $t = 0$); and two steps are needed, since the line has no maximum or minimum.

Similarly, all ${\uparrow}\mathbf{R}^n$ are coarsely d-contractible in two steps (for $n > 0$): one can take as a past deformation retract $X = {\uparrow}]-\infty, 0]^n$, moving all points of the complement to the boundary of X along lines parallel to the main diagonal $x_1 = \cdots = x_n$.

Consider now the following subspaces of the ordered plane, V and the *infinite stairway* W

$$\begin{aligned}
V &= ([0, 1] \times \{0\}) \cup (\{0\} \times [0, 1])) \subset {\uparrow}\mathbf{R}^2, \\
W &= \bigcup_{k \in \mathbf{Z}} (([k, k+1] \times \{-k\}) \cup (\{k\} \times [-k, 1-k])),
\end{aligned} \qquad (1.80)$$

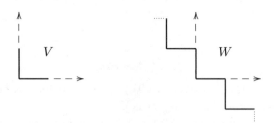

V is past contractible (to the origin). W is not even coarsely d-contractible: it is easy to see that the existence of a homotopy $f \to \mathrm{id}W$ or $\mathrm{id}W \to f$ implies $f(W) = W$. But a *finite stairway* consisting of $2n$ or $2n-1$ consecutive segments of W is coarsely d-contractible in n steps (in the even case, each step contracts the first and last segment; in the odd one, the first step contracts one of them).

1.3.5 Homotopy pushouts

Let $f\colon X \to Y$ and $g\colon X \to Z$ be two morphisms with the same domain, in the dI1-category $\mathbf{A}$. The *standard homotopy pushout*, or *h-pushout*, *from f to g* is a four-tuple $(A; u, v; \lambda)$, as in the left diagram below, where $\lambda\colon uf \to vg\colon X \to A$ is a homotopy satisfying the following universal property (of *cocomma squares*)

$$
\begin{array}{ccc}
X \xrightarrow{\ g\ } Z & \qquad & X \xrightarrow{\ \mathrm{id}\ } X \\
f\downarrow \ \ {}_{\lambda}\nearrow\ \downarrow v & & \mathrm{id}\downarrow \ \ {}_{\lambda}\nearrow\ \downarrow \partial^{+} \\
Y \xrightarrow[\ u\ ]{} A & & X \xrightarrow[\ \partial^{-}\ ]{} IX
\end{array}
\qquad (1.81)
$$

For every $\lambda'\colon u'f \to v'g\colon X \to A'$, there is precisely one map $h\colon A \to A'$ such that $u' = hu$, $v' = hv$, $\lambda' = h \circ \lambda$.

The existence of the solution depends on $\mathbf{A}$, of course; its uniqueness *up to isomorphism* is obvious. The object A, a 'double mapping cylinder', will be denoted as $I(f, g)$. When g or f is an identity, one has a *mapping cylinder*, $I(f, X)$ or $I(X, g)$.

The reflection $rX\colon IRX \to RIX$ induces an isomorphism

$$
r^{I}\colon I(Rg, Rf) \to RI(f, g),
$$

which will be called the *reflection of h-pushouts*.

As shown in the right diagram above, the cylinder IX itself is the h-pushout of the pair $(\mathrm{id}X, \mathrm{id}X)$: equipped with the obvious *structural* homotopy ∂ (*cylinder evaluation*, represented by the identity of the cylinder)

$$
\partial\colon \partial^{-} \to \partial^{+}\colon X \to IX, \qquad \hat{\partial} = \mathrm{id}(IX); \qquad (1.82)
$$

it establishes a bijection between maps $h\colon IX \to W$ and homotopies $h \circ \partial\colon h\partial^{-} \to h\partial^{+}\colon X \to W$, because of the very definition of homotopy.

On the other hand, every homotopy pushout can be constructed using the cylinder and the ordinary colimit of the following diagram, which – as shown on the right-hand side – amounts to three ordinary pushouts (or two):

$$
\begin{array}{ccc}
\begin{array}{ccc}
X & \xrightarrow{\ g\ } & Z \\
{}_{\partial^{+}}\downarrow & & \big| \\
X \ {}^{\partial^{-}}\!\!\to IX & & \big|v \\
f\downarrow & \ \ {}_{\lambda}\searrow & \downarrow \\
Y & \xrightarrow[\ u\ ]{} & I(f,g)
\end{array}
& \qquad &
\begin{array}{ccc}
X & \xrightarrow{\ g\ } & Z \\
{}_{\partial^{+}}\downarrow & & \big| \\
X \xrightarrow{\partial^{-}} IX & \dashrightarrow & I(X,g) \\
f\downarrow & & \downarrow \\
Y \to I(f,X) & \dashrightarrow & I(f,g)
\end{array}
\end{array}
\qquad (1.83)
$$

Therefore, a dl1-category **A** has homotopy pushouts if and only if it has *cylindrical colimits*, i.e. by definition, the colimit $I(f,g)$ of each diagram of the previous type.

The existence of ordinary pushouts in **A** is sufficient for this; but interestingly is *not* necessary. For instance, the category Ch.**D** of chain complexes on an *additive* category **D** (Section A4.6), equipped with ordinary homotopies, has all h-pushouts (which can be constructed *with the additive structure*), but it has cokernels (and pushouts) if and only if **D** has. Chain complexes of *free abelian groups* are a relevant case where all h-pushouts exist, while ordinary pushouts do not, see Section 1.3.6(d). (A non-reversible example can be constructed with the category of *directed* chain complexes of free abelian groups, cf. Section 4.4.5.)

Higher properties of h-pushouts need a richer structure on the cylinder functor, which will be studied in Chapter 4.

1.3.6 Examples

(a) The construction of the double mapping cylinder $I(f,g)$, in p**Top** is made clear by the following picture (where f, g are embeddings, for the sake of simplicity)

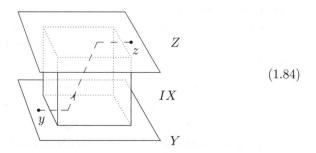

$$(1.84)$$

The space $I(f,g)$, as the colimit described above in (1.83), results of the pasting of the cylinder IX with the spaces Y, Z, under the following identifications (for $x \in X$):

$$I(f,g) = (Y + IX + Z)/\sim, \qquad [x,0] = [f(x)], \quad [x,1] = [g(x)]. \quad (1.85)$$

Thus, the structural maps $u\colon Y \to I(f,g)$ and $v\colon Z \to I(f,g)$ are always injective, while $\lambda\colon IX \to I(f,g)$ is necessarily injective *outside* of the bases $\partial^\alpha X$ (it is injective 'everywhere' if and only if both f and g are).

The preorder induced on $I(f,g)$ is also evident, and an (increasing) path from $y \in Y$ to $z \in Z$ (for instance) results from the concatenation of three paths in Y, IX and Z, as below, with $f(x_1) = f(x')$ and $g(x'') = g(x_3)$

$$a_1 : y \to f(x_1) \text{ in } Y, \qquad a_2 : (x', 0) \to (x'', 1) \text{ in } IX,$$
$$a_3 : g(x_3) \to z \text{ in } Z, \qquad ([f(x_1)] = [x', 0], \ [x'', 1] = [g(x_3)]). \tag{1.86}$$

(b) In **Cat**, $I(f,g)$ is called a *cocomma object*, and the left diagram (1.81) is called a *cocomma square*. The category $I(f,g)$ is, again, a quotient of a sum (as in (1.85)), modulo a generalised congruence of categories (also working on objects, as considered in [BBP]).

(c) For the category dCh.**Ab** of directed chain complexes of abelian groups, see Section 4.4.5.

(d) We have mentioned above (Section 1.3.5) that the category of free abelian groups and homomorphisms, say f**Ab**, lacks (some) pushouts, or equivalently (some) cokernels. Showing this fact is less obvious than one might think, since, for instance, the 'free cokernel' of multiplication by 2 in **Z** exists and is the null group. More generally, in the finite-dimensional case, taking the quotient modulo the torsion subgroup 'creates free cokernels', out of the ordinary ones.

Our example (which is not needed for the sequel) will be based on the fact that the countable power $A = \mathbf{Z}^{\mathbf{N}}$, also called the Specker group, is *not* free abelian, as proved by Specker [Sp], but has a jointly monic family of homomorphisms $p_i : A \to \mathbf{Z}$, its projections.

Take a free presentation (k, p) of A in **Ab**, and suppose for a contradiction that the monomorphism $k : F_1 \to F_0$ has a cokernel $q : F_0 \to F$ in f**Ab**.

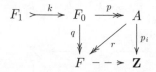

This homomorphism q is surjective (since its image in F must be free) and factorises through the ordinary cokernel $p : F_0 \to A$, which yields a surjective homomorphism $r : A \to F$. All projections $p_i : A \to \mathbf{Z}$ must factorise through r, since $p_i p : F_0 \to \mathbf{Z}$ is in f**Ab** and must factorise through q. It follows that the kernel of r is null and r is an isomorphism, which is absurd, because A is not free.

1.3.7 Functoriality

Assuming that the dI1-category **A** has h-pushouts, these form a functor $\mathbf{A}^\vee \to \mathbf{A}$, where $\vee$ is the *formal-span* category: $\bullet \leftarrow \bullet \to \bullet$. In fact, a morphism $(x, y, z): (f, g) \to (f', g')$ in $\mathbf{A}^\vee$ is a commutative diagram

$$
\begin{array}{ccccc}
Y & \xleftarrow{\;f\;} & X & \xrightarrow{\;g\;} & Z \\
{\scriptstyle y}\downarrow & & {\scriptstyle x}\downarrow & & \downarrow{\scriptstyle z} \\
Y' & \xleftarrow[f']{} & X' & \xrightarrow[g']{} & Z'
\end{array}
\tag{1.87}
$$

This yields a map $h = I(x, y, z): I(f, g) \to I(f', g')$, which is defined as suggested by the following diagram (where $\lambda': u'f' \to v'g'$ denotes the second h-pushout)

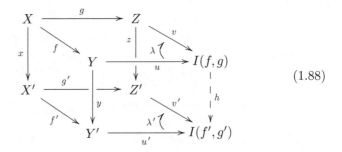

$$\tag{1.88}$$

1.3.8 Lemma (Pasting lemma for h-pushouts)

*Let **A** be a dI1-category and let λ, μ, ν be h-pushouts, as in the diagram below. Then, there is a comparison map $k: C \to D$ defined by the universal properties of λ, μ and such that:*

$$
kvu = a, \qquad kz = b, \qquad k \circ \mu = 0_{bg}, \tag{1.89}
$$

$$
\begin{array}{ccccccc}
X & \xrightarrow{\;f\;} & Y & \xrightarrow{\;g\;} & Z & = & Z \\
{\scriptstyle x}\downarrow & {\scriptstyle\lambda} & \downarrow{\scriptstyle y} & {\scriptstyle\mu} & \downarrow{\scriptstyle z} & & \downarrow{\scriptstyle b} \\
A & \xrightarrow{u} & B & \xdashrightarrow{v} & C & {\scriptstyle\nu} & \\
\| & & & & & \searrow & \downarrow \\
A & & \xrightarrow[\quad a \quad]{} & & & & D
\end{array}
$$

Again, the present setting is insufficient to get a comparison the other way round (see Theorem 4.6.2).

Proof First, we define a map $w\colon B \to C$ by the universal property of λ

$$wu = a, \qquad wy = bg, \qquad w\circ\lambda = \nu.$$

Then we define $k\colon C \to D$ by the universal property of μ, using the fact that $wy = bg$:

$$kv = w, \qquad kz = b, \qquad k\circ\mu = 0_{bg}.$$

$\square$

1.3.9 Lemma (Special pasting lemma)

Let $\mathbf{A}$ be a dI1-category. In the following diagram, λ is an h-pushout, the right-hand square is commutative and we let $\lambda' = v\lambda\colon vux \to zgf$.

$$
\begin{array}{ccccc}
X & \xrightarrow{\ f\ } & Y & \xrightarrow{\ g\ } & Z \\
\downarrow{\scriptstyle x} & {\scriptstyle \lambda} & \downarrow{\scriptstyle y} & & \downarrow{\scriptstyle z} \\
A & \xrightarrow{\ u\ } & B & \xrightarrow{\ v\ } & C
\end{array}
\tag{1.90}
$$

Then the triple (vu, z, λ') is the h-pushout of (x, gf) if and only if the right-hand square is an ordinary pushout.

Proof We give a proof based on the universal property of the h-pushout, but one could also use the cylindrical colimit (1.83) and some pasting properties of ordinary pushouts, which are well known in category theory.

First, we assume that the right square is an ordinary pushout, and prove the universal property for (vu, z, λ'). Given a triple (a, b, ν) as below

$$
\begin{array}{ccccc}
X & \xrightarrow{\ f\ } & Y & \xrightarrow{\ g\ } & Z = Z \\
\downarrow{\scriptstyle x} & {\scriptstyle \lambda} & \downarrow{\scriptstyle y} & & \downarrow{\scriptstyle z} \quad \searrow^{b} \\
A & \xrightarrow{\ u\ } & B & \dashrightarrow{\scriptstyle v} & C \quad {\scriptstyle \nu} \\
\| & & & {\scriptstyle w} & \downarrow \\
A & & \xrightarrow{\hspace{5em} a \hspace{5em}} & & D
\end{array}
\tag{1.91}
$$

we define a map $w\colon B \to C$ by the universal property of λ (as in the previous proof)

$$wu = a, \qquad wy = bg, \qquad w\circ\lambda = \nu. \tag{1.92}$$

Then we define $k\colon C \to D$ by the universal property of the pushout square:

$$kv = w, \qquad kz = b, \tag{1.93}$$

and conclude that

$$kvu = a, \qquad kz = b, \qquad k \circ \lambda' = \nu. \tag{1.94}$$

As to uniqueness, given a morphism $k\colon C \to D$ which satisfies (1.94), the composite $w = kv$ must satisfy (1.92), and is thus determined by the data (a, b, ν); then k, which satisfies (1.93), is also uniquely determined.

Conversely, let (vu, z, λ') be the h-pushout of (x, gf) and let us prove that the square $vy = zg$ is a pushout. Given a commutative square $wy = bg$, let $a = wu\colon A \to D$ and $\nu = w \circ \lambda\colon ax \to bgf$, as in diagram (1.91). Now, by the universal property of the outer h-pushout, the triple (a, b, ν) determines a unique $k\colon C \to D$ such that (1.94) holds; this implies (1.93), because the maps $kv, w\colon B \to D$ have the same composites with the h-pushout (u, y, λ). The fact that (1.93) determines $k\colon C \to D$ is proved in the same way. $\qquad\qquad\square$

1.4 Topological spaces with distinguished paths

We begin now the study of the category d**Top** of *spaces with distinguished paths*, or *d-spaces*. With respect to preordered spaces, this 'world' also contains objects having non-trivial loops, like the directed circle $\uparrow\mathbf{S}^1$ (Section 1.4.3), or point-like vortices (see Section 1.4.7). It is our main non-reversible world of a topological kind.

A preordered space X will always be viewed as a d-space by distinguishing the (weakly) increasing paths $a\colon \uparrow\mathbf{I} \to X$. But one should be warned that the canonical functor $\mathbf{d}\colon \mathrm{p}\mathbf{Top} \to \mathrm{d}\mathbf{Top}$ so defined is *not* an embedding, as it can identify different preorders of a space (Section 1.4.5).

1.4.0 Spaces with distinguished paths

The category p**Top** considered above (Section 1.1) behaves well, with respect to directed homotopy theory, but lacks interesting objects, like a 'directed circle'. Indeed, in a preordered space, every loop stays in a zone where the preorder is chaotic, and is therefore a reversible loop – even a constant one, if the preorder is antisymmetric. This deficiency can be overcome in various ways.

Our main solution, in this sense, will be the category d**Top** of d-spaces, introduced in [G8]. A *d-space* X, or *space with distinguished paths*, is a topological space equipped with a set dX of (continuous) maps $a \colon \mathbf{I} \to X$, called *distinguished paths* or *directed paths* or *d-paths*, satisfying three axioms:

(i) (*constant paths*) every constant map $\mathbf{I} \to X$ is directed;

(ii) (*partial reparametrisation*) dX is closed under composition with every (weakly) increasing map $\mathbf{I} \to \mathbf{I}$;

(iii) (*concatenation*) dX is closed under path-concatenation: if the d-paths a, b are consecutive in X (i.e. $a(1) = b(0)$), then their ordinary concatenation $a + b$ (Section 1.1.0) is also a d-path.

It is easy to see that directed paths are also closed under the *n-ary concatenation* $a_1 + \cdots + a_n$ of consecutive paths, based on the regular partition $0 < 1/n < 2/n < \cdots < 1$ of the standard interval; and, more generally, under a *generalised n-ary concatenation* based on an arbitrary partition of $[0, 1]$ in n subintervals. Note also that, in axiom (ii), we are not assuming that the increasing map $\mathbf{I} \to \mathbf{I}$ be *surjective*; therefore, any restriction of a d-path to a subinterval of $[0, 1]$ is distinguished (after reparametrisation on $\mathbf{I}$).

A *map of d-spaces*, or *d-map*, is a continuous mapping $f \colon X \to Y$ between d-spaces which preserves the directed paths: if $a \in dX$, then $fa \in dY$.

Here also, the forgetful functor $U \colon d\mathbf{Top} \to \mathbf{Top}$ has a left and a right adjoint, $D \dashv U \dashv D'$. Now, DS is the space S with the *discrete* d-structure (the finest), where the distinguished paths reduce to (all) the constant ones, while $D'S$ has the *natural* d-structure (the largest), where all (continuous) paths are distinguished. *A topological space will be viewed as a d-space by its* natural *structure* $D'S$, so that all its paths are retained; D' preserves products and subspaces.

Reversing d-paths, by the involution $r(t) = 1 - t$, yields the *opposite* d-space $RX = X^{\mathrm{op}}$, where $a \in d(X^{\mathrm{op}})$ if and only if ar is in dX. This defines the *reversor* endofunctor

$$R \colon d\mathbf{Top} \to d\mathbf{Top}, \qquad RX = X^{\mathrm{op}}. \qquad (1.95)$$

Following general terminology (Section 1.2.1), the d-space X is *reversive* if it is isomorphic to X^{op}. More particularly, we say that it is *reversible* if $X = X^{\mathrm{op}}$, i.e. if its distinguished paths are closed under reversion. *But notice that we do not want to extend such a definition to all categories with reversor*, since in **Cat** or **Cub** it would not give a reasonable notion (see the last remark in Section 1.6.4).

We shall consider in Section 1.9 other 'topological' settings for directed algebraic topology, like 'locally ordered' topological spaces, 'inequilogical spaces' and bitopological spaces in the sense of J.C. Kelly [Ky], which are rather defective with respect to the present setting. On the other hand we have already remarked that, in p**Top** and d**Top**, the directed structure is essentially 'one-dimensional' (Section 1.1.5): cubical sets and 'spaces with distinguished cubes' yield a richer directed structure, if a more complex one (Section 1.6).

1.4.1 Limits and colimits

The d-structures on a topological space S are closed under arbitrary intersection in the lattice of parts of $\mathbf{Top}(\mathbf{I}, S)$ and form therefore a complete lattice for the inclusion, or 'finer' relation (corresponding to the fact that id$S$ be directed).

A (directed) *subspace* $X' \subset X$ of a d-space X has the restricted structure, which selects those paths in X' that are directed in X; it is the less fine structure which makes the inclusion into a map. A (directed) *quotient* X/R has the quotient structure, generated by the *projected d-paths* via reparametrisation and finite concatenation (or, equivalently, via generalised finite concatenation, in the sense described above); it is the finest structure which makes the projection $X \to X/R$ into a map.

It follows easily that the category d**Top** has all limits and colimits, constructed as in **Top** and equipped with the initial or final d-structure for the structural maps. For instance a path $\mathbf{I} \to \Pi X_j$ with values in a product of d-spaces is directed if and only if all its components $\mathbf{I} \to X_j$ are, while a path $\mathbf{I} \to \Sigma X_j$ with values in a sum of d-spaces is directed if and only if it is directed in some summand X_j. Equalisers and coequalisers are realised as subspaces or quotients, in the sense described above.

If X is a d-space and $A \subset |X|$ is a *non-empty* subset, X/A will denote the d-quotient of X which identifies all points of A. More generally, for any subset A of $|X|$, one should define the quotient X/A as the following pushout of the inclusion $A \to X$

$$
\begin{array}{ccc}
A & \longrightarrow & X \\
\downarrow & \dashrightarrow & \downarrow \\
\{*\} & \longrightarrow & X/A
\end{array}
\qquad (1.96)
$$

which amounts to the previous description when $A \neq \emptyset$, but gives the sum $X + \{*\}$ when A is empty. Notice that we always get a (naturally) *pointed* d-space; these are dealt with below (Section 1.5.4).

Let now G be a group, written in additive notation (independently of commutativity). A (right) *action* of G on a d-space X is an action on the underlying topological space such that, for each $g \in G$, the induced homeomorphism

$$X \to X, \qquad x \mapsto x + g, \tag{1.97}$$

is a map of d-spaces (and therefore an isomorphism of d-spaces, with inverse $x \mapsto x - g = x + (-g)$).

The d-space *of orbits* X/G is obviously defined as the quotient d-space, modulo the (usual) equivalence relation which arises from the action. Its d-structure has a simpler description than for a general quotient.

1.4.2 Lemma (Group actions on d-spaces)

If the group G acts on a d-space X, the d-space of orbits X/G is the usual topological quotient, equipped with the projections of the distinguished paths of X.

Proof It suffices to prove that these projections are closed under partial increasing reparametrisation and concatenation. The first fact is obvious. As to the second, let $a, b \colon \mathbf{I} \to X$ be two distinguished paths whose projections are consecutive in X/G. Then there is some $g \in G$ such that $a(1) = b(0) + g$. The path $b'(t) = b(t) + g$ is distinguished in X (by (1.97)), and the concatenation $c = a + b'$ is also. Finally, writing $p \colon X \to X/G$ the canonical projection, $pb' = pb$ and $pc = pa + pb$. $\quad\square$

1.4.3 Standard models

The euclidean spaces $\mathbf{R}^n, \mathbf{I}^n, \mathbf{S}^n$ will have their *natural* (reversible) d-structure, admitting all (continuous) paths as distinguished ones. $\mathbf{I}$ will be called the *natural* interval.

The *directed real line*, or *d-line* $\uparrow\mathbf{R}$, will be the euclidean line with directed paths given by the increasing maps $\mathbf{I} \to \mathbf{R}$ (with respect to the natural orders). Its cartesian power in d**Top**, the *n-dimensional real d-space* $\uparrow\mathbf{R}^n$, is similarly described (with respect to the product order: $x \leqslant x'$ if and only if $x_i \leqslant x'_i$, for all i). The *standard d-interval* $\uparrow\mathbf{I} = \uparrow[0, 1]$

has the subspace structure of the d-line; the *standard d-cube* $\uparrow\mathbf{I}^n$ is its n-th power, and a subspace of $\uparrow\mathbf{R}^n$.

These d-spaces are reversive (i.e. isomorphic to their opposite); in particular, the canonical reflecting isomorphism

$$r\colon \uparrow\mathbf{I} \to R(\uparrow\mathbf{I}), \qquad\qquad t \mapsto 1 - t, \qquad\qquad (1.98)$$

will play a role, by reflecting paths and homotopies into the opposite d-space. (The structure of $\uparrow\mathbf{I}$ as a dIP1-interval will be analysed in Section 1.5.1.)

The *standard directed circle* $\uparrow\mathbf{S}^1$ will be the standard circle with the *anticlockwise structure*, where the directed paths $a\colon \mathbf{I} \to \mathbf{S}^1$ move this way, in the oriented plane $\mathbf{R}^2$: $a(t) = (\cos\vartheta(t), \sin\vartheta(t))$, with an increasing (continuous) argument $\vartheta\colon \mathbf{I} \to \mathbf{R}$

$$\qquad\qquad\qquad\qquad\qquad\qquad\qquad\qquad\qquad (1.99)$$

$\uparrow\mathbf{S}^1$ can be obtained as the coequaliser in d**Top** of the following pair of maps

$$\partial^-, \partial^+ \colon \{*\} \rightrightarrows \uparrow\mathbf{I}, \qquad \partial^-(*) = 0, \quad \partial^+(*) = 1. \qquad (1.100)$$

Indeed, the 'standard construction' of this coequaliser is the quotient $\uparrow\mathbf{I}/\partial\mathbf{I}$, which identifies the endpoints; the d-structure of the quotient (*generated* by the projected paths) is the required one, precisely because of the axioms on concatenation and reparametrisation of d-paths.

The directed circle can also be described as an orbit space

$$\uparrow\mathbf{S}^1 = \uparrow\mathbf{R}/\mathbf{Z}, \qquad\qquad\qquad (1.101)$$

with respect to the action of the group of integers on the directed line $\uparrow\mathbf{R}$ (by translations); therefore, the distinguished paths of $\uparrow\mathbf{S}^1$ are simply the projections of the increasing paths in the line (Lemma 1.4.2).

The *directed n-dimensional sphere* $\uparrow\mathbf{S}^n$ is defined, for $n > 0$, as the quotient of the directed cube $\uparrow\mathbf{I}^n$ modulo its (ordinary) boundary $\partial\mathbf{I}^n$, while $\uparrow\mathbf{S}^0$ has the discrete topology and the natural d-structure (obviously discrete)

$$\uparrow\mathbf{S}^n = (\uparrow\mathbf{I}^n)/(\partial\mathbf{I}^n) \quad (n > 0), \qquad\qquad \uparrow\mathbf{S}^0 = \mathbf{S}^0 = \{-1, 1\}. \quad (1.102)$$

All directed spheres are reversive; their d-structure, further analysed in Section 1.4.7, can be described by an asymmetric distance (see (6.14)).

The standard circle has another d-structure of interest, induced by the d-space $\mathbf{R} \times \uparrow\mathbf{R}$ and called the *ordered circle* $\uparrow\mathbf{O}^1$ (as motivated in Section 1.4.5)

$$\uparrow\mathbf{O}^1 \subset \mathbf{R} \times \uparrow\mathbf{R}.$$ (1.103)

Here, d-paths have an increasing second projection. $\uparrow\mathbf{O}^1$ is the quotient of $\uparrow\mathbf{I} + \uparrow\mathbf{I}$ which identifies lower and upper endpoints, separately; i.e. the coequaliser of the following two natural embeddings of $\mathbf{S}^0$

$$\mathbf{S}^0 \to \uparrow\mathbf{I} \rightrightarrows \uparrow\mathbf{I} + \uparrow\mathbf{I}.$$ (1.104)

It is thus easy to guess that the *unpointed* d-suspension of $\mathbf{S}^0$ will give $\uparrow\mathbf{O}^1$, while the *pointed* one will give $\uparrow\mathbf{S}^1$ and, by iteration, all higher $\uparrow\mathbf{S}^n$ (Sections 1.7.4 and 1.7.5).

Various d-structures on the projective plane can be constructed as directed mapping cones (see [G7]). For the disc, see Section 1.4.7.

1.4.4 Remarks

(a) *Direction should not be confused with orientation.* Every rotation of the plane preserves orientation, but only the trivial rotation preserves the directed structure of $\uparrow\mathbf{R}^2$; on the other hand, the transposition of coordinates preserves the d-structure but reverses orientation. A non-orientable surface like the Klein bottle has a natural d-structure, which is locally isomorphic to $\uparrow\mathbf{R}^2$. On the torus, the directed structure $\uparrow\mathbf{S}^1 \times \uparrow\mathbf{S}^1$ has nothing to do with its orientation, as is the case of all the directed structures considered above (Section 1.4.3), in dimension $\geqslant 2$.

(b) A line in $\uparrow\mathbf{R}^2$ inherits the canonical d-structure (isomorphic to $\uparrow\mathbf{R}$) *if and only if* its slope belongs to $[0, +\infty]$; otherwise, it acquires the *discrete* d-structure (with the euclidean topology). Similarly, the d-structure induced by $\uparrow\mathbf{R}^2$ on any circle has two d-discrete arcs, where the slope is negative: the directed circle $\uparrow\mathbf{S}^1$ *cannot be embedded in the directed plane.*

(c) The join of the d-structures of $\uparrow\mathbf{R}$ and $\uparrow\mathbf{R}^{\mathrm{op}}$ is not the natural $\mathbf{R}$, but a finer structure $\mathbf{R}^\sim$: a d-path there is a *piecewise monotone* map $[0, 1] \to \mathbf{R}$, i.e. a generalised finite concatenation of increasing and decreasing maps. The *reversible interval* $\mathbf{I}^{\sim} \subset \mathbf{R}^{\sim}$ is of interest for reversible paths (Section 1.4.6).

(d) To define a *d-topological group* G, one should require that the structural operations be directed maps $G \times G \to G$ and $G \to G^{\mathrm{op}}$ in d**Top**. This is the case of $\uparrow\mathbf{R}^n$ and $\uparrow\mathbf{S}^1$. (This structure will be examined in Section 5.4.3.) Notice that ordered groups present a similar pattern: the inverse gives a mapping $G \to G^{\mathrm{op}}$ of ordered sets.

1.4.5 Comparing preordered spaces and d-spaces

The interplay between p**Top** and d**Top** is somewhat less trivial than one might think. We have two obvious adjoint functors

$$\mathbf{p} \colon \mathrm{d}\mathbf{Top} \rightleftarrows \mathrm{p}\mathbf{Top} \colon \mathbf{d}, \qquad\qquad \mathbf{p} \dashv \mathbf{d}, \qquad\qquad (1.105)$$

where $\mathbf{d}$ equips a preordered space X with the (preorder-preserving) maps $\uparrow\mathbf{I} \to X$ as distinguished paths, while $\mathbf{p}$ provides a d-space with the *path-preorder* $x \preceq x'$ (x' is *reachable* from x), meaning that there exists a distinguished path from x to x'. Both functors preserve products and the directed interval, so that both are strict dI1-functors.

Both functors are faithful, *but* $\mathbf{d}$ *is not an embedding* (nor is $\mathbf{p}$, of course). In fact, for a preordered space $(X, \prec)$, the path-preorder of $\mathbf{pd}(X, \prec) = (X, \preceq)$ can be strictly finer than the original one; for instance, the preordered circle $(\mathbf{S}^1, \prec) \subset \mathbf{R} \times \uparrow\mathbf{R}$ gives the d-space $\mathbf{d}(\mathbf{S}^1, \prec) = \uparrow\mathbf{O}^1$ (cf. (1.103)), which has a path-*order* $(\mathbf{S}^1, \preceq)$ strictly finer than the original preorder: two points $x \neq x'$ with the same second coordinate are preorder-equivalent but cannot be joined by a directed path. Now, $\mathbf{d}(\mathbf{S}^1, \prec) = \mathbf{d}(\mathbf{S}^1, \preceq) = \uparrow\mathbf{O}^1$, which proves that $\mathbf{d}$ is not injective on objects. Moreover, $\mathbf{d}$ takes the counit $(\mathbf{S}^1, \prec) \to (\mathbf{S}^1, \preceq)$ to $\mathrm{id}(\uparrow\mathbf{O}^1)$, which shows that it is *not* full (a full and faithful functor reflects isomorphisms).

The functor $\mathbf{d} \colon \mathrm{p}\mathbf{Top} \to \mathrm{d}\mathbf{Top}$ preserves limits (as a right adjoint) but does not preserve colimits: the coequaliser of the endpoints $\{*\} \rightrightarrows \uparrow\mathbf{I}$ has the indiscrete preorder in p**Top** and a non-trivial d-structure $\uparrow\mathbf{S}^1$ in d**Top**, which is why d**Top** is more interesting.

A d-space will be said to be *of (pre)order type* if it can be obtained, as above, from a topological space with such a structure. Thus, $\uparrow\mathbf{R}^n, \uparrow\mathbf{I}^n$ are of order type; $\mathbf{R}^n, \mathbf{I}^n$ and $\mathbf{S}^n$ are of *chaotic preorder* type; and the product $\mathbf{R} \times \uparrow\mathbf{R}$ is of preorder type. The d-space $\uparrow\mathbf{S}^1$ is not of preorder type. The ordered circle $\uparrow O^1 = \mathbf{d}(\mathbf{S}^1, \prec) = \mathbf{d}(\mathbf{S}^1, \preceq)$ is *of order type*, which motivates the name we are using (even if it can also be defined by the *preorder* relation $\prec$).

1.4.6 Directed paths

A *path* in a d-space X is defined as a d-map $a \colon \uparrow\mathbf{I} \to X$ on the standard d-interval. It is easy to check that *this is the same as a distinguished path* $a \in dX$. It will also be called a *directed* path, when we want to stress the difference with ordinary paths in the underlying space UX, (which is a more general notion, of course).

The path a has two endpoints, or *faces* $\partial^-(a) = a(0)$, $\partial^+(a) = a(1)$. Every point $x \in X$ has a *degenerate path* 0_x, constant at x. A *loop* $(\partial^-(a) = \partial^+(a))$ amounts to a d-map $\uparrow\mathbf{S}^1 \to X$ (by (1.100)).

By the very definition of d-structure (Section 1.4.0), we already know that the concatenation $a + b$ of two consecutive paths $(\partial^+ a = \partial^- b)$ is directed. This amounts to saying that, in d**Top** (as for spaces and pre-ordered spaces, see Sections 1.1.0 and 1.1.4), the *standard concatenation pushout* – pasting two copies of the d-interval, one after the other – can be realised as $\uparrow\mathbf{I}$ itself, with embeddings c^α into the first or second half of the interval

$$
\begin{array}{ccc}
\{*\} & \xrightarrow{\ \partial^+\ } & \uparrow\mathbf{I} \\
{\scriptstyle \partial^-}\big\downarrow & {\scriptstyle c^-}\big\downarrow & \\
\uparrow\mathbf{I} & \xrightarrow[\ c^+\ ]{} & \uparrow\mathbf{I}
\end{array}
\qquad\qquad
\begin{array}{l}
c^-(t) = t/2, \\[2mm]
\\[2mm]
c^+(t) = (t+1)/2.
\end{array}
\qquad (1.106)
$$

Recall that the existence of a path in X from x to x' gives the *path preorder*, $x \preceq x'$ (Section 1.4.5). The equivalence relation $\simeq$ spanned by $\preceq$ gives the partition of a d-space in its *path components* and yields a functor

$$ \uparrow\Pi_0 \colon d\mathbf{Top} \to \mathbf{Set}, \qquad \uparrow\Pi_0(X) = |X|/\simeq. \qquad (1.107) $$

(As a matter of notation, we use a lower case π when dealing with pointed objects.) A non-empty d-space X is *path connected* if $\uparrow\Pi_0(X)$ is a point. Then, also the underlying space UX is path connected, while the converse is obviously false (cf. Section 1.4.4(b)). The directed spaces $\uparrow\mathbf{R}^n, \uparrow\mathbf{I}^n, \uparrow\mathbf{S}^n$ are path connected $(n > 0)$; but $\uparrow\mathbf{S}^n$ is more strongly so, because its path-preorder $\preceq$ is already chaotic: every point can reach each other.

The path a will be said to be *reversible* if also the mapping $a(1-t)$ is a directed path in X:

$$-a \colon \uparrow\mathbf{I} \to X, \qquad (-a)(t) = a(1-t) \qquad \text{(reversed path)}. \qquad (1.108)$$

Obviously, such paths are closed under concatenation. This notion should not be confused with the *reflected path* $r(a) \colon \uparrow\mathbf{I} \to X^{\mathrm{op}}$ (cf. (1.50)), which is defined by the same formula $a(1-t)$ and always exists, *but lives in the opposite d-space* X^{op}. Of course, if X itself is a reversible d-space (Section 1.4.0), i.e. $X = X^{\mathrm{op}}$, the two notions coincide.

Equivalently, a is reversible if it is a d-map $a \colon \mathbf{I}^{\sim} \to X$ on the reversible interval (Section 1.4.4(c)). (On the other hand, requiring that a be directed on the natural interval $\mathbf{I}$ is a stronger condition, not closed under concatenation: the pasting, on an endpoint, of two copies of the natural interval $\mathbf{I}$ *in* dTop is not isomorphic to $\mathbf{I}$.)

1.4.7 Vortices, discs and cones

Let X be a d-space. Loosely speaking, a *non*-reversible path in X with equal (resp. different) endpoints can be viewed as *revealing a vortex* (resp. a *stream*). If X is of preorder type (Section 1.4.5), it cannot have a vortex, since every loop must be reversible, as we have already remarked. On the other hand, the directed circle $\uparrow\mathbf{S}^1$ has non-reversible loops.

We shall say that the d-space X has a *point-like vortex* at x if every neighbourhood of x in X contains some non-reversible loop. It is easy to realise a directed disc having a point-like vortex (see below), while $\uparrow\mathbf{S}^1$ has none.

All higher directed spheres $\uparrow\mathbf{S}^n = (\uparrow\mathbf{I}^n)/(\partial\mathbf{I}^n)$, for $n \geqslant 2$, have a point-like vortex at the class $[0]$ (of the boundary points), as showed by the following sequence of non-reversible loops in $\uparrow\mathbf{S}^2$, which are 'arbitrarily small'

$$(1.109)$$

Since every point $x \neq [0]$ of $\uparrow\mathbf{S}^n$ has a neighbourhood isomorphic to $\uparrow\mathbf{R}^n$, this also shows that our higher d-spheres are not locally isomorphic to any fixed 'model'. This fact cannot be reasonably avoided, since we shall see that the d-spaces $\uparrow\mathbf{S}^n$ are determined as pointed suspensions of $\mathbf{S}^0$.

There are various d-structures of interest on the disc $\mathbf{D}^2 = C\mathbf{S}^1$, i.e. the mapping cone of $\mathrm{id}\mathbf{S}^1$ in **Top**.

In the figure below, we represent four cases which induce the natural structure $\mathbf{S}^1$ on the boundary: a directed path in these d-spaces is a map $(\rho(t)\cos\vartheta(t), \rho(t)\sin\vartheta(t))$, where the continuous function $\rho\colon \mathbf{I} \to \mathbf{I}$ is, respectively, decreasing, increasing, constant, arbitrary, while $\vartheta\colon \mathbf{I} \to \mathbf{R}$ is just continuous

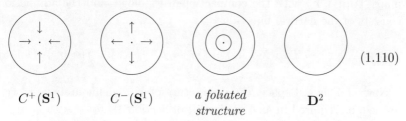

$$C^+(\mathbf{S}^1) \qquad C^-(\mathbf{S}^1) \qquad \textit{a foliated} \qquad \mathbf{D}^2 \tag{1.110}$$
$$\textit{structure}$$

All these structures are of preorder type (defined by the following relations, respectively: $\rho \geqslant \rho'; \rho \leqslant \rho'; \rho = \rho'$; chaotic). One can view the first (resp. second) as a conical peak (resp. sink) directed upwards. The first two cases will be obtained as upper or lower directed cones of $\mathbf{S}^1$ and have been named accordingly; see Section 1.7.2.

Similarly, we can consider four structures which induce $\uparrow\mathbf{S}^1$ on the boundary: take the function ρ as above, and $\vartheta\colon \mathbf{I} \to \mathbf{R}$ increasing

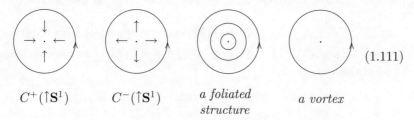

$$C^+(\uparrow\mathbf{S}^1) \qquad C^-(\uparrow\mathbf{S}^1) \qquad \textit{a foliated} \qquad \textit{a vortex} \tag{1.111}$$
$$\textit{structure}$$

These four structures have a point-like vortex at the origin. Again, the first two cases are upper or lower directed cones, of $\uparrow\mathbf{S}^1$. Four other structures can be similarly obtained from $\uparrow\mathbf{O}^1$, see (1.103).

1.4.8 Theorem (Exponentiable d-spaces)

Let $\uparrow A$ be any d-structure on a locally compact Hausdorff space A. Then $\uparrow A$ is exponentiable in $\mathrm{d}\mathbf{Top}$ (Section A4.2). More precisely, for every d-space Y

$$Y^{\uparrow A} = \mathrm{d}\mathbf{Top}(\uparrow A, Y) \subset \mathbf{Top}(A, UY), \tag{1.112}$$

*is the set of directed maps, with the compact-open topology restricted
from the topological exponential* $(UY)^A$ *and the d-structure where a path*
$c\colon \mathbf{I} \to U(Y^{\uparrow A}) \subset (UY)^A$ *is directed if and only if the corresponding
map* $\check{c}\colon \mathbf{I}\times A \to UY$ *is a d-map* $\uparrow\mathbf{I} \times \uparrow A \to Y$.

Proof We have already recalled (in Section 1.1.2) that a locally compact
Hausdorff space A is *exponentiable* in **Top**: the space Y^A is the set of
maps **Top**(A, Y) with the compact-open topology, and the adjunction
consists of the natural bijection

$$\mathbf{Top}(X, Y^A) \to \mathbf{Top}(X\times A, Y), \quad f \mapsto \check{f}, \;\; \check{f}(x, a) = f(x)(a). \quad (1.113)$$

Now, if Y is a d-space, the d-structure of $Y^{\uparrow A}$ defined above is well
formed, as required in axioms (i)–(iii) of Section 1.4.0.

(i) *Constant paths.* If $c\colon \mathbf{I} \to Y^{\uparrow A}$ is constant at the d-map $g\colon \uparrow A \to Y$,
then $\check{c}$ can be factored as $\uparrow\mathbf{I}\times\uparrow A \to \uparrow A \to Y$, and is directed as well.

(ii) *Partial reparametrisation.* For any $h\colon \uparrow\mathbf{I} \to \uparrow\mathbf{I}$, the map $(ch)\check{} =
\check{c}.(h\times\uparrow A)$ is directed.

(iii) *Concatenation.* Let $c = c_1 + c_2\colon \mathbf{I} \to U(Y^{\uparrow A})$, with $\check{c}_i\colon \uparrow\mathbf{I}\times\uparrow A \to Y$.
By the next lemma, the product $- \times \uparrow A$ preserves the concatenation
pushout (1.106). Therefore $\check{c}$, as the pasting of $\check{c}_1, \check{c}_2$ on this pushout, is
a directed map.

Finally, we must prove that the bijection (1.113) restricts to a bijection
between $d\mathbf{Top}(X, Y^{\uparrow A})$ and $d\mathbf{Top}(X\times\uparrow A, Y)$. In fact, we have a chain
of equivalent conditions

- the map $f\colon X \to Y^{\uparrow A}$ is directed;
- $\forall\, x \in dX$, the map $(fx)\check{} = \check{f}.(x \times \uparrow A)\colon \uparrow\mathbf{I}\times\uparrow A \to Y$ is directed;
- $\forall\, x \in dX, \forall\, h \in d\uparrow\mathbf{I}, \forall\, a \in d\uparrow A$, the map $\check{f}.(xh\times a)\colon \uparrow\mathbf{I}\times\uparrow\mathbf{I} \to Y$
  is directed;
- the map $\check{f}\colon X \times \uparrow A \to Y$ is directed. $\qquad\qquad\square$

1.4.9 Lemma

*For every d-space X, the functor $X \times -\colon d\mathbf{Top} \to d\mathbf{Top}$ preserves
the standard concatenation pushout (1.106), yielding the concatenation*

pushout of the cylinder functor

$$
\begin{array}{ccc}
X & \xrightarrow{\ \partial^+\ } & IX \\
{\scriptstyle \partial^-}\downarrow & \overset{-\ -}{}\downarrow{\scriptstyle c^-} & \\
IX & \xrightarrow[\ c^+\ ]{} & IX
\end{array}
\qquad
\begin{aligned}
c^-(x,t) &= (x,t/2), \\[1em]
c^+(x,t) &= (x,(t+1)/2).
\end{aligned}
\qquad (1.114)
$$

Proof In **Top**, the underlying space UX satisfies this property, because the spaces $UX \times [0,1/2]$ and $UX \times [1/2,1]$ form a finite closed cover of $UX \times \mathbf{I}$, so that each mapping defined on the latter and continuous on such closed parts is continuous.

Consider then a map $f \colon UX \times \mathbf{I} \to UY$ obtained by pasting two maps f_0, f_1 on the topological pushout $UX \times \mathbf{I}$

$$
f(x,t) = \begin{cases}
f_0(x,2t), & \text{for } 0 \leqslant t \leqslant 1/2, \\
f_1(x,2t-1), & \text{for } 1/2 \leqslant t \leqslant 1.
\end{cases}
$$

Let now $\langle a,h\rangle \colon {\uparrow}\mathbf{I} \to {\uparrow}X \times {\uparrow}\mathbf{I}$ be any directed map. If the image of h is contained in one half of $\mathbf{I}$, then $f.\langle a,h\rangle$ is certainly directed. Otherwise, there is some $t_1 \in {]}0,1{[}$ such that $h(t_1) = 1/2$, and we can assume that $t_1 = 1/2$ (up to pre-composing with an automorphism of ${\uparrow}\mathbf{I}$). Now, the path $f.\langle a,h\rangle \colon \mathbf{I} \to UY$ is directed in Y, because it is the concatenation of the following two directed paths $c_i \colon {\uparrow}\mathbf{I} \to Y$

$$
\begin{aligned}
c_0(t) &= f(a(t/2), h(t/2)) = f_0(a(t/2), 2h(t/2)), \\
c_1(t) &= f(a((t+1)/2), h((t+1)/2)) \\
&= f_1(a((t+1)/2), 2h((t+1)/2) - 1).
\end{aligned}
$$

$\square$

1.5 The basic homotopy structure of d-spaces

The ordered interval ${\uparrow}\mathbf{I}$ of p**Top** is still exponentiable in the new domain, and d**Top** is a cartesian dIP1-category, in the sense of Section 1.2.5. We begin to investigate homotopies and directed homotopy equivalence, for d-spaces and pointed d-spaces.

1.5.1 The directed structure

Directed homotopy in d**Top** is established in much the same way as for preordered spaces, in Section 1.1.4. It is based on the cartesian product and the reversor $R \colon d\mathbf{Top} \to d\mathbf{Top}$.

The standard directed interval $\uparrow\mathbf{I} = \uparrow[0, 1]$, a d-space of order type (Section 1.4.3), is exponentiable by Theorem 1.4.8. It is a cartesian dIP1-interval, with the obvious faces, degeneracy and reflection (as in p**Top**, Section 1.1.4)

$$\partial^\alpha : \{*\} \rightrightarrows \uparrow\mathbf{I} : e, \quad \partial^-(*) = 0, \ \partial^+(*) = 1, \qquad e(t) = *,$$
$$r : \uparrow\mathbf{I} \to \uparrow\mathbf{I}^{\mathrm{op}}, \qquad\qquad r(t) = 1 - t \qquad\qquad (reflection). \tag{1.115}$$

The cylinder functor and the path functor thus form a cartesian dIP1-structure (Section 1.2.5):

$$I(X) = X \times \uparrow\mathbf{I}, \qquad P(Y) = Y^{\uparrow\mathbf{I}} \qquad\qquad (I \dashv P). \tag{1.116}$$

Here, the cylinder $X \times \uparrow\mathbf{I}$ has the product topology *and* distinguished paths $\langle a, h \rangle : \uparrow\mathbf{I} \to X \times \uparrow\mathbf{I}$, where $a \in dX$ and $h : \uparrow\mathbf{I} \to \uparrow\mathbf{I}$ is continuous and order-preserving. On the other hand, the d-space $Y^{\uparrow\mathbf{I}}$ is the set of d-paths $d\mathbf{Top}(\uparrow\mathbf{I}, Y)$ with the compact-open topology (induced by the topological path-space $P(UY) = \mathbf{Top}(\mathbf{I}, UY)$) and the d-structure where a map

$$c : \mathbf{I} \to d\mathbf{Top}(\uparrow\mathbf{I}, Y) \subset \mathbf{Top}(\mathbf{I}, UY), \tag{1.117}$$

is directed if and only if, for all increasing maps $h, k : \mathbf{I} \to \mathbf{I}$, the associated path $t \mapsto c(h(t))(k(t))$ is in dY.

As for preordered spaces, the forgetful functor $U : d\mathbf{Top} \to \mathbf{Top}$ is a strict dI1-functor (Section 1.2.6) and a *lax* dP1-functor, with a non-invertible comparison $UP(Y) = U(Y^{\uparrow\mathbf{I}}) \subset (UY)^{\mathbf{I}} = PU(Y)$. Thus, also here, the right adjoint $D' : \mathbf{Top} \to p\mathbf{Top}$ inherits the structure of a *lax* dI1-functor, which is not strong.

Both adjoint functors $\mathbf{p} : d\mathbf{Top} \rightleftarrows p\mathbf{Top} : \mathbf{d}$ (Section 1.4.5) preserve products and the directed interval and commute with the reversor, whence they are both strict dI1-functors. It follows that the right adjoint **d** is a lax dP1-functor.

1.5.2 Homotopies

A (directed) *homotopy* $\varphi : f \to g : X \to Y$ of d-spaces is defined in the usual way, as a map $IX \to Y$ or equivalently $X \to PY$.

As for paths (Section 1.4.6), we say that a homotopy $\varphi : f \to g$ is *reversible* if also the mapping $(-\varphi)(x, t) = \varphi(x, 1-t)$ is a d-map $X \times \uparrow\mathbf{I} \to Y$, which yields a directed homotopy

$$-\varphi : g \to f : X \to Y.$$

The latter should not be confused with the *reflected homotopy* (1.38)

$$\varphi^{\mathrm{op}} : g^{\mathrm{op}} \to f^{\mathrm{op}} : X^{\mathrm{op}} \to Y^{\mathrm{op}},$$

which is defined by the same formula $\varphi(x, 1 - t)$ and always exists, but is concerned with the opposite d-spaces.

The relation of future (or past) homotopy equivalence of d-spaces (Section 1.3.1) implies the usual homotopy equivalence of the underlying spaces. In fact, it is strictly stronger. As a trivial example, the d-discrete structure $D\mathbf{R}$ on the real line (where all d-paths are constant, Section 1.4.0) is not (even) coarsely d-contractible (Section 1.3.3). Less trivially, within path connected d-spaces, it is easy to show that the d-spaces $\mathbf{S}^1, \uparrow\mathbf{S}^1$ and $\uparrow\mathbf{O}^1$ defined above (Section 1.4.3) are not future (or past) homotopy equivalent. In fact, a directed map $\mathbf{S}^1 \to \uparrow\mathbf{O}^1$ or $\uparrow\mathbf{S}^1 \to \uparrow\mathbf{O}^1$ must stay in *one* half of $\uparrow\mathbf{O}^1$, whence its underlying map is homotopically trivial. On the other hand, a d-map $\mathbf{S}^1 \to \uparrow\mathbf{S}^1$ is necessarily constant.

The inclusion $X_0 \subset X$ of a d-subspace is the embedding of a future deformation retract of X (Section 1.3.1) if and only if there is a d-map φ such that

$$\varphi : X \times \uparrow\mathbf{I} \to X, \qquad \varphi(x, 0) = x, \qquad \varphi(x, 1) \in X_0,$$
$$\varphi(x', 1) = x' \qquad\qquad (\text{for } x \in X, x' \in X_0). \qquad (1.118)$$

For instance, as for preordered spaces (Section 1.3.4), the cylinder $X \times \uparrow\mathbf{I}$ has a strong future deformation retract at its upper basis $\partial^+(X)$ and a strong past deformation retract at its lower basis $\partial^-(X)$. With reference to Section 1.4.7, the discs $C^+(\mathbf{S}^1)$ and $C^+(\uparrow\mathbf{S}^1)$ are future contractible to their vertex, while the 'lower' ones, labelled C^-, are past contractible.

1.5.3 Homotopy functors

Each d-space A gives a covariant 'representable' *homotopy functor* (and a contravariant one)

$$[A, -] : \mathrm{d}\mathbf{Top} \to \mathbf{Set}, \qquad ([-, A] : \mathrm{d}\mathbf{Top}^* \to \mathbf{Set}), \qquad (1.119)$$

where $[A, X]$ denotes the set of classes of maps $A \to X$, up to the equivalence relation $\sim_1$ generated by directed homotopies (see (1.74)). These functors are plainly d-homotopy invariant: if $\varphi : f \to g$ is a homotopy in $\mathrm{d}\mathbf{Top}$, then $[A, f] = [A, g]$.

In particular, $\uparrow\Pi_0(X) = [\{*\}, X]$. Also $[\uparrow\mathbf{S}^1, -], [\mathbf{S}^1, -], [\uparrow\mathbf{O}^1, -]$ express invariants of interest; the first gives the set of *homotopy classes* *of directed* free *loops*, and should not be confused with the fundamental monoid of a pointed d-space (Section 3.2.5).

1.5.4 Pointed directed spaces

A *pointed d-space* is obviously a pair (X, x_0) formed of a d-space X with a base point $x_0 \in X$; a morphism $f \colon (X, x_0) \to (Y, y_0)$ is a map of d-spaces which preserves the base points: $f(x_0) = (y_0)$.

The corresponding category will be written as $d\mathbf{Top}_\bullet$; it is pointed, with zero-object $\{*\}$ (where the base point need not be specified). Formally, $d\mathbf{Top}_\bullet$ can be identified with the 'slice' category $d\mathbf{Top} \backslash \{*\}$, whose objects are the morphisms $\{*\} \to X$ of $d\mathbf{Top}$ (see Section 5.2). The reversor of $d\mathbf{Top}_\bullet$ is obviously $R(X, x_0) = (X^{\mathrm{op}}, x_0)$.

Limits and colimits are obvious (as for pointed sets): *limits* and *quotients* are computed as in $d\mathbf{Top}$ and pointed in the obvious way, whereas *sums* are quotients (in $d\mathbf{Top}$) of the corresponding unpointed sums, under identification of the base points. Notice that this identification requires the closure of the induced d-paths under reparametrisation and concatenation, so that a sum in $d\mathbf{Top}_\bullet$ of path-connected pointed d-spaces is path connected.

The forgetful functor which forgets the base point

$$U \colon d\mathbf{Top}_\bullet \to d\mathbf{Top}, \qquad\qquad (-)_\bullet \dashv U, \qquad\qquad (1.120)$$

has a left adjoint $X \mapsto X_\bullet$, which adds an 'isolated' base point $*$. In other words, $X_\bullet$ is the sum d-space $X + \{*\}$, based at the added point; the latter is open and the only d-path through it is the constant one. This functor yields the *pointed directed interval*

$$\uparrow\!\mathbf{I}_\bullet = (\uparrow\!\mathbf{I} + \{*\}, *). \qquad\qquad (1.121)$$

1.5.5 Pointed directed homotopies

This interval $\uparrow\!\mathbf{I}_\bullet$ provides a symmetric monoidal dIP1-structure on $d\mathbf{Top}_\bullet$ (Section 1.2.5), with respect to the *smash product*:

$$(X, x_0) \wedge (Y, y_0) = ((X \times Y)/(X \times \{y_0\} \cup \{x_0\} \times Y), [x_0, y_0]). \quad (1.122)$$

Notice that the classical formula of the smash product of pointed topological spaces – which collapses all pairs (x, y_0) and (x_0, y) – is here interpreted as a *quotient of d-spaces*. The unit of the smash product is the discrete d-pace $\mathbf{S}^0 = \{-1, 1\}$, pointed at 1 (for instance).

The functor $(-)_{\bullet}: d\mathbf{Top} \to d\mathbf{Top}_{\bullet}$ transforms the cartesian structure of $d\mathbf{Top}$ into the monoidal structure of $d\mathbf{Top}_{\bullet}$ (up to natural isomorphisms)

$$(X \times Y)_{\bullet} \cong X_{\bullet} \wedge Y_{\bullet}, \qquad \{*\}_{\bullet} \cong \mathbf{S}^0. \qquad (1.123)$$

Thus, the pointed cylinder

$$\begin{aligned} & I: d\mathbf{Top}_{\bullet} \to d\mathbf{Top}_{\bullet}, \\ & I(X, x_0) = (X, x_0) \wedge \uparrow\mathbf{I}_{\bullet} \cong (IX/I\{x_0\}, [x_0, t]), \end{aligned} \qquad (1.124)$$

is the quotient of the unpointed cylinder which collapses the fibre at the base point, $I\{x_0\} = \{x_0\} \times \uparrow\mathbf{I}$. It inherits the following structural transformations

$$\begin{aligned} & \partial^\alpha: (X, x_0) \to I(X, x_0), & \partial^\alpha(x) = [x, \alpha] \quad (\alpha = 0, 1), \\ & e: I(X, x_0) \to (X, x_0), & e[x, t] = x, \\ & r: I(X^{\mathrm{op}}, x_0) \to (I(X, x_0))^{\mathrm{op}}, & r[x, t] = [x, 1 - t]. \end{aligned} \qquad (1.125)$$

Its right adjoint, the pointed cocylinder, is just the ordinary cocylinder, pointed at the constant loop at the base point: $\omega_0 = 0_{y_0}$

$$P: d\mathbf{Top}_{\bullet} \to d\mathbf{Top}_{\bullet}, \qquad P(Y, y_0) = (PY, \omega_0). \qquad (1.126)$$

We have already noted, in Section 1.3.2, that contractibility is based on the zero-object, while points are defined by the unit of the smash product, $\mathbf{S}^0 \cong \{*\}_{\bullet}$ (the *free point*, Section 1.2.4).

The forgetful functor $U: d\mathbf{Top}_{\bullet} \to d\mathbf{Top}$ is a strict dP1-functor and a lax dI1-functor, with comparison at (X, x_0) given by the canonical projection $IX \to UI(X, x_0) = IX/I\{x_0\}$.

A path in the pointed d-space (X, x_0) is the same as a path in X, and path components are the same. But the functor $\uparrow\Pi_0: d\mathbf{Top} \to \mathbf{Set}$ (defined in (1.107)) should now be enriched, with values in pointed sets

$$\uparrow\pi_0: d\mathbf{Top}_{\bullet} \to \mathbf{Set}_{\bullet}, \qquad \uparrow\pi_0(X, x_0) = (\uparrow\Pi_0(X), [x_0]). \qquad (1.127)$$

The category $p\mathbf{Top}_{\bullet}$ of *pointed preordered spaces* can be dealt with in a similar way.

1.6 Cubical sets

Cubical sets have a classical non-symmetric monoidal structure (Section 1.6.3). As a consequence, the obvious directed interval $\uparrow\mathbf{i}$, freely generated by a 1-cube (Section 1.6.4), gives rise to a *left cylinder* $\uparrow\mathbf{i} \otimes X$ and a *right cylinder* $X \otimes \uparrow\mathbf{i}$, and to two notions of directed homotopy (which are related by an endofunctor, the 'transposer' S). Most of this material, for its non-classical part, comes from [G12].

Cubical singular homology of topological spaces can be found in the texts of Massey [Ms] and Hilton and Wylie [HW].

1.6.0 The singular cubical set of a space

Every topological space X has an associated *cubical set* $\square X$, with components $\square_n X = \mathbf{Top}(\mathbf{I}^n, X)$, the set of *singular n-cubes* of X. Its faces and degeneracies (for $\alpha = 0, 1$; $i = 1, ..., n$)

$$
\begin{aligned}
\partial_i^\alpha : \square_n X \to \square_{n-1} X, &\qquad (faces), \\
e_i : \square_{n-1} X \to \square_n X &\qquad (degeneracies),
\end{aligned}
\tag{1.128}
$$

arise (contravariantly) from the faces and degeneracies of the standard topological cubes $\mathbf{I}^n$

$$
\begin{aligned}
\partial_i^\alpha &= \mathbf{I}^{i-1} \times \partial^\alpha \times \mathbf{I}^{n-i} : \mathbf{I}^{n-1} \to \mathbf{I}^n, \\
&\qquad \partial_i^\alpha(t_1, ..., t_{n-1}) = (t_1, ..., \alpha, ..., t_{n-1}), \\
e_i &= \mathbf{I}^{i-1} \times e \times \mathbf{I}^{n-i} : \mathbf{I}^n \to \mathbf{I}^{n-1}, \\
&\qquad e_i(t_1, ..., t_n) = (t_1, ..., \hat{t}_i, ..., t_n).
\end{aligned}
\tag{1.129}
$$

(Here, $\hat{t}_i$ means omit the coordinate t_i, a standard notation in algebraic topology.)

This is actually true in every symmetric monoidal dI1-category, with formal interval $\mathbf{I}$ (as we will see in Section 2.6.5). But in the present case, a cubical set *of type* $\square X$ actually has a much richer, relevant structure, obtained from the structure of the standard interval $\mathbf{I}$ as an *involutive lattice* in $\mathbf{Top}$ (Section 1.1.0); indeed, connections, reversion and transposition yield similar transformations between singular cubes of the space X (for $\alpha = 0, 1$; $i = 1, ..., n$)

$$
\begin{aligned}
g_i^\alpha : \square_n X \to \square_{n+1} X &\qquad (connections), \\
r_i : \square_n X \to \square_n X &\qquad (reversions), \\
s_i : \square_{n+1} X \to \square_{n+1} X &\qquad (transpositions).
\end{aligned}
\tag{1.130}
$$

As we have already remarked (Section 1.1.5), the group of symmetries of the n-cube, $(\mathbf{Z}/2)^n \rtimes S_n$, acts on $\square_n X$: reversions and transpositions generate, respectively, the action of the first and second factor of this semidirect product. Now, in homotopy theory, reversion (in team with connections) yields reverse homotopies and inverses in homotopy groups, while transposition yields the homotopy-preservation property of the cylinder, cone and suspension endofunctors (see Chapter 4). On the other hand, *not assigning this additional structure* allows us to break symmetries (reversion and transposition) which are intrinsic to topological spaces.

1.6.1 Cubical sets

Abstracting from the singular cubical set of a topological space recalled above, a *cubical set* $X = ((X_n), (\partial_i^\alpha), (e_i))$ is a sequence of sets X_n ($n \geqslant 0$), together with mappings, called *faces* (∂_i^α) and *degeneracies* (e_i)

$$\partial_i^\alpha = \partial_{ni}^\alpha \colon X_n \to X_{n-1},$$
$$e_i = e_{ni} \colon X_{n-1} \to X_n \qquad (\alpha = \pm; \; i = 1, ..., n), \tag{1.131}$$

satisfying the *cubical* relations

$$\partial_i^\alpha . \partial_j^\beta = \partial_j^\beta . \partial_{i+1}^\alpha \qquad (j \leqslant i),$$
$$e_j . e_i = e_{i+1} . e_j \qquad (j \leqslant i), \tag{1.132}$$

$$\partial_i^\alpha . e_j = \begin{cases} e_j . \partial_{i-1}^\alpha & (j < i), \\ \mathrm{id} & (j = i), \\ e_{j-1} . \partial_i^\alpha & (j > i). \end{cases}$$

Elements of X_n are called *n-cubes*, and *vertices* or *edges* for $n = 0$ or 1, respectively. Every n-cube $x \in X_n$ has 2^n vertices: $\partial_1^\alpha \partial_2^\beta \partial_3^\gamma (x)$ for $n = 3$. Given a vertex $x \in X_0$, the *totally degenerate n-cube at* x is obtained by applying n degeneracy operators to the given vertex, in any (legitimate) way:

$$e_1^n (x) = e_{i_n} ... e_{i_2} e_{i_1} (x) \in X_n \qquad (1 \leqslant i_j \leqslant j). \tag{1.133}$$

A *morphism* $f = (f_n) \colon X \to Y$ is a sequence of mappings $f_n \colon X_n \to Y_n$ which commute with faces and degeneracies. All this forms a category **Cub** which has all limits and colimits and is cartesian closed.

(**Cub** is the presheaf category of functors $X \colon \mathbb{I}^{\mathrm{op}} \to \mathbf{Set}$, where $\mathbb{I}$ is the subcategory of **Set** consisting of the *elementary cubes* 2^n, together with the maps $2^m \to 2^n$ which delete some coordinates and insert some 0's and 1's, without modifying the order of the remaining coordinates. See Sections 1.6.7 and A1.8; or [GM] for cubical sets with a richer structure, and further developments.)

The terminal object $\top$ is freely generated by one vertex $*$ and will also be written $\{*\}$; notice that each of its components is a singleton. The initial object is empty, i.e. all its components are; all the other cubical sets have non-empty components in every degree.

We will make use of two covariant involutive endofunctors, called *reversor* and *transposer*

$$R: \mathbf{Cub} \to \mathbf{Cub}, \qquad RX = X^{\mathrm{op}} = ((X_n), (\partial_i^{-\alpha}), (e_i)) \qquad (reversor),$$

$$S: \mathbf{Cub} \to \mathbf{Cub}, \qquad SX = ((X_n), (\partial_{n+1-i}^{\alpha}), (e_{n+1-i})) \quad (transposer),$$

$$RR = \mathrm{id}, \qquad SS = \mathrm{id}, \qquad RS = SR. \tag{1.134}$$

(The meaning of $-\alpha$, for $\alpha = \pm$, is obvious.) The first reverses the one-dimensional direction, the second the two-dimensional one; plainly, *they commute*. If $x \in X_n$, the same element viewed in X^{op} will often be written as x^{op}, so that $\partial_i^-(x^{\mathrm{op}}) = (\partial_i^+ x)^{\mathrm{op}}$.

We say that a cubical set X is *reversive* if $RX \cong X$ and *permutative* if $SX \cong X$.

1.6.2 Subobjects and quotients

A *cubical subset* $Y \subset X$ is a sequence of subsets $Y_n \subset X_n$, stable under faces and degeneracies.

An *equivalence relation* $\mathcal{E}$ in X is a cubical subset of $X \times X$ whose components $\mathcal{E}_n \subset X_n \times X_n$ are equivalence relations; then, the *quotient* $X/\mathcal{E}$ is the sequence of quotient sets $X_n/\mathcal{E}_n$, with induced faces and degeneracies. In particular, for a *non-empty* cubical subset $Y \subset X$, the quotient X/Y has components X_n/Y_n, where all cubes $y \in Y_n$ are identified.

For a cubical set X, we define the *homotopy set*

$$\Pi_0(X) = X_0/\sim, \tag{1.135}$$

where the relation $\sim$ of *connection* is the equivalence relation in X_0 generated by being vertices of a common edge. The *connected component* of X at an equivalence class $[x] \in \Pi_0(X)$ is the cubical subset formed by all cubes of X whose vertices lie in $[x]$; X is always the sum (or disjoint union) of its connected components. If X is not empty, we say that it is *connected* if it has one connected component, or equivalently if $\Pi_0(X)$ is a singleton.

One can easily see that the forgetful functor $(-)_0 \colon \mathbf{Cub} \to \mathbf{Set}$ has a left adjoint, defined by the *discrete* cubical set on a set

$$D: \mathbf{Set} \to \mathbf{Cub}, \qquad (DS)_n = S \qquad (n \in \mathbf{N}), \tag{1.136}$$

whose components are constant, while faces and degeneracies are identities. Then, *the functor* $\Pi_0 \colon \mathbf{Cub} \to \mathbf{Set}$ *is left adjoint to* D.

The forgetful functor $(-)_0$ also has a right adjoint, defined by the *indiscrete* cubical set $D'S = \mathbf{Set}(2^\bullet, S)$. In this formula, $2^\bullet$ denotes the functor $\mathbb{I} \to \mathbf{Set}$ (a *cocubical set*) given by the embedding which realises the 'formal n-cube' as 2^n; see Section 1.6.1.

More generally, there is an obvious functor of n-truncation

$$\mathrm{tr}_n \colon \mathbf{Cub} \to \mathbf{Cub}_n,$$

with values in the category of n-*truncated cubical sets* (with components in degree $\leqslant n$); and there are adjoints $\mathrm{sk}_n \dashv \mathrm{tr}_n \dashv \mathrm{cosk}_n$, called the n-*skeleton* and n-*coskeleton*, respectively. Thus, $(-)_0 = \mathrm{tr}_0$, $D = \mathrm{sk}_0$ and $D' = \mathrm{cosk}_0$.

1.6.3 Tensor product

The category $\mathbf{Cub}$ has a *non*-symmetric monoidal structure [Ka1, BH3]

$$(X \otimes Y)_n = (\Sigma_{p+q=n} \, X_p \times Y_q)/ \sim_n, \tag{1.137}$$

where $\sim_n$ is the equivalence relation generated by identifying $(e_{r+1}x, y)$ with $(x, e_1 y)$, for all $(x, y) \in X_r \times Y_s$ (where $r + s = n - 1$).

Writing $x \otimes y$ the equivalence class of (x, y), faces and degeneracies are defined as follows, when x is of degree p and y of degree q

$$\begin{aligned}
\partial_i^\alpha (x \otimes y) &= (\partial_i^\alpha x) \otimes y & (1 \leqslant i \leqslant p), \\
\partial_i^\alpha (x \otimes y) &= x \otimes (\partial_{i-p}^\alpha y) & (p+1 \leqslant i \leqslant p+q), \\
e_i (x \otimes y) &= (e_i x) \otimes y & (1 \leqslant i \leqslant p+1), \\
e_i (x \otimes y) &= x \otimes (e_{i-p} y) & (p+1 \leqslant i \leqslant p+q+1).
\end{aligned} \tag{1.138}$$

Note that $e_{p+1}(x \otimes y) = (e_{p+1}\, x) \otimes y = x \otimes (e_1 y)$ is well defined precisely because of the previous equivalence relation.

The (bilateral) identity of the tensor product is the terminal object $\{*\}$, i.e. the cubical set generated by one zero-dimensional cube; it is reversive and permutative (Section 1.6.1). The tensor product is linked to reversor and transposer (1.134) as follows

$$R(X \otimes Y) = RX \otimes RY, \tag{1.139}$$

$$s(X, Y) \colon S(X \otimes Y) \cong SY \otimes SX,$$
$$x \otimes y \mapsto y \otimes x \qquad (external\ symmetry). \tag{1.140}$$

The isomorphism (1.140) replaces the symmetry of a symmetric tensor product. (In other words, R is a strict isomorphism of the monoidal structure, while (S, s) is an anti-isomorphism.)

Therefore, reversive objects are stable under tensor products while permutative objects are stable under tensor *powers*: if $SX \cong X$, then $S(X^{\otimes n}) \cong (SX)^{\otimes n} \cong X^{\otimes n}$.

(The construction of internal homs will be recalled in (1.156).)

1.6.4 Standard models

The (elementary) *standard interval* $\uparrow\!\mathbf{i} = \mathbf{2}$ is freely generated by a 1-cube, u

$$0 \xrightarrow{\ u\ } 1 \qquad\qquad \partial_1^-(u) = 0, \quad \partial_1^+(u) = 1. \qquad (1.141)$$

This cubical set is reversive and permutative.

The (elementary) *n-cube* is its n-th tensor power

$$\uparrow\!\mathbf{i}^{\otimes n} = \uparrow\!\mathbf{i} \otimes ... \otimes \uparrow\!\mathbf{i} \qquad (\text{for } n \geqslant 0),$$

freely generated by its n-cube $u^{\otimes n}$, still reversive and permutative. (It is the representable presheaf $y(2^n) = \mathbb{I}(-, 2^n) \colon \mathbb{I}^{\mathrm{op}} \to \mathbf{Set}$, cf. Section A1.8).

The (elementary) *standard square* $\uparrow\!\mathbf{i}^2 = \uparrow\!\mathbf{i} \otimes \uparrow\!\mathbf{i}$ can be represented as follows, showing the generator $u \otimes u$ and its faces $\partial_i^\alpha(u \otimes u)$

$$
\begin{array}{ccc}
00 & \xrightarrow{\ 0 \otimes u\ } & 01 \\
{\scriptstyle u \otimes 0}\big\downarrow & {\scriptstyle u \otimes u} & \big\downarrow{\scriptstyle u \otimes 1} \\
10 & \xrightarrow[\ 1 \otimes u\ ]{} & 11
\end{array}
\qquad\qquad
\begin{array}{ccc}
\bullet & \xrightarrow{\ 2\ } & \\
\big\downarrow{\scriptstyle 1} & & \\
& &
\end{array}
\qquad (1.142)
$$

with $\partial_1^-(u \otimes u) = 0 \otimes u$ orthogonal to direction 1. Note that, for each cubical object X, $\mathbf{Cub}(\uparrow\!\mathbf{i}^{\otimes n}, X) = X_n$ (Yoneda lemma, [M3]).

The *directed (integral) line* $\uparrow\!\mathbf{z}$ is generated by (countably many) vertices $n \in \mathbf{Z}$ and edges u_n, from $\partial_1^-(u_n) = n$ to $\partial_1^+(u_n) = n + 1$. The *directed integral interval* $\uparrow\![i, j]_{\mathbf{z}}$ is the obvious cubical subset with vertices in the integral interval $[i, j]_{\mathbf{z}}$ (and all cubes whose vertices lie there); in particular, $\uparrow\!\mathbf{i} = \uparrow\![0, 1]_{\mathbf{z}}$.

The (elementary) *directed circle* $\uparrow\!\mathbf{s}^1$ is generated by one 1-cube u with equal faces

$$* \xrightarrow{\ u\ } * \qquad\qquad \partial_1^-(u) = \partial_1^+(u) = *. \qquad (1.143)$$

Similarly, the *elementary directed n-sphere* $\uparrow\mathbf{s}^n$ (for $n > 1$) is generated by one n-cube u all whose faces are totally degenerate (see (1.133)), hence equal

$$\partial_i^\alpha(u) = (e_1)^{n-1}(\partial_1^-)^n(u) \qquad (\alpha = \pm;\ i = 1, ..., n). \qquad (1.144)$$

Moreover, $\uparrow\mathbf{s}^0 = \mathbf{s}^0$ is the discrete cubical set on two vertices (1.136), say $D\{0,1\}$. The *elementary directed n-torus* is a tensor power of $\uparrow\mathbf{s}^1$

$$\uparrow\mathbf{t}^n = (\uparrow\mathbf{s}^1)^{\otimes n}. \qquad (1.145)$$

We also consider the *ordered circle* $\uparrow\mathbf{o}^1$, generated by two edges with the same faces

$$v^- \overset{u'}{\underset{u''}{\rightrightarrows}} v^+ \qquad\qquad \partial_1^\alpha(u') = \partial_1^\alpha(u''), \qquad (1.146)$$

which is a 'cubical model' of the ordered circle $\uparrow\mathbf{O}^1$ already defined in (1.103) as a d-space. (The latter is the directed geometric realisation of the former, in the sense of Section 1.6.7.)

More generally, we have the *ordered spheres* $\uparrow\mathbf{o}^n$, generated by two n-cubes u', u'' with the same boundary: $\partial_i^\alpha(u') = \partial_i^\alpha(u'')$ and further relations. Starting from $\mathbf{s}^0$, the *unpointed suspension* provides all $\uparrow\mathbf{o}^n$ (see (1.182)) while the *pointed suspension* provides all $\uparrow\mathbf{s}^n$ (see (2.55)); of course, these models have the same geometric realisation $\mathbf{S}^n$ (as a topological space) and the same homology; but their *directed* homology is different (Section 2.1.4). The models $\uparrow\mathbf{s}^n$ are more interesting, with a non-trivial order in directed homology.

All these cubical sets are reversive and permutative. Coming back to the remark on reversible d-spaces, in Section 1.4.0, let us note that $\uparrow\mathbf{s}^1$ *coincides* with its opposite cubical set. We do not want to consider this reversive object, a model of the d-space $\uparrow\mathbf{S}^1$, as a 'reversible' cubical set (which has not been defined).

1.6.5 Elementary directed homotopies

Let us start from the standard interval $\uparrow\mathbf{i}$, and work with the monoidal structure recalled above, with unit $\{*\}$ and reversor R. Recall that u denotes the one-dimensional generator of $\uparrow\mathbf{i}$, and u^{op} is the corresponding edge of $\uparrow\mathbf{i}^{\mathrm{op}}$ (Section 1.6.1).

The cubical set $\uparrow\mathbf{i}$ has an obvious structure of monoidal dI1-interval (as defined in Section 1.2.5)

$$\partial^\alpha \colon \{*\} \to \uparrow\mathbf{i}, \qquad \partial^\alpha(*) = \alpha \quad (\alpha = 0, 1),$$
$$e \colon \uparrow\mathbf{i} \to \{*\}, \qquad e(\alpha) = *, \qquad e(u) = e_1(*), \tag{1.147}$$
$$r \colon \uparrow\mathbf{i} \to \uparrow\mathbf{i}^{\mathrm{op}}, \qquad r(0) = 1^{\mathrm{op}}, \ \ r(1) = 0^{\mathrm{op}}, \ \ r(u) = u^{\mathrm{op}}.$$

Since the tensor product is not symmetric, the elementary directed interval yields a *left (elementary) cylinder* $\uparrow\mathbf{i}\otimes X$ and a *right cylinder* $X\otimes\uparrow\mathbf{i}$. But each of these functors determines the other, using the transposer S (see (1.134) and (1.140)) and the property $S(\uparrow\mathbf{i}) = \uparrow\mathbf{i}$

$$I \colon \mathbf{Cub} \to \mathbf{Cub}, \qquad IX = \uparrow\mathbf{i} \otimes X,$$
$$SIS \colon \mathbf{Cub} \to \mathbf{Cub}, \quad SIS(X) = S(\uparrow\mathbf{i} \otimes SX) = X \otimes \uparrow\mathbf{i}. \tag{1.148}$$

(The last equality is actually a canonical isomorphism, cf. (1.140).) Let us begin by considering the left cylinder, $IX = \uparrow\mathbf{i} \otimes X$. It has two faces, a degeneracy and a reflection, as follows (for $\alpha = 0, 1$)

$$\partial^\alpha = \partial^\alpha \otimes X \colon X \to IX, \qquad \partial^\alpha(x) = \alpha \otimes x,$$
$$e = e \otimes X \colon IX \to X,$$
$$e(u \otimes x) = e_1(*) \otimes x = * \otimes e_1(x) = e_1(x), \tag{1.149}$$
$$r = r \otimes RX \colon IRX \to RIX,$$
$$r(\alpha \otimes x^{\mathrm{op}}) = ((1 - \alpha) \otimes x)^{\mathrm{op}}, \qquad r(u \otimes x^{\mathrm{op}}) = (u \otimes x)^{\mathrm{op}}.$$

Moreover, I has a right adjoint, the (elementary) *left cocylinder* or *left path* functor, which has a simpler description: P shifts down all components discarding the faces and degeneracies of index 1 (which are then used to build three natural transformations, the faces and degeneracy of P):

$$P \colon \mathbf{Cub} \to \mathbf{Cub}, \qquad PY = ((Y_{n+1}), (\partial_{i+1}^\alpha), (e_{i+1})),$$
$$\partial^\alpha = \partial_1^\alpha \colon PY \to Y, \qquad e = e_1 \colon Y \to PY. \tag{1.150}$$

The adjunction is defined by the following unit and counit (for $x \in X_n$, $y \in Y_n$, $y' \in Y_{n+1}$):

$$\eta X \colon X \to P(\uparrow\mathbf{i} \otimes X), \qquad x \mapsto u \otimes x,$$
$$\varepsilon Y \colon \uparrow\mathbf{i} \otimes PY \to Y, \qquad \alpha \otimes y' \mapsto \partial_1^\alpha(y'), \quad u \otimes y \mapsto y. \tag{1.151}$$

$\mathbf{Cub}$ has thus a *left dIP1-structure* $\mathbf{Cub}_L$ consisting of the adjoint functors $I \dashv P$, their faces, degeneracy and reflection. An (elementary or immediate) *left homotopy* $f \colon f^- \to_L f^+ \colon X \to Y$ is defined as a map $f \colon IX \to Y$ with $f\partial^\alpha = f^\alpha$; or, equivalently, as a map $f \colon X \to PY$

with $\partial^\alpha f = f^\alpha$. This second expression leads immediately to a simple expression of f as a family of mappings

$$f_n: X_n \to Y_{n+1}, \quad \partial^\alpha_{i+1} f_n = f_{n-1} \partial^\alpha_i,$$
$$e_{i+1} f_{n-1} = f_n e_i, \quad \partial^\alpha_1 f_n = f^\alpha \quad (\alpha = \pm; \ i = 1, ..., n). \tag{1.152}$$

But **Cub** also has a *right dIP1-structure* **Cub**$_R$, based on the right cylinder $SIS(X) = X \otimes \uparrow\mathbf{i}$. Its right adjoint SPS, called the *right co-cylinder* or *right path* functor, shifts down all components and discards the faces and degeneracies of highest index (used again to build the corresponding three natural transformations)

$$SPS: \textbf{Cub} \to \textbf{Cub}, \quad SPS(Y) = ((Y_{n+1}), (\partial^\alpha_i), (e_i)),$$
$$\partial^\alpha: SPS(Y) \to Y, \quad \partial^\alpha = (\partial^\alpha_{n+1}: Y_{n+1} \to Y_n)_{n \geqslant 0}, \tag{1.153}$$
$$e: Y \to SPS(Y), \quad e = (e_{n+1}: Y_n \to Y_{n+1})_{n \geqslant 0}.$$

For this structure, an (elementary) *right homotopy* $f: f^- \to_R f^+$: $X \to Y$ is a map $f: X \to SPS(Y)$ with faces $\partial^\alpha f = f^\alpha$, i.e. a family (f_n) such that

$$f_n: X_n \to Y_{n+1}, \quad \partial^\alpha_i f_n = f_{n-1} \partial^\alpha_i,$$
$$e_i f_{n-1} = f_n e_i, \quad \partial^\alpha_{n+1} f_n = f^\alpha, \quad (\alpha = \pm; \ i = 1, ..., n). \tag{1.154}$$

The transposer (1.134) can be viewed as an isomorphism $S: \textbf{Cub}_L \to \textbf{Cub}_R$ between the left and the right structure: it is, actually, an invertible strict dIP1-functor (Section 1.2.6), since $RS = SR$, $IS = S(SIS)$ and $SP = (SPS)S$. One can define an *external transposition* s (corresponding to the ordinary transposition $s: P^2 \to P^2$ of spaces, in (1.16)), which is actually an identity

$$s: PSPS \to SPSP, \quad s_n = \text{id}Y_{n+2}, \tag{1.155}$$

since both functors shift down all components of two degrees, discarding the faces and degeneracies of least and greatest index.

Elementary homotopies of cubical sets (the usual ones, without connections) are a very defective notion (like intrinsic homotopies of 'face-simplicial' sets, without degeneracies): one cannot even contract the elementary interval $\uparrow\mathbf{i}$ to a vertex (a simple computation on (1.152) shows that this requires a non-degenerate 2-cube $f(u)$, with the same faces as $g_1^-(u)$ or $g_1^+(u)$, if connections exist).

Moreover, to obtain 'non-elementary' paths, which can be concatenated, and a fundamental category $\uparrow\Pi_1(X)$ one should use, instead of the elementary interval $\uparrow\mathbf{i} = \uparrow[0,1]_\mathbf{Z}$, the *directed integral line* $\uparrow\mathbf{z}$

(Section 1.6.4), as in [G6] for simplicial complexes: then, paths are parametrised on $\uparrow\mathbf{z}$, but are eventually constant at left and right, in order to have initial and terminal vertices. However, here we are interested in homology, where concatenation is surrogated by formal sums of cubes, and we will restrain ourselves to proving its invariance up to elementary homotopies, right and left. Of course, we do not want to rely on the classical geometric realisation (Section 1.6.6), which would ignore the directed structure.

The category **Cub** has left and right internal homs, which we shall not need (see [BH3] for their definition). Let us only recall that the *right* internal hom $CUB(A, Y)$ can be constructed with the *left* cocylinder functor P and its natural transformations (which give rise to a cubical object $P^\bullet Y$)

$$- \otimes A \quad \dashv \quad CUB(A, -), \qquad CUB_n(A, Y) = \mathbf{Cub}(A, P^n Y). \quad (1.156)$$

1.6.6 The classical geometric realisation

We have already recalled, in Section 1.6.0, the functor

$$\square: \mathbf{Top} \to \mathbf{Cub}, \qquad \square S = \mathbf{Top}(\mathbf{I}^\bullet, S), \qquad (1.157)$$

which assigns to a topological space S the singular cubical set of (continuous) n-cubes $\mathbf{I}^n \to S$, produced by the *cocubical* set of standard cubes $\mathbf{I}^\bullet = ((\mathbf{I}^n), (\partial_i^\alpha), (e_i))$ (see (1.129)).

As for simplicial sets, the *geometric realisation* $\mathcal{R}(K)$ of a cubical set is given by the left adjoint functor

$$\mathcal{R}: \mathbf{Cub} \rightleftarrows \mathbf{Top} :\square, \qquad \mathcal{R} \dashv \square. \qquad (1.158)$$

The topological space $\mathcal{R}(K)$ is constructed by pasting a copy of the standard cube $\mathbf{I}^n$ for each n-cube $x \in K_n$, along faces and degeneracies. This pasting (formally, the *coend* of the functor $K.\mathbf{I}^\bullet: \mathbb{I}^{\mathrm{op}} \times \mathbb{I} \to \mathbf{Top}$, see [M3]) comes with a family of structural mappings, one for each cube x, coherently with faces and degeneracies (of $\mathbf{I}^\bullet$ and K)

$$\hat{x}: \mathbf{I}^n \to \mathcal{R}(K), \qquad \hat{x}.\partial_i^\alpha = (\partial_i^\alpha x)\hat{}, \qquad \hat{x}.e_i = (e_i x)\hat{}. \qquad (1.159)$$

$\mathcal{R}(K)$ has the finest topology making all the structural mappings continuous.

This realisation is important, since it is well known that the combinatorial homology of a cubical set K coincides with the homology of the CW-space $\mathcal{R}K$ (cf. [Mu], 4.39), for the simplicial case). But we also want a finer 'directed realisation', keeping more information about the cubes of K: we shall use a d-space (Section 1.6.7) or also a set equipped with a presheaf of distinguished cubes (Section 1.6.8).

1.6.7 A directed geometric realisation

Cubical sets have a clear realisation as d-spaces, since we obviously want to realise the object with one free generator of degree n as $\uparrow\mathbf{I}^n$.

(Simplicial sets can also be realised as d-spaces, by a choice of $\uparrow\Delta^n \subset \uparrow\mathbf{R}^n$ which agrees with faces and degeneracies: the convex hull of the points $0 < e_1 < e_1 + e_2 < \cdots < e_1 + \cdots + e_n$ obtained from the canonical basis of $\mathbf{R}^n$.)

Recall that a cubical set $K = ((K_n), (\partial_i^\alpha), (e_i))$ is a functor $K \colon \mathbb{I}^{op} \to$ **Set**, for a category $\mathbb{I} \subset$ **Set** already recalled in Section 1.6.1. Its objects are the sets $2^n = \{0, 1\}^n$, its mappings are generated by the elementary faces $\partial^\alpha \colon 2^0 \to 2$ and degeneracy $e \colon 2 \to 2^0$, under finite products (in **Set**) and composition. Equivalently, the mappings of $\mathbb{I}$ are generated under composition by the following higher faces and degeneracies ($i = 1, ..., n$; $\alpha = 0, 1$; $t_i = 0, 1$)

$$\partial_i^\alpha = 2^{i-1} \times \partial^\alpha \times 2^{n-1} \colon 2^{n-1} \to 2^n,$$

$$\partial_i^\alpha(t_1, ..., t_{n-1}) = (t_1, ..., t_{i-1}, \alpha, ..., t_{n-1}),$$

$$e_i = 2^{i-1} \times e \times 2^{n-1} \colon 2^n \to 2^{n-1}, \tag{1.160}$$

$$e_i(t_1, ..., t_n) = (t_1, ..., \hat{t}_i, ..., t_n).$$

There is an obvious embedding of $\mathbb{I}$ in d**Top**, where faces and degeneracies are realised as above, with the standard $\partial^\alpha \colon \{*\} \rightrightarrows \uparrow\mathbf{I} \colon e$ (and $t_i \in \uparrow\mathbf{I}$)

$$I \colon \mathbb{I} \to \mathrm{d}\mathbf{Top}, \qquad 2^n \mapsto \uparrow\mathbf{I}^n,$$

$$\partial_i^\alpha = \uparrow\mathbf{I}^{i-i} \times \partial^\alpha \times \uparrow\mathbf{I}^{n-i} \colon \uparrow\mathbf{I}^{n-1} \to \uparrow\mathbf{I}^n, \tag{1.161}$$

$$e_i = \uparrow\mathbf{I}^{i-1} \times e \times \uparrow\mathbf{I}^{n-i} \colon \uparrow\mathbf{I}^{n+1} \to \uparrow\mathbf{I}^n.$$

Now, the *directed singular cubical set* of a d-space X and its left adjoint functor, the *directed geometric realisation* $\uparrow\mathcal{R}(K)$ of a cubical set K, can be constructed as in the classical case recalled above

$$\uparrow\mathcal{R} \colon \mathbf{Cub} \rightleftarrows \mathrm{d}\mathbf{Top} \colon \uparrow\square, \qquad \uparrow\mathcal{R} \dashv \uparrow\square,$$

$$\uparrow\square_n(X) = \mathrm{d}\mathbf{Top}(\uparrow\mathbf{I}^n, X). \tag{1.162}$$

The d-space $\uparrow\mathcal{R}(K)$ is thus the pasting in d**Top** of K_n copies of $I(2^n) = \uparrow\mathbf{I}^n$ ($n \geqslant 0$), along faces and degeneracies (again, the coend of the functor $K.I \colon \mathbb{I}^{op} \times \mathbb{I} \to \mathrm{d}\mathbf{Top}$).

In other words (since a colimit in d**Top** is the colimit of the underlying topological spaces, equipped with the relevant d-structure), one starts from the ordinary geometric realisation $\mathcal{R}K$, as a topological space, and equips it with the following d-structure $\uparrow\mathcal{R}(K)$: the distinguished paths are generated, under concatenation and increasing reparametrisation, by the mappings $\hat{x}a\colon \mathbf{I} \to \mathbf{I}^n \to \mathcal{R}K$, where $a\colon \mathbf{I} \to \mathbf{I}^n$ is an order-preserving map and $\hat{x}$ corresponds to some cube $x \in K_n$, in the colimit-construction of $\mathcal{R}K$.

The adjunction $U \dashv D'$ (Section 1.4.0) between spaces and d-spaces

$$\mathbf{Cub} \underset{\uparrow\square}{\overset{\uparrow\mathcal{R}}{\rightleftarrows}} \mathrm{d}\mathbf{Top} \underset{D'}{\overset{U}{\rightleftarrows}} \mathbf{Top} \qquad \uparrow\mathcal{R} \dashv \uparrow\square, \quad U \dashv D' \qquad (1.163)$$

gives back the ordinary realisation $\mathcal{R} = U.\uparrow\mathcal{R}\colon \mathbf{Cub} \to \mathbf{Top}$, with $\mathcal{R} \dashv \square = \uparrow\square.D'$.

Various basic objects of d**Top** are directed realisations of simple cubical sets already considered in Section 1.6.4. For instance, the directed interval $\uparrow\mathbf{I}$ realises $\uparrow\mathbf{i} = \{0 \to 1\}$; the directed line $\uparrow\mathbf{R}$ realises $\uparrow\mathbf{z}$; the ordered circle $\uparrow\mathbf{O}^1$ realises $\uparrow\mathbf{o}^1 = \{0 \rightrightarrows 1\}$; the directed circle $\uparrow\mathbf{S}^1$ realises $\uparrow\mathbf{s}^1 = \{* \to *\}$.

It is easy to prove that $\uparrow\mathcal{R}\colon \mathbf{Cub}_L \to \mathrm{d}\mathbf{Top}$ is a strong dI1-functor (Section 1.2.6). In fact, $\uparrow\mathcal{R}$ and the cylinder functors of $\mathbf{Cub}_L$ and d**Top** preserve all colimits, as left adjoints, and every cubical set is a colimit of representable presheaves $y(2^n) = \uparrow\mathbf{i}^{\otimes n}$. But, on such objects, one trivially has

$$\uparrow\mathcal{R}(I(\uparrow\mathbf{i}^{\otimes n})) = \uparrow\mathcal{R}(\uparrow\mathbf{i} \otimes \uparrow\mathbf{i}^{\otimes n}) \cong \uparrow\mathbf{I}^{n+1} = \uparrow\mathbf{I}^n \times \uparrow\mathbf{I} \cong I(\uparrow\mathcal{R}(\uparrow\mathbf{i}^{\otimes n})).$$

1.6.8 Sets with distinguished cubes

A finer (directed) *cubical* realisation $\uparrow\mathcal{C}(K)$ can be given in the category c**Set** of *sets with distinguished cubes*, or *c-sets*, which we introduce now.

A *c-set* L is a set equipped with a sub-presheaf cL of the cubical set $\mathbf{Set}(\mathbf{I}^\bullet, L)$, such that L is covered by all distinguished cubes. In other words, the structure of the set L consists of a sequence of sets of *distinguished cubes* $c_n L \subset \mathbf{Set}(\mathbf{I}^n, L)$, preserved by faces and degeneracies (of the cocubical set $\mathbf{I}^\bullet$) and satisfying the *covering condition* $L = \bigcup\mathrm{Im}(x)$ (for x varying in the set of all distinguished cubes), so that the canonical mapping $p_L\colon \mathcal{R}(cL) \to L$ is surjective. A *morphism* of c-sets $f\colon L \to L'$ is a mapping of sets which preserves distinguished cubes: if $x\colon \mathbf{I}^n \to L$ is distinguished, also $fx\colon \mathbf{I}^n \to L'$ is.

(Note that, differently from the definition of d**Top**, here we are not asking that distinguished cubes be 'closed under concatenation and reparametrisation', which would give a complicated structure. The present category c**Set** is well-adapted for directed homology, but less adequate than d**Top** for directed homotopy.)

Now, the adjunction $\uparrow\mathcal{R} \dashv \uparrow\square$ constructed above (Section 1.6.7) can be factored through c**Set**

$$\mathbf{Cub} \underset{c}{\overset{\uparrow\mathcal{C}}{\rightleftarrows}} \mathbf{cSet} \underset{(-)_\square}{\overset{\mathbf{d}}{\rightleftarrows}} \mathbf{Top} \qquad \uparrow\mathcal{C} \dashv c, \quad \mathbf{d} \dashv (-)_\square. \qquad (1.164)$$

First, if X is a d-space, its singular cubes in $\uparrow\square X$ cover the underlying set (since all constant cubes are directed). Thus, we factorise the functor $\uparrow\square \colon \mathrm{d}\mathbf{Top} \to \mathbf{Cub}$ letting $X_\square$ be the underlying *set* $|X|$ equipped with the presheaf $\uparrow\square X \subset \mathbf{Set}(\mathbf{I}^\bullet, X)$, and letting c be the forgetful functor assigning to a c-set L its structural presheaf cL. Note that the functor c is faithful (because $pL \colon \mathcal{R}(cL) \to L$ is surjective).

Then, the left adjoint of c yields the *directed realisation* $\uparrow\mathcal{C}(K)$ of a cubical set K as a c-set: it is the set RK underlying the geometric realisation $\mathcal{R}(K)$, without topology but equipped with convenient distinguished cubes. The latter arise from the n-cubes $x \in K_n$, via the associated mappings $\hat{x} \colon \mathbf{I}^n \to RK$ (cf. (1.159)), which are closed under faces and degeneracies

$$c_n(\uparrow\mathcal{R}(K)) = \{\hat{x} \mid x \in K_n\} \subset \mathbf{Set}(\mathbf{I}^n, RK). \qquad (1.165)$$

The bijection $(\uparrow\mathcal{C}(K), L) = (K, cL)$ is easy to construct: given $f \colon \uparrow\mathcal{C}(K) \to L$, we define $f_n \colon K_n \to c_n L$ letting $f_n(x) = f\hat{x}$; given $g \colon K \to cL$, we take $f = p_L.\mathcal{C}f = (\mathcal{C}(K) \to \mathcal{C}(cL) \to L)$.

Finally, the functor $\mathbf{d} \colon \mathbf{cSet} \to \mathrm{d}\mathbf{Top}$ (left adjoint to $(-)_\square$), acting on a c-set L, gives the underlying set $|L|$ equipped with the *cubical topology*, i.e. the finest topology making all distinguished cubes $\mathbf{I}^n \to L$ continuous, and further equipped with the distinguished paths generated by the distinguished 1-cubes of $c_1 L$ (via reparametrisation and concatenation). The bijection $(\mathbf{d}(L), X) = (L, X_\square)$ is obvious, since a mapping $L \to X$ is continuous for the cubical topology of L if and only if it is continuous on each distinguished n-cube $x \colon \mathbf{I}^n \to L$, if and only if each composite $f \circ x$ is an n-cube of $\square X$.

We end with some comments on the category c**Set**. Given a c-set $L = (L, cL)$, a *c-subset* $M = (M, cM)$ will be a c-set with $cM \subset cL$; in other words, we are considering a subset $M \subset L$ equipped with a sub-presheaf $cM \subset cL \cap \mathbf{Set}(\mathbf{I}^\bullet, M)$ satisfying the covering condition on M. It is a *regular subobject* (Section 1.3.1) if $cM = cL \cap \mathbf{Set}(\mathbf{I}^\bullet, M)$, that is if the distinguished cubes of M are precisely those of L whose image is contained in M; a regular subobject thus amounts to a subset $M \subset L$ which is a union of images of distinguished cubes of L (equipped with the restricted structure).

The *quotient* $L/\sim$ of a c-set modulo an equivalence relation (on the *set* L) will be the set-theoretical quotient, equipped with the projections $\mathbf{I}^n \to L \to L/\sim$ of the distinguished cubes of L (which are obviously stable under the faces and degeneracies of $\mathbf{I}^\bullet$). This easy description of quotients will be exploited in Section 2.5, as an advantage of c-sets with respect to cubical sets: *one has just to assign an equivalence relation on the underlying set.*

1.6.9 Pointed cubical sets

A strong reason for considering *pointed* cubical sets is that their homology theory behaves much better than reduced homology of the unpointed objects (as we shall see in Section 2.3).

A *pointed cubical set* is a pair (X, x_0) formed of a cubical set with a base point $x_0 \in X_0$; a morphism $f \colon (X, x_0) \to (Y, y_0)$ is a morphism of cubical sets which preserves the base points: $f(x_0) = y_0$. The corresponding category, written $\mathbf{Cub_\bullet}$, is pointed, with zero-object $\{*\}$.

As for pointed d-spaces (Section 1.5.4), the forgetful functor $\mathbf{Cub_\bullet} \to \mathbf{Cub}$ has a left adjoint $X \mapsto X_\bullet$, which adds a *discrete* base point $*$ (in the sense that the only cubes having some vertex at $*$ are totally degenerate, see (1.133)). The *pointed directed interval* is $\uparrow\mathbf{i_\bullet}$.

Again, *limits* and *quotients* are computed as in $\mathbf{Cub}$ and pointed in the obvious way, whereas *sums* are quotients of the corresponding unpointed sums, under identification of the base points.

The (left) cylinder and path endofunctors can be obtained from the pointed directed interval $\uparrow\mathbf{i_\bullet}$, via the *smash tensor product*:

$$(X, x_0) \otimes (Y, y_0) = ((X \otimes Y)/\sim, [x_0, y_0]), \qquad (1.166)$$

where $\sim$ is the equivalence relation which identifies all n-cubes $x \otimes y_0$ and $x_0 \otimes y$ with $e^n(x_0 \otimes y_0)$. Its unit is the discrete cubical set $\mathbf{s}^0 = D\{0, 1\}$ (Section 1.6.4), pointed at 0 (for instance).

However, it will be simpler to give a direct definition of these functors (Section 2.3.2).

1.7 First-order homotopy theory by the cylinder functor, II

Coming back to the general theory, we study *dI1-homotopical categories*, i.e. those dI1-categories which have a terminal object and all h-pushouts, and therefore all mapping cones and suspensions. We end by constructing, in this setting, the (lower or upper) cofibre sequence of a map (Theorem 1.7.9).

The present matter essentially appeared in [G3, G7]. Its classical counterpart for topological spaces is the well-known Puppe sequence [Pu], whose study in categories with an abstract cylinder functor was developed in Kamps' dissertation [Km1].

1.7.0 Homotopical categories via cylinders

Let **A** be a dI1-category with a *terminal* object $\top$, so that every object X has a unique morphism $p_X : X \to \top$.

Note that, since homotopies are defined by a cylinder functor, the object $\top$ is automatically 2-*terminal*, in the sense that each map $p_X : X \to \top$ has precisely one endohomotopy, the trivial one (represented by the unique map $IX \to \top$). Moreover, $R\top \cong \top$ (since R is an automorphism) and we shall assume, for simplicity, that $R\top = \top$, so that $R(p_X) = p_{RX}$.

A map $X \to Y$ which factors through $\top$ is said to be *constant*. Recall (from Section 1.3.2) that an object X is said to be *future contractible* if it is future homotopy equivalent to $\top$, or equivalently if there is a map $i\colon \top \to X$ which is the embedding of a future deformation retract, i.e. admits a homotopy $1_X \to ip_X$.

We say that **A** is a *dI1-homotopical category* if it is a dI1-category with all h-pushouts (Section 1.3.5) and a terminal object $\top$. In the rest of the section, we assume that this is the case.

Every dI1-category with a terminal object and ordinary pushouts is dI1-homotopical (by Section 1.3.5). But we have already seen, implicitly, that a dI1-homotopical category need not have all pushouts (Section 1.3.5).

The categories

$$\textbf{pTop, dTop, Cub, Cat, pTop}_\bullet\textbf{, dTop}_\bullet$$

are dI1-homotopical, with the cylinder functor already described above; in fact, all of them are complete and cocomplete. Other examples include: inequilogical spaces (Section 1.9.1), 'convenient' locally preordered spaces (Section 1.9.3), pointed cubical sets (Section 2.3), chain complexes (Section 4.4), and various constructions on the previous categories, like functor categories or categories of algebras (see Chapter 5).

A *dI1-homotopical functor*

$$H = (H, i, h) \colon \mathbf{A} \to \mathbf{X} \qquad (i \colon RH \to HR, \quad h \colon IH \to HI),$$

will be a *strong* dI1-functor (Section 1.2.6) between dI1-homotopical categories which preserves the terminal object and satisfies the following equivalent properties

(i) H preserves every h-pushout;
(ii) H preserves every cylindrical colimit (Section 1.3.5), as a colimit.

The equivalence is proved in the next lemma. Notice that property (i) makes sense for all lax dI1-functors, since they act on homotopies (1.58), and is obviously closed under composition, while (ii) makes sense for arbitrary functors, but, then, is not closed under composition.

The directed geometric realisation functor $\uparrow\mathcal{R} \colon \mathbf{Cub} \to \mathrm{d}\mathbf{Top}$ (Section 1.6.7), which preserves all colimits as a left adjoint, is a dI1-homotopical functor.

1.7.1 Lemma (The preservation of h-pushouts)

A strong dI1-functor $H = (H, i, h) \colon \mathbf{A} \to \mathbf{X}$ between dI-categories preserves an h-pushout of $\mathbf{A}$ if and only if it preserves the corresponding cylindrical colimit (Section 1.3.5) as a colimit.

Proof Consider an h-pushout of $\mathbf{A}$, the left diagram below

$$(1.167)$$

This diagram is transformed by H into the right diagram above, where, as defined in (1.58), the homotopy $H\lambda$ is represented by the map

$$(H\lambda)\hat{\ } = H(\hat{\lambda}).hX \colon IHX \to HIX \to H(I(f,g)). \qquad (1.168)$$

We have seen in Section 1.3.5 that the h-pushout λ can be expressed by the ordinary colimit in the left diagram below, which is transformed by $H\colon \mathbf{A} \to \mathbf{B}$ into the right diagram

$$
\begin{array}{ccc}
X \xrightarrow{\ g\ } Z & \qquad & HX \xrightarrow{\ Hg\ } HZ
\end{array}
\tag{1.169}
$$

Now, since the natural transformation $h\colon IH \to HI$ is invertible, we can equivalently replace HIX with IHX, the map $H\hat{\lambda}$ with $(H\lambda)\widehat{\ } = H(\hat{\lambda}).(hX)$ and the maps $H\partial^\alpha$ with

$$
\partial^\alpha H = (hX)^{-1}.H\partial^\alpha \colon HX \to HIX \to IHX,
$$

Therefore, the right diagram (1.167) is an h-pushout if and only if the right diagram (1.169) is a colimit. $\qquad\square$

1.7.2 Mapping cones, cones and suspension

Let $\mathbf{A}$ be a dI1-homotopical category, as defined above.

(a) A map $f\colon X \to Y$ has an *upper mapping cone* $C^+ f = I(f, p_X)$, or *upper h-cokernel*, defined as the h-pushout below, at the left

$$
\begin{array}{ccc}
X \xrightarrow{\ p\ } \top & \qquad & X \xrightarrow{\ f\ } Y
\end{array}
\tag{1.170}
$$

Its structural maps are the *lower basis* $u = \mathrm{hcok}^+(f)\colon Y \to C^+ f$ and the *upper vertex* $v^+ \colon \top \to C^+ f$; furthermore, we have a *structural homotopy* $\gamma\colon uf \to v^+ p\colon X \to C^+ f$, which links f to a constant map, in a universal way. The h-cokernel is a functor $C^+ \colon \mathbf{A}^2 \to \mathbf{A}$, defined on the category of morphisms of $\mathbf{A}$.

Symmetrically, f has a *lower mapping cone* $C^- f = I(p_X, f)$ defined as the right h-pushout above, with a *lower vertex* v^- and an *upper basis* u. As a particular case of the reflection r^I of h-pushouts (Section 1.3.5), we have a *reflection* isomorphism for mapping cones

$$
r^C \colon C^+(f^{\mathrm{op}}) \to (C^- f)^{\mathrm{op}},
$$

induced by the reflection $rX\colon IRX \to RIX$.

Notice that the name we are choosing for these constructions, *upper* or *lower*, agrees with the vertex and not with the basis. This choice of terminology (in contrast with [G3]) comes from a relationship of cones with *future* and *past* contractibility, which is examined below (Lemma 1.7.3).

(b) In particular, every object X has an *upper cone* $C^+ X = C^+ (\mathrm{id} X) = I(\mathrm{id} X, p_X)$, defined as the left h-pushout below; or, equivalently, as the ordinary pushout on the right (with $u^- = \gamma . \partial^- X$)

$$\begin{array}{ccc} X & \overset{p}{\longrightarrow} & \top \\ {\scriptstyle \mathrm{id}}\downarrow & {\scriptstyle \gamma}\searrow & \downarrow{\scriptstyle v^+} \\ X & \underset{u^-}{\longrightarrow} & C^+ X \end{array} \qquad\qquad \begin{array}{ccc} X & \overset{p}{\longrightarrow} & \top \\ {\scriptstyle \partial^+}\downarrow & \dashrightarrow & \downarrow{\scriptstyle v^+} \\ IX & \underset{\gamma}{\longrightarrow} & C^+ X \end{array} \qquad (1.171)$$

This defines a functor $C^+ : \mathbf{A} \to \mathbf{A}$. Notice: given a map $f : X \to Y$, one should not confuse the mapping cone $C^+ f = I(f, p_X)$, which is an *object* of $\mathbf{A}$, with the *map* $C^+ [f] : C^+ X \to C^+ Y$ induced by f between the cones of X and Y. We will use square brackets in the second case, even though the context should already be sufficient to distinguish these things.

Symmetrically, there is a *lower cone* functor $C^- : \mathbf{A} \to \mathbf{A}$, with $C^- X = C^- (\mathrm{id} X) = I(p_X, \mathrm{id} X)$ and a lower vertex v^-. The two cones are linked by a reflection isomorphism $r^C : C^+ (X^{\mathrm{op}}) \to (C^- X)^{\mathrm{op}}$.

If $\mathbf{A}$ is pointed, $\top$ is the zero-object and the right diagram above shows that the map $\gamma : IX \to C^+ X$ is the ordinary cokernel of $\partial^+ : X \to IX$, whence a normal *epimorphism*. But, in the unpointed case, γ need not be epi (see Section 1.7.4).

In Proposition 4.8.5, under stronger hypotheses on $\mathbf{A}$, we will show that an upper cone $C^+ X$ is always strongly future contractible, to its vertex v^+.

(c) Finally, the *suspension* of an object

$$\Sigma X = C^+ (p_X) = C^- (p_X) = I(p_X, p_X),$$

is an upper and a lower cone, at the same time. It is obtained by collapsing, independently, the bases of IX to an upper and a lower vertex, v^+ and v^-

$$\begin{array}{ccc} X & \overset{p}{\longrightarrow} & \top \\ {\scriptstyle p}\downarrow & {\scriptstyle \sigma}\searrow & \downarrow{\scriptstyle v^+} \\ \top & \underset{v^-}{\longrightarrow} & \Sigma X \end{array} \qquad (1.172)$$

There is now a reflection isomorphism for the suspension, $r^\Sigma \colon \Sigma(X^{\mathrm{op}}) \to (\Sigma X)^{\mathrm{op}}$.

For the terminal object $\top$, $p \colon \top \to \top$ is an isomorphism, and $\gamma\top, \sigma\top$ are also invertible. Identifying

$$I\top = C^+\top = C^-\top = \Sigma\top,$$

the faces $\partial^\alpha \colon \top \to I\top$ are also the faces and vertices of the cones and suspension.

1.7.3 Lemma (Contractible objects and cones)

In a dI1-homotopical category **A**, *an object* X *is* future *contractible if and only if the basis* $u \colon X \to C^+X$ *of its upper* cone *has a retraction* $h \colon C^+X \to X$.

Proof We use the notation of (1.171). If $hu = \mathrm{id}X$, then the map $h\gamma \colon IX \to X$ is a homotopy from $h\gamma\partial^-X = hu = \mathrm{id}X$ to $h\gamma\partial^+X = hv^+p_X \colon X \to X$, and the latter is a constant endomap.

Conversely, if there is a homotopy $\varphi \colon IX \to X$ with $\varphi\partial^-X = \mathrm{id}X$ and $\varphi\partial^+X = ip_X \colon X \to X$, we define $h \colon C^+X \to X$ as the unique map such that $h\gamma = \varphi \colon IX \to X$ and $hv^+ = i \colon \top \to X$, and we obtain $hu = h\gamma.\partial^-X = \varphi.\partial^-X = \mathrm{id}X$. □

1.7.4 Cones and suspension for d-spaces

Let X be a d-space. Its upper cone C^+X, given by the right-hand pushout in (1.171), can be computed as the quotient

$$C^+X = (IX + \{*\})/(\partial^+X + \{*\}), \tag{1.173}$$

where the upper basis of the cylinder is collapsed to an upper vertex $v^+ = v^+(*)$, while the lower basis $\partial^- \colon X \to IX \to C^+X$ 'subsists'.

Note that $C^+\emptyset = \{*\}$: the cone C^+X is a quotient of the cylinder IX *only for* $X \neq \emptyset$.

Dually, the *lower cone* C^-X is obtained by collapsing the lower basis of IX to a lower vertex $v^- = v^-(*)$.

We have already described, in Section 1.4.7, the upper and lower cones of $\mathbf{S}^1$ and $\uparrow\mathbf{S}^1$, in d**Top**:

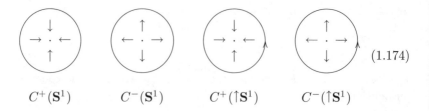

$$C^+(\mathbf{S}^1) \qquad C^-(\mathbf{S}^1) \qquad C^+(\uparrow\mathbf{S}^1) \qquad C^-(\uparrow\mathbf{S}^1) \tag{1.174}$$

The suspension ΣX is the colimit of the diagram

$$\begin{array}{ccc}
 & X \longrightarrow \{*\} & \\
 & \downarrow{\scriptstyle\partial^+} & \\
X \underset{\partial^-}{\to} IX & & v^+ \\
\downarrow & \quad\hat{\sigma}\searrow & \downarrow \\
\{*\} & \xrightarrow{\quad v^- \quad} & \Sigma X
\end{array} \tag{1.175}$$

which collapses, independently, the bases of IX to a lower and an upper vertex, v^- and v^+. Note that $\Sigma\emptyset = \mathbf{S}^0$ (with the unique d-structure), and $\Sigma\mathbf{S}^0 = \uparrow\mathbf{O}^1$ (1.103).

As happens for topological spaces, the suspension ΣX of a d-space can also be obtained by the following pushout: the pasting of both cones $C^\alpha X$ of X, along their bases

$$\begin{array}{ccc}
X & \xrightarrow{\quad u^- \quad} & C^+ X \\
{\scriptstyle u^+}\downarrow & \quad\vdash\dashv\quad & \downarrow \\
C^- X & \longrightarrow & \Sigma X
\end{array} \tag{1.176}$$

This coincidence is *not* a general fact, within directed algebraic topology. It rests on the fact that, in d**Top** (or in **Top**), pasting two copies of the standard interval one after the other (a particular case of (1.176), for $X = \{*\}$), we get an object isomorphic to the standard interval (which is $\Sigma\{*\}$). But this is not true for **Cat** (where the corresponding pasting gives the ordinal **3**, nor for cubical sets (where we similarly get three vertices), nor for chain complexes (directed or not).

It is easy to prove that a future cone $C^+ X$ is *past* contractible if and only if X is empty or past contractible.

1.7.5 Cones and suspension for pointed d-spaces

Let us recall, from (1.124), that, in the dI1-homotopical category d**Top.** of pointed d-spaces, the cylinder $I(X, x_0)$ can be expressed as the quotient of the unpointed cylinder IX which collapses the fibre at the base point x_0 (thus providing the new base point)

$$I(X, x_0) = (IX/I\{x_0\}, [x_0, t]). \tag{1.177}$$

As a consequence, the upper cone and suspension of a pointed d-space can also be obtained from the corresponding unpointed constructions, by collapsing the fibre of the base point

$$C^+(X, x_0) = I(X, x_0)/\partial^+(X, x_0) = (C^+X/I\{x_0\}, [x_0, t]),$$
$$\Sigma(X, x_0) = (\Sigma X/I\{x_0\}, [x_0, t]). \tag{1.178}$$

In particular, pointing the directed sphere $\uparrow\mathbf{S}^n = (\uparrow\mathbf{I}^n)/(\partial\mathbf{I}^n)$ (cf. (1.102)) at $x_n = [(0, ..., 0)]$, we have, for $n \geqslant 0$

$$\Sigma(\uparrow\mathbf{S}^n, x_n) \cong (\uparrow\mathbf{S}^{n+1}, x_{n+1}). \tag{1.179}$$

1.7.6 Cones and suspension for cubical sets

Let X be a cubical set. Its *left* upper cone C^+X (relative to the *left* cylinder $IX = \uparrow\mathbf{i} \otimes X$, cf. Section 1.6.5) can again be computed as the quotient (1.173).

Analytically, we can describe C^+X by saying that it is generated by $(n+1)$-dimensional cubes $u \otimes x \in IX$ (for each $x \in X_n$) plus a vertex v^+, under the relations arising from X together with the identification

$$1 \otimes x = e_1^n(v^+) \qquad (x \in X_n). \tag{1.180}$$

The *left suspension* ΣX is the colimit in **Cub** of the diagram (1.175). Thus, the suspension of $\mathbf{s}^0 = D\{0, 1\}$ yields the 'ordered circle' $\uparrow\mathbf{o}^1$ (1.146)

$$v^- \mathrel{\substack{\xrightarrow{u'} \\[-0.3em] \xrightarrow[u'']{}}} v^+ \qquad u' = [0 \otimes u], \qquad u'' = [1 \otimes u], \tag{1.181}$$

(the square brackets denote equivalence classes in the colimit (1.175)). More generally, we define

$$\Sigma^n(\mathbf{s}^0) = \uparrow\mathbf{o}^n. \tag{1.182}$$

But we are more interested in the *pointed suspension* (i.e. the suspension of pointed cubical sets) which will be studied in Section 2.3, and yields the directed spheres $\uparrow\mathbf{s}^n$ (as in the previous result for d-spaces, (1.179)).

1.7.7 Differential and comparison

Let us come back to the general theory. In the dl1-homotopical category $\mathbf{A}$, every map $f\colon X \to Y$ has a *lower differential* $d = d^-(f)$ defined by the universal property of the lower h-cokernel $C^- f$ (see the diagram below)

$$
\begin{aligned}
&d = d^-(f)\colon C^- f \to \Sigma X && \text{(\emph{lower differential})}, \\
&dv^- = v^-\colon \mathsf{T} \to \Sigma X, && du = v^+ p_Y \colon Y \to \Sigma X, && (1.183)\\
&d\circ\gamma = \sigma\colon v^- p_X \to v^+ p_X \colon X \to \Sigma X && (u = \mathrm{hcok}^-(f)).
\end{aligned}
$$

Now, we go on with a construction which only depends on f and makes an *alternate* use of lower and upper h-cokernels; the construction is called 'lower', simply because it begins by the lower h-cokernel of f.

Following the pattern of the pasting lemma 1.3.8 for h-pushouts

$$ (1.184) $$

we can compare the pasting of the *lower* h-cokernel $C^- f$ *and the upper* h-cokernel $C^+ u$, *with* the global h-pushout ΣX.

We thus obtain a *lower comparison map* $k = k^-(f)$, defined by the universal property of $C^+ u$

$$
\begin{aligned}
&k = k^-(f)\colon C^+ u \to \Sigma X && \text{(\emph{lower comparison map})}, \\
&ku' = d, \quad kv^+ = v^+, \qquad k\circ\gamma' = 0\colon du \to v^+ p_Y \colon Y \to \Sigma X.
\end{aligned}
$$
$$ (1.185) $$

This yields a diagram, the *reduced lower cofibre diagram of* f

$$ (1.186) $$

which commutes, except (possibly) at the $\sharp$-marked square; the latter will be called the *lower comparison square* of f.

This diagram will be extended to the lower cofibre diagram of f, in Theorem 1.7.9. In a stronger setting, we will be able to prove that the comparison square commutes up to homotopy and that k is a future homotopy equivalence (see Theorem 4.7.5).

Both d^- and k^- are natural in f. More precisely, d^- and k^- are natural transformations of functors defined on $\mathbf{A}^2$, the category of morphisms of $\mathbf{A}$

$$d^- : C^- \to \Sigma.\mathrm{Dom} : \mathbf{A}^2 \to \mathbf{A},$$
$$k^- : C^+.\mathrm{hcok}^- \to \Sigma.\mathrm{Dom} : \mathbf{A}^2 \to \mathbf{A}. \tag{1.187}$$

By reflection duality, we also have an *upper differential* $d^+(f)$ and an *upper comparison map* $k^+(f)$ (natural in f), which are characterised by the following conditions (see the diagram below):

$$
\begin{aligned}
&d = \partial^+(f) : C^+ f \to \Sigma X && \text{(\emph{upper differential})},\\
&du = v^- p_Y : Y \to \Sigma X, && dv^+ = v^+ : \top \to \Sigma X, && \text{(1.188)}\\
&d \circ \gamma = \sigma : v^- p_X \to v^+ p_X : X \to \Sigma X && (u = \mathrm{hcok}^+(f)).
\end{aligned}
$$

$$
\begin{aligned}
&k = k^+(f) : C^- u \to \Sigma X && \text{(\emph{upper comparison map})},\\
&kv^- = v^-, \quad ku' = d, \quad k \circ \gamma' = 0 : v^- p_Y \to du : Y \to \Sigma X. && \text{(1.189)}
\end{aligned}
$$

1.7.8 The cofibre sequences of a map

In a dI1-homotopical category, every map $f : X \to Y$ has a natural *lower cofibre sequence* (or lower Puppe sequence)

$$X \xrightarrow{f} Y \xrightarrow{u} C^- f \xrightarrow{d} \Sigma X \xrightarrow{\Sigma f} \Sigma Y \xrightarrow{\Sigma u} \Sigma C^- f \xrightarrow{\Sigma d} \Sigma^2 X \ldots$$

$$u = \mathrm{hcok}^-(f), \qquad\qquad d = d^-(f), \tag{1.190}$$

formed by its lower h-cokernel u, its lower differential d (Section 1.7.7) and the suspension functor.

Symmetrically, we have the *upper cofibre sequence* (or upper Puppe sequence) of f

$$X \xrightarrow{f} Y \xrightarrow{u} C^+f \xrightarrow{d} \Sigma X \xrightarrow{\Sigma f} \Sigma Y \xrightarrow{\Sigma u} \Sigma C^+f \xrightarrow{\Sigma d} \Sigma^2 X \dots$$

$$u = \mathrm{hcok}^+(f), \qquad\qquad d = d^+(f), \qquad\qquad (1.191)$$

The upper sequence of f^{op} is equivalent to the R-dual of the lower sequence of f, via the isomorphisms $r^C \colon C^+(f^{\mathrm{op}}) \to (C^-f)^{\mathrm{op}}$ and $r^\Sigma \colon \Sigma(X^{\mathrm{op}}) \to (\Sigma X)^{\mathrm{op}}$ (Section 1.7.2)

$$
\begin{array}{ccccccccc}
X^{\mathrm{op}} & \rightarrow & Y^{\mathrm{op}} & \rightarrow & C^+(f^{\mathrm{op}}) & \rightarrow & \Sigma(X^{\mathrm{op}}) & \rightarrow & \Sigma(Y^{\mathrm{op}}) \dots \\
\| & & \| & & \downarrow{r} & & \downarrow{r} & & \downarrow{r} \\
X^{\mathrm{op}} & \rightarrow & Y^{\mathrm{op}} & \rightarrow & (C^-f)^{\mathrm{op}} & \rightarrow & (\Sigma X)^{\mathrm{op}} & \rightarrow & (\Sigma Y)^{\mathrm{op}} \dots
\end{array}
\qquad (1.192)
$$

1.7.9 Theorem and definition (The cofibre diagram)

In the dI1-homotopical category **A**, *every map* $f \colon X \to Y$ *has a* lower cofibre diagram

$$
\begin{array}{ccccccccccc}
X & \xrightarrow{f} & Y & \xrightarrow{u} & C^-f & \xrightarrow{d} & \Sigma X & \xrightarrow{\Sigma f} & \Sigma Y & \xrightarrow{\Sigma u} & \Sigma C^-f \dots \\
\| & & \| & & \| & & \uparrow{h_1} & \natural & \uparrow{h_2} & \natural & \uparrow{h_3} \\
X & \xrightarrow[f]{} & Y & \xrightarrow[u]{} & C^-f & \xrightarrow[u_2]{} & C^+u & \xrightarrow[u_3]{} & C^-u_2 & \xrightarrow[u_4]{} & C^+u_3 \dots
\end{array}
\qquad (1.193)
$$

This diagram links the lower cofibre sequence of f *to a sequence of it-*erated h-cokernels *of* f, *where each map is*, alternately, *the* lower or upper *h-cokernel of the preceding one. The squares marked with* $\natural$ *do not* commute, *generally*.

Note. In a stronger setting, we will prove that these squares are commutative up to the homotopy congruence $\sim_1$ and that every h_i is a composition of past and future homotopy equivalences; see Theorem 4.7.5.

Proof We will make repeated use of the lower and upper comparison maps, defined above (Section 1.7.7).

Let us begin by noting that the lower comparison $k_1 = k^-(f) \colon C^+u \to \Sigma X$ links the lower sequence of $u_0 = f$ to the upper sequence of $u =$

$u_1 = \mathrm{hcok}^-(f)$

$$
\begin{array}{ccccccccccc}
X & \xrightarrow{f} & Y & \xrightarrow{u} & C^-f & \xrightarrow{d} & \Sigma X & \xrightarrow{\Sigma f} & \Sigma Y & \xrightarrow{\Sigma u} & \Sigma C^-f \ \ldots \\
& & \| & & \| & & k_1 \uparrow & & \sharp & & \| \\
& & Y & \xrightarrow[u_1]{} & C^-f & \xrightarrow[u_2]{} & C^+u_1 & \xrightarrow[d_2]{} & \Sigma Y & \xrightarrow[\Sigma u_1]{} & \Sigma C^-f \ \ldots
\end{array}
$$
(1.194)

where the square(s) marked with $\sharp$ need not commute.

Now, we iterate this procedure (and its R-dual). Setting $\alpha_n = -$ for n odd and $\alpha_n = +$ for n even, we obtain the *expanded lower cofibre diagram* of f

$$
\begin{array}{ccccccccccccc}
X & \xrightarrow{f} & Y & \xrightarrow{u} & C^-f & \xrightarrow{d} & \Sigma X & \xrightarrow{\Sigma f} & \Sigma Y & \xrightarrow{\Sigma u} & \Sigma C^-f & \xrightarrow{\Sigma d} & \Sigma^2 X \ \ldots \\
& & \| & & \| & & k_1\uparrow & & \sharp & & \| & & \Sigma k_1\uparrow \quad \sharp \\
& & Y & \xrightarrow[u_1]{} & C^-f & \xrightarrow[u_2]{} & C^+u_1 & \xrightarrow[d_2]{} & \Sigma Y & \xrightarrow[\Sigma u_1]{} & \Sigma C^-f & \xrightarrow[\Sigma d_2]{} & \Sigma(C^+u_1)\ldots \\
& & & & \| & & \| & & k_2\uparrow & & \sharp & & \| \\
& & & & C^-f & \xrightarrow[u_2]{} & C^+u_1 & \xrightarrow[u_3]{} & C^-u_2 & \xrightarrow[d_3]{} & \Sigma C^-f & \xrightarrow[\Sigma u_2]{} & \Sigma(C^+u_1)\ldots \\
& & & & & & \| & & \| & & k_3\uparrow & & \sharp \quad \| \\
& & & & & & C^+u_1 & \xrightarrow[u_3]{} & C^-u_2 & \xrightarrow[u_4]{} & C^+u_3 & \xrightarrow[d_4]{} & \Sigma(C^+u_1)\ldots
\end{array}
$$
(1.195)

$$
\begin{aligned}
u_n &= \mathrm{hcok}^{\alpha_n}(u_{n-1}), \quad u_0 = f && (n \geqslant 1), \\
d_n &= d^{\alpha_n}(u_{n-1}), \qquad k_n = k^{\alpha_n}(u_{n-1}).
\end{aligned}
$$
(1.196)

Finally, we compose all columns and get diagram (1.193), letting

$$
\begin{aligned}
h_n &= k_n & (n = 1, 2, 3), \\
h_n &= (\Sigma k_{n-3}).k_n & (n = 4, 5, 6), \\
h_n &= (\Sigma^2 k_{n-6}).(\Sigma k_{n-3}).k_n & (n = 7, 8, 9), \ldots
\end{aligned}
$$
(1.197)

$\square$

1.8 First-order homotopy theory by the path functor

Working now on a directed path functor, we dualise Sections 1.3 and 1.7, introducing dP1-homotopical categories and the (lower or upper) fibre sequences of a map. We end by combining both results in the pointed self-dual case, i.e. pointed dIP1-homotopical categories (Sections 1.8.6 and 1.8.7).

1.8.1 Homotopy pullbacks

Dualising Section 1.3, let us assume that $\mathbf{A}$ is a dP1-category. Given two arrows f, g with the same codomain, the *standard homotopy pullback*, or *h-pullback, from f to g* is a four-tuple $(A; u, v; \lambda)$ as in the left diagram below, where $\lambda\colon fu \to gv\colon A \to X$ is a homotopy satisfying the following universal property (of *comma squares*)

$$
\begin{array}{ccc}
A & \xrightarrow{\ v\ } & Z \\
\scriptstyle u \downarrow & \nearrow \scriptstyle \lambda & \downarrow \scriptstyle g \\
Y & \xrightarrow[f]{} & X
\end{array}
\qquad
\begin{array}{ccc}
PX & \xrightarrow{\ \partial^+\ } & X \\
\scriptstyle \partial^- \downarrow & \nearrow \scriptstyle \lambda & \downarrow \scriptstyle \mathrm{id} \\
X & \xrightarrow[\mathrm{id}]{} & X
\end{array}
\tag{1.198}
$$

For every $\lambda'\colon fu' \to gv'\colon A' \to X$, there is precisely one map $h\colon A' \to A$ such that $u' = uh, v' = vh, \lambda' = \lambda h$.

The solution $(A; u, v; \lambda)$ is determined up to isomorphism (when it exists). We write $A = P(f, g)$; we have: $P(g, f) = R(P(Rf, Rg))$.

In particular, the path object $PX = P(\mathrm{id}X, \mathrm{id}X)$, displayed in the right diagram above, comes equipped with the obvious *structural* homotopy

$$
\partial\colon \partial^- \to \partial^+\colon PX \to X, \qquad \check{\partial} = \mathrm{id}(PX).
\tag{1.199}
$$

The homotopy pullback can be constructed with the path-object and the ordinary limit of the left diagram below, which amounts to three ordinary pullbacks

$$
\begin{array}{ccc}
P(f, g) & \xrightarrow{\ v\ } & Z \\
\Big\downarrow \ \ \searrow{\scriptstyle \lambda} & & \Big\downarrow \scriptstyle g \\
\scriptstyle u \ \ \ \ PX & \xrightarrow{\partial^+} & X \\
\Big\downarrow \quad \scriptstyle \partial^- \downarrow & & \\
Y & \xrightarrow[f]{} & X
\end{array}
\qquad
\begin{array}{ccccc}
P(f, g) & \dashrightarrow & P(X, g) & \dashrightarrow & Z \\
\downarrow & & \downarrow & & \Big\downarrow \scriptstyle g \\
P(f, X) & \dashrightarrow & PX & \xrightarrow{\partial^+} & X \\
\downarrow & & \partial^- \downarrow & & \\
Y & & \xrightarrow[f]{} & & X
\end{array}
\tag{1.200}
$$

The dP1-category $\mathbf{A}$ has homotopy pullbacks if and only if it has *cocylindrical limits*, i.e. the limits of all diagrams of the previous type. The existence of all ordinary pullbacks in $\mathbf{A}$ is a stronger condition.

Standard homotopy pullbacks, when they exist 'everywhere', form a functor $\mathbf{A}^\wedge \to \mathbf{A}$ defined on the cospans of $\mathbf{A}$ (i.e. the diagrams of $\mathbf{A}$ based on the *formal-cospan* category $\wedge$). Pasting works in the obvious way, dualising Lemma 1.3.8.

The construction of the homotopy pullback $P(f, g)$ in pTop (or dTop) is obvious from the diagram above

$$P(f, g) = \{(y, a, z) \in Y \times PX \times Z \mid a(0) = fy, \ a(1) = gz\} \atop \subset Y \times PX \times Z, \tag{1.201}$$

with the preorder (or d-structure) of a regular subobject of the product $Y \times PX \times Z$.

Recall that the structure of PX is described in Section 1.1.4 for pTop and in (1.117) for dTop; in the first case $PX = X^{\uparrow \mathbf{I}}$ is the set of increasing paths $a \colon \uparrow\mathbf{I} \to X$, with the compact-open topology and the pointwise preorder.

In pTop. (pointed preordered spaces) or dTop. (pointed d-spaces), one gets again the object (1.201), pointed at the triple (y_0, ω_0, z_0) formed of the base points of Y, PX and Z; of course, $\omega_0 = 0_{x_0}$ is the constant loop at the base point of X (1.126).

Similarly, in **Cat**, one obtains the usual construction of the comma category $(f \downarrow g)$ [M3], with objects (y, a, z) where $y \in Y$ (i.e. y is an object of Y), $z \in Z$ and $a \colon f(y) \to g(z)$ is a map of X. In **Cat.** (pointed small categories) one gets the same comma category pointed at the obvious object $(y_0, 0_{x_0}, z_0)$, with 0_{x_0} the identity of x_0.

1.8.2 Homotopical categories via cocylinders

We say that **A** is a *dP1-homotopical category* if it is a dP1-category with all h-pullbacks and an initial object $\bot$. In the rest of this section, we assume that this is the case.

Now, every object X has a unique morphism $i_X \colon \bot \to X$. The initial object is automatically 2-initial and we can assume that $R(\bot) = \bot$. A map $X \to Y$ which factors through $\bot$ will be said to be *co-constant*, and an object X is future co-contractible (resp. past co-contractible) if there is a map $p \colon X \to \bot$ and a homotopy $1_X \to i_X p$ (resp. $i_X p \to 1_X$).

Every dP1-category with a terminal object and ordinary pullbacks is dP1-homotopical.

The *non-pointed* dI1-categories pTop, dTop, **Cub**, **Cat** considered in the previous section are also dP1-homotopical. But we have already noted that their initial object – the 'empty object', in the appropriate sense – is an *absolute initial object*: every morphism $X \to \bot$ is invertible (in fact, the identity of $X = \bot$, in all these cases). It follows that the only (future or past) co-contractible object is the initial object itself and *the fibre sequence of every map degenerates* (cf. Section 1.8.5). As in classical algebraic topology, one should rather consider the corresponding *pointed* categories ($\mathbf{pTop_\bullet}$, etc.) to find non-trivial fibre sequences.

But of course there exist *non-pointed* dP1-homotopical categories with non-trivial fibre sequences, for instance the opposite categories $\mathbf{A}^* = \mathbf{pTop}^*$, $\mathbf{dTop}^*$, etc. with path functor $P = I^* \colon \mathbf{A}^* \to \mathbf{A}^*$. Differential graded algebras give a more natural example, which will be studied in Chapter 5.

A *dP1-homotopical functor*

$$K = (K, i, k) \colon \mathbf{A} \to \mathbf{X} \qquad (i \colon KR \to RK, \quad k \colon KP \to PK),$$

is a strong dP1-functor (Section 1.2.6) between dP1-homotopical categories which preserves the initial object and h-pullbacks (see the dual notion in Section 1.7.0).

1.8.3 Cocones and loop objects

Let $\mathbf{A}$ be always a dP1-homotopical category.

(a) A map $f \colon X \to Y$ has an *upper mapping cocone* $E^+ f = P(f, i_X)$, or *upper h-kernel*, defined as the left h-pullback below; its structural maps are $u = \mathrm{hker}^+(f) \colon E^+ f \to X$ and the *upper co-vertex* $v^+ \colon E^+ f \to \bot$; they come with a homotopy $\gamma \colon fu \to iv^+ \colon E^+ f \to X$

$$\begin{array}{ccc}
E^+ f \xrightarrow{\;v^+\;} \bot & \qquad & E^- f \xrightarrow{\;u\;} X \\
{\scriptstyle u}\downarrow \quad {\scriptstyle \gamma} \nearrow \quad \downarrow {\scriptstyle i} & & {\scriptstyle v^-}\downarrow \quad {\scriptstyle \gamma}\nearrow \quad \downarrow {\scriptstyle f} \\
X \xrightarrow{\;f\;} Y & & \bot \xrightarrow{\;i\;} Y
\end{array} \qquad (1.202)$$

The upper mapping cocone is a functor $E^+ \colon \mathbf{A}^2 \to \mathbf{A}$. Symmetrically, there is a *lower mapping cocone* $E^- f = P(i_X, f)$, equipped with a lower co-vertex v^-; it is defined as the right h-pullback above, and $(E^- f)^{\mathrm{op}} = E^+(f^{\mathrm{op}})$.

(b) Every object X has an *upper cocone* $E^+ X = E^+(\mathrm{id}X) = P(\mathrm{id}X, i_X)$ and a *lower cocone* $E^- X = E^-(\mathrm{id}X) = P(i_X, \mathrm{id}X)$. They determine each other, by R-duality: $(E^- X)^{\mathrm{op}} = E^+(X^{\mathrm{op}})$.

(c) Finally, the *loop object*

$$\Omega X = E^+(i_X) = E^-(i_X) = P(i_X, i_X),$$

is an upper and a lower cocone, at the same time

$$
\begin{array}{ccc}
\Omega X & \xrightarrow{v^+} & \bot \\
{\scriptstyle v^-}\downarrow & {}^{\omega}\nearrow & \downarrow{\scriptstyle i} \\
\bot & \xrightarrow{i} & X
\end{array}
\tag{1.203}
$$

The loop object thus has a structural homotopy $\omega\colon i_X v^- \to i_X v^+\colon \Omega X \to X$, universal within homotopies $A \to_1 X$ whose faces are co-constant. It forms a functor $\Omega\colon \mathbf{A} \to \mathbf{A}$ which commutes with the reversor: $(\Omega X)^{\mathrm{op}} = \Omega(X^{\mathrm{op}})$.

If the initial object is absolute, all these notions become trivial: $E^\pm f$, $E^\pm X$, and ΩX always coincide with the initial object itself.

1.8.4 Examples of cocones and loop objects

In p**Top.** and d**Top.**, the object X, pointed at x_0, has the following upper cocone

$$
\begin{aligned}
E^+(X) &= P(\mathrm{id}X, i_X) \\
&= \{(x, a) \in X \times PX \mid a(0) = x,\ a(1) = x_0\} \subset X \times PX,
\end{aligned}
\tag{1.204}
$$

with the structure of a regular subobject of $X \times PX$, and in particular the same base point: $(x_0, 0_{x_0})$.

The lower cocone $E^-(X) = P(i_X, \mathrm{id}X)$ and the loop object $\Omega(X)$ are

$$
\begin{aligned}
E^-(X) &= \{(x, a) \in X \times PX \mid a(0) = x_0,\ a(1) = x\}, \\
\Omega(X) &= P(i_X, i_X) = \{a \in PX \mid a(0) = x_0 = a(1)\} \subset PX.
\end{aligned}
\tag{1.205}
$$

In **Cat.** one gets similar comma categories.

As in Section 1.7.4, we can note that in p**Top.** and d**Top.** (but *not* in **Cat.**) the loop object ΩX can also be obtained as the following pullback

$$
\begin{array}{ccc}
\Omega X & \longrightarrow & E^+ X \\
\downarrow & & \downarrow{\scriptstyle u^-} \\
E^- X & \xrightarrow{u^+} & X
\end{array}
\tag{1.206}
$$

1.8.5 Fibre sequences

Dualising Section 1.7.7, every map $f \colon X \to Y$ has a *lower differential* $d = d^-(f) \colon \Omega Y \to E^- f$ and a *lower (fibre) comparison* $h = h^-(f) \colon \Omega Y \to E^+ v$, with values in the upper cocone of $v = \mathrm{hker}^-(f)$.

In the *reduced lower fibre diagram of* f, all squares commute, except (possibly) the $\sharp$-marked one

$$
\begin{array}{ccccccccc}
\Omega X & \xrightarrow{\Omega f} & \Omega Y & \xrightarrow{d} & E^- f & \xrightarrow{v} & X & \xrightarrow{f} & Y \\
\| & & {}^{\sharp}\Big\downarrow{}^{h} & & \| & & \| & & \| \\
\Omega X & \xrightarrow{d'} & E^+ v & \xrightarrow{v'} & E^- f & \xrightarrow{v} & X & \xrightarrow{f} & Y
\end{array}
\qquad (1.207)
$$

$$
v = \mathrm{hker}^-(f), \qquad d = d^-(f), \qquad v' = \mathrm{hker}^+(v), \qquad d' = d^+(v).
$$

Dualising Section 1.7.8, the map $f \colon X \to Y$ has a natural *lower fibre sequence*

$$
\ldots \Omega^2 Y \xrightarrow{\Omega d} \Omega E^- f \xrightarrow{\Omega v} \Omega X \xrightarrow{\Omega f} \Omega Y \xrightarrow{d} E^- f \xrightarrow{v} X \xrightarrow{f} Y
\qquad (1.208)
$$

formed by its lower h-kernel v, its lower differential d and the loop functor.

Dualising Section 1.7.9, this sequence can be linked to a sequence of *iterated h-kernels* of f, where each map is, *alternately*, the *lower or upper* h-kernel of the following one

$$
\begin{array}{ccccccccccc}
\ldots \Omega E^- f & \xrightarrow{\Omega v} & \Omega X & \xrightarrow{\Omega f} & \Omega Y & \xrightarrow{d} & E^- f & \xrightarrow{v} & X & \xrightarrow{f} & Y \\
{}^{k_3}\Big\downarrow & {}^{\sharp} & {}^{k_2}\Big\downarrow & {}^{\sharp} & {}^{k_1}\Big\downarrow & & \| & & \| & & \| \\
\ldots E^+ v_3 & \xrightarrow{v_4} & E^- v_2 & \xrightarrow{v_3} & E^+ v & \xrightarrow{v_2} & E^- f & \xrightarrow{v_1} & X & \xrightarrow{f} & Y
\end{array}
\qquad (1.209)
$$

This will be called the *lower fibre diagram* of f. Again, the squares marked with $\sharp$ do not commute, generally.

By R-duality, we have the *upper fibre sequence* of f (and an upper fibre diagram)

$$
\ldots \Omega^2 Y \xrightarrow{\Omega d} \Omega E^+ f \xrightarrow{\Omega v} \Omega X \xrightarrow{\Omega f} \Omega Y \xrightarrow{d} E^+ f \xrightarrow{v} X \xrightarrow{f} Y
\qquad (1.210)
$$

$$
v = \mathrm{hker}^+(f), \qquad d = d^+(f).
$$

(Again, if the initial object is absolute, the fibre sequence of every map becomes trivial: all the objects at the left of X in the diagrams above coincide with the initial object.)

One can find in [G3], Section 3.7, a 'concrete' motivation for using, alternately, lower and upper h-kernels. Working in $\mathbf{Top_{\bullet}}$, we showed that a uniform use of lower h-kernels in the second row of diagram (1.209) would give a homotopically *anti*-commutative diagram, with respect to loop-reversion. Thus, in a non-reversible situation, the diagram would get 'out of control'.

1.8.6 Pointed homotopical categories

The results of Section 1.3, for dI1-homotopical categories, and the preceding ones, for the dP1-homotopical case, can be combined in a useful way *in the pointed case*, i.e. when the terminal and initial objects coincide, yielding the zero-object $\top = \bot = 0$.

We say that $\mathbf{A}$ is a *pointed dIP1-homotopical category* if it is a dIP1-category (Section 1.2.2), has all h-pushouts and h-pullbacks, and a zero-object 0.

Then, after the adjunction $I \dashv P$, we also have canonical adjunctions

$$C^\alpha \dashv E^\alpha, \qquad \Sigma \dashv \Omega \qquad\qquad (\alpha = \pm). \qquad (1.211)$$

Indeed, a map $C^-X \to Y$ amounts to a morphism $f \colon X \to Y$ together with a homotopy $\varphi \colon f \to 0 \colon X \to Y$, and thus corresponds to a map $X \to E^-Y$. Similarly, a map $\Sigma X \to Y$ amounts to an endohomotopy $\varphi \colon 0 \to 0 \colon X \to Y$, and to a map $X \to \Omega Y$.

Obvious examples of (non-reversible) pointed dIP1-homotopical categories are $\mathbf{pTop_{\bullet}}$, $\mathbf{dTop_{\bullet}}$, $\mathbf{Cub_{\bullet}}$, $\mathbf{cSet_{\bullet}}$, $\mathbf{Cat_{\bullet}}$. Directed chain complexes will be studied in Section 2.1.

1.8.7 The fibre–cofibre sequence

Let $\mathbf{A}$ be again a pointed dIP1-homotopical category. Putting together the previous results, every map $f \colon X \to Y$ has a natural *lower fibre–cofibre sequence*, unbounded in both directions

$$\ldots \Omega^2 Y \overset{\Omega d}{\Rightarrow} \Omega E^- f \overset{\Omega v}{\Rightarrow} \Omega X \xrightarrow{\Omega f} \Omega Y \overset{d}{\to} E^- f \overset{v}{\to} X$$

$$Y \underset{u}{\leftarrow} C^- f \underset{d}{\to} \Sigma X \underset{\Sigma f}{\to} \Sigma Y \underset{\Sigma u}{\Rightarrow} \Sigma C^- f \underset{\Sigma d}{\Rightarrow} \Sigma^2 X \ldots$$

$$v = \mathrm{hker}^-(f), \qquad\qquad u = \mathrm{hcok}^-(f). \qquad (1.212)$$

It is formed by the lower h-kernel v and the lower h-cokernel u of f, their differentials and the action of the adjoint functors $\Sigma \dashv \Omega$.

This sequence can be linked to a sequence of iterated h-kernels and h-cokernels of f (again of alternating type, lower and upper), forming the lower fibre–cofibre diagram of f

$$
\begin{array}{ccccccccc}
\ldots \Omega Y & \xrightarrow{d} & E^- f & \xrightarrow{v} & X & \xrightarrow{f} & Y & \xrightarrow{u} & C^- f & \xrightarrow{d} & \Sigma X \ldots \\
k_1 \downarrow & & \| & & \| & & \| & & \| & & \uparrow h_1 \\
\ldots E^+ v & \xrightarrow[v_2]{} & E^- f & \xrightarrow[v_1]{} & X & \xrightarrow[f]{} & Y & \xrightarrow[u_1]{} & C^- f & \xrightarrow[u_2]{} & C^+ u \ldots
\end{array}
\tag{1.213}
$$

We end this section by writing down the dual of Lemma 1.3.8, and then applying it to **Cat**.

1.8.8 Pasting lemma for h-pullbacks

Let **A** *be a dP1-category and let* λ, μ, ν *be h-pullbacks, as in the diagram below. Then, there is a comparison map* $i \colon Z \to Y$ *defined by the universal properties of* λ, μ *and such that:*

$$
pp'i = p'', \qquad\qquad q'i = q'', \qquad\qquad \mu \circ i = 0_{hq''},
\tag{1.214}
$$

1.8.9 Homotopy pullbacks of categories

Also as a preparation for Chapter 3, we remark that the necessity of notions of directed homotopy in **Cat** *already appears in the general theory of categories*, for instance in the diagrammatic properties of (co)comma squares. (This motivation mostly makes sense for a reader already acquainted with these constructions and their importance in category theory; it is not necessary for the sequel.)

We have already observed that, in **Cat**, a comma square $X = (f \downarrow g)$ is the h-pullback of the functors f, g (Section 1.8.1). Consider now the pasting of two comma squares $X = (f \downarrow g)$, $Y = (q \downarrow h)$, as in Lemma 1.8.8, and the 'global' comma $Z = (f \downarrow gh)$.

The comparison functor i given by the previous lemma is computed as follows (with obvious notation: $a \in A$, $d \in D$, the morphism u is in C)

$$i\colon Z \to Y, \quad i(a,d;u\colon f(a) \to gh(d)) \\ = (a,h(d),d;u\colon f(a) \to gh(d), 1_{h(d)}). \tag{1.215}$$

It is a full embedding, and is *not* an equivalence of categories, generally. In fact, Z is a *full reflective subcategory* of Y (Section A3.4), with reflector p and unit η

$$p\colon Y \to Z, \quad p(a,b,d;u\colon f(a) \to g(b); v\colon b \to h(d)) = \\ = (a,d;g(v){\circ}u\colon f(a) \to gh(d)), \\ \eta\colon 1 \to ip, \quad \eta(a,b,d;u,v) = (1_a,v,1_d)\colon \\ (a,b,d;u,v) \to (a,hd,d;g(v){\circ}u,1_{hd}). \tag{1.216}$$

(This is in accord with a stronger pasting theorem for h-pullbacks, whose dual will be given later, in a suitable setting: Theorem 4.6.2.)

Reversing the 'direction' of these comma categories ($X = (g \downarrow f)$, $Y = (h \downarrow q)$, $Z = (gh \downarrow f)$), the global comma Z becomes a full *coreflective* subcategory of Y. Similar results hold for other diagrammatic properties of comma (or cocomma) squares. A general treatment should be based on the universal properties of the latter, to take advantage of duality and avoid the complicated construction of cocomma categories.

We should therefore be prepared to consider a full reflective *or* coreflective subcategory $Z \subset Y$ as 'equivalent' to Y, in some sense related to directed homotopy in **Cat**. Indeed, being full reflective (resp. coreflective) subcategories of a common one will amount to the notion of 'future equivalence' (resp. 'past equivalence') studied in Chapter 3, as a coherent version of future (or past) homotopy equivalence in **Cat**. Future and past equivalences are thus natural tools to describe the diagrammatic properties of comma and cocomma categories.

1.9 Other topological settings

After Sections 1.1, 1.4 and 1.5, we discuss here other topological settings for directed algebraic topology, in the light of some general, seemingly reasonable requirements expounded in Section 1.9.0. Some of these settings present problems for path-concatenation, like inequilogical spaces (Section 1.9.1), c-sets (Section 1.9.4) and generalised metric spaces (Section 1.9.6). Others do not even satisfy the requirements of Section 1.9.0: for instance, bitopological spaces lack a cocylinder functor (Section 1.9.5).

Locally preordered spaces, when defined in a convenient way (Section 1.9.3), give a good setting. But, as a disadvantage with respect to d-spaces, they are not sufficiently 'fine' to establish a good relationship with non-commutative geometry (Chapter 2).

1.9.0 General principles

The topological settings we are interested in will be constructed starting from the category **Top** of topological spaces, and compared with it.

As we have already considered, **Top** is itself a *reversible* dIP1-category (Section 1.2.2). This structure arises from the cartesian product and the standard interval $\mathbf{I} = [0, 1]$ (with euclidean topology), an exponentiable object, and yields the ordinary homotopies.

Now, a 'good topological setting' **A** for directed algebraic topology should satisfy some *general principles*:

(i) **A** is a (non-reversible) dIP1-category with all limits and colimits;

(ii) **A** is equipped with a *forgetful* strict dI1-functor $U \colon \mathbf{A} \to \mathbf{Top}$, which is automatically a *lax* dP1-functor (Section 1.2.6);

(iii) U has a right adjoint $D' \colon \mathbf{A} \to \mathbf{Top}$, which is automatically a lax dI1-functor, and can therefore be extended to homotopies (Section 1.2.6).

As a consequence, U preserves colimits, which means that **A**-colimits are computed as in **Top** and provided with the suitable additional structure. On the other hand, classical algebraic topology can be viewed within the directed one, via the right adjoint D'. The existence of a left adjoint $D \dashv U$ would say that limits in **A** are also computed as in **Top**; but D cannot be extended to homotopies.

We have already seen that such principles are satisfied by our basic setting p**Top** (Section 1.1) as well as by the richer setting d**Top** (Sections 1.4 and 1.5). Other settings, perhaps less convenient, will be briefly examined below.

1.9.1 Inequilogical spaces

The category p**Eql** of *inequilogical spaces* was introduced in [G11], as a *directed version* of D. Scott's equilogical spaces [Sc, BBS, Ro, G13]. Notwithstanding various points of interest, this setting will only be used here in marginal way, because it makes concatenation complicated.

An object $X = (X^\sharp, \sim_X)$ of p**Eql** is a preordered topological space $X^\sharp$ equipped with an equivalence relation $\sim_X$. A morphism $[f]: X \to Y$ is an equivalence class of preorder-preserving continuous mappings $X^\sharp \to Y^\sharp$ which respect the given equivalence relations, such mappings being equivalent if they induce the same mapping, modulo these relations. A preordered topological space A is viewed as an inequilogical one with the equality relation: $A = (A, =)$.

This category p**Eql** *is cartesian closed.* As for equilogical spaces, the proof is not trivial, but we only need the fact that the directed interval $\uparrow I = \uparrow[0,1]$ is exponentiable. Now, as proved in [G11], Theorem 1.8, if $A = (A, =)$ is a preordered topological space with a locally compact, Hausdorff topology and $Y = (Y^\sharp, \sim_Y)$ is any inequilogical space, the *inequilogical exponential* Y^A is computed as:

$$Y^A = ((Y^\sharp)^A, \sim_E),$$
$$h' \sim_E h'' \text{ if } (\forall a \in A, \; h'(a) \sim_Y h''(a)), \tag{1.217}$$

where $(Y^\sharp)^A$ is the exponential in p**Top** (with compact-open topology and pointwise preorder, cf. Section 1.1.3), and $\sim_E$ is the pointwise equivalence relation of maps $h: A \to Y^\sharp$, made explicit above.

Thus p**Eql** has a structure of cartesian dIP1-category. The directed homology of inequilogical spaces was studied in [G11]. The forgetful functor is

$$U: p\mathbf{Eql} \to \mathbf{Top},$$
$$X = (X^\sharp, \sim_X) \mapsto |X| = X^\sharp / \sim_X, \tag{1.218}$$

where $X^\sharp / \sim_X$ has the induced topology. Its right adjoint $D': \mathbf{Top} \to$ p**Eql** equips a space with the indiscrete preorder and the equality relation. U does not have a left adjoint.

There are various models of the directed circle, all equivalent up to 'local homotopy' (see [G11]), but the simplest (or the nicest) is perhaps the inequilogical space $\uparrow\overline{\mathbf{S}}^1_e = (\uparrow\mathbf{R}, \equiv_{\mathbf{Z}})$, i.e. the quotient (in this category) of the *ordered* topological line $\uparrow\mathbf{R}$ modulo the action of the group $\mathbf{Z}$ ([G11], Section 1.7).

More generally, we have the inequilogical sphere

$$\uparrow\overline{\mathbf{S}}^n_e = (\uparrow\mathbf{R}^n, \sim_n), \tag{1.219}$$

where the equivalence relation $\sim_n$ is generated by congruence modulo $\mathbf{Z}^n$ and by identifying all points $(t_1, ..., t_n)$ where at least one coordinate belongs to $\mathbf{Z}$.

The powers of this directed circle $\uparrow\overline{\mathbf{S}}_e^1$ in p**Eql** give the *inequilogical tori* $(\uparrow\overline{\mathbf{S}}_e^1)^n = (\uparrow\mathbf{R}^n, \equiv \mathbf{Z}^n)$, where directed paths have to turn 'counterclockwise in each variable'.

1.9.2 Locally transitive spaces

The usual notions of topological space equipped with a 'local preorder' do not allow one to construct mapping cones and suspension, and are thus inadequate to develop homotopy theory.

To show this, let us say that a *locally transitive topological space*, or *lt-space* $X = (X, \prec)$, is a topological space equipped with a *precedence* relation $\prec$ which is reflexive and *locally transitive*, i.e. transitive on a suitable neighbourhood of each point. (A similar, stronger notion was used in [FGR2, GG] and called a *local order*: the space is equipped with an open covering and a coherent system of closed orders on such open subsets; then, the join of such relations is locally transitive and gives back the given orderings, by restriction.) A *map* $f\colon X \to Y$ is required to be *locally increasing*, i.e. to preserve $\prec$ on some neighbourhood of every $x \in X$. Their category will be written lt**Top**.

Note that on a given space, infinitely many precedence relations may give *equivalent* lt-structures, isomorphic via the identity. This is a minor problem: it can be mended, replacing the precedence relation by its *germ*, or equivalence class, in the same way as a manifold structure is often defined as an equivalence class of atlases; or it can be ignored, since our mending would just replace lt**Top** with an equivalent category.

This category lt**Top** has obvious limits and sums, but not all colimits, as proved in [G8], Section 4.6. The point is that an lt-space $(X, \prec)$ cannot have a point-like vortex (Section 1.4.7): each point $x \in X$ has a neighbourhood V on which $\prec$ is transitive, so that any loop $a\colon \uparrow\mathbf{I} \to V$ is necessarily reversible. But, loosely speaking, a cone on a directed circle must have a point-like vortex at its vertex.

The forgetful functor $\mathbf{d}\colon$ lt**Top** $\to$ d**Top** is defined by taking as distinguished paths of an lt-space X the locally increasing paths $a\colon \uparrow[0,1] \to X$, on the ordered interval. By compactness of $[0,1]$ and local transitivity of $\prec$, this amounts to a continuous mapping, preserving precedence on each subinterval $[t_{i-1}, t_i]$ of a suitable decomposition $0 = t_0 < t_1 < \ldots < t_n = 1$.

Now, the directed circle $\uparrow\mathbf{S}^1$ is of locally transitive type, i.e. it can be obtained as $\mathbf{d}(\mathbf{S}^1, \prec)$ with some obvious precedence relation (not uniquely determined). But the higher directed spheres $\uparrow\mathbf{S}^n$ are *not* of lt-type, for $n \geqslant 2$, because they have a point-like vortex.

1.9.3 Locally preordered spaces

There are better ways of localising preorders, studied in a recent paper by S. Krishnan [Kr].

The main subject of this paper is a 'stream', which consists of a topological space X with a 'circulation', i.e. a family $\prec = (\prec_U)$ of preorder relations, one on each open subset U of X, which satisfies the following cosheaf condition:

(i) if $U = \bigcup_i U_i$ is a union of open subsets of X, the preorder $\prec_U$ is the join of the preorders $\prec_{U_i}$.

Actually we prefer a weaker notion, called a 'precirculation' in [Kr] and a *local preorder* here, where (i) is replaced with a weaker co*pre*sheaf condition:

(ii) if $U \subset V$ and $x \prec_U x'$, then $x \prec_V x'$.

In fact, this notion has a better relationship with p**Top**: every preordered space has a local preorder, defined by restricting its relation to the open subsets; but this family need not satisfy the cosheaf condition.

A *locally preordered topological space*, or *lp-space*, will be a topological space X equipped with a local preorder. An *lp-map $f\colon X \to Y$* will be a continuous mapping between lp-spaces which preserves the local preorder, in the sense that, for every U open in X and V open in Y, if $f(U) \subset V$ and $x \prec_U x'$, then $f(x) \prec_V f(x')$. This defines the category lp**Top**. (Maps of streams are defined in the same way, in [Kr].)

Both notions, streams and lp-spaces, yield a 'good topological setting' for directed algebraic topology, in the sense of Section 1.9.0. (The left and right adjoint to the forgetful functor lp**Top** $\to$ **Top** are defined as for preordered spaces.) But none of them is adequate to establish a relationship with non-commutative geometry, as will be developed with cubical sets and d-spaces, in the next chapter: the quotient of the ordered line $\uparrow\mathbf{R}$ modulo the action of a dense subgroup, in both the previous frameworks, is an indiscrete topological space with the indiscrete preorder.

1.9.4 Sets with distinguished cubes

The category c**Set** of sets with distinguished cubes, or c-sets, has been introduced in Section 1.6.8. It is complete and cocomplete, and has a cartesian dIP1-structure defined by the obvious directed interval $\uparrow\mathbf{I}$, which is here the set $\mathbf{I}$ equipped with the presheaf of increasing continuous mappings $\mathbf{I}^n \to \mathbf{I}$.

The forgetful functor

$$U \colon \mathbf{cSet} \to \mathbf{Top}, \qquad (1.220)$$

equips a c-set L with the *cubical topology*, i.e. the finest topology making all distinguished cubes $\mathbf{I}^n \to L$ continuous (as for the functor $\mathbf{d} \colon \mathbf{cSet} \to \mathbf{dTop}$ defined in Section 1.6.8).

The adjoints $D \dashv U \dashv D'$ are obvious: for a topological space S, the c-set DS is equipped with the constant cubes of S, and $D'S$ with all the continuous ones. Finally, U is a strict dI1-functor, since it preserves products and satisfies $U(\uparrow\mathbf{I}) = \mathbf{I}$.

This setting has some disadvantages for directed homotopy: the obvious paths (and homotopies), based on $\uparrow\mathbf{I}$, cannot be concatenated because the distinguished cubes are not assumed to be closed under the various concatenation procedures. But, for directed homology, c-sets offer the same advantages as cubical sets, with respect to d-spaces (see Section 2.2.5 and Proposition 2.2.6).

1.9.5 Bitopological spaces

A *bitopological space* (a notion introduced by J.C. Kelly [Ky]) is a set equipped with a pair of topologies $X = (X, \tau^-, \tau^+)$, which we will call the *past* and the *future* topology, respectively. Their category b$\mathbf{Top}$, with the obvious maps – continuous with respect to past and future topologies, separately – has all limits and colimits, calculated as in $\mathbf{Set}$ and, separately, on past and future topologies (calculated in $\mathbf{Top}$). The reversor turns past into future, and vice versa.

The forgetful functor lp$\mathbf{Top} \to$ b$\mathbf{Top}$ is easily defined. Given an lp-space X (Section 1.9.3), a fundamental system of past or future neighbourhoods at x_0 arises from any fundamental system $\mathcal{V}$ of open neighbourhoods of the original topology, letting (for $(V \in \mathcal{V})$

$$V^- = \{x \in V \mid x \prec_V x_0\}, \qquad V^+ = \{x \in V \mid x_0 \prec_V x\}. \qquad (1.221)$$

$\uparrow\mathbf{I}$ thus inherits the left and right euclidean topologies.

Problems for establishing directed homotopy in b$\mathbf{Top}$ originate from pathologies of (say) left euclidean topologies; in fact, for a fixed Hausdorff space A, the product $- \times A$ preserves quotients (if and) only if A is locally compact ([Mi], Theorem 2.1 and footnote (5)). Thus, the cylinder endofunctor $- \times \uparrow\mathbf{I}$ in b$\mathbf{Top}$ does not preserve colimits and has no right adjoint: the path-object is missing (and homotopy pullbacks as well, while homotopy pushouts have poor properties; cf. [G7]).

1.9.6 Metrisability

Directed spaces can be defined by 'asymmetric distances'. A generalised metric space X in the sense of Lawvere [Lw1], called here a *directed metric space* or *δ-metric space*, is a set X equipped with a *δ-metric* $δ: X \times X \to [0, \infty]$, satisfying the axioms

$$\delta(x, x) = 0, \qquad \delta(x, y) + \delta(y, z) \geqslant \delta(x, z). \qquad (1.222)$$

This structure is natural within the theory of enriched categories, as showed in [Lw1]; see also Section 6.1. (If the value ∞ is forbidden, $δ$ is usually called a *quasi-pseudo-metric*, cf. [Ky]; but including it has various structural advantages, e.g. the existence of all limits and colimits.)

$δ\mathbf{Mtr}$ will denote the category of such $δ$-metric spaces, with (weak) *contractions* $f: X \to Y$, satisfying the condition $\delta(x, x') \geqslant \delta(f(x), f(x'))$. Limits and colimits exist and are calculated as in **Set**.

A product $\Pi_i X_i$ has the l_∞-type $δ$-metric (always defined because ∞ is included):

$$\delta(\mathbf{x}, \mathbf{y}) = \sup \delta_i(x_i, y_i) \qquad (\mathbf{x} = (x_i),\ \mathbf{y} = (y_i)). \qquad (1.223)$$

An equaliser has the restricted $δ$-metric. A sum $\Sigma_i X_i$ has the obvious $δ$-metric (which needs ∞ also in the binary case):

$$\delta((x, i), (y, i)) = \delta_i(x, y), \qquad \delta((x, i), (y, j)) = \infty \qquad (i \neq j). \qquad (1.224)$$

A coequalisers has the $δ$-metric induced on a quotient X/R:

$$\delta(\xi, \eta) = \inf_{\mathbf{x}} (\Sigma_j\ \delta(x_{2j-1}, x_{2j})) \\ (\mathbf{x} = (x_1, ..., x_{2p});\ x_1 \in \xi;\ x_{2j} R x_{2j+1};\ x_{2p} \in \eta). \qquad (1.225)$$

The *opposite* $δ$-metric space $R(X) = X^{op}$ has the opposite $δ$-metric, $\delta^{op}(x, y) = \delta(y, x)$. A *symmetric* $δ$-metric ($δ = δ^{op}$) is the same as an *écart* in Bourbaki [Bk].

A $δ$-metric space $X = (X, δ)$ has an associated bitopological space (X, τ^-, τ^+). At the point $x_0 \in X$, the *past* topology τ^- (resp. the *future* topology τ^+) has a canonical system of fundamental neighbourhoods consisting of the *past* discs D^- (resp. *future* discs D^+) centred at x_0

$$D^-(x_0, \varepsilon) = \{x \in X \mid \delta(x, x_0) < \varepsilon\}, \\ D^+(x_0, \varepsilon) = \{x \in X \mid \delta(x_0, x) < \varepsilon\} \qquad (\varepsilon > 0). \qquad (1.226)$$

This describes a forgetful functor to bitopological spaces

$$\delta\mathbf{Mtr} \to \mathbf{bTop}, \qquad (X, \delta) \mapsto (X, \tau^-, \tau^+). \qquad (1.227)$$

Homotopies of δ-metric spaces will be studied in Chapter 6, where we also construct a forgetful functor $\delta\mathbf{Mtr} \to \mathbf{dTop}$ which uses a 'symmetric' topology, instead of the previous ones (Section 6.1.9).

2

Directed homology and its relation with non-commutative geometry

Homology theories of directed 'spaces' will take values in *directed* algebraic structures. We will use *preordered* abelian groups, letting the preorder express most (not all) of the information codified in the original distinguished directions. One could also use abelian *monoids*, following a procedure developed by A. Patchkoria [P1, P2] for the homology semimodule of a 'chain complex of semimodules', but this would give here less information (see Section 2.1.4).

Directed homology of cubical sets, the main subject of this chapter, has interesting features, also related to non-commutative geometry.

Indeed, it may happen that the quotient $S/\sim$ of a topological space has a trivial topology, while the corresponding quotient of its singular cubical set $\square S$ keeps relevant topological information, identified by its homology and agreeing with the interpretation of such 'quotients' in non-commutative geometry. This relationship, briefly explored here, should be further clarified.

Let us start from the classical results on the homology of an orbit space S/G, for a group G acting *properly* on a space S; these results can be extended to *free* actions if we replace S with its singular cubical set and take the *quotient cubical set* $(\square S)/G$ (Corollary 2.4.4 and Theorem 2.4.5). Thus, for the group $G_\vartheta = \mathbf{Z} + \vartheta\mathbf{Z}$ (ϑ irrational), the orbit space $\mathbf{R}/G_\vartheta$ has a *trivial topology* (the indiscrete one), but can be replaced with a non-trivial cubical set, $X = (\square\mathbf{R})/G_\vartheta$, whose homology is the same as the homology of the group $G_\vartheta \cong \mathbf{Z}^2$, and coincides with the homology of the torus $\mathbf{T}^2$ (see (2.76)). *Algebraically*, all this is in accord with Connes' interpretation of $\mathbf{R}/G_\vartheta$ as a 'non-commutative space', i.e. a non-commutative C^*-algebra A_ϑ [C1, C2, C3, Ri1, Bl]; however, our $\uparrow H_n(X)$ has a trivial preorder, for $n > 0$, independently of ϑ.

To enhance this similarity, one can modify the quotient $(\square \mathbf{R})/G_\vartheta$, replacing $\square \mathbf{R}$ with the cubical set $\square \uparrow \mathbf{R}$ of all *order-preserving* maps $\mathbf{I}^n \to \mathbf{R}$. Algebraically, the homology groups are unchanged, but now $\uparrow H_1((\square \uparrow \mathbf{R})/G_\vartheta) \cong \uparrow G_\vartheta$ as a (totally) *ordered* subgroup of $\mathbf{R}$ (Theorem 2.5.8): thus the *irrational rotation cubical sets* $C_\vartheta = (\square \uparrow \mathbf{R})/G_\vartheta$ have the same classification *up to isomorphism* (Theorem 2.5.9) as the C^*-algebras A_ϑ *up to strong Morita equivalence* [PV, Ri1]: ϑ is determined up to the action of the group $\mathrm{PGL}(2, \mathbf{Z})$.

This example shows that the ordering of directed homology can carry much finer information than its algebraic structure. Furthermore, a comparison with the stricter classification of the algebras A_ϑ *up to isomorphism* (Section 2.5.1) shows that cubical sets provide a sort of 'non-commutative topology', without the metric character of non-commutative geometry (cf. Section 2.5.2). To take into account also this aspect, one should move to the richer domain of *weighted* algebraic topology (Chapter 6).

The reader can gain a quick overview of these motivations, reading Sections 2.5.1–2.5.3, on rotation structures and foliations of tori; Section 2.5 contains other results on higher dimensional tori.

This chapter follows rather closely the paper [G12], except for the last section which is new. It defines a directed reduced homology theory on a dI1-homotopical category, by three axioms based on: homotopy invariance, stability under suspension and exactness on the cofibre sequence of a map.

2.1 Directed homology of cubical sets

Combinatorial homology of cubical sets is a simple theory, with evident proofs. We study here its enrichment with a natural preorder: the items of the cubical set generate the positive chains, and positive cycles induce positive homology classes.

2.1.1 Directed chain complexes

Given a cubical set X, we begin by enriching the usual chain complex of abelian groups $\mathrm{Ch}_+(X)$ with an order on each component; the boundary homomorphisms *will not preserve* this order relation, but a chain morphism $f_\sharp \colon \mathrm{Ch}_+(X) \to \mathrm{Ch}_+(Y)$ induced by a map $f \colon X \to Y$ *will preserve it*. Motivated by all this, we define the category $\mathrm{dCh}_+ \mathbf{Ab}$ of directed chain complexes of abelian groups (indexed on natural numbers).

Let us start from the category p**Ab** of *preordered abelian groups*: an object $\uparrow L$ is an abelian group equipped with a preorder $\lambda \leqslant \lambda'$ preserved by the sum; or, equivalently, with a submonoid, the *positive cone* $L^+ = \{\lambda \in L \mid \lambda \geqslant 0\}$. A morphism is a preorder-preserving homomorphism.

Plainly, the category p**Ab** has all limits and colimits, computed as in **Ab** and equipped with the required preorder. It is enriched on abelian monoids and has finite biproducts (Section A4.6). Notice also that a bijective morphism (which is mono and epi) need not be an isomorphism.

The symmetric monoidal closed structure of abelian groups can be easily lifted to p**Ab**: the positive cone of $\uparrow L \otimes \uparrow M$ is the submonoid generated by the tensors $\lambda \otimes \mu$, for $\lambda \in L^+$, $\mu \in M^+$, while the internal hom $\text{Hom}(\uparrow M, \uparrow N)$ is the abelian group $\text{Hom}(M, N)$ of *all* algebraic homomorphisms, with positive cone given by the increasing ones

$$\begin{aligned} (\text{Hom}(\uparrow M, \uparrow N))^+ &= \text{p}\mathbf{Ab}(\uparrow M, \uparrow N) \\ &= \{f \in \text{Hom}(M, N) \mid f(M^+) \subset N^+\}. \end{aligned} \tag{2.1}$$

The unit of the tensor product is the ordered group of integers, $\uparrow \mathbf{Z}$. The forgetful functor p**Ab** $\to$ **Ab**, written $\uparrow L \mapsto L$, has left adjoint $\uparrow_d A$ and right adjoint $\uparrow_c A$, respectively enriching an abelian group A with the discrete order ($A^+ = \{0\}$) or with the chaotic preorder ($A^+ = A$).

On the other hand, the forgetful functor

$$\text{p}\mathbf{Ab}(\uparrow \mathbf{Z}, -) = (-)^+ : \text{p}\mathbf{Ab} \to \mathbf{Set}$$

represented by the monoidal unit $\uparrow \mathbf{Z}$ has (only) a left adjoint

$$\uparrow \mathbf{Z}.(-) \colon \mathbf{Set} \to \text{p}\mathbf{Ab}, \qquad S \mapsto \uparrow \mathbf{Z}.S. \tag{2.2}$$

Here, the *free ordered abelian group* $\uparrow \mathbf{Z}.S$ is the usual free abelian group $\mathbf{Z}S$, of formal $\mathbf{Z}$-linear combinations of elements of S, with positive cone consisting of the submonoid $\mathbf{N}S$ of positive combinations (with coefficients in $\mathbf{N}$).

We now introduce the category dCh$_+$**Ab** of *directed chain complexes of abelian groups*. An object $\uparrow A = ((\uparrow A_n), (\partial_n))$ is a chain complex of abelian groups, where every component is a *preordered* abelian group, but the differentials are just algebraic homomorphisms, *not* assumed to preserve the preorder (the dot-marked arrow below is meant to recall this fact)

$$\partial_n \colon \uparrow A_n \dashrightarrow \uparrow A_{n-1}. \tag{2.3}$$

A *chain morphism* $f = (f_n) \colon {\uparrow}A \to {\uparrow}B$ between directed chain complexes is a chain morphism in the usual sense, where every component $f_n \colon {\uparrow}A_n \to {\uparrow}B_n$ preserves preorders (i.e. lives in p**Ab**).

Again, the category dCh$_+$**Ab** has all limits and colimits, is enriched on abelian monoids and has finite biproducts (Section A4.6).

The *directed homology* of a directed chain complex is defined as a sequence of *preordered abelian groups*

$${\uparrow}H_n \colon \mathrm{dCh}_+\mathbf{Ab} \to \mathrm{p}\mathbf{Ab}, \tag{2.4}$$

where ${\uparrow}H_n({\uparrow}A)$ is the ordinary homology group $\mathrm{Ker}\partial_n/\mathrm{Im}\partial_{n+1}$, equipped with the preorder induced by ${\uparrow}A_n$ on this subquotient. Notice that, even when ${\uparrow}A_n$ is ordered, the induced preorder need not be antisymmetric.

Similarly, we have the category of directed *co*chain complexes of abelian groups, dCh$^+$**Ab**, and its directed cohomology functor

$${\uparrow}H^n \colon \mathrm{dCh}^+\mathbf{Ab} \to \mathrm{p}\mathbf{Ab}.$$

Directed homotopies in dCh$_+$**Ab** are not needed now; their d-structure will be studied in Section 4.4.

2.1.2 Directed homology of cubical sets

Let us start by recalling the usual construction of the homology groups of cubical sets. Every cubical set X determines a collection

$$\mathrm{Deg}_n X \;=\; \bigcup_i \mathrm{Im}(e_i \colon X_{n-1} \to X_n), \tag{2.5}$$

of subsets of *degenerate elements* (with $\mathrm{Deg}_0 X = \emptyset$); this collection is not a cubical subset (unless X is empty), but satisfies weaker properties (for all $i = 1, ..., n$)

$$\begin{aligned} x \in \mathrm{Deg}_n X \;\;\Rightarrow\;\; (\partial_i^\alpha x \in \mathrm{Deg}_{n-1}X \;\text{ or }\; \partial_i^- x = \partial_i^+ x), \\ e_i(\mathrm{Deg}_{n-1}X) \subset \mathrm{Deg}_n X. \end{aligned} \tag{2.6}$$

The cubical set X determines a (*normalised*) *chain complex* of free abelian groups

$$\begin{aligned} \mathrm{Ch}_n(X) = (\mathbf{Z}X_n)/(\mathbf{Z}\mathrm{Deg}_n X) \cong \mathbf{Z}\overline{X}_n \;\; (\overline{X}_n = X_n \setminus \mathrm{Deg}_n X), \\ \partial_n(\hat{x}) = \textstyle\sum_{i,\alpha}(-1)^{i+\alpha}(\partial_i^\alpha x)\hat{} \qquad (x \in X_n), \end{aligned} \tag{2.7}$$

where $\mathbf{Z}S$ is the free abelian group on the set S, and $\hat{x}$ is the class of the n-cube x up to degenerate cubes; but we shall generally write the normalised class $\hat{x}$ as x, identifying all degenerate cubes with 0. (The first property in (2.6) shows that $\partial_n(\hat{x})$ is well defined.)

Coming now to our enrichment, each component $\mathrm{Ch}_n(X)$ can be ordered by the positive cone of *positive chains* $\mathbf{N}\overline{X}_n$, and will be written as $\uparrow\mathrm{Ch}_n(X)$ when thus enriched; notice that *the positive cone is not preserved by the differential* $\partial_n\colon \uparrow\mathrm{Ch}_n(X) \to \uparrow\mathrm{Ch}_{n-1}(X)$, which is just a homomorphism of the underlying abelian groups (as stressed, again, by marking its arrow with a dot).

On the other hand, a morphism of cubical sets $f\colon X \to Y$ induces a sequence of *order-preserving* homomorphisms $f_{\sharp n}\colon \uparrow\mathrm{Ch}_n(X) \to \uparrow\mathrm{Ch}_n(Y)$. We have thus defined a covariant functor

$$\uparrow\mathrm{Ch}_+\colon \mathbf{Cub} \to \mathrm{dCh}_+\mathbf{Ab}, \tag{2.8}$$

with values in the category $\mathrm{dCh}_+\mathbf{Ab}$ of directed chain complexes of abelian groups, defined above (Section 2.1.1).

This functor gives the *directed homology* of a cubical set, as a sequence of preordered abelian groups

$$\uparrow H_n\colon \mathbf{Cub} \to \mathrm{pAb}, \qquad \uparrow H_n(X) = \uparrow H_n(\uparrow\mathrm{Ch}_+X), \tag{2.9}$$

where $\uparrow H_n(\uparrow\mathrm{Ch}_+X)$ (defined in Section 2.1.1) is the ordinary homology of the underlying chain complex, equipped with the preorder induced by the ordered group $\uparrow\mathrm{Ch}_n(X)$ on the subquotient $\mathrm{Ker}\partial_n/\mathrm{Im}\partial_{n+1}$.

When we forget preorders, the usual chain and homology functors will be written as usual

$$\mathrm{Ch}_+\colon \mathbf{Cub} \to \mathrm{Ch}_+\mathbf{Ab}, \qquad H_n\colon \mathbf{Cub} \to \mathbf{Ab}. \tag{2.10}$$

We will also apply the functors $\uparrow\mathrm{Ch}_+$ and $\uparrow H_n$ to a *c-set* L (Section 1.6.8), by letting them act on the cubical set cL of distinguished cubes of L

$$\uparrow\mathrm{Ch}_+(L) = \uparrow\mathrm{Ch}_+(cL), \qquad \uparrow H_n(L) = \uparrow H_n(cL). \tag{2.11}$$

2.1.3 Preordered coefficients

The (obvious) symmetric monoidal closed structure of preordered abelian groups has been recalled in Section 2.1.1. Let $\uparrow L$ be a preordered abelian group; the functors $- \otimes \uparrow L\colon \mathrm{pAb} \to \mathrm{pAb}$ and $\mathrm{Hom}(-, \uparrow L)\colon \mathrm{pAb}^* \to \mathrm{pAb}$ have obvious extensions to $\mathrm{dCh}_+\mathbf{Ab}$.

Using these tools, we get the directed combinatorial homology of cubical sets, *with coefficients in the preordered abelian group* $\uparrow L$

$$
\begin{aligned}
\uparrow \mathrm{Ch}_+(-; \uparrow L) &: \mathbf{Cub} \to \mathrm{dCh}_+ \mathbf{Ab}, \\
\uparrow \mathrm{Ch}_+(X; \uparrow L) &= \uparrow \mathrm{Ch}_+(X) \otimes \uparrow L, \\
\uparrow H_n(-; \uparrow L) &: \mathbf{Cub} \to \mathrm{pAb}, \\
\uparrow H_n(X; \uparrow L) &= \uparrow H_n(\uparrow \mathrm{Ch}_+(X; \uparrow L)),
\end{aligned}
\tag{2.12}
$$

and the directed combinatorial cohomology

$$
\begin{aligned}
\uparrow \mathrm{Ch}^+(-; \uparrow L) &: \mathbf{Cub}^* \to \mathrm{dCh}^+ \mathbf{Ab}, \\
\uparrow \mathrm{Ch}^+(X; \uparrow L) &= \mathrm{Hom}(\uparrow \mathrm{Ch}_+(X), \uparrow L), \\
\uparrow H^n(-; \uparrow L) &: \mathbf{Cub}^* \to \mathrm{pAb}, \\
\uparrow H^n(X; \uparrow L) &= \uparrow H^n(\uparrow \mathrm{Ch}^+(X; \uparrow L)).
\end{aligned}
\tag{2.13}
$$

Therefore, $\uparrow H_n(X) = \uparrow H_n(X; \uparrow \mathbf{Z})$ will also be called *directed combinatorial homology with ordered integral coefficients*; below, we generally consider this case, but the extension to arbitrary preordered coefficients is easy.

The algebraic part of the universal coefficient theorems holds, with the usual proof; the preorder aspect should be examined, but we shall restrict to considering *rational* and *real* coefficients. First, it is easy to verify that, for the ordered group of rationals $\uparrow \mathbf{Q}$, the canonical algebraic isomorphism

$$
\uparrow H_n(X) \otimes \uparrow \mathbf{Q} \to \uparrow H_n(X; \uparrow \mathbf{Q}), \qquad [z] \otimes \lambda \mapsto [z \otimes \lambda],
\tag{2.14}
$$

which obviously preserves preorder, *also reflects it*. In fact, a positive chain in $\uparrow \mathrm{Ch}_n(X; \uparrow \mathbf{Q})$ can always be written as $c = \lambda.c'$ where $\lambda > 0$ is rational and c' is a positive chain with *integral* coefficients; further, if c is a cycle, also c' is, and $[c] = [c'] \otimes \lambda$ belongs to the positive cone of $\uparrow H_n(X) \otimes \uparrow \mathbf{Q}$.

As a consequence, the same property holds for the ordered group $\uparrow \mathbf{R}$ of real numbers: it suffices to take a positive basis of the reals on the rationals.

But one can also give a more elementary argument. A positive chain in $\uparrow \mathrm{Ch}_n(X; \uparrow \mathbf{R})$ can be rewritten as a finite linear combination $c = \Sigma \lambda_i c_i$ where the $\lambda_i > 0$ are real numbers, linearly independent on the rationals, and all c_i are positive chains with integral coefficients; since each boundary $\lambda_i(\partial_n c_i)$ still has coefficients in $\lambda_i \mathbf{Q}$, we arrive at the same conclusion as above: if c is a cycle, so are all c_i and $[c] = \Sigma[c_i] \otimes \lambda_i$ belongs to the positive cone of $\uparrow H_n(X) \otimes \uparrow \mathbf{R}$.

2.1.4 Elementary computations

The homology of a sum $X = \Sigma X_i$ is a direct sum $\uparrow H_n X = \oplus_i \uparrow H_n X_i$ (and every cubical set is the sum of its connected components, Section 1.6.2).

It is also easy to see that, if X is connected (non-empty), then $\uparrow H_0(X) \cong \uparrow \mathbf{Z}$; the isomorphism is induced by the augmentation $\partial_0 \colon \uparrow Ch_0 X = \uparrow \mathbf{Z} X_0 \to \uparrow \mathbf{Z}$, which takes each vertex $x \in X_0$ to $1 \in \mathbf{Z}$.

Thus, for every cubical set X

$$\uparrow H_0(X) = \uparrow \mathbf{Z}.\Pi_0 X, \tag{2.15}$$

is the free ordered abelian group generated by the homotopy set $\Pi_0 X$ (Section 1.6.2).

In particular, $\uparrow H_0(\uparrow \mathbf{s}^0) = \uparrow \mathbf{Z}^2$. Now, it is easy to see that, for $n > 0$

$$\uparrow H_n(\uparrow \mathbf{s}^n) = \uparrow \mathbf{Z}, \tag{2.16}$$

is the group of integers with the natural order: a normalised n-chain ku (where u is the n-dimensional generator of $\uparrow \mathbf{s}^n$, as in Section 1.6.4) is positive when $k \geqslant 0$ (and is always a cycle).

Similarly, one proves that the n-dimensional elementary torus $\uparrow \mathbf{t}^n = (\uparrow \mathbf{s}^1)^{\otimes n}$ has directed homology:

$$\uparrow H_k(\uparrow \mathbf{t}^n) = \uparrow \mathbf{Z}^{\binom{n}{k}}, \tag{2.17}$$

where a power of $\uparrow \mathbf{Z}$ has the product order.

For instance, for $k = 1$, a normalised 1-chain can be written as follows (where u is the one-dimensional generator of $\uparrow \mathbf{s}^1$ and $*$ its unique vertex)

$$h_1(u \otimes * \otimes \ldots \otimes *) + \ldots + h_n(* \otimes \ldots \otimes * \otimes u) \qquad (h_i \in \mathbf{Z}), \tag{2.18}$$

and is positive when all its coefficients h_i are $\geqslant 0$. Again, it is always a cycle, because so is u.

On the other hand, the ordered sphere $\uparrow \mathbf{o}^n$ (Section 1.6.4) has $\uparrow H_n(\uparrow \mathbf{o}^n) = \uparrow_d \mathbf{Z}$, with the discrete order: the positive cone is reduced to 0. In fact, a normalised n-chain $hu' + ku''$ (notation of Section 1.6.4) is a cycle when $h + k = 0$, and a positive chain for $h \geqslant 0$, $k \geqslant 0$. (Notice that a directed homology with values in monoids, defined along the same lines as in Patchkoria [P1, P2] for his notion of 'chain complex of semimodules', would give here the positive cone of $\uparrow_d \mathbf{Z}$, which is null, missing the existence of non-positive cycles.)

The graded preordered abelian group of a cubical set X will also be written as a *formal polynomial*

$$\uparrow H_\bullet(X) = \Sigma_i \ \sigma^i . \uparrow H_i(X), \tag{2.19}$$

whose coefficients are preordered abelian groups, while the 'indeterminate' σ displays the homology degree. One can think of σ^i as a power of the suspension operator of chain complexes, acting here on a preordered abelian group, embedded in $\mathrm{dCh}_+ \mathbf{Ab}$ (in degree 0): then the expression (2.19) is a direct sum of graded preordered abelian groups (each of them concentrated in one degree). Notice also that the direct sum of graded preordered abelian groups amounts to the sum of the corresponding polynomials, computed – in the obvious way – by means of the direct sum of their coefficients.

In this notation, the directed homology of the elementary torus $\uparrow \mathbf{t}^n = (\uparrow \mathbf{s}^1)^{\otimes n}$, computed above, becomes the polynomial

$$\begin{aligned}
\uparrow H_\bullet(\uparrow \mathbf{t}^n) &= (\uparrow \mathbf{Z} + \sigma . \uparrow \mathbf{Z})^{\otimes n} \\
&= \uparrow \mathbf{Z} + \sigma . \uparrow \mathbf{Z}^{\binom{n}{1}} + \sigma^2 . \uparrow \mathbf{Z}^{\binom{n}{2}} + \cdots + \sigma^n . \uparrow \mathbf{Z}.
\end{aligned} \tag{2.20}$$

2.1.5 Relative directed homology

Relative homology of cubical sets is defined in the usual way.

A *cubical pair* (X, A) consists of a cubical set X and a cubical subset $A \subset X$; a *morphism* $f \colon (X, A) \to (Y, B)$ is a map $f \colon X \to Y$ which sends A into B. An (elementary) *left relative homotopy* $f \colon f^- \to_L f^+ \colon (X, A) \to (Y, B)$ is a map $f \colon X \to PY$ with $\partial^\alpha f = f^\alpha$, which sends A into PB; it can be described as a family of mappings of sets $f_n \colon X_n \to Y_{n+1}$ which send A_n into B_{n+1} and satisfy the equations of (1.152).

The embedding $i \colon A \to X$ induces a map $i_\sharp \colon \uparrow \mathrm{Ch}_+ A \to \uparrow \mathrm{Ch}_+ X$ of directed chain complexes, which is also injective (a cube in A is degenerate in X if and only if it is already so in A). We obtain the *relative directed chains* of (X, A) by the usual short exact sequence of chain complexes, interpreted in $\mathrm{dCh}_+ \mathbf{Ab}$

$$0 \to \uparrow \mathrm{Ch}_+ A \to \uparrow \mathrm{Ch}_+ X \to \uparrow \mathrm{Ch}_+ (X, A) \to 0 \tag{2.21}$$

i.e. letting each component $\uparrow \mathrm{Ch}_n(X, A)$ have the induced preorder (it is again a free ordered abelian group). Now, the *relative directed homology* is the homology of the quotient

$$\uparrow H_n(X, A) = \uparrow H_n(\uparrow \mathrm{Ch}_+(X, A)). \tag{2.22}$$

The exact sequence of the pair (X, A) comes from the exact homology sequence of (2.21), with differential $\Delta_n[c] = [\partial_n c]$; *the latter need not preserve the preorder* (and its arrow is dot-marked)

$$\cdots \dashrightarrow \uparrow H_n A \longrightarrow \uparrow H_n X \longrightarrow \uparrow H_n(X, A) \overset{\Delta}{\dashrightarrow} \uparrow H_{n-1} A \cdots$$

$$\cdots \dashrightarrow \uparrow H_0 A \longrightarrow \uparrow H_0 X \longrightarrow \uparrow H_0(X, A) \longrightarrow 0$$

(2.23)

(For pairs of *pointed* cubical sets, there is a more effective way of defining relative directed homology, based on the upper or lower mapping cone of the embedding; algebraically, the result is the same but the new preorder is different and preserved by the differential. See Section 2.3.7.)

Obviously, $\uparrow \mathrm{Ch}_+(X, \emptyset) = \uparrow \mathrm{Ch}_+(X)$ and $\uparrow H_n(X, \emptyset) = \uparrow H_n(X)$.

More generally, given a *cubical triple* (X, A, B), consisting of cubical subsets $B \subset A \subset X$, the snake lemma gives a short exact sequence of chain complexes

$$0 \to \uparrow \mathrm{Ch}_+(A, B) \to \uparrow \mathrm{Ch}_+(X, B) \to \uparrow \mathrm{Ch}_+(X, A) \to 0$$

providing the exact homology sequence of the triple.

Tensoring by $\uparrow L$ our chain complexes (with free ordered components), one gets *relative directed homology with arbitrary coefficients*.

2.1.6 Cohomology

The (normalised) *cochain complex* $\uparrow \mathrm{Ch}^+(X; \uparrow L) = \mathrm{Hom}(\uparrow \mathrm{Ch}_+(X); \uparrow L)$ of a cubical set X, with coefficients in a preordered abelian group $\uparrow L$ (Section 2.1.3), has a simple description

$$\mathrm{Ch}^n(X; \uparrow L) = \{\lambda \colon X_n \to L \mid \lambda(\mathrm{Deg}_n X) = 0\},$$

$$(d_n \lambda)(a) = \Sigma_{i,\alpha}(-1)^{i+\alpha} \lambda(\partial_i^\alpha a) \qquad (a \in X_{n+1}),$$

(2.24)

with components preordered by the cones of *positive cochains*, $\lambda \colon X_n \to L^+$, again not preserved by the differential.

Forgetting preorders and *assuming that L is a ring*, the cochain complex $\mathrm{Ch}^+(X; L)$ has a natural structure of differential graded coalgebra, by the cup product (cf. [HW], 9.3)

$$(\lambda \cup \mu)(a) = \Sigma_{HK}(-1)^{\rho(HK)} \lambda(\partial_H^- a).\mu(\partial_K^+ a)$$

$$(\lambda \in \mathrm{Ch}^p(X; L), \quad \mu \in \mathrm{Ch}^q(X; L), \quad a \in X_{p+q}),$$

(2.25)

where (H, K) varies among all partitions of $\{1, ..., n\}$ in two subsets of p and q elements, respectively, $\rho(HK)$ is the class of this permutation,

$\partial_H^- a$ is the *lower H-face* of a and $\partial_K^+ a$ its *upper K-face*. Thus, $H^\bullet(X; L)$ is a graded algebra, isomorphic to $H^\bullet(\mathcal{R}X; L)$ (and graded commutative).

This definition shows that *the product of positive cochains need not be positive*. Moreover, the *graded* commutativity of $H^\bullet(X; L)$ (for a commutative ring L) says that positive cohomology classes can hardly be closed under product.

For an actual counterexample, we use graded commutativity in odd degree, where $[\lambda] \cup [\mu] = -[\mu] \cup [\lambda]$, looking for a case where cohomology is *ordered* (not just preordered) and $[\lambda], [\mu], [\lambda] \cup [\mu]$ are strictly positive (whence $[\mu] \cup [\lambda]$ is not).

As we have seen in Section 2.1.4, the torus $\uparrow \mathbf{t}^2 = \uparrow \mathbf{s}^1 \otimes \uparrow \mathbf{s}^1$ has one 0-cube $(*)$, two non-degenerate 1-cubes $(u \otimes *$ and $* \otimes u)$ and one non-degenerate 2-cube $(u \otimes u)$, which provide the positive generators of $\uparrow H_\bullet(\uparrow \mathbf{t}^2)$. Similarly, in cohomology, we have an *ordered* object

$$\uparrow H^\bullet(\uparrow \mathbf{t}^2) = \uparrow \mathbf{Z} + \sigma.\uparrow \mathbf{Z}^2 + \sigma^2.\uparrow \mathbf{Z}, \qquad (2.26)$$

and the positive generators in degree 1, 2 come from the following cocycles (zero elsewhere)

$$\lambda(u \otimes *) = 1, \qquad \mu(* \otimes u) = 1, \qquad (\lambda \cup \mu)(u \otimes u) = 1. \qquad (2.27)$$

2.2 Main properties of the directed homology of cubical sets

The new aspects of directed homology deal, of course, with the homology preorder. We prove that it is preserved and reflected by excision (Theorem 2.2.3), preserved by tensor product (Theorem 2.2.4), but not preserved by the differential of the Mayer–Vietoris exact sequences (Theorem 2.2.2).

We also define here the directed homology of d-spaces, using their directed singular cubical set. In this case, the preorder is only relevant in degree 1 (Sections 2.2.5–2.2.7), in accord with fact that the directed structure of d-spaces is essentially one-dimensional (as already remarked in Section 1.1.5).

2.2.1 Theorem (Homotopy invariance)

The homology functor $\uparrow H_n$: $\mathbf{Cub} \to \mathrm{pAb}$ is invariant for elementary left (or right) homotopies: given a left homotopy $f: f^- \to_L f^+ : X \to Y$, then $\uparrow H_n(f^-) = \uparrow H_n(f^+)$. Similarly for relative homology and elementary left (or right) relative homotopies (Section 2.1.5).

As a consequence, the functor $\uparrow H_n$ is invariant up to homotopy congruence, and a fortiori up to coarse d-homotopy equivalence (Section 1.3.3).

Proof We can forget about preorders, since the thesis is merely algebraic: $\uparrow H_n(f^-) = \uparrow H_n(f^+)$. However, we write down the proof because the homology theory of cubical sets is much less well known than that of simplicial sets.

By (1.152), the left homotopy $f \colon f^- \to_L f^+ \colon X \to Y$ satisfies

$$
\begin{aligned}
&f_n \colon X_n \to Y_{n+1}, \quad \partial_{i+1}^\alpha f_n = f_{n-1} \partial_i^\alpha, \\
&\partial_1^\alpha f_n = f^\alpha, \qquad f_n e_i = e_{i+1} f_{n-1} \quad (\alpha = \pm; \ i = 1, ..., n),
\end{aligned}
\tag{2.28}
$$

and gives rise to a homotopy of the associated (normalised, non-directed) chain complexes

$$
\begin{aligned}
&f_{\sharp n} \colon \mathrm{Ch}_n X \to \mathrm{Ch}_{n+1} Y, \qquad f_{\sharp n}(\mathrm{Deg}_n X) \subset \mathrm{Deg}_{n+1} Y, \\
&\partial_{n+1} f_{\sharp n} = \partial_1^+ f_{\sharp n} - \partial_1^- f_{\sharp n} - \sum_{i\alpha} (-1)^{i+a} \partial_{i+1}^\alpha f_{\sharp n} \\
&\phantom{\partial_{n+1} f_{\sharp n}} = f_n^+ - f_n^- - f_{\sharp n-1} \partial_n.
\end{aligned}
\tag{2.29}
$$

The relative case is similar. It will be useful to note that the thesis also holds for a *generalised left homotopy*, replacing the condition $f_n e_i = e_{i+1} f_{n-1}$ with $f_n(\mathrm{Deg}_n X) \subset \mathrm{Deg}_{n+1} Y$. $\qquad\square$

2.2.2 Theorem (The Mayer–Vietoris sequence)

Let the cubical set X be covered by its subobjects U, V, i.e. $X = U \cup V$. Then we have an exact sequence

$$
\begin{aligned}
\cdots \longrightarrow \ &\uparrow H_n(U \cap V) \xrightarrow{(i_*, j_*)} \uparrow H_n U \oplus \uparrow H_n V \\
&\xrightarrow{[u_*, -v_*]} \uparrow H_n(X) \xrightarrow{\ \Delta_n\ } \uparrow H_{n-1}(U \cap V) \longrightarrow \cdots
\end{aligned}
\tag{2.30}
$$

with the obvious meaning of brackets.

The maps $u \colon U \to X$, $v \colon V \to X$, $i \colon U \cap V \to U$, $j \colon U \cap V \to X$ are inclusions and the connective Δ *(which need not preserve preorder) is:*

$$
\Delta_n[c] = [\partial_n a], \qquad c = a + b \quad (a \in \uparrow \mathrm{Ch}_n(U), \ b \in \uparrow \mathrm{Ch}_n(V)).
\tag{2.31}
$$

The sequence is natural, for a cubical map $f \colon X \to X' = U' \cup V'$, which restricts to $U \to U'$, $V \to V'$.

Proof The proof is similar to the topological one, simplified by the fact that here no subdivision is needed.

Given two cubical subsets $U, V \subset X$, their union $U \cup V$ (resp. intersection $U \cap V$) just consists of the union (resp. intersection) of all components. Therefore, $\uparrow\mathrm{Ch}_+$ takes subobjects of X to directed chain subcomplexes of $\uparrow\mathrm{Ch}_+ X$, preserving joins and meets

$$\uparrow\mathrm{Ch}_+ (U \cup V) = \uparrow\mathrm{Ch}_+ U + \uparrow\mathrm{Ch}_+ V,$$
$$\uparrow\mathrm{Ch}_+ (U \cap V) = \uparrow\mathrm{Ch}_+ U \cap \uparrow\mathrm{Ch}_+ V. \tag{2.32}$$

Now, it is sufficient to apply the algebraic theorem of the exact homology sequence to the following sequence of directed chain complexes

$$0 \to \uparrow\mathrm{Ch}_+ (U \cap V) \xrightarrow{f_\sharp} \uparrow\mathrm{Ch}_+ U \oplus \uparrow\mathrm{Ch}_+ V \xrightarrow{g_\sharp} \uparrow\mathrm{Ch}_+ (X) \to 0$$

$$f = (i, j), \qquad g = [u, -v]. \tag{2.33}$$

Its exactness needs one non-trivial verification. Take $a \in \uparrow\mathrm{Ch}_n U$, $b \in \uparrow\mathrm{Ch}_n V$ and assume that $u_\sharp(a) = v_\sharp(b)$; therefore, each cube really appearing in a (and b) belongs to $U \cap V$; globally, there is (one) normalised chain $c \in \uparrow\mathrm{Ch}_n (U \cap V)$ such that $i_\sharp(c) = a$, $i_\sharp(c) = b$. □

2.2.3 Theorem and definition (Excision)

Let a cubical set X be given, with subobjects $B \subset Y \cap A$. The inclusion map $i \colon (Y, B) \to (X, A)$ is said to be excisive *whenever $Y_n \backslash B_n = X_n \backslash A_n$, for all n. Or equivalently:*

$$Y \cup A = X, \qquad Y \cap A = B \tag{2.34}$$

in the lattice of subobjects of X.

Then the inclusion i induces isomorphisms in homology, preserving and reflecting preorder.

Proof After (2.32), the proof reduces to a Noether isomorphism for directed chain complexes

$$\begin{aligned}
\uparrow\mathrm{Ch}_+ (Y, B) &= (\uparrow\mathrm{Ch}_+ Y)/(\mathrm{Ch}_+ (Y \cap A)) \\
&= (\uparrow\mathrm{Ch}_+ Y)/(\mathrm{Ch}_+ Y \cap \mathrm{Ch}_+ A) \\
&\cong (\uparrow\mathrm{Ch}_+ Y + \uparrow\mathrm{Ch}_+ A)/(\mathrm{Ch}_+ A) \\
&= \uparrow\mathrm{Ch}_+ (Y \cup A)/(\mathrm{Ch}_+ A) = \uparrow\mathrm{Ch}_+ (X, A).
\end{aligned}$$

□

2.2.4 Theorem (The homology of a tensor product)

Given two cubical sets X, Y, there is a natural isomorphism and a natural monomorphism

$$\uparrow\!\mathrm{Ch}_+(X) \otimes \uparrow\!\mathrm{Ch}_+(Y) = \uparrow\!\mathrm{Ch}_+(X \otimes Y),$$
$$\uparrow\!H_\bullet(X) \otimes \uparrow\!H_\bullet(Y) \rightarrowtail \uparrow\!H_\bullet(X \otimes Y). \tag{2.35}$$

Proof It suffices to prove the first part, and apply the Künneth formula. First, the canonical (positive) basis of the free ordered abelian group

$$\uparrow\!\mathrm{Ch}_p(X) \otimes \uparrow\!\mathrm{Ch}_q(Y)$$

is $\overline{X}_p \times \overline{Y}_q$ (as in Section 2.1.2, with $\overline{X}_p = X_p \setminus \mathrm{Deg}_p X$). Recall now that the set $(X \otimes Y)_n$ is a quotient of the set $\sum_{p+q=n} X_p \times Y_q$ modulo an equivalence relation which only identifies pairs where a term is degenerate (see (1.137)); moreover, a class $x \otimes y$ is degenerate if and only if x or y is degenerate (see (1.138)).

Therefore, the canonical positive basis of $\uparrow\!\mathrm{Ch}_n(X \otimes Y)$ is precisely the sum (disjoint union) of the preceding sets $\overline{X}_p \times \overline{Y}_q$, for $p + q = n$. We can identify the ordered abelian groups

$$\uparrow\!\mathrm{Ch}_n(X \otimes Y) = \bigoplus_{p+q=n} \uparrow\!\mathrm{Ch}_p(X) \otimes \uparrow\!\mathrm{Ch}_q(Y),$$

respecting the canonical positive bases.

Finally, the differential of an element $x \otimes y$, with $(x, y) \in \overline{X}_p \times \overline{Y}_q$, is the same in both chain complexes

$$\sum_{i,\alpha} (-1)^{i+\alpha} \, \partial_i^\alpha (x \otimes y)$$
$$= \sum_{i \leqslant p, \alpha} (-1)^{i+\alpha} (\partial_i^\alpha x) \otimes y + \sum_{j \leqslant q, \alpha} (-1)^{p+j+\alpha} x \otimes (\partial_j^\alpha y)$$
$$= (\partial_p x) \otimes y + (-1)^p x \otimes (\partial_q y).$$

$\square$

2.2.5 Directed homology of d-spaces

We end this section by defining directed homology for d-spaces. In this setting, where the directed structure is essentially one-dimensional (being defined by distinguished paths, i.e. 1-cubes), we will see that preorder is only relevant in $\uparrow\!H_1$.

First, let us recall a well-known fact: if S is a topological space, its singular homology can be defined by the singular *cubical* set $\square S$ (as for instance in Massey's text [Ms])

$$H_n(S) = H_n(\square S). \tag{2.36}$$

The equivalence with the simplicial definition can be proved by acyclic models, see [HW].

Of course, we can equip these groups with the preorder $\uparrow H_n(S) = \uparrow H_n(\square S)$, but this would have no interest whatsoever. In fact, $\uparrow H_0(\square S)$ has an order generated by the homology classes of points (cf. (2.15)), which conveys no information; and we prove below that, for $n \geqslant 1$, *the preorder of* $\uparrow H_n(\square S)$ *is chaotic* (Corollary 2.2.7).

More interestingly, starting from a *d-space* S, we define its *directed singular homology* by letting $\uparrow H_n$ act on the singular cubical set $\uparrow \square S \subset \square S$ (see (1.162))

$$\uparrow H_n : \mathrm{d}\mathbf{Top} \to \mathrm{p}\mathbf{Ab}, \qquad \uparrow H_n(S) = \uparrow H_n(\uparrow \square S). \qquad (2.37)$$

Here, the preorder is relevant in degree 1, but becomes chaotic for $n \geqslant 2$ (Corollary 2.2.7).

The proof of the first fact, for **Top**, will be based on the presence in this category of the maps of *reversion* and *lower connection* (Section 1.1.0)

$$\begin{aligned} r: \mathbf{I} \to \mathbf{I}, \qquad & r(t) = 1 - t & (reversion), \\ g^-: \mathbf{I}^2 \to \mathbf{I}, \qquad & g^-(t_1, t_2) = \max(t_1, t_2) & (lower\ connection). \end{aligned} \qquad (2.38)$$

The proof of the second fact, for d**Top**, will be based on *transposition* (which can only act on dimension $\geqslant 2$) and, again, lower connection (which also preserves the order)

$$s: \uparrow \mathbf{I}^2 \to \uparrow \mathbf{I}^2, \quad s(t_1, t_2) = (t_2, t_1) \quad (transposition). \qquad (2.39)$$

All this is proved below, in the more general situation of c-sets (Proposition 2.2.6(b), (c)). It could be further extended to *abstract* cubical sets equipped with symmetries and connections (cf. [GM]) but we prefer to avoid this heavy structure.

2.2.6 Proposition (Symmetry versus preorder)

We use the same notation as above (Section 2.2.5).

(a) Let K be a cubical set and $n \geqslant 1$. Suppose that for every n-cube $a \in K_n$ there exists some n-cube a' such that the chain $a + a' \in \uparrow \mathrm{Ch}_n(K)$ is a boundary. Then $\uparrow H_n(K)$ has a chaotic preorder.

(b) Let X be a c-set and $n \geqslant 1$. The following conditions imply that $\uparrow H_n(X)$ (defined in Section 2.1.2) has a chaotic preorder:

- the set $c_n X$ is closed under pre-composition with the involution $r \times$ $\mathbf{I}^{n-1} : \mathbf{I}^n \to \mathbf{I}^n$;
- if $a \in c_n X$, then $a.((g^-(r \times \mathbf{I})) \times \mathbf{I}^{n-1}) \in c_{n+1} X$.

(Actually, the first condition is redundant, as will be evident from the proof.)

(c) *Let X be a c-set and $n \geqslant 2$. The following conditions imply that $\uparrow H_n(X)$ has a chaotic preorder:*

- the set $c_n X$ is closed under pre-composition with the involution $s \times$ $\mathbf{I}^{n-2} : \mathbf{I}^n \to \mathbf{I}^n$;
- if $a \in c_n X$, then $a.((g^- \times \mathbf{I}^{n-1})(\mathbf{I} \times s \times \mathbf{I}^{n-2})) \in c_{n+1} X$.

(Again, the first condition is redundant.)

Proof (a) Every n-cycle $\sum_i \lambda_i.a_i$ is equivalent modulo boundaries to a cycle $\sum_i \mu_i.b_i$ where all coefficients are $\geqslant 0$, replacing $\lambda_i.a_i$ with $(-\lambda_i).a_i'$ when $\lambda_i < 0$. Thus, all homology classes in $\uparrow H_n(K)$ are positive. (Note that the hypothesis is only possible for $n > 0$, unless K is empty.)

The rest of the proof ensues from this point, making use of the following (easy) computation of differentials:

$$\begin{aligned}
\partial((g^-(r \times \mathbf{I})) \times \mathbf{I}^{n-1}) &= \text{id} + r \times \mathbf{I}^{n-1}, \\
\partial((g^- \times \mathbf{I}^{n-1})(\mathbf{I} \times s \times \mathbf{I}^{n-2})) &= -\text{id} - s \times \mathbf{I}^{n-2}.
\end{aligned} \tag{2.40}$$

(b) For every distinguished n-cube $a \colon \mathbf{I}^n \to X$, the cubes

$$\begin{aligned}
a' &= a.(r \times \mathbf{I}^{n-1}) \colon \mathbf{I}^n \to X, \\
a'' &= a.((g^-(r \times \mathbf{I})) \times \mathbf{I}^{n-1}) \colon \mathbf{I}^{n+1} \to X,
\end{aligned} \tag{2.41}$$

are also distinguished. Since, by (2.40), $\partial_n a'' = a + a'$ in $\uparrow \text{Ch}_n(X)$, the thesis follows from (a).

(c) For every distinguished n-cube $a \colon \mathbf{I}^n \to X$, the cubes

$$\begin{aligned}
a' &= a.(s \times \mathbf{I}^{n-2}) \colon \mathbf{I}^n \to X, \\
a'' &= a.((g^- \times \mathbf{I}^{n-1})(\mathbf{I} \times s \times \mathbf{I}^{n-2})) \colon \mathbf{I}^{n+1} \to X,
\end{aligned} \tag{2.42}$$

are distinguished. Now, $\partial_n a'' = -a - a'$ and the thesis follows again from (a). $\qquad \square$

2.2.7 Corollary (Chaotic preorders)

If S is a d-space, then the directed singular homology $\uparrow H_n(S) = \uparrow H_n(\uparrow\square S)$ (defined in Section 2.2.5) has a chaotic preorder for all $n \geqslant 2$. If S is a topological space, this holds for all $n \geqslant 1$.

Proof Follows from points (c) and (b) of the previous proposition. $\square$

2.3 Pointed homotopy and homology of pointed cubical sets

Pointed suspension and the pointed cofibre sequence have a good relationship with *directed pointed homology*; the latter can also be viewed as a form of reduced homology, well adapted to preorder, while the ordinary reduced homology is not.

2.3.1 The interest of pointed objects in the directed case

We have introduced in Section 1.6.9 the category **Cub.** of pointed cubical sets, whose homotopy and homology will be studied below. Working on such objects with directed algebraic tools one often gets better results than with the corresponding unpointed ones (while, forgetting about directions, we would get equivalent solutions).

First, from the point of view of *directed homotopy theory*, the advantage is evident from the following example: within cubical sets, suspension gives rise to the *ordered spheres* $\Sigma^n(\mathbf{s}^0) = \uparrow\mathbf{o}^n$ (1.182), while we show below (Section 2.3.2) that, within pointed cubical sets, (pointed) suspension gives the *elementary directed spheres*: $\Sigma^n(\mathbf{s}^0, 1) = (\uparrow\mathbf{s}^n, *)$, which are more interesting. Note that their classical geometric realisations are homeomorphic spaces.

Second, from the point of view of *directed homology theory*, let us compare reduced homology (of unpointed objects) and pointed homology (of the pointed ones). Classically, one introduces the reduced homology of a cubical set (or a topological space) X as the kernel of the homomorphism induced by the terminal map $X \to \{*\}$

$$\tilde{H}_n(X) = \mathrm{Ker}(H_n(X) \to H_n(\{*\})). \tag{2.43}$$

Reduced homology has a suspension isomorphism

$$h_n : \tilde{H}_n(X) \to \tilde{H}_{n+1}(\Sigma(X)), \tag{2.44}$$

and yields, for every map $f: X \to Y$, an exact cofibre sequence

$$\ldots \tilde{H}_n(X) \xrightarrow{f_{*n}} \tilde{H}_n(Y) \xrightarrow{u_{*n}} \tilde{H}_n(C^+f) \xrightarrow{d_n} \tilde{H}_{n-1}(X)$$

$$\ldots \tilde{H}_0(X) \xrightarrow{f_{*0}} \tilde{H}_0(Y) \xrightarrow{u_{*0}} \tilde{H}_0(C^+f) \longrightarrow 0. \tag{2.45}$$

Here, $u: Y \to C^+f$ is the upper homotopy cokernel of f, while the differential comes from the differential $d(f): C^+f \to \Sigma X$ of the cofibre sequence of f (Section 1.7.8)

$$d_n = (h_{n-1})^{-1}(d(f))_{*n} : \tilde{H}_n(C^+f) \to \tilde{H}_n(\Sigma X) \to \tilde{H}_{n-1}(X), \tag{2.46}$$

(which amounts to $(h_{n-1})^{-1}$ when $Y = \{*\}$ and $C^+f \cong \Sigma X$).

One can obtain similar results with the pointed homology of pointed cubical sets (or pointed topological spaces), which can be defined, very simply, as a particular case of relative homology

$$H_n(X, x_0) = H_n(X, \{x_0\}) \cong \tilde{H}_n(X). \tag{2.47}$$

Algebraically, the two approaches are more or less equivalent – even if the latter has some formal advantage: pointed homology (of pointed objects) *preserves sums* while reduced homology (of unpointed objects) does not.

But, *introducing preorders*, reduced homology and pointed homology give different results, and the latter behaves much better.

Indeed, $\uparrow\tilde{H}_0(X)$ has a trivial preorder, the discrete one, since the positive cone $\Sigma\lambda_i[x_i]$ ($\lambda_i \in \mathbf{N}$) of $\uparrow H_0(X)$ has a null trace on $\mathrm{Ker}(H_0(X) \to \mathbf{Z}$. This is not true for pointed homology (see (2.59)).

We thus have

$$\uparrow\tilde{H}_0(\mathbf{s}^0) = \mathrm{Ker}(\uparrow H_0(\mathbf{s}^0) \to \uparrow H_0(\{*\})) \cong \uparrow_d\mathbf{Z},$$

$$\uparrow\tilde{H}_1(\uparrow\mathbf{o}^1) \cong \uparrow_d\mathbf{Z} \tag{2.48}$$

(where $\uparrow_d\mathbf{Z}$ has the discrete order), while pointed homology and suspension will yield an isomorphism of non-trivial ordered abelian groups (see Theorem 2.3.5).

2.3.2 Pointed homotopies

Let us recall (from Section 1.6.9) that, in the category **Cub.** of pointed cubical sets, *limits* and *quotients* are computed as for cubical sets and pointed in the obvious way, whereas *sums* are quotients of the corresponding unpointed sums, under identification of the base points (as for pointed sets).

The (elementary, left) *pointed cylinder* is

$$I: \mathbf{Cub_{\bullet}} \to \mathbf{Cub_{\bullet}}, \qquad I(X, x_0) = (IX/I\{x_0\}, [0 \otimes x_0]),$$
$$\partial^\alpha: (X, x_0) \to I(X, x_0), \quad \partial^\alpha(x) = [\alpha \otimes x] \qquad (\alpha = 0, 1), \qquad (2.49)$$
$$e: I(X, x_0) \to (X, x_0), \qquad e[u \otimes x] = e_1(x).$$

Its right adjoint, the (elementary, left) *pointed cocylinder*, is

$$P: \mathbf{Cub_{\bullet}} \to \mathbf{Cub_{\bullet}},$$
$$P(Y, y_0) = (PY, \omega_0), \quad \omega_0 = e_1(y_0) \in Y_1. \qquad (2.50)$$

An (elementary, left) *pointed homotopy* $f: f^- \to_L f^+: (X, x_0) \to (Y, y_0)$ is a map $f: I(X, x_0) \to (Y, y_0)$ with $f\partial^\alpha = f^\alpha$, or, equivalently, a map $f: (X, x_0) \to P(Y, y_0)$ with $\partial^\alpha f = f^\alpha$, which – as we have seen in (1.152) – amounts to a family

$$f_n: X_n \to Y_{n+1},$$
$$\partial^\alpha_{i+1} f_n = f_{n-1} \partial^\alpha_i, \qquad \partial^\alpha_1 f_n = f^\alpha, \qquad (2.51)$$
$$e_{i+1} f_{n-1} = f_n e_i, \qquad f_0(x_0) = \omega_0 \qquad (\alpha = \pm; i = 1, ..., n).$$

The (left) pointed upper cone $C^+(X, x_0)$ is a quotient of the pointed cylinder

$$\begin{array}{ccc} (X, x_0) & \xrightarrow{\ p\ } & \top \\ \partial^+ \Big\downarrow & \dashrightarrow v^+ & \Big\downarrow \\ I(X, x_0) & \xrightarrow{\ \gamma\ } & C^+(X, x_0) \end{array} \qquad \begin{array}{l} C^+(X, x_0) = \\[4pt] (IX)/(I\{x_0\} \cup \partial^+ X). \quad (2.52) \end{array}$$

The (left) pointed suspension is the quotient

$$\Sigma(X, x_0) = (IX)/(\partial^- X \cup I\{x_0\} \cup \partial^+ X). \qquad (2.53)$$

Thus, the pointed suspension of $(\mathbf{s}^0, 0)$ yields the elementary directed circle $\uparrow\mathbf{s}^1$

$$(2.54)$$

and, more generally

$$(\uparrow\mathbf{s}^n, *) = \Sigma^n(\mathbf{s}^0, 0). \qquad (2.55)$$

2.3.3 Pointed homology

A pointed cubical set (X, x_0) naturally gives a directed chain complex $\uparrow\mathrm{Ch}_+(X, x_0)$.

Its higher components, for $n > 0$, coincide with the unpointed ones, $\uparrow\mathrm{Ch}_n(X)$, while the 0-component is the free preordered abelian group generated by the *pointed* set (X, x_0), so that the base point is annihilated

$$\uparrow\mathrm{Ch}_0(X, x_0) = \uparrow\mathbf{Z}(X_0, x_0) = (\uparrow\mathbf{Z}X_0)/(\mathbf{Z}x_0). \tag{2.56}$$

The functor $\uparrow\mathbf{Z}(-, -)$ which we are applying is left adjoint to the forgetful functor

$$\mathbf{pAb} \to \mathbf{Set}_{\scriptscriptstyle\bullet}, \qquad A \mapsto (A^+, 0).$$

Moreover, we can identify the directed chain complex $\uparrow\mathrm{Ch}_+(X, x_0)$ with the corresponding relative complex

$$\uparrow\mathrm{Ch}_+(X, \{x_0\}) \; = \; \uparrow\mathrm{Ch}_+(X)/\uparrow\mathrm{Ch}_+(\{x_0\}).$$

We thus have the *pointed directed homology* of a pointed cubical set, a particular case of relative directed homology

$$\uparrow H_n \colon \mathbf{Cub}_{\scriptscriptstyle\bullet} \to \mathbf{pAb},$$
$$\uparrow H_n(X, x_0) = \uparrow H_n(\uparrow\mathrm{Ch}_+(X, x_0))$$
$$= \uparrow H_n(\uparrow\mathrm{Ch}_+(X, \{x_0\})) = \uparrow H_n(X, \{x_0\}). \tag{2.57}$$

On the other hand, it is also true that relative homology *of cubical sets* is determined by pointed homology. In fact, the projection $p \colon (X, A) \to (X/A, \{*\})$ induces an isomorphism $p_\sharp$ of directed chain complexes and isomorphisms p_{*n} in directed homology

$$p_\sharp \colon \uparrow\mathrm{Ch}_+(X, A) \; \to \; \uparrow\mathrm{Ch}_+(X/A, \{*\}),$$
$$p_{*n} \colon \uparrow H_n(X, A) \; \cong \; \uparrow H_n(X/A, *), \tag{2.58}$$

Pointed homology only differs from the unpointed one in degree zero, where

$$\uparrow H_0(X, x_0) \; \cong \; \uparrow\mathbf{Z}.\pi_0(|X|, x_0) \tag{2.59}$$

is the free ordered abelian group generated by the *pointed set* of connected components of (X, x_0), or equivalently by the (unpointed) *set* of components different from the component of the base point.

A pair $((X, x_0), (A, x_0))$ of pointed cubical sets (with $x_0 \in A \subset X$) has a relative homology which coincides with the unpointed one (in every degree)

$$
\begin{aligned}
\uparrow H_n((X, x_0), (A, x_0)) &= \uparrow H_n(\uparrow \mathrm{Ch}_+(X, x_0)/\mathrm{Ch}_+(A, x_0)) \\
&= \uparrow H_n(X, A). \qquad (2.60)
\end{aligned}
$$

The *exact homology sequence* of this pair is just the homology sequence of the triple $(X, A, \{x_0\})$ of cubical sets (see Section 2.1.5).

2.3.4 Theorem (Homotopy invariance)

The pointed homology functor $\uparrow H_n \colon \mathbf{Cub}_\bullet \to \mathrm{pAb}$ is invariant for elementary left (or right) pointed homotopies (Section 2.3.2): given $f \colon f^- \to_L f^+ \colon X \to Y$, then $\uparrow H_n(f^-) = \uparrow H_n(f^+)$.

Proof Follows from the invariance of relative homology (Section 2.2.1), since pointed homology is a particular instance of relative homology (see (2.57)) and a left pointed homotopy is the same as a left relative homotopy (Section 2.1.5) between pointed cubical sets. $\qquad \square$

2.3.5 Theorem (Homology of a suspension)

There is a natural isomorphism of preordered abelian groups

$$
\begin{aligned}
h_n \colon \uparrow H_n(X, x_0) &\to \uparrow H_{n+1}(\Sigma(X, x_0)), \\
[\textstyle\sum_k \lambda_k x_k] &\mapsto [\textstyle\sum_k \lambda_k \langle u \otimes x_k \rangle] \qquad (n \geqslant 0),
\end{aligned}
\qquad (2.61)
$$

where $\langle ... \rangle$ denotes an equivalence class in the suspension $\Sigma(X, x_0)$ as a quotient of $I(X, x_0)$, and u is the generator of the elementary interval $\uparrow \mathbf{i}$.

In particular, one finds again the ordered homology groups $\uparrow H_n(\uparrow \mathbf{s}^n) = \uparrow \mathbf{Z}$, for $n > 0$; cf. Section 2.1.4.

Proof First, let us note that, for $x \in X_n$, we have the following relation in $\uparrow \mathrm{Ch}_{n+1}(\Sigma(X, x_0))$

$$
\begin{aligned}
\partial \langle u \otimes x \rangle &= \langle 1 \otimes x - 0 \otimes x \rangle - \textstyle\sum_{i,\alpha}(-1)^{i+\alpha} \langle u \otimes \partial_i^\alpha x \rangle \\
&= -\langle u \otimes \partial x \rangle.
\end{aligned}
\qquad (2.62)
$$

Now, the isomorphism of the thesis is induced by the following inverse isomorphisms of preordered abelian groups, which *anti*-commute with differentials

$$f_n \colon {\uparrow}\mathrm{Ch}_n(X, x_0) \to {\uparrow}\mathrm{Ch}_{n+1}(\Sigma(X, x_0)),$$

$$f(x) = \langle u \otimes x \rangle \qquad\qquad\qquad (x \in X_n),$$

$$\partial f(x) = \partial \langle u \otimes x \rangle = -\langle u \otimes \partial x \rangle = -f(\partial x), \quad f(x_0) = \langle u \otimes x_0 \rangle = 0,$$

$$f(e_k y) = \langle u \otimes e_k y \rangle = \langle e_{k+1}(u \otimes y) \rangle = 0 \qquad (\text{for } n > 0, \, y \in X_{n-1});$$

$$g_n \colon {\uparrow}\mathrm{Ch}_{n+1}(\Sigma(X, x_0)) \to {\uparrow}\mathrm{Ch}_n(X, x_0),$$

$$g\langle u \otimes x \rangle = x, \qquad g\partial\langle u \otimes x \rangle = -g\langle u \otimes \partial x \rangle = -\partial x = -\partial g\langle u \otimes x \rangle.$$

$$\square$$

2.3.6 Theorem (The homology cofibre sequence of a map)

Given a map $f \colon (X, x_0) \to (Y, y_0)$ of pointed cubical sets, there is a homology upper cofibre sequence (obtained from the upper cofibre sequence of f, in Section 1.7.8) which is exact and natural, with preorder-preserving homomorphisms:

$$\ldots {\uparrow}H_n(X, x_0) \xrightarrow{f_{*n}} {\uparrow}H_n(Y, y_0) \xrightarrow{u_{*n}} {\uparrow}H_n(C^+ f, [y_0]) \xrightarrow{d_n}$$

$$\uparrow H_{n-1}(X, x_0) \longrightarrow \qquad \cdots \qquad \longrightarrow {\uparrow}H_0(C^+ f, [y_0]) \longrightarrow 0,$$

$$\begin{aligned} u &= \mathrm{hcok}^+(f), \\ d_n &= (h_{n-1})^{-1}(d^+(f))_{*n} \colon {\uparrow}H_n(C^+ f) \to {\uparrow}H_{n-1}(X, x_0). \end{aligned} \qquad (2.63)$$

The same holds for the lower cofibre sequence.

Proof All the homomorphisms above are preorder-preserving, also because of the previous result on suspension (Theorem 2.3.5).

The rest will be proved showing that our sequence amounts to the homology sequence of the pair $(C, j(X, x_0))$ of pointed cubical sets (Section 2.3.3), where $C = I(f, \mathrm{id}(X, x_0))$ is a mapping cylinder (Section 1.3.5) in **Cub.** and j is the embedding of (X, x_0) into its upper basis

$$j \colon (X, x_0) \to I(X, x_0) \to I(f, \mathrm{id}(X, x_0)), \qquad j(x) = [x, 1].$$

Indeed, the pointed homology ${\uparrow}H_n(C^+ f, [y_0])$ coincides with the relative directed homology ${\uparrow}H_n(C, j(X, x_0))$ (by (2.58)). Moreover (X, x_0)

is isomorphic to $(jX, [y_0])$, while (Y, y_0) is past homotopy equivalent to $(C, [y_0])$, actually a past deformation retract (Section 1.3.1)

$i\colon (Y, y_0) \subset (C, [y_0]),$

$p\colon (C, [y_0]) \to (Y, y_0), \qquad p[x, t] = f(x), \quad p[y] = y,$

$\psi\colon ip \to 1_C, \qquad\qquad \psi([x, t], t') = [x, tt'], \quad \psi([y], t') = y.$

$\square$

2.3.7 Upper relative pointed homology

For a pair $((X, x_0), (A, x_0))$ of pointed cubical sets, one can also define an upper (or lower) relative pointed homology, which – for preorder – behaves better than the relative directed homology $\uparrow H_n((X, x_0), (A, x_0)) = \uparrow H_n(X, A)$, already considered above (in (2.60)).

By definition, the *upper* (or *lower*) *relative pointed homology* of our pointed pair is based on the *upper* (or *lower*) *pointed cone* of the embedding $j\colon (A, x_0) \to (X, x_0)$

$$\uparrow H_n^\alpha((X, x_0), (A, x_0)) = \uparrow H_n(C^\alpha((A, x_0) \to (X, x_0))) \quad (\alpha = \pm). \quad (2.64)$$

Using the homology upper (or lower) cofibre sequence of the embedding j (Theorem 2.3.6), we get an exact sequence with a differential *which does preserve preorder* (and can reflect it, or not, according to cases, see the example below):

$$\ldots \uparrow H_n(A, x_0) \xrightarrow{j_{*n}} \uparrow H_n(X, x_0) \xrightarrow{u_{*n}} \uparrow H_n^\alpha((X, x_0), (A, x_0)) \xrightarrow{d_n}$$

$$\ldots \uparrow H_{n-1}(A, x_0) \longrightarrow \quad \cdots \quad \longrightarrow \uparrow H_0^\alpha((X, x_0), (A, x_0)) \to 0,$$

$$u = \mathrm{hcok}^+(j). \quad (2.65)$$

2.3.8 An elementary example

Consider the pointed pair $(\uparrow\mathbf{i}, \mathbf{s}^0)$, with (unwritten) base point at 0 and inclusion $j\colon \mathbf{s}^0 \to \uparrow\mathbf{i}$. The cones of j are:

$$C^+(j) \qquad\qquad C^-(j) \cong \uparrow\mathbf{o}^1.$$

$$(2.66)$$

The upper and lower homology of the pointed pair $(\uparrow\mathbf{i}, \mathbf{s}^0)$ are the group of integers (generated by the homology class of the chain w defined below), with the natural or discrete order, respectively:

$$\begin{aligned}
\uparrow H_1^+(\uparrow\mathbf{i}, \mathbf{s}^0) &= \uparrow\mathbf{Z}[w], & w = w_1 + w_2 & \quad\text{(a positive chain)},\\
\uparrow H_1^-(\uparrow\mathbf{i}, \mathbf{s}^0) &= \uparrow_d\mathbf{Z}[w], & w = w_1 - w_2 & \quad\text{(a chain)}.
\end{aligned} \tag{2.67}$$

Therefore:

(a) in the sequence of *upper* relative homology, the differential preserves order and also reflects it:

$$\begin{aligned}
d_1 &: \uparrow H_1^+(\uparrow\mathbf{i}, \mathbf{s}^0) \to \uparrow H_0(\mathbf{s}^0, 0),\\
d_1 &: \uparrow\mathbf{Z}[w] \to \uparrow\mathbf{Z}[1], & d_1([w]) = [1];
\end{aligned} \tag{2.68}$$

(b) in the sequence of *lower* relative homology, the differential preserves order but does not reflect it:

$$\begin{aligned}
d_1 &: \uparrow H_1^-(\uparrow\mathbf{i}, \mathbf{s}^0) \to \uparrow H_0(A, 0),\\
d_1 &: \uparrow_d\mathbf{Z}[w] \to \uparrow\mathbf{Z}[1], & d_1([w]) = [1].
\end{aligned} \tag{2.69}$$

Finally in the exact sequence of ordinary (unpointed) relative homology of the pair $(\uparrow\mathbf{i}, \mathbf{s}^0)$, the differential *does not even preserve* the order relation:

$$\begin{aligned}
\Delta_1 &: \uparrow H_1(\uparrow\mathbf{i}, \mathbf{s}^0) \to \uparrow H_0(\mathbf{s}^0),\\
\Delta_1 &: \uparrow\mathbf{Z}[u] \to \uparrow\mathbf{Z}\{[0], [1]\}, & \Delta_1([u]) = [1] - [0].
\end{aligned} \tag{2.70}$$

2.4 Group actions on cubical sets

The classical theory of *proper* actions on topological spaces, culminating in a spectral sequence, is extended here to *free* actions on cubical sets. G is a group, written in additive notation (independently of commutativity). The action of an operator $g \in G$ on an element x is written as $x + g$.

2.4.1 Basics

Take a cubical set X and a group G *acting on it*, on the right. In other words, we have an action $x + g$ (for $x \in X_n$, $g \in G$) on each component, consistently with faces and degeneracies (or, equivalently, a cubical object in the category of G-sets).

Plainly, there is a cubical set of orbits X/G, with components X_n/G and induced structure; and a natural projection $p\colon X \to X/G$ in **Cub**.

One says that the action is *free* if G acts freely on each component, i.e. the relation $x = x + g$, for some $x \in X_n$ and $g \in G$ implies $g = 0$. This is equivalent to saying that *G acts freely on the set of vertices X_0* (because $x = x + g$ implies that their first vertices coincide).

It is now easy to extend to *free actions on cubical sets* the classical results of actions of groups on topological spaces ([M1], IV.11), which hold for groups acting *properly* on a space, a much stronger condition than acting freely on the space (it means that every point has an open neighbourhood U such that all subsets $U + g$ are disjoint). Note, however, that all results below which involve the homology of the group G *ignore preorder*, necessarily (cf. Section 2.5.6).

Of course, an *action of G on a c-set* (X, cX) (Section 1.6.8) is defined to be an action on the set X coherent with the structural presheaf cX: for every distinguished cube $x \colon \mathbf{I}^n \to X$, all mappings $x + g$ are also distinguished. Thus, for a topological space S, a G-action on the space gives an action on the c-set $S_\square = (|S|, \square S)$ and on the cubical set $\square S$.

2.4.2 Lemma (Free actions)

(a) If G acts freely on the cubical set X, then $\uparrow\mathrm{Ch}_+(X)$ is a complex of free right G-modules, and one can choose a (positive) basis $B_n \subset X_n$ of $\uparrow\mathrm{Ch}_n(X)$ which projects bijectively onto $\overline{X}_n/G$, the canonical basis of $\uparrow\mathrm{Ch}_n(X/G)$.

(b) Moreover, if $\uparrow L$ is a preordered abelian group, viewed as a trivial G-module, then the canonical projection $p \colon X \to X/G$ induces an isomorphism of directed (co)chain complexes, and hence an isomorphism in (co)homology

$$
\begin{aligned}
p_\sharp &\colon \uparrow\mathrm{Ch}_+(X) \otimes_G \uparrow L \to \uparrow\mathrm{Ch}_+(X/G; \uparrow L), \\
p_{*n} &\colon H_n(\uparrow\mathrm{Ch}_+(X) \otimes_G \uparrow L) \to \uparrow H_n(X/G; \uparrow L), \\
p^\sharp &\colon \uparrow\mathrm{Ch}^+(X/G; \uparrow L) \to \mathrm{Hom}_G(\uparrow\mathrm{Ch}_+(X), \uparrow L), \\
p^{*n} &\colon \uparrow H^n(X/G; \uparrow L) \to H_n(\mathrm{Hom}_G(\uparrow\mathrm{Ch}_+(X), \uparrow L).
\end{aligned}
\tag{2.71}
$$

Proof This lemma adapts [M1], IV.11.2-4. It is sufficient to prove (a), which implies (b).

The action of G on X_n extends to a right action on the free abelian group $\mathbf{Z}X_n$, which is consistent with faces and degeneracies and preserves the canonical basis; it thus induces an obvious action on

$\uparrow \mathrm{Ch}_n(X) = \uparrow \mathbf{Z}.\overline{X}_n$, consistent with the positive cone and the differential

$$(\textstyle\sum \lambda_i x_i) + g = \textstyle\sum \lambda_i (x_i + g), \qquad \partial(\textstyle\sum \lambda_i x_i) + g = \partial(\textstyle\sum \lambda_i x_i + g).$$

Thus $\uparrow \mathrm{Ch}_n(X)$ is a complex of G-modules, whose components are preordered G-modules. Take now a subset $B_0 \subset X_0$ choosing exactly one point in each orbit; then B_0 is a G-basis of $\uparrow \mathrm{Ch}_0(X)$. Letting $B_n \subset X_n$ be the subset of those non-degenerate n-cubes x whose 'initial vertex' $\partial_1^- ... \partial_n^- x$ belongs to B_0, we have more generally a G-basis of $\uparrow \mathrm{Ch}_n(X)$ which satisfies our requirements.   □

2.4.3 Theorem (Free actions on acyclic cubical sets)

Let X be an acyclic *cubical set (which means that it has the homology of the point). Let G be a group acting freely on X and L an abelian group with trivial G-structure.*

Then, with respect to coefficients in L and forgetting preorder (cf. Section 2.5.6), the combinatorial (co)homology of the cubical set of orbits is isomorphic to the (co)homology of the group G:

$$H_{\bullet}(X/G; L) \cong H_{\bullet}(G; L), \qquad H^{\bullet}(X/G; L) \cong H^{\bullet}(G; L). \qquad (2.72)$$

Proof As in [M1], IV.11.5, the augmented sequence

$$... \to \mathrm{Ch}_1(X) \to \mathrm{Ch}_0(X) \to \mathbf{Z} \to 0$$

is exact, since X is acyclic.

By Lemma 2.4.2(a), this sequence forms a G-free resolution of the G-trivial module $\mathbf{Z}$. Therefore, applying the definition of the group homology $H_n(G; L)$ and the isomorphism (2.71), we get the thesis for homology (and similarly for cohomology)

$$H_n(G; L) = H_n(\mathrm{Ch}_+(X) \otimes_G L) \cong H_n(X/G; L).$$

    □

2.4.4 Corollary (Free actions on acyclic spaces)

Let S be an acyclic *topological space (with the singular homology of the point) and G a group acting freely on it.*

Then $H_{\bullet}((\square S)/G) \cong H_{\bullet}(G)$, and $\uparrow H_n((\square S)/G)$ has a chaotic preorder for $n \geqslant 1$. The same holds in cohomology.

Proof It suffices to apply the preceding theorem to the singular cubical set $\square S$ of continuous cubes of S. This cubical set has the same homology as S, and G acts obviously on it, by $(x + g)(t) = x(t) + g$ (for $t \in \mathbf{I}^n$).

Moreover, the action is free because it is on the set of vertices, S. Finally, the remark on the preorder of $\uparrow H_n((\square S)/G)$ follows from Proposition 2.2.6(b). $\qquad\qquad\square$

2.4.5 Theorem (The spectral sequence of a G-free cubical set)

Let X be a connected cubical set, G a group acting freely on it and L a G-module. Then there is a spectral sequence

$$E_{p,q}^2 = H_p(G; H_q(X; L)) \;\Rightarrow_p\; H_n(X/G; L). \qquad (2.73)$$

Proof This result extends Corollary 2.4.4, without assuming X acyclic. It will not be used below.

The proof is the same as in [M1], XI.7.1, where X is a path-connected topological space with a *proper* G-action. One computes the terms $E_{p,q}^2$ of the two spectral sequences of the double complex of components

$$K_{pq} = L \otimes \mathrm{Ch}_p(X) \otimes_G B_q(G),$$

where $B_\bullet(G)$ is a G-free resolution of $\mathbf{Z}$ as a trivial G-module. The argument depends on the fact that $\mathrm{Ch}_+(X)$ is a chain complex of free G-modules with $\mathrm{Ch}_+(X) \otimes_G L \cong \mathrm{Ch}_+(X/G; L)$, which is also true in our case (Lemma 2.4.2). $\qquad\qquad\square$

2.5 Interactions with non-commutative geometry

We compute now the directed homology of various cubical sets, which simulate – in a natural way – 'virtual spaces' of non-commutative geometry, the irrational rotation C*-algebras A_ϑ and non-commutative tori of dimension $\geqslant 2$; ϑ is always an irrational real number.

The classification of our irrational rotation cubical sets C_ϑ (Section 2.5.2) will be based on the order of $\uparrow H_1(C_\vartheta)$, much as the classification of the C*-algebras A_ϑ (up to Morita equivalence) is based on the order of $K_0(A_\vartheta)$.

These results, which first appeared in [G12], show that the preorder structure of directed homology can carry much stronger information than the algebraic structure.

2.5.1 Rotation algebras

Let us begin by recalling some well-known 'non-commutative spaces'.

First, take the topological line $\mathbf{R}$ and its additive subgroup $G_\vartheta = \mathbf{Z} + \vartheta\mathbf{Z}$, acting on the line by translations. Since ϑ is irrational, G_ϑ is a dense subgroup of the real line; therefore, in **Top**, the orbit space $\mathbf{R}/G_\vartheta = \mathbf{S}^1/\vartheta\mathbf{Z}$ is trivial: an uncountable set with the indiscrete topology.

Second, consider the *Kronecker foliation* F of the torus $\mathbf{T}^2 = \mathbf{R}^2/\mathbf{Z}^2$, with slope ϑ (explicitly recalled below, in Section 2.5.3), and the set $\mathbf{T}_\vartheta^2 = \mathbf{T}^2/\equiv_F$ of its leaves. It is well known, and easy to see, that the sets $\mathbf{R}/G_\vartheta$ and $\mathbf{T}_\vartheta^2$ have a canonical bijective correspondence (cf. Section 2.5.3). Again, ordinary topology gives no information on $\mathbf{T}_\vartheta^2$ since the quotient $\mathbf{T}^2/\equiv_F$ in **Top** is indiscrete.

In non-commutative geometry, both these sets are 'interpreted' as the (non-commutative) C*-algebra A_ϑ generated by two unitary elements u, v under the relation $vu = \exp(2\pi i\vartheta).uv$. A_ϑ is called the *irrational rotation C*-algebra* associated with ϑ, or also a *non-commutative torus* [C1, C2, C3, Ri1, Bl]. Both its complex K-theory groups are two-dimensional.

An interesting achievement of K-theory, which combines results of Pimsner and Voiculescu [PV] and Rieffel [Ri1], classifies these algebras. In fact, it is proved that $K_0(A_\vartheta) \cong \mathbf{Z} + \vartheta\mathbf{Z}$ as an *ordered subgroup of* $\mathbf{R}$, and that the traces of the projections of A_ϑ cover the set $G_\vartheta \cap [0, 1]$. It follows that A_ϑ and $A_{\vartheta'}$ are *isomorphic* if and only if $\vartheta' \in \pm\vartheta + \mathbf{Z}$ ([Ri1], Theorem 2) and *strongly Morita equivalent* if and only if ϑ and ϑ' are equivalent modulo the *fractional* action (on the irrationals) of the group $\mathrm{GL}(2, \mathbf{Z})$ of invertible integral 2×2 matrices ([Ri1], Theorem 4)

$$\begin{pmatrix} a & b \\ c & d \end{pmatrix}.t = \frac{at + b}{ct + d} \qquad (a, b, c, d \in \mathbf{Z};\ ad - bc = \pm 1) \qquad (2.74)$$

(or, equivalently, modulo the action of the projective general linear group $\mathrm{PGL}(2, \mathbf{Z})$ on the projective line).

Since $\mathrm{GL}(2, \mathbf{Z})$ is generated by the matrices

$$R = \begin{pmatrix} 0 & 1 \\ 1 & 0 \end{pmatrix}, \qquad T = \begin{pmatrix} 1 & 1 \\ 0 & 1 \end{pmatrix}, \qquad (2.75)$$

the orbit of ϑ is its closure $\{\vartheta\}_{RT}$ under the transformations $R(t) = t^{-1}$ and $T^{\pm 1}(t) = t \pm 1$ (which act on $\mathbf{R} \setminus \mathbf{Q}$).

A similar result, based on the 1-cohomology of an associated etale topos, can be found in [Ta].

We show now how one can obtain similar results with cubical sets naturally arising from the previous situations: the point is to replace a topologically trivial orbit space S/G with the corresponding quotient of the singular cubical set $\square S$, which identifies the cubes $I^n \to S$ modulo the action of the group G.

2.5.2 *Irrational rotation structures*

(a) As a first step on this route, instead of considering the trivial quotient $\mathbf{R}/G_\vartheta$ of topological spaces, we replace $\mathbf{R}$ with the singular cubical set $\square\mathbf{R}$ (on which G_ϑ acts freely) and consider the cubical set $(\square\mathbf{R})/G_\vartheta$.

Or, equivalently, we replace $\mathbf{R}$ with the c-set $\mathbf{R}_\square = (|\mathbf{R}|, \square\mathbf{R})$ (1.164) and take the quotient $\mathbf{R}_\square/G_\vartheta$, i.e. the set $\mathbf{R}/G_\vartheta$ equipped with the projections of the (continuous) cubes of $\mathbf{R}$. (In fact, if the cubes $x, y\colon I^n \to \mathbf{R}$ coincide when projected to $\mathbf{R}/G_\vartheta$, their difference $g = x - y\colon I^n \to \mathbf{R}$ takes values in the totally disconnected subset $G_\vartheta \subset \mathbf{R}$, and is constant; therefore, x and y also coincide in $(\square\mathbf{R})/G_\vartheta$.)

Then, applying Corollary 2.4.4, we find that the c-set $\mathbf{R}_\square/G_\vartheta$ (or the cubical set $(\square\mathbf{R})/G_\vartheta$) has the same homology as the group $G_\vartheta \cong \mathbf{Z}^2$, which coincides with the ordinary homology of the torus $\mathbf{T}^2$

$$H_\bullet(\mathbf{R}_\square/G_\vartheta) = H_\bullet(G_\vartheta) = H_\bullet(\mathbf{T}^2) = \mathbf{Z} + \sigma.\mathbf{Z}^2 + \sigma^2.\mathbf{Z} \qquad (2.76)$$

(the second 'equality' follows, for instance, from the classical version of Theorem 2.4.3 ([M1], IV.11.5), applied to the proper action of the group $\mathbf{Z}^2$ on the acyclic space $\mathbf{R}^2$). We also know that directed homology only gives the chaotic preorder on $\uparrow H_1(\mathbf{R}_\square/G_\vartheta)$ (again by Corollary 2.4.4). In cohomology, we have the same graded group.

Algebraically, all this is in accord with the K-theory of the rotation algebra A_ϑ, since both $H^{\mathrm{even}}(\mathbf{R}_\square/G_\vartheta)$ and $H^{\mathrm{odd}}(\mathbf{R}_\square/G_\vartheta)$ are two-dimensional.

(b) But a much more interesting result (and accord) can be obtained with the c-structure $\uparrow\mathbf{R}_\square$ of the line given by topology *and natural order*, where $c_n(\uparrow\mathbf{R}_\square)$ is the set of continuous order-preserving mappings $I^n \to \mathbf{R}$. The quotient

$$C_\vartheta = \uparrow\mathbf{R}_\square/G_\vartheta = \uparrow\mathbf{S}^1_\square/\vartheta\mathbf{Z} \qquad (2.77)$$

will be called an *irrational rotation c-set* and we want to classify its isomorphism classes, for $\vartheta \notin \mathbf{Q}$.

(We have already used the symbol C_ϑ for the *cubical set* $(\square\uparrow\mathbf{R})/G_\vartheta$, which consists precisely of the distinguished cubes of the c-set $\uparrow\mathbf{R}_\square/G_\vartheta$

which we are considering now. But there is no real need to introduce different symbols for these two closely related structures, which, by definition, have the same directed homology.)

We prove below (Theorems 2.5.8 and 2.5.9) that $\uparrow H_1(\uparrow\mathbf{R}_\square/G_\vartheta) \cong \uparrow G_\vartheta$, *as an ordered subgroup of the line* and that the c-sets C_ϑ have the same classification *up to isomorphism* as the rotation algebras A_ϑ *up to strong Morita equivalence*: while the algebraic homology of C_ϑ is the same as in (a), *independent of* ϑ, the (pre)order of directed homology determines ϑ up to the equivalence relation $\uparrow G_\vartheta \cong \uparrow G_{\vartheta'}$, which amounts to ϑ and ϑ' being conjugate under the action of the group $\mathrm{GL}(2, \mathbf{Z})$.

Note that the stronger classification of rotation algebras up to isomorphism (recalled in Section 2.5.1) has no analogue here: *cubical sets lack the 'metric information'* contained in C*-algebras. This can be recovered in a richer setting, in the domain of weighted algebraic topology (see Chapter 6).

The *irrational rotation d-space*

$$D_\vartheta = \uparrow\mathbf{R}/G_\vartheta = \uparrow\mathbf{S}^1/\vartheta\mathbf{Z}, \tag{2.78}$$

i.e. the quotient *in* d**Top** of the directed line modulo the action of the group G_ϑ, would give the same classification as C_ϑ. This could be proved here, with the directed homology of d-spaces. But we will get this result for free from the richer classification of w-spaces mentioned above (Section 6.7.6).

2.5.3 The non-commutative two-dimensional torus

Consider now the *Kronecker foliation F* of the torus $\mathbf{T}^2 = \mathbf{R}^2/\mathbf{Z}^2$, with irrational slope ϑ, and the set $\mathbf{T}^2_\vartheta = \mathbf{T}^2/ \equiv_F$ of its leaves. The foliation F is induced by the following foliation $\hat{F} = (F_\lambda)$ of the plane

$$F_\lambda = \{(x, y) \in \mathbf{R}^2 \mid y = \vartheta x + \lambda\} \qquad (\lambda \in \mathbf{R}). \tag{2.79}$$

Therefore $\equiv_F$ is induced by the following equivalence relation $\equiv$ on the plane (containing the congruence modulo $\mathbf{Z}^2$)

$$(x, y) \equiv (x', y') \iff y + k - \vartheta(x + h) = y' + k' - \vartheta(x' + h') \tag{2.80}$$

for some $h, k, h', k' \in \mathbf{Z}$.

Now, we replace the set $\mathbf{T}^2_\vartheta$ with the quotient c-set $\mathbf{T}^2_\square/ \equiv_F$, i.e. the set $\mathbf{T}^2_\vartheta$ equipped with the projection of the cubes of the torus (or of the plane). This is proved below to be isomorphic to the previous c-set $K = \mathbf{R}_\square/G_\vartheta$ (Section 2.5.2(a)), whose directed (co)homology has been computed above, in accordance (*algebraically*) with the complex K-theory groups of A_ϑ.

Indeed, the isomorphism we want can be realised with two inverse c-maps $i' \colon K \to \mathbf{T}_\vartheta^2$ and $p' \colon \mathbf{T}_\vartheta^2 \to K$, respectively induced by the following maps (in **Top**):

$$i \colon \mathbf{R} \to \mathbf{R}^2, \qquad i(t) = (0, t),$$
$$p \colon \mathbf{R}^2 \to \mathbf{R}, \qquad p(x, y) = y - \vartheta x. \tag{2.81}$$

First, the induction on quotients is legitimate because, for $t \equiv t + h + k\vartheta$ in $\mathbf{R}$ and $(x, y) \equiv (x', y')$ in $\mathbf{R}^2$ (as in (2.80))

$$i(t + h + k\vartheta) = (0, t + h + k\vartheta) \equiv (1, t + \vartheta) \equiv (0, t) = i(t),$$
$$p(x, y) - p(x', y') = (y - \vartheta x) - (y' - \vartheta x')$$
$$= k' - \vartheta h' - k + \vartheta h \in \mathbf{Z} + \vartheta \mathbf{Z}.$$

Second, $pi = \mathrm{id}\mathbf{R}$, while $i'p' = \mathrm{id}(\mathbf{T}_\vartheta^2)$ because

$$ip(x, y) = (0, y - \vartheta x) \equiv (x, y) \qquad (y - \vartheta x - \vartheta.0 = y - \vartheta x).$$

Finally, the following diagram shows that i', p' preserve distinguished cubes, since i and p preserve singular cubes

$$
\begin{array}{ccccccc}
\mathbf{I}^n & \xrightarrow{a} & \mathbf{R} & \underset{p}{\overset{i}{\rightleftarrows}} & \mathbf{R}^2 & \xleftarrow{b} & \mathbf{I}^n \\
& & \downarrow & & \downarrow & & \\
\mathbf{R}/G_\vartheta & =\!\!= & K & \underset{p'}{\overset{i'}{\rightleftarrows}} & \mathbf{R}^2/\!\equiv & =\!\!= & \mathbf{T}_\vartheta^2
\end{array}
$$

2.5.4 Higher foliations of codimension 1

(a) Extending Sections 2.5.2(a) and 2.5.3, take an n-tuple of real numbers $\vartheta = (\vartheta_1, ..., \vartheta_n)$, linearly independent on the rationals, and consider the additive subgroup

$$G_\vartheta = \Sigma_j\, \vartheta_j \mathbf{Z} \cong \mathbf{Z}^n, \tag{2.82}$$

which acts freely on $\mathbf{R}$. (The previous case corresponds to the pair $(\vartheta_1, \vartheta_2) = (1, -\vartheta)$, the minus sign being introduced to simplify the computations below.)

Now, the c-set $\mathbf{R}_\square/G_\vartheta$ has the homology (or cohomology) of the n-dimensional torus $\mathbf{T}^n$ (notation as in Section 2.1.4)

$$H_\bullet(\mathbf{R}_\square/G_\vartheta) = H_\bullet(G_\vartheta) = H_\bullet(\mathbf{T}^n)$$
$$= \mathbf{Z} + \sigma.\mathbf{Z}^{\binom{n}{1}} + \sigma^2.\mathbf{Z}^{\binom{n}{2}} + \cdots + \sigma^n.\mathbf{Z}. \tag{2.83}$$

Again, this coincides with the homology of a c-set $\mathbf{T}^n_{\square}/\equiv_F$ arising from the foliation F of the n-dimensional torus $\mathbf{T}^n = \mathbf{R}^n/\mathbf{Z}^n$ induced by the hyperplanes $\sum_j \vartheta_j x_j = \lambda$ of $\mathbf{R}^n$. (In the previous proof, one can replace the maps i, p of (2.81) with $i(t) = (t/\vartheta_1, 0, ..., 0)$ and $p(x_1, ..., x_n) = \sum_j \vartheta_j x_j$.)

(b) Extending now Section 2.5.2(b) (and Theorem 2.5.8), the c-set $\uparrow \mathbf{R}_{\square}/G_\vartheta$ has a more interesting directed homology, with a total order in degree 1:

$$\uparrow H_1(\uparrow \mathbf{R}_{\square}/G_\vartheta) = \uparrow G_\vartheta = \uparrow(\textstyle\sum_j \vartheta_j \mathbf{Z}) \qquad (G_\vartheta^+ = G_\vartheta \cap \mathbf{R}^+). \qquad (2.84)$$

2.5.5 Higher foliations

As a further generalisation of Section 2.5.4(a), let us replace the hyperplane $\sum_j \vartheta_j x_j = 0$ with a linear subspace $H \subset \mathbf{R}^n$ of codimension k $(0 < k < n)$, such that $H \cap \mathbf{Z}^n = \{0\}$.

Let $\hat{F}$ be the foliation of $\mathbf{R}^n$ whose leaves are all the $(n - k)$-dimensional planes $H + x$, parallel to H. These can be parametrised by letting x vary in some convenient k-dimensional subspace transverse to H; equivalently, choose a projector $e \colon \mathbf{R}^n \to \mathbf{R}^n$ with $H = \text{Ker}(e)$ and an epi-mono (linear) factorisation of e through $\mathbf{R}^k$

$$\mathbf{R}^n \xrightarrow{\ p\ } \mathbf{R}^k \xrightarrow{\ i\ } \mathbf{R}^n \qquad\qquad ip = e, \ \ pi = \text{id}, \qquad (2.85)$$

so that the leaves F_λ of $\hat{F}$ are bijectively parametrised by p on $\mathbf{R}^k$

$$F_\lambda = \{x \in \mathbf{R}^n \mid p(x) = \lambda\} \qquad\qquad (\lambda \in \mathbf{R}^k).$$

The projection $\mathbf{R}^n \to \mathbf{T}^n = \mathbf{R}^n/\mathbf{Z}^n$ is injective on each leaf F_λ (because $\text{Ker}(p) \cap \mathbf{Z}^n = H \cap \mathbf{Z}^n = \{0\}$). Therefore, $\hat{F}$ induces a foliation F of $\mathbf{T}^n$ of codimension k, and an equivalence relation $\equiv_F$ (namely, to belong to the same leaf). The set of leaves $\mathbf{T}^n/\equiv_F$ can be identified with the quotient $\mathbf{R}^n/\equiv$, modulo the equivalence relation $\equiv$ generated by the congruence modulo $\mathbf{Z}^n$ and the equivalence relation $x \hat{\equiv} y$ of the original foliation (i.e., $p(x) = p(y)$):

$$x \equiv x' \text{ in } \mathbf{R}^n \ \ \Leftrightarrow \ \ p(x) - p(x') \in p(\mathbf{Z}^n). \qquad (2.86)$$

Note that $G_p = p(\mathbf{Z}^n)$ is an additive subgroup of $\mathbf{R}^k$ *isomorphic to* $\mathbf{Z}^n$, because $\text{Ker}(p) \cap \mathbf{Z}^n = \{0\}$, again. Now, we are interested in the c-set $T^n_{\square}/\equiv_F$, isomorphic to $\mathbf{R}^n_{\square}/\equiv$. Because of (2.86), the maps p, i in

(2.85) induce a bijection of sets

$$\mathbf{R}^n/\equiv \quad \xrightarrow{\ p'\ } \quad \mathbf{R}^k/G_p \quad \xrightarrow{\ i'\ } \quad \mathbf{R}^n/\equiv$$

and an isomorphism of c-sets

$$\mathbf{T}^n_\square/\equiv_F \ \cong\ \mathbf{R}^n_\square/\equiv\ \cong\ \mathbf{R}^k_\square/G_p.$$

Since the cubical set $\square\mathbf{R}^k$ is acyclic and $G_p \cong \mathbf{Z}^n$, we conclude by Theorem 2.4.3 (together with its classical version) that the homology of $\mathbf{T}^n_\square/\equiv_F$ is the same as the ordinary homology of the torus $\mathbf{T}^n$ (cf. Section 2.1.4)

$$H_\bullet(\mathbf{T}^n_\square/\equiv_F) = H_\bullet(\mathbf{R}^k_\square/G_p) = H_\bullet(G_p) = H_\bullet(\mathbf{T}^n). \tag{2.87}$$

It should be interesting to compare these results with the analysis of the general *n-dimensional non-commutative torus* A_Θ [Ri2]. This is the C*-algebra generated by n unitary elements $u_1, ..., u_n$ under the relations $u_k u_h = \exp(2\pi i.\vartheta_{hk}).u_h u_k$ given by an antisymmetric matrix $\Theta = (\vartheta_{hk})$; it has the same K-groups as $\mathbf{T}^n$.

2.5.6 Remarks

The previous results also show that *it is not possible to preorder group-homology* so that the isomorphism $H_\bullet(G) \cong H_\bullet(X/G)$ (in (2.72)) be extended to $\uparrow H_\bullet(X/G)$: a group G can act *freely* on two *acyclic* cubical sets X_i producing *different preorders* on the groups $\uparrow H_n(X_i/G_\vartheta)$.

In fact, it is sufficient to take $G_\vartheta = \mathbf{Z} + \vartheta\mathbf{Z}$, as above, and recall that $\uparrow H_1(\mathbf{R}_\square/G_\vartheta)$ has a *chaotic preorder* (Corollary 2.4.4) while $\uparrow H_1(\uparrow\mathbf{R}_\square/G_\vartheta) = \uparrow G_\vartheta$ is totally ordered (Theorem 2.5.8).

We end this section by proving the main results on the directed homology of the rotation c-set $C_\vartheta = \uparrow\mathbf{R}_\square/G_\vartheta$, already announced in Section 2.5.2.

2.5.7 Lemma

Let ϑ, ϑ' be irrationals. Then $G_\vartheta = G_{\vartheta'}$, as subsets of $\mathbf{R}$, if and only if $\vartheta' \in \pm\vartheta + \mathbf{Z}$. Moreover the following conditions are equivalent

(a) $\uparrow G_\vartheta \cong \uparrow G_{\vartheta'}$ *as ordered groups;*
(b) *ϑ and ϑ' are conjugate under the action of* $\mathrm{GL}(2, \mathbf{Z})$ *(Section 2.5.1);*
(c) *ϑ' belongs to the closure $\{\vartheta\}_{RT}$ of $\{\vartheta\}$ under the transformations* $R(t) = t^{-1}$ *and* $T^{\pm1}(t) = t \pm 1$.

Further, these conditions imply the following one (which will be proved to be equivalent to them in Theorem 2.5.9)

(d) $\uparrow \mathbf{R}_\square / G_\vartheta \cong \uparrow \mathbf{R}_\square / G_{\vartheta'}$ *as c-sets.*

Note. The equivalence of the first three conditions is well known, within the classification of the C*-algebras A_ϑ up to strong Morita equivalence.

Proof First, if $G_\vartheta = G_{\vartheta'}$, then $\vartheta = a + b\vartheta'$ and $\vartheta' = c + d\vartheta$, whence $\vartheta = a + bc + bd\vartheta$ and $d = \pm 1$; the converse is obvious.

We have already seen, in Section 2.5.1, that (b) and (c) are equivalent, because the group $GL(2, \mathbf{Z})$ is generated by the matrices R, T (see (2.75)), which give the transformations $R(t) = t^{-1}$ and $T^k(t) = t + k$, on $\mathbf{R} \setminus \mathbf{Q}$ (for $k \in \mathbf{Z}$).

To prove that (c) implies (a) and (d), it suffices to consider the cases $\vartheta' = \vartheta + k$ and $\vartheta' = \vartheta^{-1}$. In the first case, $\uparrow G_\vartheta$ and $\uparrow G_{\vartheta'}$ coincide (as well as their action on $\uparrow \mathbf{R}_\square$); in the second, the isomorphism of c-sets

$$f \colon \uparrow \mathbf{R}_\square \to \uparrow \mathbf{R}_\square, \qquad f(t) = |\vartheta|.t,$$

restricts to an isomorphism $f' \colon \uparrow G_\vartheta \to \uparrow G_{\vartheta'}$, obviously consistent with the actions $(f(t + g) = f(t) + f'(g))$, and induces an isomorphism $\uparrow \mathbf{R}_\square / G_\vartheta \to \uparrow \mathbf{R}_\square / G_{\vartheta'}$.

We are left with proving that (a) implies (c). Let us begin by noting that any irrational number ϑ defines an algebraic isomorphism $\mathbf{Z}^2 \cong G_\vartheta$, which becomes an order isomorphism for the following ordered abelian group $\uparrow_\vartheta \mathbf{Z}^2$

$$\uparrow_\vartheta \mathbf{Z}^2 \to G_\vartheta, \qquad (a, b) \mapsto a + b\vartheta,$$
$$(a, b) \geqslant_\vartheta 0 \Leftrightarrow a + b\vartheta \geqslant 0.$$

Conversely, the number ϑ is (completely) determined by this order, as an upper bound in $\mathbf{R}$

$$\vartheta = \sup\{-a/b \mid a, b \in \mathbf{Z}, \, b > 0, \, (a, b) >_\vartheta 0\}. \qquad (2.88)$$

Take now an algebraic isomorphism $f \colon \mathbf{Z}^2 \to \mathbf{Z}^2$. Since $GL(2, \mathbf{Z})$ is generated by the matrices R and T, this isomorphism can be factorised as $f = f_n ... f_1$, with factors f_R, f_T^k

$$f_R(a, b) = (b, a), \qquad f_T^k(a, b) = (a + kb, b).$$

Let us replace ϑ with a positive representative in $\{\vartheta\}_{RT}$, which leaves $\uparrow_\vartheta \mathbf{Z}^2$ invariant up to isomorphism. Then f_R (resp. f_T^k) is an *order isomorphism* $\uparrow_\vartheta \mathbf{Z}^2 \to \uparrow_\zeta \mathbf{Z}^2$ with $\zeta = R(\vartheta)$ (resp. $\zeta = T^{-k}(\vartheta)$), still belonging to $\{\vartheta\}_{RT}$

$$(a,b) >_\vartheta 0 \iff a + b\vartheta > 0 \iff b + a\vartheta^{-1} > 0$$
$$\iff (b,a) >_\zeta 0 \qquad (\zeta = \vartheta^{-1}),$$
$$(a,b) >_\vartheta 0 \iff a + b\vartheta > 0 \iff a + kb + b(\vartheta - k) > 0$$
$$\iff (a + kb, b) >_\zeta 0 \qquad (\zeta = \vartheta - k).$$

Thus, $f = f_n...f_1$ can be viewed as an isomorphism $\uparrow_\vartheta \mathbf{Z}^2 \to \uparrow_\zeta \mathbf{Z}^2$ where ζ belongs to the closure $\{\vartheta\}_{RT}$.

Finally, given two irrational numbers ϑ, ϑ', an isomorphism $\uparrow G_\vartheta \cong \uparrow G_{\vartheta'}$ yields an isomorphism $\uparrow_\vartheta \mathbf{Z}^2 \to \uparrow_{\vartheta'} \mathbf{Z}^2$; but we have seen that the *same* algebraic isomorphism is an order isomorphism $\uparrow_\vartheta \mathbf{Z}^2 \to \uparrow_\zeta \mathbf{Z}^2$ where ζ belongs to the closure of $\{\vartheta\}$; by (2.88), $\vartheta' = \zeta$ and the thesis holds. □

2.5.8 Theorem (The homology of irrational rotation c-sets)

The c-set $\uparrow \mathbf{R}_\square$ (Section 2.5.2(b)) is acyclic.

The directed homology of the irrational rotation c-set $C_\vartheta = \uparrow \mathbf{R}_\square / G_\vartheta$ is the homology of $\mathbf{T}^2$, with a total order on $\uparrow H_1$ and a chaotic preorder on $\uparrow H_2$

$$\uparrow H_1(C_\vartheta) = \uparrow G_\vartheta = \uparrow(\mathbf{Z} + \vartheta\mathbf{Z}) \qquad (G_\vartheta^+ = G_\vartheta \cap \mathbf{R}^+),$$
$$\uparrow H_2(C_\vartheta) = \uparrow_c \mathbf{Z}, \tag{2.89}$$

and obviously $\uparrow H_0(C_\vartheta) = \uparrow \mathbf{Z}$.

The first isomorphism above has a simple description on the positive cone $G_\vartheta \cap \mathbf{R}^+$

$$j: \uparrow G_\vartheta \to \uparrow H_1(C_\vartheta), \qquad j(\rho) = [pa_\rho] \quad (\rho \in G_\vartheta \cap \mathbf{R}^+),$$
$$a_\rho: \mathbf{I} \to \mathbf{R}, \qquad a_\rho(t) = \rho t, \tag{2.90}$$

where $p: \uparrow \mathbf{R}_\square \to \uparrow \mathbf{R}_\square / G_\vartheta$ is the canonical projection.

Proof　First, let us consider the c-subset $\uparrow[x, +\infty[$ of $\uparrow \mathbf{R}_\square$ (for $x \in \mathbf{R}$) and the following left homotopy f of cubical sets (cf. (1.152); noting that it *does* preserve directed cubes)

$$f_n \colon c_n(\uparrow[x, +\infty[) \to c_{n+1}(\uparrow[x, +\infty[),$$

$$f_n(a)(t_1, ..., t_{n+1}) = x + t_1.(a(t_2, ..., t_{n+1}) - x),$$

$$\partial^\alpha_{i+1} f_n = f_{n-1} \partial^\alpha_i, \qquad f_n e_i = e_{i+1} f_{n-1}.$$

Computing its faces in direction 1, f is a homotopy from the map $f^- = \partial^-_1 f$, which is constant at x, to the identity $f^+ = \partial^+_1 f = \mathrm{id}\uparrow[x, +\infty[$. This proves that every c-set $\uparrow[x, +\infty[$ is past contractible (to its minimum x), hence acyclic. Since cubes of $\uparrow\mathbf{R}_\square$ have a compact image in the line, it follows easily that also $\uparrow\mathbf{R}_\square$ is acyclic.

Now, Theorem 2.4.3 proves that the cubical homology of $\uparrow\mathbf{R}_\square/G_\vartheta$ coincides, algebraically, with the homology of the group G_ϑ, or equivalently of the torus $\mathbf{T}^2$. It also proves that $H_1(C_\vartheta)$ is generated by the homology classes $[pa_1]$ and $[pa_\vartheta]$. Since $[pa_{\rho+\rho'}] = [pa_\rho] + [pa_{\rho'}]$, the mapping j in (2.90) is an algebraic isomorphism. By construction, it preserves preorders, and we still have to prove that it reflects it.

To simplify the argument, a 1-chain z of $\uparrow\mathbf{R}_\square$ which projects to a cycle $p_\sharp(z)$ in C_ϑ, or a boundary, will be called a *pre-cycle* or a *pre-boundary*, respectively. (Note that, since $p_\sharp$ is surjective, the homology of C_ϑ is isomorphic to the quotient of pre-cycles modulo pre-boundaries.)

Let $z = \sum_i \lambda_i a_i$ be a positive pre-cycle, with all $\lambda_i > 0$; let us call the sum $\lambda = \sum_i \lambda_i$ its *weight*. We have to prove that z is equivalent to a positive combination of pre-cycles of type a_ρ ($\rho \in G^+_\vartheta$), modulo pre-boundaries.

Let us decompose $z = z' + z''$, putting in z' all the summands $\lambda_i a_i$ which are pre-cycles themselves, and replace any such a_i, up to pre-boundaries, with a_{ρ_i} (see (2.90)), where $\rho_i = \partial^+ a_i - \partial^- a_i \in G^+_\vartheta$. If $z'' = 0$ we are done, otherwise $z'' = z - z'$ is still a pre-cycle; let us act on it. Reorder its paths a_i so that a_1 (the first) has a minimal coefficient λ_1 (strictly positive); since $\partial^+ a_1$ has to annihilate in $\partial p_\sharp(z')$, there is some a_i ($i > 1$) with $\partial^+ a_1 - \partial^- a_i \in G_\vartheta$. By a G_ϑ-translation of a_i (leaving pa_i unaffected), we can assume that $\partial^- a_i = \partial^+ a_1$, and then replace (modulo pre-boundaries) $\lambda_1 a_1 + \lambda_i a_i$ with $\lambda_1 \hat{a}_1 + (\lambda_i - \lambda_1)a_i$ where $\hat{a}_1 = a_1 * a_i$ is a concatenation (and $\lambda_i - \lambda_1 \geqslant 0$). Now, the new weight is $\lambda - \lambda_1 < \lambda$, strictly less than the previous one.

Continuing this way, the procedure ends in a finite number of steps; this means that, modulo pre-boundaries, we have modified z into a positive combination of pre-cycles of the required form, a_ρ.

Finally, we already know that the group $\uparrow H_2(C_\vartheta) = \mathbf{Z}$ gets the chaotic preorder, by Proposition 2.2.6(c).   □

2.5.9 Theorem (Classifying the irrational rotation c-sets)

The c-sets $C_\vartheta = \uparrow \mathbf{R}_\square/G_\vartheta$ and $C_{\vartheta'}$ are isomorphic if and only if the ordered groups $\uparrow G_\vartheta$ and $\uparrow G_{\vartheta'}$ are isomorphic, if and only if ϑ and ϑ' are conjugate under the action of $\mathrm{GL}(2,\mathbf{Z})$ (see (2.74)), if and only if ϑ' belongs to the closure $\{\vartheta\}_{RT}$ (see (2.75)).

Proof Follows immediately from Lemma 2.5.7 and Theorem 2.5.8, which gives the missing implication of the lemma: if our c-sets are isomorphic, also their ordered groups $\uparrow H_1$ are, whence $\uparrow G_\vartheta \cong \uparrow G_{\vartheta'}$.   □

2.6 Directed homology theories

We briefly consider directed homology for inequilogical spaces (Section 2.6.1), and its 'defective' character with respect to preorder, as for d-spaces (cf. Sections 2.2.5–2.2.7).

We end with the axioms for a *perfect directed homology theory*, which have already been verified above for pointed cubical sets. In stronger settings, we will be able to reduce this axiomatic system to a simpler, equivalent one (Theorem 4.7.6).

2.6.1 Directed singular homology of inequilogical spaces

For the category p**Eql** of inequilogical spaces (Section 1.9.1), we have a directed singular homology, defined in a similar way as for d-spaces (Section 2.2.5)

$$\uparrow\square: \text{p}\mathbf{Eql} \to \mathbf{Cub}, \qquad \uparrow\square_n X = \text{p}\mathbf{Eql}(\uparrow\mathbf{I}^n, X),$$
$$\uparrow H_n: \text{p}\mathbf{Eql} \to \text{p}\mathbf{Ab}, \qquad \uparrow H_n(X) = \uparrow H_n(\uparrow\square X). \tag{2.91}$$

Of course, $\uparrow\mathbf{I}^n$ is now viewed in p**Eql**: the ordered n-cube $\uparrow\mathbf{I}^n$ with the equality relation.

This theory has been studied in [G11], showing interactions with non-commutative geometry similar to those of Section 2.5.

As for cubical sets, homotopy invariance holds; there is an exact Mayer–Vietoris sequence, whose differential does *not* preserve preorders (cf. Theorem 2.2.2), while excision gives an isomorphism of preordered abelian groups (cf. Theorem 2.2.3).

Furthermore, as for d-spaces, the existence of the transposition symmetry $s\colon \uparrow\mathbf{I}^2 \to \uparrow\mathbf{I}^2$ and of the lower connection $g^-\colon \mathbf{I}^2 \to \mathbf{I}$ entails the fact that the preorder of $\uparrow H_n(X)$ is always chaotic, for $n \geqslant 2$ (Proposition 2.2.6(c)). As a consequence, also here suspension cannot agree with the preorder of homology.

As shown in [G11], 3.5, the directed homology of the inequilogical spheres $\uparrow\overline{\mathbf{S}}_e^n$ (see (1.219)) yields the usual algebraic groups. Their $\uparrow H_0$ is always $\uparrow\mathbf{Z}$, for $n > 0$, and:

$$\uparrow H_1(\overline{\mathbf{S}}_e^1) = \uparrow\mathbf{Z}.$$

But, for all $n \geqslant 2$, $\uparrow H_n(\overline{\mathbf{S}}_e^n)$ is the group of integers with the *chaotic* preorder.

As we have seen, these drawbacks are directly related to the fact that the transposition symmetry s subsists in p**Eql**: as for d-spaces, the directed structure of inequilogical spaces distinguishes directed *paths* in an effective way, but can only distinguish *higher cubes* through directed paths; this is not sufficient to get good results for $\uparrow H_n$, with $n > 1$.

2.6.2 Axioms for directed homology

A *theory of (reduced) directed homology* on a dI1-homotopical category **A**, with values in the category p**Ab** of preordered abelian groups (2.1.1), will be a pair $\uparrow H = ((\uparrow H_n), (h_n))$ subject to the following axioms.

(dhlt.0) (*naturality*) The data consist of functors $\uparrow H_n$ and natural transformations h_n involving the suspension Σ of **A**

$$\uparrow H_n\colon \mathbf{A} \to \mathrm{p}\mathbf{Ab}, \qquad h_n\colon \uparrow H_n \to \uparrow H_{n+1}\Sigma \qquad (n \in \mathbf{Z}). \tag{2.92}$$

$\uparrow H_n$ is called a (reduced) *homology* functor; the preorder-preserving homomorphism associated to a map f is generally written as $f_{*n} = \uparrow H_n(f)$, or just f_*.

(dhlt.1) (*homotopy invariance*) If there is a homotopy $f \to g$ in **A**, then $f_{*n} = g_{*n}$ (for all n).

(dhlt.2) (*algebraic stability*) Every component

$$h_n X\colon \uparrow H_n X \to \uparrow H_{n+1}(\Sigma X)$$

is an algebraic isomorphism (of abelian groups).

(dhlt.3) (*exactness*) For every $f\colon X \to Y$ in **A** and every n, the following sequence is *exact* in p**Ab**

$$\uparrow H_n X \xrightarrow{f_*} \uparrow H_n Y \xrightarrow{u_*} \uparrow H_n C^- f \xrightarrow{\delta_*} \uparrow H_n \Sigma X \xrightarrow{(\Sigma f)_*} \uparrow H_n \Sigma Y \tag{2.93}$$

where $u\colon Y \to C^-f$ is the lower homotopy cokernel of f and $\delta = d_-f\colon C^-f \to \Sigma X$ denotes the differential of the corresponding Puppe sequence (Section 1.7.8).

Then, the exactness of the corresponding upper sequence of f comes from the exactness of the lower sequence of f^{op}, see (1.192). Notice that the components $h_n X\colon {\uparrow} H_n X \to {\uparrow} H_{n+1}\Sigma X$ are *algebraic* isomorphisms which preserve preorder but need not reflect it.

If the dl1-structure of **A** is concrete (Section 1.2.4), with representative object E, the homology theory ${\uparrow} H_n$ is said to be *ordinary* if it satisfies a further axiom

(dhlt.4) (*dimension*) ${\uparrow} H_n (E) = 0$, for all $n \neq 0$.

In this case, ${\uparrow} H_0 (E)$ is called the *preordered* (abelian) *group of coefficients*.

Forgetting everything about preorders, we get the classical notion of a (reduced) *homology theory* on **A**.

2.6.3 Homology sequences and perfect theories

As a consequence of these axioms, every map $f : X \to Y$ has a lower (and an upper) exact *homology sequence* of preordered abelian groups

$$\ldots\ {\uparrow} H_n(X) \xrightarrow{f_*} {\uparrow} H_n(Y) \xrightarrow{u_*} {\uparrow} H_n(C^-f) \xrightarrow{d_n} {\uparrow} H_{n-1}(X)$$

$$\ldots\ {\uparrow} H_0(X) \longrightarrow {\uparrow} H_0(Y) \longrightarrow {\uparrow} H_0(C^-f) \longrightarrow 0. \tag{2.94}$$

The differential d_n is the following *algebraic homomorphism* (which need *not* preserve preorders)

$$d_n = (h_{n-1}X)^{-1}.(d^-f)_{*n}\colon {\uparrow} H_n(C^-f) \to {\uparrow} H_n(\Sigma X) \to {\uparrow} H_{n-1}(X).$$

Exactness of (2.94) (a merely algebraic condition) is proved by the following diagram with exact rows

$$
\begin{array}{ccccccc}
{\uparrow} H_n X & \xrightarrow{f_*} & {\uparrow} H_n Y & \xrightarrow{u_*} & {\uparrow} H_n C^-f & \xrightarrow{\delta_*} & {\uparrow} H_n \Sigma X & \longrightarrow & {\uparrow} H_n \Sigma Y \\
& & & & {\scriptstyle d_n}\nwarrow & & \uparrow{\scriptstyle h} & & \uparrow{\scriptstyle h} \\
& & & & & {\uparrow} H_{n-1}X & \underset{f_*}{\to} & {\uparrow} H_{n-1}Y
\end{array}
\tag{2.95}
$$

We speak of a *perfect* theory of directed homology when the components $h_n X$ also reflect preorder, i.e. are isomorphisms in **pAb**, or equivalently:

(dhlt.2′) (*full stability*) $h_n: \uparrow H_n \to \uparrow H_{n+1}\Sigma: \mathbf{A} \to p\mathbf{Ab}$ is a functorial isomorphism.

Then also the differentials d_n and the exact homology sequence (2.94) are in $p\mathbf{Ab}$.

2.6.4 Examples

We have encountered only one perfect directed homology theory, the directed homology of *pointed cubical sets* (Section 2.3.3), with respect to the left d-structure of $\mathbf{Cub}_\bullet$, defined by the left cylinder (Section 2.3.2). The axioms have already been verified: *homotopy invariance* in Theorem 2.3.4, *stability* in Theorem 2.3.5, *exactness* in Theorem 2.3.6. (The right d-structure gives the same results.)

This theory is ordinary, with *ordered integral coefficients*: $\uparrow H_0(\mathbf{s}^0, 0) = \uparrow \mathbf{Z}$. As in Section 2.1.3, one can deduce a perfect theory with coefficients in an arbitrary preordered abelian group $\uparrow L$

$$\uparrow \mathrm{Ch}_+(-; \uparrow L): \mathbf{Cub}_\bullet \to d\mathrm{Ch}_+ \mathbf{Ab},$$
$$\uparrow \mathrm{Ch}_+(X, x_0; \uparrow L) = \uparrow \mathrm{Ch}_+(X, x_0) \otimes \uparrow L,$$
$$\uparrow H_n(-; \uparrow L): \mathbf{Cub}_\bullet \to p\mathbf{Ab},$$
$$\uparrow H_n(X, x_0; \uparrow L) = \uparrow H_n(\uparrow \mathrm{Ch}_+(X, x_0; \uparrow L)).$$

Also the reduced directed homology $\uparrow \tilde{H}_n: \mathbf{Cub} \to p\mathbf{Ab}$, defined in Section 2.3.1, is a directed homology theory for cubical sets, and likely a perfect one. But we have seen that its preorder is much less interesting.

The reduced homology of d-spaces (Section 2.2.5) and inequilogical spaces (Section 2.6.1), and their pointed analogues, are non-perfect directed homology theories.

2.6.5 The singular cubical set

Let $\mathbf{A}$ be a concrete dI1-category, with (representable) forgetful functor $U = \mathbf{A}(E, -): \mathbf{A} \to \mathbf{Set}$. We have already seen that the object $\mathbf{I} = I(E)$ acquires the structure of a dI1-interval (1.2.4). Its faces, degeneracy and reflection are written also here as:

$$\partial^\alpha: E \rightrightarrows \mathbf{I}: e, \qquad r: \mathbf{I} \to \mathbf{I}^{\mathrm{op}} \qquad (\alpha = \pm).$$

But actually we have a cocubical object in $\mathbf{A}$, extending the left diagram above

$$\mathbf{I}^{(n)} = I^n(E), \qquad \partial_i^\alpha = I^{i-1}\partial^\alpha I^{n-i} \colon \mathbf{I}^{(n-1)} \to \mathbf{I}^{(n)},$$
$$e_i = I^{i-1}e I^{n-i} \colon \mathbf{I}^{(n)} \to \mathbf{I}^{(n-1)}, \qquad (\alpha = \pm; \; i = 1, ..., n). \tag{2.96}$$

Therefore, applying the contravariant functor $\mathbf{A}(-, X)\colon \mathbf{A}^* \to \mathbf{Set}$, every object X has a cubical set $\square X$ of *singular cubes*, of which points and paths form the components of degree 0 and 1, respectively

$$\square_n X = \mathbf{A}(\mathbf{I}^{(n)}, X),$$
$$\partial_i^\alpha X = \mathbf{A}(\partial_i^\alpha, X), \qquad e_i X = \mathbf{A}(e_i, X). \tag{2.97}$$

This defines a functor $\square \colon \mathbf{A} \to \mathbf{Cub}$, which one can use to define the singular directed homology of $\mathbf{A}$

$$\uparrow H_n \colon \mathbf{A} \to \mathrm{pAb}, \qquad \uparrow H_n(X) = \uparrow H_n(\square X). \tag{2.98}$$

If, moreover, $\mathbf{A}$ is a pointed category, then every hom-set $\mathbf{A}(A, X)$ can be pointed at the zero-map, yielding contravariant functors $\mathbf{A}(-, X)\colon \mathbf{A}^* \to \mathbf{Set}_{\bullet}$. Now, the cocubical object (2.96) yields functors

$$\mathbf{A} \to \mathbf{Cub}_{\bullet},$$

which can be composed with the directed homology of pointed cubical sets (Section 2.3.3), a perfect theory.

3

Modelling the fundamental category

In classical algebraic topology, homotopy equivalence between 'spaces' gives rise to a plain equivalence of their fundamental groupoids; therefore, the categorical skeleton provides a minimal model of the latter.

But the study of homotopy invariance in directed algebraic topology is far richer and more complex. Our directed structures have a fundamental category $\uparrow\Pi_1(X)$, and this must be studied up to appropriate notions of *directed* homotopy equivalence of categories, which are more general than categorical equivalence.

We shall use two (dual) directed notions, which take care, respectively, of variation 'in the future' or 'from the past': a *future equivalence* in **Cat** is a future homotopy equivalence (Section 1.3.1) satisfying two conditions of coherence; it can also be seen as a symmetric version of an adjunction, with two units. Its dual, a *past equivalence*, has two counits. Then we study how to combine these two notions, so as to take into account both kinds of invariance. *Minimal models* of a category, up to these equivalences, are then introduced to better understand the 'shape' and properties of the category we are analysing, as well as of the process it represents.

Within category theory, the study of future (and past) equivalences is a sort of 'variation on adjunctions': they compose as the latter (Section 3.3.3) and, moreover, two categories are future homotopy equivalent if and only if they can be embedded as full reflective subcategories of a common one (Theorem 3.3.5); therefore, a property is invariant for future equivalences if and only if it is preserved by full reflective embeddings and by their reflectors.

After introducing the fundamental category of a d-space in Section 3.2, Section 3.3 introduces and studies future and past equivalences.

Then, in the next three sections, we combine such equivalences, dealing with injective and projective models. Section 3.7 investigates *future invariant properties*, like future regular points and future branching ones.

In the next two sections, these properties are used to define and study *pf-spectra*, which give a minimal injective model and an associated projective one. In Section 3.9, we compute these invariants for the fundamental category of various ordered spaces (or preordered, in Section 3.9.5). Hints to possible applications outside of concurrency can be found in Section 3.9.9.

The material of this chapter essentially comes from [G8, G14]. A study of higher fundamental categories, begun in [G15, G16, G17], is still at the level of research and will not be dealt with here.

3.1 Higher properties of homotopies of d-spaces

After the basic properties of the cylinder and path functors of d**Top**, developed in Section 1.5, we examine here their higher structure, which will be used below to define the fundamental category of a d-space. This structure will be formalised in an abstract setting for directed algebraic topology, in Chapter 4.

3.1.1 An example

An elementary example will give some idea of the analysis which will be developed in this chapter. Let us consider the 'square annulus' $X \subset \uparrow[0,1]^2$ represented below, i.e. the ordered compact subspace of the standard ordered square $\uparrow[0,1]^2$, which is the complement of the *open* square $]1/3, 2/3[^2$ (marked with a cross)

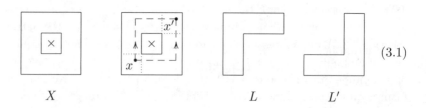

$$X \qquad\qquad L \qquad L' \qquad\qquad (3.1)$$

As we will see, the fundamental category $C = \uparrow\Pi_1(X)$ has *some* arrow $x \to x'$ provided that $x \leqslant x'$ and both points are in L or L' (the closed subspaces represented above). Precisely, there are *two* arrows when

$x \leqslant p = (1/3, 1/3)$ and $x' \geqslant q = (2/3, 2/3)$ (as in the second figure above), and *one* otherwise. (This evident fact can be easily proved with a 'van Kampen' type theorem, using precisely the subspaces L, L'; see Section 3.2.7(f)).

Thus, the whole category C is easy to visualise and 'essentially represented' by the full subcategory E on four vertices $0, p, q, 1$ (the central cell does not commute)

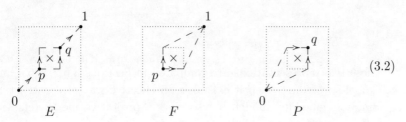

$$(3.2)$$

But E is far from being equivalent to C, as a category: C *is already a skeleton*, in the ordinary sense (Section 3.6.1), and one cannot reduce it up to category equivalence. On the other hand, it obviously contains a huge amount of redundant information, which we want to reduce to some essential model.

The procedure which we are to establish, in order to model C, begins by determining the least *full reflective* subcategory F of C, so that F is future equivalent to C and minimal as such; in this example, its objects are a *future branching* point p (where one must choose between different ways out of it) and a *maximal* point 1 (where one cannot further progress); they form the *future spectrum* $\mathrm{sp}^+(C) = \{p, 1\}$. Similarly, we determine the *past spectrum* P, i.e. the least *full coreflective* subcategory, whose objects form the *past spectrum* $\mathrm{sp}^-(C) = \{0, q\}$.

E is now the full subcategory of C on $\mathrm{sp}(C) = \mathrm{sp}^-(C) \cup \mathrm{sp}^+(C)$, called the *spectral injective model* of X. It is a minimal embedded model, in a sense which will be made precise.

The situation can now be analysed as follows, in E:

- the action begins at 0, from where we move to p;
- p is an (effective) future branching point, where we have to choose between two paths;
- which join at q, an (effective) past branching point;
- from where we can only move to 1, where the process ends.

An alternative description will be obtained with the associated *projective model* M, the full subcategory of the category C^2 (of morphisms of C) on the four maps $\lambda, \mu, \sigma, \tau$, obtained from a canonical factorisation of the composed adjunction $P \rightleftarrows C \rightleftarrows F$ (cf. Theorem 3.5.7)

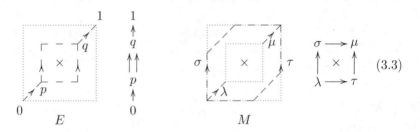

$$(3.3)$$

These two representations are compared in Sections 3.6.2, 3.6.4 and 3.9. A pf-spectrum (when it exists) is an effective way of constructing a *minimal* embedded model; it also gives a projective model (cf. Section 3.8.5 and Theorem 3.8.8)

This study has similarities with other recent ones, using categories of fractions [FRGH] or generalised quotients of categories [GH], for the same goal: to construct a 'minimal model' of the fundamental category; the models obtained in these two papers are often similar to the *projective* models considered here. See also [Ra2].

Notice: as already warned at the end of Section 1.2.1, one should not confuse the 'reflector' $p\colon C \to C_0$ of a full reflective subcategory $C_0 \subset C$ (left adjoint to the inclusion functor) with the 'reflection' of a cylinder (or cocylinder) functor, which is a natural transformation $r\colon IR \to RI$ (or $r\colon RP \to PR$).

3.1.2 Directed homotopy invariance

Let us summarise the problem we want to analyse.

In algebraic topology, the fundamental groupoid $\Pi_1(X)$ of a topological space is homotopy invariant in a clear sense: a homotopy $\varphi\colon f \to g\colon X \to Y$ gives an isomorphism of the associated functors $f_*, g_*\colon \Pi_1(X) \to \Pi_1(Y)$, so that a homotopy equivalence $X \simeq Y$ yields *an equivalence of groupoids* $\Pi_1(X) \simeq \Pi_1(Y)$. Thus, a one-dimensional homotopy model of the space is its fundamental groupoid, *up to groupoid-equivalence*; if we want a *minimal model*, we can always take a skeleton of the latter (choosing one point in each path component of the space).

In directed algebraic topology, homotopy invariance requires a deeper analysis. A d-space has a fundamental category $\uparrow\!\Pi_1(X)$, and a homotopy

$\varphi \colon f \to g \colon X \to Y$ only gives a natural transformation between the associated functors

$$\varphi_* \colon f_* \to g_* \colon \uparrow\!\Pi_1(X) \to \uparrow\!\Pi_1(Y),$$
$$\varphi_* x = [\varphi(x, -)] \colon f(x) \to g(x) \qquad (x \in X), \tag{3.4}$$

which, generally, is *not* invertible, because the paths $\varphi(x, -) \colon \uparrow\!\mathbf{I} \to Y$ need not be reversible.

Thus, a future homotopy equivalence of spaces only becomes a *future homotopy equivalence of categories*.

Ordinary equivalence of categories (Section A1.5) is not, by far, sufficient to 'link' categories having – loosely speaking – the same aspect; and the problem of defining and constructing *minimal models* is important, both theoretically, for directed algebraic topology, and in applications.

3.1.3 The higher structure of the cylinder

The directed interval $\uparrow\!\mathbf{I} = \uparrow[0,1]$ has a rich structure in p**Top** and d**Top**, which extends the basic structure already seen in Chapter 1, for preordered spaces (Section 1.1.4) and d-spaces (Section 1.5.1). (From a formal point of view, the extended structure will be defined and studied in Section 4.2.)

Thus, after faces (∂^-, ∂^+), degeneracy (e) and reflection (r), we have a second-order structure which applies to the standard square $\uparrow\!\mathbf{I}^2$, namely two *connections* or main operations (g^-, g^+) and a *transposition* (s), as already considered in **Top** (Section 1.1.0)

$$\{*\} \overset{\partial^\alpha}{\underset{e}{\rightrightarrows}} \uparrow\!\mathbf{I} \overset{g^\alpha}{\longleftarrow} \uparrow\!\mathbf{I}^2 \qquad \uparrow\!\mathbf{I} \overset{r}{\longrightarrow} \uparrow\!\mathbf{I}^{\mathrm{op}} \qquad \uparrow\!\mathbf{I}^2 \overset{s}{\longrightarrow} \uparrow\!\mathbf{I}^2$$

$$\partial^\alpha(*) = \alpha, \quad g^-(t, t') = \max(t, t'), \quad g^+(t, t') = \min(t, t'),$$
$$r(t) = 1 - t, \quad s(t, t') = (t', t). \tag{3.5}$$

(Recall that the reflection is expressed via the reversor $R \colon \mathrm{d}\mathbf{Top} \to \mathrm{d}\mathbf{Top}$, $R(X) = X^{\mathrm{op}}$ yielding the opposite d-space, with the reversed distinguished paths.)

As a consequence, the (directed) *cylinder* endofunctor

$$I \colon \mathrm{d}\mathbf{Top} \to \mathrm{d}\mathbf{Top}, \qquad I(-) = -\times\uparrow\!\mathbf{I}, \tag{3.6}$$

has natural transformations, written as above

$$1 \overset{\partial^\alpha}{\underset{e}{\rightrightarrows}} I \overset{g^\alpha}{\longleftarrow} I^2 \qquad IR \overset{r}{\longrightarrow} RI \qquad I^2 \overset{s}{\longrightarrow} I^2 \tag{3.7}$$

linked by algebraic equations which will be considered later (Section 4.2.1).

Consecutive homotopies will be pasted via the *concatenation pushout* of the cylinder functor (cf. Lemma 1.4.9)

$$
\begin{array}{ccc}
X & \xrightarrow{\partial^+} & IX \\
\partial^- \downarrow & \dashrightarrow{\ c^-} & \downarrow c^- \\
IX & \xrightarrow{c^+} & IX
\end{array}
\qquad
\begin{aligned}
c^-(x,t) &= (x, t/2), \\[2ex]
c^+(x,t) &= (x, (t+1)/2).
\end{aligned}
\qquad (3.8)
$$

3.1.4 The higher structure of the path functor

As a consequence of Theorem 1.4.8, the directed interval $\uparrow\!I$ is exponentiable: the cylinder functor $I = -\times\uparrow\!I$ has a right adjoint, the (directed) *path functor*, or *cocylinder* P. Explicitly, in this functor

$$
P\colon \mathrm{d}\mathbf{Top} \to \mathrm{d}\mathbf{Top}, \qquad P(Y) = Y^{\uparrow\!I}, \qquad (3.9)
$$

the d-space $Y^{\uparrow\!I}$ is the set of d-paths $\mathrm{d}\mathbf{Top}(\uparrow\!I, Y)$, equipped with the compact-open topology (induced by the topological path-space $P(UY) = \mathbf{Top}(\mathbf{I}, UY)$) and the following d-structure.

A path

$$
c\colon \mathbf{I} \to \mathrm{d}\mathbf{Top}(\uparrow\!I, Y) \subset \mathbf{Top}(\mathbf{I}, UY),
$$

is directed if and only if, for all increasing maps $h, k\colon \mathbf{I} \to \mathbf{I}$, the consequent path $t \mapsto c(h(t))(k(t))$ is in dY. Notice that $P^2(Y) = Y^{\uparrow\!I^2}$, by the closure of adjunction under composition.

The lattice structure of $\uparrow\!I$ in $\mathrm{d}\mathbf{Top}$ gives – contravariantly – a dual structure on P; its natural transformations, mates to the transformations of I (Section A5.3), are denoted as the latter, but go in the opposite direction

$$
1 \underset{e}{\overset{\partial^\alpha}{\rightleftarrows}} P \overset{g^\alpha}{\rightrightarrows} P^2 \qquad RP \xrightarrow{r} PR \qquad P^2 \xrightarrow{s} P^2 \qquad (3.10)
$$

$$
\begin{aligned}
\partial^\alpha(a) &= a(\alpha), & e(x)(t) &= x, \\
g^-(a)(t,t') &= a(\max(t,t')), & g^+(a)(t,t') &= a(\min(t,t')), \\
r(a)(t) &= a(1-t), & s(a)(t,t') &= a(t',t).
\end{aligned}
$$

Again, the *concatenation pullback* (the object of pairs of consecutive paths) can be realised as PY

$$
\begin{array}{ccc}
PY & \xrightarrow{c^+} & PY \\
c^- \downarrow\,\text{-- -}\!\! & & \downarrow \partial^- \\
PY & \xrightarrow[\partial^+]{} & Y
\end{array}
\qquad
\begin{aligned}
c^-(a)(t) &= a(t/2), \\[1ex]
c^+(a)(t) &= a((t+1)/2).
\end{aligned}
\tag{3.11}
$$

We defer to the end of this section the (easy) verification that (3.11) is indeed a pullback (Lemma 3.1.7(b)). One can also deduce this fact from general results on 'mates' and (co)limits (Section A5.4).

3.1.5 Concatenation of paths

Recall that, in a d-space X, a path is a map $a\colon {\uparrow}\mathbf{I} \to X$ and amounts to a distinguished path of the structure, i.e. an element of dX (Section 1.4.6). Their concatenation is the usual one, described in Section 1.4.6: given a consecutive path b

$$
(a+b)(t) = \begin{cases} a(2t) & \text{for } 0 \leqslant t \leqslant 1/2, \\ b(2t-1), & \text{for } 1/2 \leqslant t \leqslant 1. \end{cases}
\tag{3.12}
$$

The concatenation of paths is actually 'written' inside the concatenation pullback (3.11).

Let us also recall that the path $a\colon {\uparrow}\mathbf{I} \to X$ is said to be *reversible* (see (1.108)) if also $(-a)(t) = a(1-t)$ is a directed path in X, or equivalently if $a\colon \mathbf{I}^{\sim} \to X$ is a d-map on the reversible interval (Section 1.4.4(c)). Such paths are closed under concatenation.

Since concatenation is not (strictly) associative, and the constant paths $0_x\colon x \to x$ are not neutral for it, we need two-dimensional homotopies of paths, which will be analysed in the next section.

3.1.6 Homotopies of d-spaces

As we have already seen in Section 1.2, a (directed) *homotopy* $\varphi\colon f \to g\colon X \to Y$ is defined as a d-map $\hat{\varphi}\colon IX = X \times {\uparrow}\mathbf{I} \to Y$ whose two faces, $\partial^\alpha(\varphi) = \hat{\varphi}.\partial^\alpha\colon X \to Y$ are f and g, respectively. Equivalently, it is a map $X \to PY = Y^{{\uparrow}\mathbf{I}}$, with faces as above. A path is a homotopy between two points, $a\colon x \to x'\colon \{*\} \to X$.

After *trivial homotopies, reflection of homotopies* (Section 1.2.1) and *whisker composition* of maps and homotopies (Section 1.2.3), we also have the *concatenation of (consecutive) homotopies*: $\varphi + \psi \colon f \to h$, defined in the usual way, by means of the concatenation pushout (3.8)

$$(\varphi + \psi)c^- = \varphi, \qquad (\varphi + \psi)c^+ = \psi \qquad (\partial^+ \varphi = \partial^- \psi),$$

$$(\varphi + \psi)(x, t) = \begin{cases} \varphi(x, 2t), & \text{for } 0 \leqslant t \leqslant 1/2, \\ \psi(x, 2t - 1), & \text{for } 1/2 \leqslant t \leqslant 1. \end{cases} \tag{3.13}$$

The endofunctors I and P can be extended to homotopies, *via their transposition*: for $\varphi \colon f \to g \colon X \to Y$, let

$$(I\varphi)\hat{} = I(\hat{\varphi}).sX \colon I^2(X) \to I(Y),$$
$$(P\varphi)\hat{} = sY.P(\hat{\varphi}) \colon P(X) \to P^2(Y),$$
$$(I\varphi)\hat{}(\partial^- IX) = (\hat{\varphi} \times \uparrow\mathbf{I}).(X \times s).(X \times \uparrow\mathbf{I} \times \partial^-) \tag{3.14}$$
$$= (\hat{\varphi} \times \uparrow\mathbf{I}).(X \times \partial^- \times \uparrow\mathbf{I}) = f \times \uparrow\mathbf{I} = I(f).$$

More formally, this comes from the fact that the cylinder functor $I \colon \mathrm{d}\mathbf{Top} \to \mathrm{d}\mathbf{Top}$ is a strong dl1-functor (see Section 1.2.6) via the natural transformations $i = r^{-1} \colon RI \to IR$ and $s \colon I^2 \to I^2$ (see Section 4.1.5).

Recall that homotopy of d-spaces $\varphi \colon f \to g \colon X \to Y$ is said to be reversible (Section 1.5.2) if also the mapping $(-\varphi)(x, t) = \varphi(x, 1 - t)$ is a d-map $X \times \uparrow\mathbf{I} \to Y$. This gives a directed homotopy

$$-\varphi \colon g \to f \colon X \to Y,$$

not to be confused with the *reflected homotopy* $\varphi^{\mathrm{op}} \colon g^{\mathrm{op}} \to f^{\mathrm{op}} \colon X^{\mathrm{op}} \to Y^{\mathrm{op}}$ (1.38), which always exist.

3.1.7 Lemma (From pushouts to pullbacks)

(a) Let us assume that the left diagram below is a pushout in $\mathrm{d}\mathbf{Top}$, *preserved by all products* $X \times -$

$$
\begin{array}{ccc}
A \xrightarrow{\ f\ } B & \qquad & Y^A \xleftarrow{\ f^*\ } Y^B \\
\end{array}
\tag{3.15}
$$

$$
\begin{array}{ccccc}
A & \xrightarrow{f} & B & & Y^A & \xleftarrow{f^*} & Y^B \\
\downarrow{g} & & \downarrow{h} & & \uparrow{g^*} & & \uparrow{h^*} \\
C & \xrightarrow{k} & D & & Y^C & \xleftarrow{k^*} & Y^D
\end{array}
$$

If its objects A, B, C, D *have a locally compact Hausdorff topology, and* Y *is an arbitrary d-space, the contravariant functor* $Y^{(-)}$ *transforms the pushout into a pullback, the right square above.*

(b) (Concatenation pullback for the cocylinder of d-spaces) *The diagram* (3.11) *is a pullback.*

Proof (a) It is an easy consequence of Theorem 1.4.8 on exponentiable d-spaces.

Given two maps $u\colon X \to Y^B$, $v\colon X \to Y^C$ in d**Top** such that $f^*u = g^*v$, the exponential law yields two maps $u'\colon X \times B \to Y$, $v'\colon X \times C \to Y$ such that $u'(X \times f) = v'(X \times g)$; by hypothesis, there is one map $w'\colon X \times D \to Y$ such that $w'(X \times h) = u'$ and $w'(X \times k) = v'$, which means precisely one map $w\colon X \to Y^D$ which solves the pullback problem: $h^*w = u$, $k^*w = v$.

(b) Apply the previous point to the standard concatenation pushout (1.106), which pastes two copies of $\uparrow\mathbf{I}$ one after the other; the preservation hypothesis holds, by Lemma 1.4.9. $\qquad\square$

.

3.2 The fundamental category of a d-space

We introduce the fundamental category of a d-space. Computations are based on a van Kampen-type theorem (Theorem 3.2.6), similar to R. Brown's version for the fundamental groupoid of spaces [Br].

3.2.1 Double homotopies and 2-homotopies

A (directed) *double homotopy* of d-spaces is a map

$$\Phi\colon X \times \uparrow\mathbf{I}^2 = I^2X \to Y,$$

(or, equivalently, $X \to P^2Y$). Roughly speaking, double homotopies (and double paths, in particular) behave as in **Top**, as long as we work on the ordered square $\uparrow\mathbf{I}^2$ via increasing maps.

The *second-order cylinder* $I^2X = X \times \uparrow\mathbf{I}^2$ has four one-dimensional faces, written

$$\begin{aligned}
\partial_1^\alpha &= I\partial^\alpha = (X \times \partial^\alpha) \times \uparrow\mathbf{I}\colon IX \to I^2X, &\quad \partial_1^\alpha(x,t) &= (x,\alpha,t),\\
\partial_2^\alpha &= \partial^\alpha I = (X \times \uparrow\mathbf{I}) \times \partial^\alpha\colon IX \to I^2X, &\quad \partial_2^\alpha(x,t) &= (x,t,\alpha).
\end{aligned} \tag{3.16}$$

Thus, a double homotopy $\Phi\colon X \times \uparrow\mathbf{I}^2 \to Y$ has four faces, which will be drawn as below

$$\begin{aligned}
\partial_1^\alpha(\Phi) &= \Phi.\partial_1^\alpha = \Phi.(X \times \partial^\alpha \times \uparrow\mathbf{I}),\\
\partial_2^\alpha(\Phi) &= \Phi.\partial_2^\alpha = \Phi.(X \times \uparrow\mathbf{I} \times \partial^\alpha),
\end{aligned} \tag{3.17}$$

$$\begin{array}{ccc} & \xrightarrow{\partial_2^-(\Phi)} & \\ f & \xrightarrow{\quad} & h \\ \partial_1^-(\Phi)\Big\downarrow & & \Big\downarrow \partial_1^+(\Phi) \\ k & \xrightarrow[\partial_2^+(\Phi)]{} & g \end{array} \qquad\qquad \begin{array}{c} \bullet \xrightarrow{\; 1\;} \\ \Big\downarrow 2 \\ \; \end{array}$$

The faces are homotopies linking the four vertices, the maps $f = \partial^-\partial_1^-(\Phi) = \partial^-\partial_2^-(\Phi)$, etc.

The concatenation, or pasting, of double homotopies *in direction* 1 or 2 is defined as usual (under the obvious boundary conditions) and satisfies a strict *middle-four interchange property*

$$(A +_1 B) +_2 (C +_1 D) = (A +_2 C) +_1 (B +_2 D), \qquad (3.18)$$

$$\begin{array}{ccccc} \bullet & \rightarrow & \bullet & \rightarrow & \bullet \\ \downarrow & A & \downarrow & B & \downarrow \\ \bullet & \rightarrow & \bullet & \rightarrow & \bullet \\ \downarrow & C & \downarrow & D & \downarrow \\ \bullet & \rightarrow & \bullet & \rightarrow & \bullet \end{array} \qquad\qquad \begin{array}{c} \bullet \xrightarrow{\; 1\;} \\ \Big\downarrow 2 \\ \; \end{array}$$

In particular, a (directed) *2-homotopy*

$$\Phi\colon \varphi \to \psi\colon f \to g\colon X \to Y$$

is a double homotopy whose faces ∂_1^α are degenerate, while the faces ∂_2^α are φ, ψ (the symmetric choice is equivalent, by transposition)

$$\begin{array}{ccc} f & \xrightarrow{\;\varphi\;} & g \\ 0_f\Big\downarrow & \Phi & \Big\downarrow 0_g \\ f & \xrightarrow[\psi]{} & g \end{array} \qquad \begin{array}{ll} \partial_2^-(\Phi) = \varphi, & \partial_2^+(\Phi) = \psi, \\[4pt] \partial^-\varphi = f = \partial^-\psi, & \partial^+\varphi = g = \partial^+\psi, \\[4pt] \partial_1^-(\Phi) = 0_f, & \partial_1^+(\Phi) = 0_g. \end{array} \qquad (3.19)$$

Such particular double homotopies are closed under pasting in both directions (also because $0_f + 0_f = 0_f$). The preorder $\varphi \preceq_2 \psi$ (i.e. there is a 2-homotopy $\varphi \to \psi$) spans an equivalence relation $\simeq_2$; two homotopies which satisfy the relation $\varphi \simeq_2 \psi$ are said to be *2-homotopic*.

3.2.2 Constructing double homotopies

(a) Two 'horizontally' consecutive homotopies

$$\varphi\colon f^- \to f^+\colon X \to Y, \qquad \psi\colon h^- \to h^+\colon Y \to Z,$$

can be composed, to form a *double* homotopy $\Phi = \psi \circ \varphi$

$$
\begin{array}{ccc}
h^- f^- & \xrightarrow{\ h^- \circ \varphi\ } & h^- f^+ \\
{\scriptstyle \psi \circ f^-}\downarrow & \psi \circ \varphi & \downarrow{\scriptstyle \psi \circ f^+} \\
h^+ f^- & \xrightarrow[\ h^+ \circ \varphi\ ]{} & h^+ f^+
\end{array}
\qquad
\begin{array}{c}
(\psi \circ \varphi)\hat{} = \hat{\psi}.(\hat{\varphi} \times \mathop{\uparrow}\mathbf{I}): \\[4pt]
X \times \mathop{\uparrow}\mathbf{I}^2 \ \to \ Z,
\end{array}
\qquad (3.20)
$$

$$
\partial_1^\alpha (\Phi) = \psi.(\varphi \partial^\alpha \times \mathop{\uparrow}\mathbf{I}) = \psi \circ f^\alpha, \qquad \partial_2^\alpha (\Phi) = \psi.(\varphi \times \partial^\alpha) = h^\alpha \circ \varphi.
$$

Notice that, together with the whisker composition in Section 1.2.3, this is a particular instance of the cubical enrichment given by the cylinder functor: composing a p-uple homotopy $\Phi \colon I^p X \to Y$ with a q-uple homotopy $\Psi \colon I^q Y \to Z$ gives a $(p+q)$-uple homotopy $\Psi \circ \Phi = \Psi.I^q \Phi$. (In the cocylinder approach, one would have: $\Psi \circ \Phi = P^p \Psi.\Phi$.)

(b) *Acceleration.* For every homotopy $\varphi \colon f \to g$, there are *acceleration* 2-homotopies

$$
\Theta' \colon 0_f + \varphi \to \varphi, \qquad \Theta'' \colon \varphi \to \varphi + 0_g, \qquad (3.21)
$$

but *not* the other way round: slowing down conflicts with direction.

To construct them, it suffices to consider the particular case $\varphi = \mathrm{id}\mathop{\uparrow}\mathbf{I}$ (and compose it with an arbitrary homotopy); thus, $\Theta'' \colon \mathop{\uparrow}\mathbf{I}^2 \to \mathop{\uparrow}\mathbf{I}$ is defined as follows

$$
\begin{array}{ccc}
f & \xrightarrow{\ \varphi\ } & g \\
{\scriptstyle 0_f}\downarrow & \Theta'' & \downarrow{\scriptstyle 0_g} \\
f & \xrightarrow[\ \psi + 0\ ]{} & g
\end{array}
\qquad
\begin{array}{l}
\varphi = \mathrm{id}\mathop{\uparrow}\mathbf{I}, \qquad f = \partial^-, \qquad g = \partial^+, \\[4pt]
\varphi(t) = t, \quad (\varphi + 0_g)(t) = \min(2t, 1), \\[4pt]
\Theta''(t, t') = (1 - t').t + t'.\min(2t, 1).
\end{array}
\qquad (3.22)
$$

In fact, Θ'' is an affine interpolation (in t') *from φ to $\varphi + 0_g$*; since $\varphi(t) \leqslant (\varphi + 0_g)(t)$, the mapping Θ'' preserves the order of the square and is a d-map $\mathop{\uparrow}\mathbf{I}^2 \to \mathop{\uparrow}\mathbf{I}$.

(c) *Folding.* A double homotopy $\Phi \colon A \times \mathop{\uparrow}\mathbf{I}^2 \to X$ with faces $\varphi, \psi, \sigma, \tau$ (as below) gives rise to a 2-homotopy Ψ, by pasting Φ with two double homotopies of connection (denoted by $\sharp$)

$$
\begin{array}{ccccccc}
f & \xrightarrow{\ 0_f\ } & f & \xrightarrow{\ \sigma\ } & h & \xrightarrow{\ \psi\ } & g \\
{\scriptstyle 0_f}\downarrow & \sharp & {\scriptstyle \varphi}\downarrow & \Phi & {\scriptstyle \psi}\downarrow & \sharp & \downarrow{\scriptstyle 0_g} \\
f & \xrightarrow[\ \varphi\ ]{} & k & \xrightarrow[\ \tau\ ]{} & g & \xrightarrow[\ 0_g\ ]{} & g
\end{array}
\qquad (3.23)
$$

$$
\Psi \colon (0_f + \sigma) + \psi \ \to \ (\varphi + \tau) + 0_g \colon f \to g.
$$

(Together with accelerations and Theorem 3.2.4, this will show that $\sigma + \psi \simeq_2 \varphi + \tau$, in the equivalence relation defined at the end of Section 3.2.1.)

3.2.3 The fundamental category

Directed paths (reviewed in Section 3.1.5) will now be considered modulo 2-homotopy, i.e. homotopy with fixed endpoints.

A *double path* in X is a d-map $A\colon \uparrow\mathbf{I}^2 \to X$. It is the elementary instance of a double homotopy (Section 3.2.1), defined on the point, and the previous results apply; its four faces are paths in X, linking four vertices. A *2-path* is a double path whose faces ∂_1^α are degenerate, that is a 2-homotopy $A\colon a \preceq_2 b\colon x \to x'$ between its faces ∂_2^α, which have the same endpoints. A class of paths $[a]$ up to 2-homotopy is a class of the equivalence relation $\simeq_2$ spanned by the preorder $\preceq_2$ (Section 3.2.1).

The *fundamental category* $\uparrow\Pi_1(X)$ of a d-space has for objects the points of X; for arrows $[a]\colon x \to x'$ the 2-homotopy classes of paths from x to x', as defined above. Composition – written additively – is induced by concatenation of consecutive paths, and identities are induced by degenerate paths

$$[a] + [b] = [a + b], \qquad\qquad 0_x = [e(x)] = [0_x]. \qquad (3.24)$$

We prove below that $\uparrow\Pi_1(X)$ is indeed a category and that the obvious action on arrows defines a functor $\uparrow\Pi_1\colon \mathrm{d}\mathbf{Top} \to \mathbf{Cat}$, with values in the category of small categories

$$\uparrow\Pi_1(f)(x) = f(x), \qquad\qquad \uparrow\Pi_1(f)[a] = f_*[a] = [fa]. \qquad (3.25)$$

The fundamental category of X is linked to the fundamental groupoid of the underlying space UX, by the obvious *comparison* functor

$$\uparrow\Pi_1(X) \to \Pi_1(UX), \qquad\qquad x \mapsto x, \qquad [a] \mapsto [[a]],$$

which is the identity on objects and sends 2-homotopy classes of (directed) paths in X to 2-homotopy classes of paths UX. This functor need not be full (obviously) nor faithful (Section 3.2.8). Of course, if X is a topological space with the natural d-structure, which distinguishes all paths (i.e. $X = D'UX$), then $\uparrow\Pi_1(X) = \Pi_1(UX)$.

3.2.4 Theorem (The fundamental category)

(a) For every d-space X, $\uparrow\Pi_1(X)$ is a category and the previous formulas (3.25) do define a functor, which preserves sums and products. The opposite d-space gives the opposite category, $\uparrow\Pi_1(RX) = (\uparrow\Pi_1(X))^{\mathrm{op}}$.

(b) If $a\colon x \to x'$ is a reversible path (Section 1.4.6), its class $[a]$ is an invertible arrow in $\uparrow\Pi_1(X)$.

(c) The functor $\uparrow\Pi_1\colon \mathrm{d\mathbf{Top}} \to \mathbf{Cat}$ is homotopy invariant, in the following sense: a homotopy $\varphi\colon f \to g\colon X \to Y$ induces a natural transformation (i.e. a directed homotopy of categories, see Section 1.1.6)

$$\begin{aligned} &\varphi_*\colon f_* \to g_*\colon \uparrow\Pi_1(X) \to \uparrow\Pi_1(Y),\\ &\varphi_*(x) = [\varphi(x)]\colon f(x) \to g(x), \end{aligned} \qquad (3.26)$$

where $\varphi(x)$ is the path in Y given by the representative map $X \to PY$. Therefore, $\uparrow\Pi_1$ transforms a future equivalence of d-spaces into a future equivalence of categories. (The latter will be studied in the next section.)

(d) A reversible homotopy φ induces an invertible transformation φ_*. Therefore $\uparrow\Pi_1$ transforms a reversible homotopy equivalence of d-spaces into an equivalence of categories.

Proof (a) Composition is well defined, in (3.24). In fact, given 2-homotopies $A\colon a \preceq_2 a'\colon x \to x'$ and $B\colon b \preceq_2 b'\colon x' \to x''$, the pasting

$$A +_1 B\colon a + b \preceq_2 a' + b'\colon x \to x'',$$

shows that $[a + b] = [a' + b']$. The general case, for the equivalence relation $\simeq_2$, follows by taking, in A or B, a trivial 2-homotopy and applying transitivity.

In $\uparrow\Pi_1(X)$, constant paths yield (strict) identities, because of the acceleration 2-homotopies $0_x + a \to a \to a + 0'_x$ (see (3.21)).

Associativity, on three consecutive paths a, b, c in X, follows from considering a 2-homotopy, constructed as follows, pasting double paths obtained from degeneracies and connections (all of them are denoted with $\sharp$)

$$B\colon (0 + a) + (b + c) \to (a + b) + (c + 0), \qquad (3.27)$$

$$
\begin{array}{ccccccccc}
x & \xrightarrow{\;0\;} & x & \xrightarrow{\;a\;} & y & \xrightarrow{\;b\;} & z & \xrightarrow{\;c\;} & w \\
{\scriptstyle 0}\downarrow & {\scriptstyle\sharp} & {\scriptstyle a}\downarrow & {\scriptstyle\sharp} & {\scriptstyle 0}\downarrow & {\scriptstyle\sharp} & {\scriptstyle 0}\downarrow & {\scriptstyle\sharp} & {\scriptstyle 0}\downarrow \\
x & \xrightarrow{\;-a\;} & y & \xrightarrow{\;-0\;} & y & \xrightarrow{\;-b\;} & z & \xrightarrow{\;-c\;} & w \\
{\scriptstyle 0}\downarrow & {\scriptstyle\sharp} & {\scriptstyle 0}\downarrow & {\scriptstyle\sharp} & {\scriptstyle b}\downarrow & {\scriptstyle\sharp} & {\scriptstyle 0}\downarrow & {\scriptstyle\sharp} & {\scriptstyle 0}\downarrow \\
x & \xrightarrow{\;-a\;} & y & \xrightarrow{\;-b\;} & z & \xrightarrow{\;-0\;} & z & \xrightarrow{\;-c\;} & w \\
{\scriptstyle 0}\downarrow & {\scriptstyle\sharp} & {\scriptstyle 0}\downarrow & {\scriptstyle\sharp} & {\scriptstyle 0}\downarrow & {\scriptstyle\sharp} & {\scriptstyle c}\downarrow & {\scriptstyle\sharp} & {\scriptstyle 0}\downarrow \\
x & \xrightarrow{\;a\;} & y & \xrightarrow{\;b\;} & z & \xrightarrow{\;c\;} & w & \xrightarrow{\;0\;} & w
\end{array}
\tag{3.28}
$$

The argument is concluded by two other 2-homotopies, which come forth from accelerations and cannot be pasted with the previous 2-homotopy B, because of conflicting directions

$$
A \colon (0+a)+(b+c) \to a+(b+c), \qquad C \colon (a+b)+c \to (a+b)+(c+0).
$$

The fact that a d-map $f \colon X \to Y$ gives a well-defined transformation $\uparrow\Pi_1(f)[a] = [fa]$ is also obvious: a 2-homotopy $A \colon a \preceq_2 a'$ gives a 2-homotopy $fA \colon fa \preceq_2 fa'$.

We thus have a functor $\uparrow\Pi_1$. Its preservation of sums and cartesian products is proved in the same (easy) way as in the ordinary case.

(b) The reversible path a can be interpreted as a d-map $\mathbf{I}^\sim \to X$, defined on the reversible interval $\mathbf{I}^\sim$ (Section 1.4.6). The double path

$$
\begin{array}{ccc}
x' & \xrightarrow{\;0\;} & x' \\
{\scriptstyle -a}\downarrow & A & \downarrow{\scriptstyle 0} \\
x & \xrightarrow{\;a\;} & x'
\end{array}
\qquad\qquad
A = ag^-.(\mathbf{I}^\sim \times r) \colon \mathbf{I}^{\sim 2} \to X,
$$

is directed with respect to the reversible structures; in fact, given two piecewise monotone real functions h, k, also $h \vee k$ is so (if h is increasing and k decreasing on some interval $[t_0, t_1]$, and $h(t) = k(t)$ at some intermediate point, then $h \vee k$ coincides with k on $[t_0, t]$, with h on $[t, t_1]$).

Finally, by folding (Section 3.2.2(c)), and recalling that $\uparrow\mathbf{I}$ has a finer structure than $\mathbf{I}^\sim$, we get a 2-path showing that $-a+a \simeq_2 0$

$$
A' \colon \uparrow\mathbf{I}^2 \to \mathbf{I}^{\sim 2} \to \mathbf{I}^\sim \to X, \qquad A' \colon 0 \to (-a+a)+0 \colon \uparrow\mathbf{I} \to X.
$$

(c) The naturality of the transformation associated to $\varphi \colon f \to g$ on the arrow $[a] \colon x \to x'$ in $\uparrow\Pi_1(X)$ amounts to the relation $[fa] + [\varphi(x')] = [\varphi(x)] + [ga]$. This follows from the existence of the double path $\Phi = \varphi \circ a$ (cf. Section 3.2.2(a)), together with folding (cf. Section 3.2.2(c)) and the previous arguments

$$
\begin{array}{ccc}
f(x) & \xrightarrow{\; fa \;} & f(x') \\
{\scriptstyle \varphi(x)}\big\downarrow & {\scriptstyle \varphi \circ a} & \big\downarrow{\scriptstyle \varphi(x')} \\
g(x) & \xrightarrow[\; ga \;]{} & g(x')
\end{array}
\qquad
\begin{array}{l}
\Phi = \varphi \circ a = \varphi.(a \times {\uparrow}\mathbf{I}), \\[2ex]
fa + \varphi(x') \simeq_2 \varphi(x) + ga.
\end{array}
\qquad (3.29)
$$

(d) Is a straightforward consequence of (b) and (c). $\qquad\qquad\square$

3.2.5 Homotopy monoids

The *fundamental monoid* ${\uparrow}\pi_1(X, x)$ of the d-space X at the point x is the monoid of endo-arrows $[c] \colon x \to x$ in ${\uparrow}\Pi_1(X)$. It yields a functor from the category $\mathrm{d}\mathbf{Top}_{\bullet}$ of pointed d-spaces (Section 1.5.4) to the category of monoids

$$
{\uparrow}\pi_1 \colon \mathrm{d}\mathbf{Top}_{\bullet} \to \mathbf{Mon}, \qquad {\uparrow}\pi_1(X, x) = {\uparrow}\Pi_1(X)(x, x), \qquad (3.30)
$$

which is *strictly* homotopy invariant: a *pointed homotopy* $\varphi \colon f \to g \colon (X, x) \to (Y, y)$ has, by definition, a trivial path at the base point ($\varphi(x) = 0_y$), whence $f_* = g_*$ (see (3.29)).

Similarly, we have a functor from the slice category $\mathrm{d}\mathbf{Top}\backslash\mathbf{S}^0$ of *bi-pointed d-spaces* (Section 5.2)

$$
{\uparrow}\pi_1 \colon \mathrm{d}\mathbf{Top}\backslash\mathbf{S}^0 \to \mathbf{Set}, \qquad {\uparrow}\pi_1(X, x, x') = {\uparrow}\Pi_1(X)(x, x'), \qquad (3.31)
$$

which is strictly homotopy invariant, up to *bipointed homotopies* (leaving fixed each base point).

One can view (3.30) and (3.31) as representable homotopy functors (Section 1.5.3) on $\mathrm{d}\mathbf{Top}_{\bullet}$ and $\mathrm{d}\mathbf{Top}\backslash\mathbf{S}^0$, which accounts for their strict invariance. Moreover, both can be computed by the methods developed below for ${\uparrow}\Pi_1 X$. (For the homotopy structure of slice categories see Chapter 5.)

The existence of a *reversible* path from x to x' implies that their fundamental monoids are isomorphic (by Theorem 3.2.4(b)); without reversibility, this need not be true (cf. Section 3.2.8). However, in a *homogeneous* d-space, where the group $\mathrm{Aut}(X)$ acts transitively, all ${\uparrow}\pi_1(X, x)$ are isomorphic; this applies, for instance, to the directed circle ${\uparrow}\mathbf{S}^1$, whose only reversible paths are the constant ones.

3.2.6 Pasting theorem

('Seifert–van Kampen' for fundamental categories of d-spaces) *Let X be a d-space; let X_1, X_2 be two d-subspaces, and $X_0 = X_1 \cap X_2$.*

(a) If $X = \text{int}(X_1) \cup \text{int}(X_2)$, the following diagram of categories and functors (induced by inclusions) is a pushout in **Cat**

$$\begin{array}{ccc} \uparrow\Pi_1 X_0 & \overset{u_1}{\longrightarrow} & \uparrow\Pi_1 X_1 \\ {\scriptstyle u_2}\downarrow & & \downarrow{\scriptstyle v_1} \\ \uparrow\Pi_1 X_2 & \underset{v_2}{\longrightarrow} & \uparrow\Pi_1 X \end{array} \qquad (3.32)$$

(b) More generally, the same fact holds if there exist two d-subspaces $w_i \colon Y_i \subset X_i$ with retractions $p_i \colon Y_i \to X_i$ (d-maps with $p_i w_i = \text{id}Y_i$; no deformation is required) such that:

$$X = \text{int}(Y_1) \cup \text{int}(Y_2), \quad p_1 \text{ and } p_2 \text{ coincide on } Y_0 = Y_1 \cap Y_2. \quad (3.33)$$

Proof (a) We shall use the n-ary concatenation of consecutive d-paths, written $a_1 + \cdots + a_n$ (Section 1.4.0).

Let C be a small category (in additive notation, again) and take two functors $F_i \colon \uparrow\Pi_1 X_i \to C$ which coincide on $\uparrow\Pi_1 X_0$ ($F_1 u_1 = F_2 u_2$); we have to prove that they have a unique 'extension' $F \colon \uparrow\Pi_1 X \to C$.

On the objects, this is obvious since $|X| = |X_1| \cup |X_2|$ and $|X_0| = |X_1| \cap |X_2|$.

Let then $a \colon x \to y \colon \{*\} \to X$ be a path. By the Lebesgue covering lemma, there is a finite decomposition $0 < 1/n < 2/n \cdots < 1$ of the standard interval such that each subinterval $[(i-1)/n, i/n]$ is mapped by a into X_1 or X_2 (possibly both); let us call it a *suitable* decomposition for our data. Thus, $a = a_1 + \cdots + a_n$ where each $a_i \colon [0,1] \to X$ is a directed path (by partial increasing reparametrisation) contained in some X_{k_i} (with $k_i = 1, 2$), hence a d-path there. Define

$$F[a] = F_{k_1}[a_1] + \cdots + F_{k_n}[a_n] \in C(F(x), F(x')).$$

First, this morphism $F[a]$ does not depend on the choice of k_i: if $\text{Im}(a_i) \subset X_1 \cap X_2 = X_0$, then $F_1 u_1 = F_2 u_2$ shows that $F_1[a_i] = F_2[a_i]$.

Second, $F[a]$ does not depend on the choice of n: if also m gives a suitable partition, one can use the partition arising from mn to prove that they give the same result.

Third, $F[a]$ does not depend on the representative path a. It is sufficient to show this for a second path $a' \colon x \to x'$, linked to the first by a 2-path $A \colon a \to a'$; in other words, $A \colon \uparrow\mathbf{I}^2 \to X$ has degenerate 1-faces, and 2-faces coinciding with a, a'. Again by the Lebesgue covering lemma, applied to the compact metric square $[0,1]^2$, there is some integer $n > 0$ such that each elementary square

$$[(i-1)/n, i/n] \times [(j-1)/n, j/n] \qquad (i, j = 1, ..., n)$$

is mapped by A into X_1 or X_2. A can be obtained as an '$(n \times n)$-pasting' of its reparametrised restrictions to these squares, $A_{ij} \colon \uparrow\mathbf{I}^2 \to X_{k(i,j)} \subset X$

$$A = (A_{11} +_1 A_{21} +_1 \cdots +_1 A_{n1}) +_2 \cdots +_2 (A_{1n} +_1 A_{2n} +_1 \cdots +_1 A_{nn}).$$

Every 2-cube $B = A_{ij}$ yields, by folding (Section 3.2.2(c)), a 2-homotopy relation in $X_{k(i,j)}$

$$\partial_2^- B + \partial_1^+ B \simeq_2 \partial_1^- B + \partial_2^+ B. \tag{3.34}$$

Therefore, using the fact that all the 1-directed faces on the boundary (namely $\partial_1^- A_{1i}$, $\partial_1^+ A_{ni}$) are degenerate, and the coincidence of faces between contiguous little squares, we can gradually move from a to a'

$$
\begin{aligned}
F[a] \ &= F_{k(1,1)}[\partial_2^- A_{11}] + \cdots + F_{k(n,1)}[\partial_2^- A_{n1}] \\
&= F_{k(1,1)}[\partial_2^- A_{11}] + \cdots + F_{k(n,1)}[\partial_2^- A_{n1}] + F_{k(n,1)}[\partial_1^+ A_{n1}] \\
&\qquad\qquad \text{(by degeneracy)} \\
&= F_{k(1,1)}[\partial_2^- A_{11}] + \cdots + F_{k(n,1)}[\partial_1^- A_{n1}] + F_{k(n,1)}[\partial_2^+ A_{n1}] \\
&\qquad\qquad \text{(by (3.34))} \\
&= F_{k(1,1)}[\partial_2^- A_{11}] + \cdots + F_{k(n,1)}[\partial_1^+ A_{n-1,1}] + F_{k(n,2)}[\partial_2^- A_{n2}] \\
&\qquad\qquad \text{(by contiguity)} \\
&= \cdots = F_{k(1,2)}[\partial_2^- A_{21}] + \cdots + F_{k(n,2)}[\partial_2^- A_{n1}] \\
&= \cdots = F_{k(1,n)}[\partial_n^- A_{n1}] + \cdots + F_{k(n,n)}[\partial_n^- A_{n1}] \ = \ F[a'].
\end{aligned}
$$

Thus, $F \colon \uparrow\Pi_1 X \to C$ is also well defined on arrows. To show that it preserves composition just note that, if two consecutive d-paths a, b have a suitable decomposition on n subintervals, then $a+b$ inherits a suitable decomposition $a + b = a_1 + \cdots + a_n + b_1 + \cdots + b_n$ which keeps the original paths apart. Finally, the uniqueness of the functor F is obvious.

(b) By (a), the square which comes from Y_0, Y_1, Y_2, and $Y = X$ is a pushout of categories.

Also the inclusion $w_0 \colon X_0 \subset Y_0$ has a retraction p_0, the common restriction of p_1 and p_2 to Y_0. Therefore, all w_i and p_i form a retraction in the category of commutative squares of d**Top**

$$w = (w_0, w_1, w_2, \mathrm{id}X) \colon \mathbf{X} \to \mathbf{Y} \colon \mathbf{2} \times \mathbf{2} \to d\mathbf{Top},$$
$$p = (p_0, p_1, p_2, \mathrm{id}X) \colon \mathbf{Y} \to \mathbf{X} \colon \mathbf{2} \times \mathbf{2} \to d\mathbf{Top} \qquad (pw = \mathrm{id}X).$$

The functor $\uparrow\!\Pi_1$ takes all this into a retraction in the category of commutative squares of **Cat**

$$w_* \colon \uparrow\!\Pi_1 \mathbf{X} \rightleftarrows \uparrow\!\Pi_1 \mathbf{Y} \colon p_*.$$

Since $\uparrow\!\Pi_1 \mathbf{Y}$ is a pushout, also its retract $\uparrow\!\Pi_1 \mathbf{X}$ is (as can be easily checked, or seen in [Br], 6.6.7). $\qquad\square$

3.2.7 Elementary computations

Say that a d-space X is *1-simple* if its fundamental category is a preorder, or equivalently if $\uparrow\!\Pi_1 X = \mathrm{cat}(X, \preceq)$, the category associated with the path preorder of X. In other words, $(\uparrow\!\Pi_1 X)(x, x')$ has precisely one arrow when $x \preceq x'$, and no arrows otherwise.

(a) Every *convex* subspace X of $\mathbf{R}^n$, with the order structure induced by $\uparrow\!\mathbf{R}^n$, is 1-simple. More generally, the same fact holds for a subspace X of $\mathbf{R}^n$ such that, whenever $x \leqslant x'$ in X, the line segment joining x, x' is contained in X.

In fact, if $x \leqslant x'$ in X, there is a d-path in X from x to x' along that segment, e.g. the affine interpolation $a(t) = (1 - t).x + t.x'$ (the converse being obvious). Moreover, given two increasing paths $a, b \colon \uparrow\!\mathbf{I} \to X$ from x to x', we can always replace them with 2-homotopic paths $a' \preceq_2 b'$: we replace the first with $a' = 0_x + a \preceq_2 a$, the second with $b' = b + 0_{x'} \succeq_2 b$. Then, the interpolation 2-path

$$A(t, t') = (1 - t').a'(t) + t'.b'(t),$$

preserves the order of $\uparrow\!\mathbf{I}^2$ and provides a directed 2-homotopy $A \colon a' \to b'$ which stays in the convex subset X.

Note that we have actually constructed three directed 2-homotopies:

$$a' = 0_x + a \to a, \qquad\qquad b \to b' = b + 0_{x'}, \qquad\qquad a' \to b',$$

and that two paths with the same endpoints are 'rarely' linked by a 'one step' directed 2-homotopy $a \to b$. Furthermore, in an *ordered* topological space, one can have two directed 2-homotopies $a \to b \to a$ only if $a = b$.

(b) It follows that a d-space X is certainly 1-simple whenever the following condition holds: if $x' \preceq x''$, then the d-subspace

$$\{x \in X \mid x' \preceq x \preceq x''\},$$

is isomorphic to some convex d-subspace of $\uparrow \mathbf{R}^n$.

(c) The following objects are 1-simple: any interval of $\uparrow \mathbf{R}$; any product of such in $\uparrow \mathbf{R}^n$; the preordered subspaces $V, W \subset \uparrow \mathbf{R}^2$ considered in (1.80); any 'fan' formed by the union of (finitely or infinitely many) line segments or half-lines spreading from a point, in some $\uparrow \mathbf{R}^n$. (Here, one should not confuse the path-order with the order induced by $\uparrow \mathbf{R}^n$, which is coarser and of less interest.)

(d) We consider now some elementary cases which are not 1-simple, starting from the directed circle $\uparrow \mathbf{S}^1$. Let us apply the 'van Kampen' Theorem 3.2.6(a) in the obvious way: choose two arcs X_1, X_2 isomorphic to $\uparrow \mathbf{I}$, which satisfy the hypothesis, with $X_0 \cong \uparrow \mathbf{I} + \uparrow \mathbf{I}$. The resulting pushout in **Cat** shows that $\uparrow \Pi_1 \uparrow \mathbf{S}^1$ is the subcategory of the groupoid $\Pi_1 \mathbf{S}^1$ formed by the classes of anticlockwise paths (with respect to the embedding in the oriented plane). In particular, each monoid $\uparrow \pi_1 (\uparrow \mathbf{S}^1, x)$ is isomorphic to the additive monoid **N** of natural numbers.

(e) Consider now the ordered circle $\uparrow \mathbf{O}^1 \subset \mathbf{R} \times \uparrow \mathbf{R}$ (see (1.103)); let us write $x^- = (0, -1)$ and $x^+ = (0, 1)$ the minimum and maximum, and $a, b \colon x^- \to x^+$ the two obvious d-paths moving around the left and right half-circles. Applying 'van Kampen', we get that there are precisely two arrows $[a] \neq [b]$ from x^- to x^+

$$\uparrow \Pi_1 \uparrow \mathbf{O}^1 (x^-, x^+) = \{[a], [b]\}.$$

Moreover, if $x \neq x^-$ or $x' \neq x^+$, there is precisely one arrow from x to x' when $x \prec x'$ and they both lie either in the left or in the right half-circle, none otherwise. This can also be proved directly, noting that any d-path in $\uparrow \mathbf{O}^1$ stays either in the left half-circle or in the right one.

(f) Finally, for the 'square annulus' $X = \uparrow [0,1]^2 \setminus]1/3, 2/3[^2$ (Section 3.1.1)

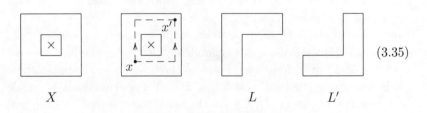

$$X \qquad\qquad L \qquad\qquad L' \tag{3.35}$$

applying the 'van Kampen' theorem to the closed subspaces L, L' (which are 1-simple), we find the description of the fundamental category $\uparrow\Pi_1(X)$ given above: there is some arrow $x \to x'$ provided that $x \leqslant x'$ and both points are in L or L'. Precisely, there are two of them when $x \leqslant p = (1/3, 1/3)$ and $x' \geqslant q = (2/3, 2/3)$ (as in the second figure above), and one otherwise.

(The fact that L and L' are 1-simple – if not accepted as obvious – can also be proved with 'van Kampen' and point (a).)

3.2.8 Remarks

For a future (or past) homotopy equivalence $p: X \to Y$, the induced functor $p_*: \uparrow\Pi_1 X \to \uparrow\Pi_1 Y$ need not be *full* nor *faithful* (even when p is a strong future deformation retraction, as defined in Section 1.3.1).

For the first case, just consider the fact that $\uparrow\mathbf{I}$ is future contractible to 1, and yet $\uparrow\Pi_1(\uparrow\mathbf{I}) = \mathrm{cat}(\mathbf{I}, \leqslant)$ keeps the information of the (path) order; thus $p_*: \uparrow\Pi_1(\uparrow\mathbf{I}) \to \uparrow\Pi_1\{1\}$ is not full.

The failure of faithfulness can give rise to even more unusual effects:

(i) a future contractible object X can have loops c which are not homotopically trivial ($[c] \neq 0$);

(ii) such loops are then annihilated by the deformation retraction $p: X \to \{*\}$ (p_* is not faithful);

(iii) such loops are 'loop-homotopic' to the constant loop, without being 2-homotopic to it.

In fact, take the disc with the structure $X = C^+(\mathbf{S}^1)$ (see (1.110)), which is future contractible to its centre $0 = v^+$, and recall that any path in this d-space moves towards the centre, in the weak sense.

Any concentric circle C inherits the natural structure of $\mathbf{S}^1$, and no path in X between two of its points can leave it; thus, the restriction of $\uparrow\Pi_1 X$ to the points of C coincides with the fundamental groupoid of the circle and has d-loops $c: \mathbf{S}^1 \to X$ with $[c] \neq 0$. Any deformation $\varphi: X \times \uparrow\mathbf{I} \to X$ with $\varphi(x, 0) = x$, $\varphi(x, 1) = v^+$ yields a loop-homotopy $\varphi.(c \times \uparrow\mathbf{I}): \mathbf{S}^1 \times \uparrow\mathbf{I} \to X$ from c to 0_{v^+}. Note also that, if c is a loop at x_0, the homotopy class of the path $a = \varphi.(x_0, -): x_0 \to v^+$ is not cancellable in $\uparrow\Pi_1 X$ (it is not a monomorphism): $[c] + [a] = [a]$.

Similar arguments allow us to completely determine the fundamental category of $C^+(\mathbf{S}^1)$: from each point to the origin $0 = v^+$ there is precisely *one* arrow (this will also follow from Proposition 3.3.6: v^+ must be a terminal object in $\uparrow\Pi_1 X$). On the other hand, if $||x_1|| \geqslant ||x_2|| > 0$

(in the euclidean norm of the disc) there are infinitely many arrows $x_1 \to x_2$, determined by their 'winding number' around the origin (an integer)

$$(\uparrow\Pi_1 X)(x_1, x_2) = \Pi_1 \mathbf{S}^1(q(x_1), q(x_2)),$$

where $q(x) = x/\|x\|$. There are no other arrows. Concatenation of maps $x_1 \to x_2 \to x_3$ (with $\|x_1\| \geqslant \|x_2\| \geqslant \|x_3\| > 0$) works by adding the winding numbers, and is trivially determined when $x_3 = 0$.

The fundamental category of $C^+(\uparrow\mathbf{S}^1)$ has a similar description, with winding numbers in $\mathbf{N}$.

All these remarks show that the fundamental category contains information which can disappear modulo future or past homotopy equivalence (and even more modulo coarse d-homotopy). This is why, in the next sections, we will study finer equivalence relations, *taking into account at the same time past and future*.

Thus, the minimal injective model of $\uparrow\Pi_1(C^+(\mathbf{S}^1))$, in Section 3.6.6, will be a small, countable model on *two* objects (the vertex and any other point), but will keep all the relevant information.

3.3 Future and past equivalences of categories

A future equivalence of categories (defined in Section 3.3.1) is a coherent instance of a future homotopy equivalence of categories, and a symmetric version of the notion of adjunction. It is meant to identify *future invariant* properties. Directed homotopy equivalence of categories is thus introduced in two dual forms, future and past equivalences, which will be combined in the next sections to give a finer analysis.

A motivation for the study of these relations, based on comma categories, has already been given in Section 1.8.9. Future equivalence tends to be of little relevance when applied to the usual (large) categories of structured sets, since all categories with a terminal object are future equivalent (to the terminal category **1**, see Proposition 3.3.6). However, standard constructions performed on such categories, like comma categories, can give rise to categories without a terminal object, for which the results of Section 1.8.9 may be relevant.

In the rest of this chapter we will generally use the usual 'multiplicative' notation for composition in categories and, as a consequence, for vertical composition of natural transformations.

3.3.0 *Review of directed homotopy in* Cat

Let us recall – with a few additions – the basic notions of directed homotopy in **Cat**, introduced in Section 1.1.6 and based on its cartesian closed structure. (We shall occasionally use the same notions for large categories.)

The reversor $R(X) = X^{\mathrm{op}}$ takes a category to the opposite one. The *directed interval* $\uparrow\mathbf{i} = \mathbf{2} = \{0 \to 1\}$ is a cartesian dIP1-interval (Section 1.2.5); it is equipped with the obvious *faces* $\partial^{\pm}\colon \mathbf{1} \to \mathbf{2}$, defined on the terminal category $\mathbf{1} = \{*\}$, and the *reflection* isomorphism $r\colon \mathbf{2} \to \mathbf{2}^{\mathrm{op}}$. The *cylinder* functor is $IX = X \times \mathbf{2}$, with right adjoint, $PY = Y^{\mathbf{2}}$ (the category of morphisms of Y).

A (directed) *homotopy* $\varphi\colon f \to g\colon X \to Y$ is the same as a natural transformation between functors, and their concatenation is by vertical composition, strictly associative and unitary.

A *point* $x\colon \mathbf{1} \to X$ of a small category X 'is' an object of the latter, which we write as $x \in X$. A (directed) *path* $a\colon \mathbf{2} \to X$ from x to x' is an arrow $a\colon x \to x'$ of X, and amounts to a homotopy $a\colon x \to x'\colon \mathbf{1} \to X$; their concatenation is by composition in X. A *reversible path* is an isomorphism. A double path $\mathbf{2} \times \mathbf{2} \to X$ is a commutative square, while a 2-path $A\colon a \to a'$ is necessarily degenerate, with $a = a'$.

Therefore $\uparrow\Pi_1 X$, defined as above for d-spaces, just *coincides* with X, and $\uparrow\Pi_1 f = f$, for every (small) functor f.

The existence of a map $x \to x'$ in X (a path) produces the *path preorder* $x \preceq x'$ (x *reaches* x') on the points of X; the resulting *path equivalence* relation, meaning that there are maps $x \rightleftarrows x'$, will be written as $x \simeq x'$. For the path preorder, a point x is

- *maximal* if it can only reach the points $\simeq x$;
- a *maximum* if it can be reached from every point of X.

(The latter is the same as a *weak terminal* object, and is only determined up to path equivalence.) If the category X 'is' a preorder, the path preorder coincides with the original relation.

3.3.1 *Future equivalences*

According to a general definition in directed homotopy theory (Section 1.3.1), a *future homotopy equivalence* $(f, g; \varphi, \psi)$ between the categories X, Y consists of a pair of functors and a pair of natural transformations (i.e. directed homotopies), the *units*

$$f \colon X \rightleftarrows Y \colon g \qquad \varphi \colon 1_X \to gf, \qquad \psi \colon 1_Y \to fg, \qquad (3.36)$$

which go *from* the identities of X, Y *to* the composed functors.

This four-tuple will be called a *future equivalence* if it is *coherent*, i.e. satisfies:

$$f\varphi = \psi f \colon f \to fgf, \qquad \varphi g = g\psi \colon g \to gfg \qquad (\textit{coherence}). \qquad (3.37)$$

In **Cat**, we shall only use such 'coherent' equivalences, which would be too restricted for d-spaces. (Note that these coherence conditions are not required for homotopy equivalence of ordinary spaces, and only one coherence condition is required for *strong* deformation retracts.)

A property (making sense *in* a category, or *for* a category) will be said to be *future invariant* if it is preserved by future equivalences. Some elementary examples will be discussed in Lemma 3.3.8, and other more interesting ones in Section 3.7.

A future equivalence is a 'symmetric variation' of the notion of adjunction, and some aspects of the theory will be similar. However, in a future equivalence, f need not determine g (see Section 3.3.9). Our data give rise to two natural transformations between hom-functors (which will often be used implicitly in what follows)

$$\begin{aligned}
\Phi &\colon Y(fx, y) \to X(x, gy), & \Phi(b) &= gb.\varphi x, \\
\Psi &\colon X(gy, x) \to Y(y, fx), & \Psi(a) &= fa.\psi y.
\end{aligned} \qquad (3.38)$$

One can also note that an adjunction $f \dashv g$ with *invertible* counit $\varepsilon \colon fg \cong 1$ amounts to a future equivalence with invertible unit $\psi = \varepsilon^{-1}$; this case, a *split* future equivalence, will be treated later (Section 3.3.4).

A future equivalence $(f, g; \varphi, \psi)$ will be said to be *faithful* if the functors f and g are faithful and, moreover, all the components of φ and ψ are *epi* and *mono*. (Motivations for the latter condition will arise in Section 3.3.4 and Theorem 3.3.5.) The next lemma (similar to classical properties of adjunctions) will prove that it suffices to know that *one* of the following equivalent conditions holds:

(i) all the components of φ and ψ are mono;

(ii) f and g are faithful and all the components of φ and ψ are epi.

Plainly, all future homotopy equivalences between preordered sets (viewed as categories) are coherent and faithful. There are non-faithful future equivalences where all unit-components are epi (see Section 3.3.9(d)). A faithful future equivalence between *balanced* categories (where every map which is mono *and* epi is an isomorphism) is an equivalence of categories. But a faithful future equivalence can link a balanced category with a non-balanced one (see Section 3.4.7).

Dually, a *past equivalence* has natural transformations, called *counits*, in the opposite direction, *from* the composed functors *to* the identities

$$f \colon X \rightleftarrows Y \colon g \qquad \varphi \colon gf \to 1, \qquad \psi \colon fg \to 1,$$
$$f\varphi = \psi f \colon fgf \to f, \quad \varphi g = g\psi \colon gfg \to g \quad (coherence). \tag{3.39}$$

An adjoint equivalence of categories (Section A1.5) is at the same time a future and a past equivalence. Future equivalences will be shown to be related to reflective subcategories and idempotent monads (Section 3.3.4); they will generally be given 'priority' over the dual case, which is related to *co*reflective subcategories and *co*monads.

We will see that each of these two notions of directed homotopy equivalence distinguishes new 'shapes', in **Cat**. Each of them is weaker than ordinary equivalence, which corresponds to *reversible homotopies*, based on the groupoid **i** (1.30). But *each of them implies ordinary homotopy equivalence of the classifying spaces*, the geometric realisations of the simplicial nerves (which is a non-directed notion).

Indeed, a natural transformation $\varphi \colon f \to g \colon C \to D$ gives, under the nerve functor $N \colon \mathbf{Cat} \to \mathbf{Smp}$ with values in the category of simplicial sets, a simplicial homotopy $N\varphi \colon Nf \to Ng \colon NC \to ND$ (because $N(C \times \mathbf{2}) = NC \times N\mathbf{2}$ and $N\mathbf{2}$ is the simplicial interval $\{0 \to 1\}$); then, by ordinary geometric realisation, $N\varphi$ yields a homotopy between the classifying spaces of C and D. (See [My], Section 16; [Cu], 1.29.)

3.3.2 Lemma (Cancellation properties of future equivalences)

Let $(f, g; \varphi, \psi)$ be a future equivalence (Section 3.3.1).

(a) *If all the components $\varphi x \colon x \to gfx$ are mono, then all of them are epi and f is a faithful functor.*

(b) *The natural transformation φ is invertible if and only if all its components are split mono; in this case f is right adjoint to g, full and faithful.*

(c) *If g is faithful and all the components of φ are epi, then f preserves all epis.*

(d) *If gf is faithful and all the components of φ are epi, then they are also mono.*

(e) The conditions (i) and (ii) of Section 3.3.1 are equivalent; when they hold, f and g preserve all epis.

Proof (a) Assume that all the components φx are mono, and take two arrows a_1, a_2 with $a_i.\varphi x = a$

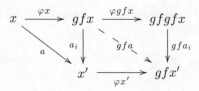

Since $\varphi g f = g f \varphi$ (by coherence) we have $\varphi x'.a_i = g f a_i.\varphi g f x = g f(a_i.\varphi x) = g f(a)$, and, cancelling $\varphi x'$, we deduce that $a_1 = a_2$. Now, the faithfulness of gf (hence of f) works as in adjunctions: given $a_i \colon x \to x'$ with $g f a_1 = g f a_2$, we get $\varphi x'.a_1 = g f a_i.\varphi x = \varphi x'.a_2$ and we cancel $\varphi x'$.

(b) The first assertion follows from (a). Then, assuming that φ is invertible, we have an adjunction $g \dashv f$ with an invertible counit $\varphi^{-1} \colon g f \to 1$, which implies that f is full and faithful (Section A3.3(d)).

(c) Assume that g is faithful and that all the components of φ are epi. Given an epimorphism $a \colon x \to x'$, we have that $g f a.\varphi x = \varphi x'.a$ is also epi, whence $g f(a)$ is epi, and finally $f(a)$ is too.

(d) Assume that gf is faithful and that all the components of φ are epi. Let $\varphi x.a_i = a \colon x' \to g f x$; then $g f a_i.\varphi x' = \varphi x.a_i = a$; cancelling $\varphi x'$ we have $g f a_1 = g f a_2$, and $a_1 = a_2$.

(e) The equivalence of (i) and (ii) follows from (a) and (d); the last point from (c). $\qquad\square$

3.3.3 The relation of future equivalence

Future equivalences can be composed (much in the same way as adjunctions, cf. Section A3.3(b)), which shows that *being future equivalent categories* is an equivalence relation. This depends on the fact that **Cat** is a 2-category, and cannot be extended to dh1-categories.

Indeed, given $(f, g; \varphi, \psi)$, as in (3.36), and a second future equivalence

$$h \colon Y \rightleftarrows Z \colon k, \qquad \vartheta \colon 1_Y \to kh, \quad \zeta \colon 1_Z \to hk,$$
$$h\vartheta = \zeta h \colon h \to hkh, \qquad \vartheta k = k\zeta \colon k \to khk,$$

their *composite* will be:

$$hf: X \rightleftarrows Z : gk, \qquad g\vartheta f.\varphi: 1 \to gk.hf, \qquad h\psi k.\zeta: 1 \to hf.gk. \quad (3.40)$$

Its coherence is proved by the following computation, where $fg\vartheta.\psi = \psi kh.\vartheta$

$$hf(g\vartheta f.\varphi) = h(fg\vartheta f.f\varphi) = h(fg\vartheta f.\psi f) = h(fg\vartheta.\psi)f,$$
$$(h\psi k.\zeta)hf = (h\psi kh.\zeta h)f = (h\psi kh.h\vartheta)f = h(\psi kh.\vartheta)f.$$

This composition is easily seen to be associative, with obvious identities.

Faithful future equivalences are closed under composition. Indeed, using the form (ii) of Section 3.3.1, it suffices to note that the general component $g(\vartheta fx).\varphi x$ is epi (also because g preserves epis, by Lemma 3.3.2(e)).

Two categories will be said to be *past and future equivalent* if they are both past equivalent and future equivalent. Generally, one needs *different* pairs of functors for these two notions (see Section 3.3.7); finer relations, where the past and future structure are linked together, will be introduced later and give more interesting results. Marginally, we also consider *coarse equivalence* of categories, as the equivalence relation generated by past equivalence and future equivalence.

3.3.4 Full reflective subcategories as future retracts

We deal now with a special case of future equivalence, which is important for its own sake, but will also be shown to generate the general case, in the next theorem.

A *split* future equivalence *of F into X* (or *of X onto F*) will be a future equivalence $(i, p; 1, \eta)$ where the unit $1 \to pi$ is an identity

$$
\begin{array}{llll}
i: F \rightleftarrows X : p, & \eta: 1_X \to ip & \text{(the \emph{main unit})}, & \\
pi = 1_F, & p\eta = 1_p, & \eta i = 1_i & (p \dashv i).
\end{array}
\quad (3.41)
$$

We also say that F is a *future retract* of X. Recall that the notion of a *strong future deformation retract*, as defined in Section 1.3.1 (in any dI1-category) only requires one coherence condition, namely $\eta i = 1_i$, and therefore is weaker than the present notion.

In (3.41), the functor p is left adjoint to i, which is full and faithful. (Note also that $(i, p; 1, \eta)$ is a split mono in the category of future equivalences, with retraction $(p, i; \eta, 1)$.)

As in Section 3.3.1, we say that this future equivalence is *faithful* – and that F is a *faithful future retract* of X – if all the components of η are mono; because of the adjunction, *this is equivalent to saying that p is faithful* (and implies that all the components of η are epi).

The structure we are considering means that the category F is (iso-morphic to) a *full reflective subcategory* of X, i.e. that there is a full embedding $i\colon F \to X$ with a left adjoint $p\colon X \to F$. Then p is essentially determined by i, and – via the universal property of the unit – can always be constructed so that the counit $pi \to 1_F$ is an identity, as we are assuming.

Equivalently, one can assign a *strictly idempotent monad* (e, η) on X

$$e\colon X \to X, \qquad \eta\colon 1_X \to e, \qquad ee = e, \qquad e\eta = 1_e = \eta e. \qquad (3.42)$$

Indeed, given $(i, p; \eta)$, we take $e = ip$; given (e, η), we factor $e = ip$ splitting e through the subcategory F of X formed of the objects and arrows which e leaves fixed.

Dually, a *split past equivalence, of P into X* (or *of X onto P*) is a past equivalence $(i, p; 1, \varepsilon)$ where the counit $pi \to 1_P$ is an identity

$$
\begin{aligned}
&i\colon P \rightleftarrows X :p, && \varepsilon\colon ip \to 1_X && \text{(the } \textit{counit}\text{)}, \\
&pi = 1_P, && p\varepsilon = 1_p, && \varepsilon i = 1_i && (i \dashv p).
\end{aligned}
\qquad (3.43)
$$

This amounts to saying that $i(P)$ is a *full coreflective subcategory* of X (with a choice of the coreflector making the unit $1 \to pi$ an identity); P will also be called a *past retract* of X.

3.3.5 Theorem
(Future equivalence and reflective subcategories)

(a) A future equivalence $(f, g; \varphi, \psi)$ *between* X *and* Y *(Section 3.3.1) has a canonical factorisation into two split future equivalences*

$$X \underset{p}{\overset{i}{\rightleftarrows}} W \underset{j}{\overset{q}{\rightleftarrows}} Y \qquad (\eta\colon 1_W \to ip, \ \eta'\colon 1_W \to jq), \qquad (3.44)$$

so that X *and* Y *are full reflective subcategories of* W.

(It is a mono-epi *factorisation in the category of future equivalences, through a sort of 'graph' of* $(f, g; \varphi, \psi)$.*)*

(b) Two categories are future equivalent if and only if they are full reflective subcategories of a third.

(c) Two categories are faithfully *future equivalent if and only if they are faithful* future retracts *of a third.*

(d) A property is future invariant if and only if it is preserved by all em-beddings of full reflective subcategories and by their reflectors. Similarly in the faithful case.

Proof (a). First, we construct the category W.

(i) An object is a four-tuple $(x, y; u, v)$ such that:

$$u\colon x \to gy \text{ (in } X), \quad v\colon y \to fx \text{ (in } Y), \quad gv.u = \varphi x, \quad fu.v = \psi y,$$

$$\text{(3.45)}$$

(ii) A morphism is a pair $(a, b)\colon (x, y; u, v) \to (x', y'; u', v')$ such that:

$$a\colon x \to x' \text{ (in } X), \quad b\colon y \to y' \text{ (in } Y), \quad gb.u = u'.a, \quad fa.v = v'.b,$$

$$\text{(3.46)}$$

Then, we have a split future equivalence of X into W:

$$
\begin{aligned}
&i\colon X \rightleftarrows W \colon p, && \eta\colon 1_W \to ip, \\
&i(x) = (x, fx; \varphi x, 1_{fx}), && i(a) = (a, fa), \\
&p(x, y; u, v) = x, && p(a, b) = a, \\
&\eta(x, y; u, v) = (1_x, v)\colon (x, y; u, v) \to (x, fx; \varphi x, 1_{fx}).
\end{aligned}
\qquad \text{(3.47)}
$$

The correctness of the definitions is easily verified, as well as the co-herence conditions: $pi = 1_W$, $p\eta = 1_p$, $\eta i = 1_i$ (in particular, i is well defined because the given equivalence is coherent).

Symmetrically, there is a split future equivalence of Y into W:

$$
\begin{aligned}
&j\colon Y \rightleftarrows W \colon q, && \eta'\colon 1_W \to jq, \\
&j(y) = (gy, y; 1_{gy}, \psi y), && j(b) = (gb, b), \\
&q(x, y; u, v) = y, && q(a, b) = b, \\
&\eta'(x, y; u, v) = (u, 1_y)\colon (x, y; u, v) \to (gy, y; 1_{gy}, \psi y).
\end{aligned}
$$

Finally, composing the two equivalences (3.44), as defined in (3.40), gives back the original future equivalence $(f, g; \varphi, \psi)$

$$
\begin{aligned}
qi(x) &= f(x), && qi(a) = f(a), \\
p\eta'i(x) &= p\eta'(x, fx; \varphi x, 1_{fx}) = p(\varphi x, 1_{fx}) = \varphi x.
\end{aligned}
$$

Now, (b) follows immediately from (a).

For (c), it suffices to modify the previous construction: if $(f, g; \varphi, \psi)$ is faithful, we use the full subcategory $W_0 \subset W$ on the objects $(x, y; u, v)$ where u and v are *mono*. Then, the functor i take values in W_0 (as $i(x) = (x, fx; \varphi x, 1_{fx}))$; we restrict p, η and get a future retract which is faithful, since the general component $\eta(x, y; u, v) = (1_x, v)$ is obviously mono. Symmetrically for j, q, η'. (One can also use a smaller full subcategory W_1, requiring that u, v be mono *and* epi.)

Finally, (d) is an obvious consequence. □

3.3.6 Definition and proposition (Strong contractibility)

The category X will be said to be strongly future contractible *if it satisfies the following equivalent conditions:*

(a) *the terminal object $\mathbf{1}$ (i.e. the singleton category $\{*\}$) is a strong future deformation retract of X (i.e. we have functors $t\colon \mathbf{1} \rightleftarrows X\colon p$ and a natural transformation $\eta\colon 1_X \to tp$ such that $\eta t = 1_t$);*

(b) *X is future equivalent to $\mathbf{1}$ (i.e. with the previous notations, we also have $p\eta = 1_p$);*

(c) *X has a terminal object.*

The fact that X be future contractible *(Section 1.3.2), i.e. future homotopy equivalent to $\mathbf{1}$ (without requiring coherence) is a strictly weaker condition.*

Symmetrically, a category is strongly past contractible *if and only if it has an initial object.*

Proof (a) $\Rightarrow$ (b). The remaining coherence condition $p\eta = 1\colon p \to p\colon X \to \mathbf{1}$ is automatically satisfied. (In other words, $\mathbf{1}$ is a 2-terminal object in **Cat**.)

(b) $\Rightarrow$ (c). A future equivalence $t\colon \mathbf{1} \rightleftarrows X\colon p$ necessarily splits, with unit $pt = 1$; thus $t\colon \mathbf{1} \to X$ is right adjoint to p and preserves the terminal object. (More analytically: every object x has a map $\eta x\colon x \to t(*)$; and indeed a unique one: given $a\colon x \to t(*)$, the naturality of η, *together with* the condition $\eta i = 1$, implies that $a = \eta x$.)

(c) $\Rightarrow$ (a). If X has a terminal object, we have a strong future deformation retract $t\colon \mathbf{1} \rightleftarrows X\colon p$, where $\eta x\colon x \to t(*)$ is the unique map to the terminal object of X.

As to the last assertion, the idempotent two-element monoid $M = \{1, a\}$, viewed as a category on one formal object $*$, has no terminal object but is future contractible, with $\eta(*) = a$. $\qquad\square$

3.3.7 Other notions of contractibility

Faithful strong contractibility is much more restrictive than strong contractibility. In fact, *the* functor $p\colon X \to \mathbf{1}$ *is faithful if and only if each hom-set of X has at most one element, which means that X 'is' a preordered set.* Therefore, a category is *faithfully* strongly future contractible – i.e. faithfully future equivalent to $\mathbf{1}$ – if and only if it is a *preordered set with a maximum*; and dually for the past.

Finally, a category is *past and future strongly contractible* (i.e. past and future equivalent to $\mathbf{1}$) if and only if it has an initial *and* a terminal object. Then, the *future embedding* $(t\colon \mathbf{1} \to X)$ and the *past* one $(i\colon \mathbf{1} \to X)$ can only coincide if X has a zero object (this will amount to contractibility for the finer relation of injective equivalence studied later, see Section 3.6.4).

Marginally, we also use the notion of *coarse contractibility*, meaning coarse equivalent to $\mathbf{1}$ (Section 3.3.3). Examples for all these cases will be considered in Section 3.3.9.

The *future cone $C^+ X$*, obtained by freely adding a terminal object to the category X, is future strongly contractible; it is also past strongly contractible if and only if X is past strongly contractible or empty.

3.3.8 Lemma (Maximal points)

The following properties of an object $x \in X$ are future invariant:

(a) x *is the* terminal *object of the category X;*
(b) x *is a* weak terminal *object of X, i.e. a* maximum *for the path preorder $\preceq$ (Section 3.3.0);*
(c) x *is* maximal *in X, for the path preorder;*
(d) x *does not reach a maximal point z.*

Proof Let $f\colon X \rightleftarrows Y\colon g$ be a future equivalence of small categories.

(a) Follows immediately from Proposition 3.3.6: if we compose the future equivalence $t\colon \mathbf{1} \rightleftarrows X\colon p$ produced by the terminal object x with the future equivalence $X \rightleftarrows Y$, we get a composite $ft\colon \mathbf{1} \rightleftarrows Y\colon pg$, which shows that $ft(*) = f(x)$ is terminal in Y.

(b) If x is a maximum in X, then, for every $y \in Y$: $g(y) \preceq x$ and $y \preceq fg(y) \preceq f(x)$.

(c) Let x be maximal in X, and $f(x) \preceq y$ in Y. Then $x \preceq gf(x) \preceq g(y)$ and all these points are equivalent, whence $f(x) \simeq fg(y)$. But $f(x) \preceq y \preceq fg(y)$ and $y \simeq f(x)$.

(d) Since z is maximal, from $z \preceq gf(z)$ we deduce that $z \simeq gf(z)$. Therefore, if $f(x) \preceq f(z)$ in Y, we have $x \preceq gf(x) \preceq gf(z) \simeq z$, and $x \preceq z$ in X. $\qquad\qquad\square$

3.3.9 Elementary examples

(a) Let us begin with some examples consisting of finite or countable ordered sets.

For preordered sets, viewed as categories, a future equivalence consists of a pair of preorder-preserving mappings $f \colon X \rightleftarrows Y \colon g$ such that $1_X \leqslant gf$ and $1_Y \leqslant fg$, and *is necessarily faithful*. We already know that future contractibility (necessarily strong) means having a maximum. Therefore:

(past and future contractible)

(just future contractible)

(just past contractible)

(just coarse-contractible).

(b) Consider again (as in the third set of examples, above) the ordered set **n** of natural numbers, as a category. (Not to be confused with the monoid **N**, a quite different category on one object.) There are future equivalences

$$f \colon \mathbf{n} \rightleftarrows \mathbf{n} \colon g, \qquad f(x) = x, \qquad g(x) = \max(x, x_0),$$
$$\varphi(x) = \psi(x) \colon x \leqslant g(x),$$

where $x_0 \in \mathbf{n}$ is arbitrary (and coherence automatically holds, since our categories are preorders).

This proves that, in a future equivalence, the functor f does *not* determine g. Note also (in relation with a previous result, Lemma 3.3.2(b)) that all components $\psi(x)$ are mono and epi, but g is not full, i.e. does not reflect the preorder (when $x_0 > 0$).

(c) Now we consider some finite categories, generated by the directed graphs drawn below; the cross-marked cells do not commute and these categories are not preorders. The category represented in (3.48)

$$0 \longrightarrow a \qquad \times \qquad b \longrightarrow 1 \qquad\qquad (3.48)$$

is (faithfully) future equivalent to the first of the following list, past equivalent to the second, past and future equivalent to the third and coarse-equivalent to the last

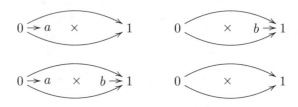

This shows a situation of interest in concurrency (see, for instance, [FGR1, FGR2]). There is a given *starting point* 0, which is minimal (Section 3.3.0), but *not* initial *nor* the unique minimal point (generally); and a given *ending point* 1, which is maximal. Moreover:

- 0 is also a *future branching* point, where one has to choose among different ways of going *forward*; being such is a future invariant property (as will be proved in Theorem 3.7.7);
- a is a *deadlock*, i.e. a maximal *unsafe vertex* (from where one cannot reach 1); this is again a future invariant property, as already proved in Lemma 3.3.8(d);
- b is a minimal *unreachable vertex* (which cannot be reached from 0); being such is a past invariant property (according to the dual of Lemma 3.3.8);
- 1 is a *past branching* point, preserved by past equivalences (Theorem 3.7.7).

The 'past and future model' above *preserves all these properties*, while the coarse one only says that there are two paths from 0 to 1.

(d) Finally, the following category (described by generators and relations)

$$
\begin{array}{ccc}
0 \xrightarrow{\ h\ } a & & uh = vh = h', \\
\end{array}
\qquad (3.49)
$$

(diagram with $0 \xrightarrow{h} a$, h' and $u \Downarrow v$ to b, k' to 1, $b \xrightarrow{k} 1$)

$$
ku = kv = k',
$$

has an initial object (0) and a terminal one (1): it is past and future strongly contractible, but *not* faithfully so. Note also that, in the future contraction, all the components $x \to 1$ of the unit are epimorphisms.

3.4 Bilateral directed equivalences of categories

We have already considered categories which are 'separately' past and future equivalent (e.g. in Section 3.3.7). However, an *unrelated* pair formed of a past equivalence and a future equivalence between the same categories is not an effective tool.

A better system, which we call a *pf-equivalence*, consists of a past and a future equivalence *which share one functor*. A particular case has been studied in category theory: essential localisations (Section 3.4.7).

In this section, g^- and g^+ denote functors; connections are not used.

3.4.1 Pf-equivalences

Let X, Y be (small) categories. A *pf-equivalence from X to Y* will be a pair formed of a past equivalence $(f, g^-; \varepsilon_X, \varepsilon_Y)$ and a future equivalence $(f, g^+; \eta_X, \eta_Y)$ sharing the same functor $f \colon X \to Y$, and also satisfying a further *pf-coherence* condition (3.51) which links the two pairs:

$$
\begin{aligned}
&f \colon X \to Y, & &g^-, g^+ \colon Y \to X, \\
&\varepsilon_X \colon g^- f \to 1_X, & &\varepsilon_Y \colon f g^- \to 1_Y, \\
&f \varepsilon_X = \varepsilon_Y f \colon f g^- f \to f, & &\varepsilon_X g^- = g^- \varepsilon_Y \colon g^- f g^- \to g^-, \quad (3.50) \\
&\eta_X \colon 1_X \to g^+ f, & &\eta_Y \colon 1_Y \to f g^+, \\
&f \eta_X = \eta_Y f \colon f \to f g^+ f, & &\eta_X g^+ = g^+ \eta_Y \colon g \to g^+ f g^+,
\end{aligned}
$$

$$
\begin{array}{ccc}
g^- & \xrightarrow{\ g^- \eta_Y\ } & g^- f g^+ \\
{\scriptstyle \eta_X g^-} \downarrow & = & \downarrow {\scriptstyle \varepsilon_X g^+} \\
g^+ f g^- & \xrightarrow[\ g^+ \varepsilon_Y\ ]{} & g^+
\end{array}
\qquad \textit{(pf-coherence)}. \qquad (3.51)
$$

This yields a natural transformation, the *comparison* from past to future

$$g: g^- \to g^+ : Y \to X, \qquad g = \varepsilon_X g^+ . g^- \eta_Y = g^+ \varepsilon_Y . \eta_X g^-, \qquad (3.52)$$

which, when convenient, will be seen as a functor $g: Y \to X^{\mathbf{2}}$ with values in the category of morphisms of X

$$g: Y \to X^{\mathbf{2}}, \qquad gy: g^- y \to g^+ y, \qquad g(b) = (g^- b, g^+ b). \qquad (3.53)$$

A pf-equivalence will often be written in one of the follwing forms

$$f: X \rightleftharpoons Y, \qquad f: X \rightleftharpoons Y : g^\alpha,$$

leaving the rest understood.

It will be said to be *faithful* if both the past and the future equivalence which compose it are faithful. By Section 3.3.1, this is the case if and only if our data satisfy these equivalent conditions:

(i) all the components of η_X, η_Y are mono and all the components of $\varepsilon_X, \varepsilon_Y$ are epi;

(ii) the functors f, g^-, g^+ are faithful; all the components of η_X, η_Y are epi and all the components of $\varepsilon_X, \varepsilon_Y$ are mono.

Two dual types of pf-equivalences, where g^-, g^+ are 'split' adjoint to f, will be treated below.

3.4.2 Composition of pf-equivalences

A pf-equivalence is not a symmetric structure. But they compose, by the composition of past equivalences and future equivalences (Section 3.3.3).

Thus, given $f: X \rightleftharpoons Y : g^\alpha$ (as in (3.50)) and a second pf-equivalence

$$h: Y \rightleftharpoons Z : k^-, k^+,$$
$$\sigma_Y : k^- h \to 1_Y, \qquad\qquad \sigma_Z : hk^- \to 1_Z, \qquad (3.54)$$
$$\zeta_Y : 1_Y \to k^+ h, \qquad\qquad \zeta_Z : 1_Z \to hk^+,$$

their composite is:

$$hf: X \rightleftharpoons Z : g^\alpha k^\alpha \qquad\qquad (\alpha = \pm),$$
$$\varepsilon_X . g^- \sigma_Y f: \ g^- k^- . hf \to g^- f \to 1_X, \qquad (3.55)$$
$$\sigma_Z . h\varepsilon_Y k^-: \ hf.g^- k^- \to hf.g^- k^- \to 1_Z, \dots$$

The following diagram shows that coherence holds (functors are replaced with dots, in the labels of arrows)

$$
\begin{array}{ccccc}
g^-k^- & \xrightarrow{\ \cdot\zeta_Z\ } & g^-k^-hk^+ & \xrightarrow{\ \cdot\eta_Y\ \cdot\ } & g^-k^-hfg^+k^+ \\
\| & & \downarrow{\scriptstyle \cdot\eta_Y\cdot} & & \downarrow{\scriptstyle \cdot\sigma_Y\cdot} \\
g^-k^- \xrightarrow{\cdot\eta_Y\cdot} g^-fg^+k^- & \xrightarrow{\cdot\zeta_Z} & g^-fg^+k^-hk^+ & \xrightarrow{\ \cdot\sigma_Y\cdot\ } & g^-fg^+k^+ \\
\downarrow{\scriptstyle \eta_X\cdot} \quad \square \quad \downarrow{\scriptstyle \cdot\varepsilon_X} & & \downarrow{\scriptstyle \varepsilon_X\cdot} & & \\
g^+fg^-k^- \xrightarrow[\cdot\varepsilon_Y\cdot]{} g^+k^- & \xrightarrow{\ \cdot\zeta_Z\ } & g^+k^-hk^+ & & \downarrow{\scriptstyle \varepsilon_X\cdot} \\
\downarrow{\scriptstyle \cdot\zeta_Y\cdot} \qquad \downarrow{\scriptstyle \cdot\zeta_Y\cdot} \quad \square \quad \downarrow{\scriptstyle \cdot\sigma_Y\cdot} & & & & \\
g^+k^+hfg^-k^- \underset{\cdot\varepsilon_Y\cdot}{\rightrightarrows} g^+k^+hk^- & \xrightarrow[\cdot\sigma_Z]{} & g^+k^+ & = & g^+k^+
\end{array}
$$

In fact, the outer square commutes, because all the inner ones do, either by pf-coherence of the data (when marked with a box) or by middle-four interchange.

In particular, the commutativity of the right upper square comes from applying middle-four interchange twice:

$$
\begin{array}{ccc}
g^-k^-hk^+ & \xrightarrow{\ \cdot\eta_Y\cdot\ } & g^-k^-hfg^+k^+ \\
\downarrow{\scriptstyle \cdot\eta_Y\cdot} \quad \overset{\cdot\sigma_Y\cdot}{\nearrow} \ g^-k^+ & & \downarrow{\scriptstyle \cdot\sigma_Y\cdot} \\
& \overset{\cdot\eta_Y\cdot}{\searrow} & \\
g^-fg^+k^-hk^+ & \xrightarrow[\ \cdot\sigma_Y\cdot\ ]{} & g^-fg^+k^+
\end{array}
$$

3.4.3 Lemma (Pf-coherence)

In a pf-equivalence $f \colon X \rightleftarrows Y \colon g^{\alpha}$, *the condition of pf-coherence is redundant (i.e. follows from the other axioms) whenever* f *is faithful or surjective on objects.*

Proof Indeed, post- and pre-composing the diagram (3.51) with the functor f, we get two diagrams whose commutativity follows from the other coherence conditions, together with middle-four interchange

$$
\begin{array}{ccc}
fg^- \xrightarrow{fg^-\eta_Y} fg^-fg^+ & \qquad & g^-f \xrightarrow{g^-\eta_Y f} g^-fg^+f \\
\downarrow{\scriptstyle f\eta_X g^-} \ \ {\scriptstyle 1_Y} \ \ \downarrow{\scriptstyle f\varepsilon_X g^+} & & \downarrow{\scriptstyle \eta_X g^-f} \ \ {\scriptstyle 1_X} \ \ \downarrow{\scriptstyle \varepsilon_X g^+ f} \\
fg^+fg^- \xrightarrow[fg^+\varepsilon_Y]{} fg^+ & & g^+fg^-f \xrightarrow[g^+\varepsilon_Y]{} g^+f
\end{array}
\tag{3.56}
$$

Since a faithful functor is left-cancellable with respect to *parallel* natural transformations ($f\varphi = f\psi$ implies $\varphi = \psi$), while a functor surjective on objects is right-cancellable, the thesis follows.   □

3.4.4 Injections and projections

(a) A pf-equivalence $f\colon X \rightleftharpoons Y : g^{\alpha}$ will be called a *pf-injection*, or *pf-embedding*, if the functor f is a *full embedding* (i.e. full, faithful and injective on objects). Pf-embeddings compose, with the composition of pf-equivalences (Section 3.4.2); they will give rise to the notion of an 'injective model' of a category (Section 3.5.1).

It is easy to see that a pf-embedding $f\colon X \rightleftharpoons Y : g^{\alpha}$ amounts to these three functors together with the *two* natural transformations *at Y*, satisfying the conditions below

$$\varepsilon_Y : fg^- \to 1_Y \qquad \text{(the *main counit*),}$$
$$\eta_Y : 1_Y \to fg^+ \qquad \text{(the *main unit*),} \qquad (3.57)$$
$$fg^-\varepsilon_Y = \varepsilon_Y fg^-, \qquad fg^+\eta_Y = \eta_Y fg^+.$$

In fact, these data can be uniquely completed to a pf-injection: there is a unique natural transformation $\eta_X : 1_X \to g^+ f$ (the *secondary unit*) such that $f\eta_X = \eta_Y f : f \to fg^+ f$ (because the latter transformation lives in the *full image* of f in Y, i.e. the full subcategory of Y determined by the objects which are reached by the functor f). The coherence relation $\eta_X g^+ = g^+ \eta_Y$ comes from cancelling f in $f(\eta_X g^+) = \eta_Y fg^+ = f(g^+ \eta_Y)$. Similarly, there is one $\varepsilon_X : g^- f \to 1_X$ such that $f\varepsilon_X = \varepsilon_Y f$ (and it is coherent with ε_Y). Finally, the global pf-coherence condition (3.51) automatically holds, by the previous lemma.

(b) A pf-equivalence $f\colon X \rightleftharpoons Y : g^{\alpha}$ will be called a *pf-surjection* if the functor f is surjective on objects, and a *pf-projection* if, moreover, the associated functor $g\colon Y \to X^2$ (3.53) is a full embedding. The latter structure will give a 'projective model' of the category X (Section 3.5.1).

We already know that, in a pf-surjection, pf-coherence is automatic (Lemma 3.4.3); it is also obvious that the transformations *at Y* are determined by the transformations at X (since $\varepsilon_Y f = f\varepsilon_X$, $\eta_Y f = f\eta_X$), but here it seems to be less easy to deduce the former from the latter.

3.4.5 Theorem (The middle model)

A pf-equivalence $f \colon X \rightleftarrows Y \colon g^\alpha$ has an associated pf-surjection *and an* associated pf-injection

$$p \colon X \rightleftarrows Z \colon r^\alpha, \qquad i \colon Z \rightleftarrows Y \colon h^\alpha, \qquad (3.58)$$

where $f = ip$. This determines Z (the middle model*), i and p up to category isomorphism; and one can always take for Z the full subcategory of Y on the objects fx ($x \in X$). Moreover, if the given pf-equivalence is faithful, so are the two associated ones.*

(In general, this is not a factorisation: the composition of these two pf-equivalences does not *give back the original one. Furthermore, the pf-surjection need not be a pf-projection, but this will be true in the cases of interest, below.)*

Proof Let us write the units and counits of $f \colon X \rightleftarrows Y \colon g^\alpha$ as in (3.50). The functor $f \colon X \to Y$ has an essentially unique factorisation $f = ip$ where p is surjective on objects and $i \colon Z \to Y$ is a full embedding: take as Z the full image of f (cf. Section 3.4.4).

Then, we define four functors ($\alpha = \pm$)

$$r^\alpha = g^\alpha i \colon Z \to X, \qquad h^\alpha = pg^\alpha \colon Y \to Z, \qquad (3.59)$$

so that:

$$r^\alpha p = g^\alpha f, \qquad ih^\alpha = fg^\alpha,$$
$$pr^\alpha = pg^\alpha i = h^\alpha i, \qquad r^\alpha h^\alpha = g^\alpha ipg^\alpha = g^\alpha fg^\alpha.$$

(Here we can already note that $r^\alpha h^\alpha$ need not be g^α.) Now, for the pf-injection (i, h^α), we only need to observe that the original natural transformations ε_Y, η_Y work as *main* counit and unit (cf. (3.57))

$$\varepsilon_Y \colon ih^- = fg^- \to 1_Y, \qquad \eta_Y \colon 1_Y \to ih^+ = fg^+,$$

since we already know that they commute with $ih^- = fg^-$ and $ih^+ = fg^+$, respectively.

On the other hand, the first pf-equivalence (p, r^α) is completed with the natural transformations

$$\varepsilon_X \colon r^- p = g^- f \to 1_X, \qquad \eta_X \colon 1_X \to r^+ p = g^+ f,$$
$$\varepsilon_Z \colon pr^- \to 1_Z, \qquad i\varepsilon_Z = \varepsilon_Y i \colon ipr^- = fg^- i \to i,$$
$$\eta_Z \colon 1_Z \to pr^+, \qquad i\eta_Z = \eta_Y i \colon i \to ipr^+ i = fg^+ i,$$

where ε_X, η_X are the original ones; ε_Z is a restriction of ε_Y (justified by the fact that $\varepsilon_Y i: fg^- i = ipr \to i$ lives in the full subcategory Z); and, similarly, η_Z is a restriction of η_Y.

Its coherence is deduced below, in brackets, from the analogous properties of the original data (recall that the pf-coherence relation need not be checked, by Lemma 3.4.3)

$$
\begin{aligned}
p\varepsilon_X &= \varepsilon_Z p & (ip\varepsilon_X &= f\varepsilon_X = \varepsilon_Y f = \varepsilon_Y ip = i\varepsilon_Z p), \\
\varepsilon_X r^- &= r^- \varepsilon_Z & (\varepsilon_X r^- &= \varepsilon_X g^- i = g^- \varepsilon_Y i = g^- i\varepsilon_Z = r^- \varepsilon_Z), \\
p\eta_X &= \eta_Z p & (ip\eta_X &= f\eta_X = \eta_Y f = \eta_Y ip = i\eta_Z p), \\
\eta_X r^+ &= r^+ \eta_Z & (\eta_X r^+ &= \eta_X g^+ i = g^+ \eta_Y i = g^+ i\eta_Z = r^+ \eta_Z).
\end{aligned}
$$

Finally, let us assume that the original pf-equivalence is faithful (Section 3.4.1). We know that all the components of η_X, η_Y are mono, whence this is also true of the components of $i\eta_Z = \eta_Y i$, and then of the ones of η_Z, because i is faithful. Dually for counits. $\qquad\square$

3.4.6 Split pf-injections

A *split pf-injection*, or *adjoint reflexive graph*, will be a pf-equivalence $i: E \rightleftarrows X : p^\alpha$ where the natural transformations $\varepsilon_E: p^- i \to 1_E$ and $\eta_E: 1_E \to p^+ i$ are identities. We show below that this essentially means that E is *a full subcategory, reflective and coreflective,* of X.

In fact, a split pf-injection consists of three functors $i: E \rightleftarrows X : p^\alpha$ and two natural transformations ε and η such that:

$$
\begin{aligned}
&i: E \to X, & &p^+ \dashv i \dashv p^-, \\
&\varepsilon: ip^- \to 1_X & &\text{(the *past counit*)}, \\
&\eta: 1_X \to ip^+ & &\text{(the *future unit*)}, \\
&p^- i = 1_E = p^+ i, \ \ p^- \varepsilon = 1, & &\varepsilon i = 1, \ \ p^+ \eta = 1, \ \ \eta i = 1.
\end{aligned}
\tag{3.60}
$$

Note that i is a full embedding (because the adjunction $p^+ \dashv i$ has an invertible counit, η_E^{-1}), so that we do have a pf-injection. Furthermore, the functor i essentially determines the rest of the structure: it embeds E as a full subcategory, reflective and coreflective, with reflector p^+ and coreflector p^-. Conversely, given a full subcategory $E \subset X$, which is reflective and coreflective, we can always choose the reflector so that the counit is an identity, and the coreflector so that the unit is an identity.

Because of pf-coherence, there is a canonical comparison $p\colon p^- \to p^+$ from the right adjoint to the left:

$$p = p^-\eta = p^+\varepsilon\colon p^- \to p^+ \colon X \to E$$
$$(\eta.\varepsilon = ip^-\eta = ip^+\varepsilon\colon ip^- \to ip^+). \tag{3.61}$$

The equations in brackets follow from middle-four interchange or (3.56), and can also be useful.

Examples related to the present notions will be given in Section 3.6.

Forgetting about smallness, there is an elegant example in homological algebra which presents homology in a symmetric way. Start from the embedding $i\colon \mathbf{G_\bullet Ab} \to \mathbf{Ch_\bullet Ab}$ of (unbounded) graded abelian groups into chain complexes, as complexes with a null differential. The left and right adjoints are computed on a chain complex $A = (A_\bullet, \partial_\bullet)$, as

$$p^+A = \mathrm{Cok}(\partial_\bullet) = A_\bullet/\partial_\bullet(A_\bullet), \qquad p^-A = \mathrm{Ker}(\partial_\bullet). \tag{3.62}$$

Now, the graded group $H_\bullet(A)$ can be defined as the image of the comparison $pA\colon p^-A \to p^+A$.

3.4.7 Split pf-projections

The dual notion of *split pf-projection* is well-known in category theory: it has been studied under the name of *essential localisation* [KL, BK], or 'unity and identity of adjoint opposites' [Lw2]; more recently, the term of 'adjoint reflexive cograph' has also been used by F.W. Lawvere.

It can be presented as a pf-equivalence $p\colon X \;\rightleftarrows\; M : i^\alpha$ where the natural transformations $\varepsilon_M\colon pi^- \to 1_M$ and $\eta_M\colon 1_M \to pi^+$ are identities. The structure thus consists of three functors (p, i^-, i^+) and two natural transformations satisfying:

$$
\begin{aligned}
&p\colon X \to M, && i^- \dashv p \dashv i^+, \\
&\varepsilon\colon i^-p \to 1_X && (\text{the } \textit{past counit}), \\
&\eta\colon 1_X \to i^+p && (\text{the } \textit{future unit}), \\
&pi^- = 1_M = pi^+, \quad p\varepsilon = 1, && \varepsilon i^- = 1, \quad p\eta = 1, \quad \eta i^+ = 1.
\end{aligned}
\tag{3.63}
$$

Thus, p is surjective on objects (and maps as well), so that pf-coherence is automatic (by Lemma 3.4.3) and we have a pf-equivalence, actually a pf-surjection, which we prove below to be a pf-projection (Proposition 3.4.8).

Again, by pf-coherence, we have *one* comparison $i \colon i^- \to i^+$

$$i = \eta i^- = \varepsilon i^+ \colon i^- \to i^+ \colon M \to X,$$
$$(\eta . \varepsilon = \eta i^- p = \varepsilon i^+ p \colon i^- p \to i^+ p). \tag{3.64}$$

Examples will be given in Section 3.6. But we can already note that (forgetting about smallness) the forgetful functor $p \colon \mathbf{Top} \to \mathbf{Set}$ from topological spaces to sets has such a structure, with left (resp. right) adjoint provided by the discrete (resp. indiscrete) topology

$$p \colon \mathbf{Top} \; \rightleftarrows \; \mathbf{Set} \colon i^\alpha, \qquad i^- \dashv p \dashv i^+ \quad (\varepsilon \colon i^- p \to 1, \; \eta \colon 1 \to i^+ p),$$

(so that $\mathbf{Set}$ is a faithful projective model of $\mathbf{Top}$, as defined in Section 3.5.1).

3.4.8 Proposition (Split pf-projections)

The structure described in (3.63) is a pf-projection (Section 3.4.4(b)).

Proof We have already observed that p is surjective on objects, so that our data define a pf-equivalence, and actually a pf-surjection (Section 3.4.4(b)).

Moreover, the embeddings i^- and i^+ are full and faithful, because the past unit and the future counit are invertible. (Here, one may recall that, starting from a pair of adjunctions $i^- \dashv p \dashv i^+$ in a 2-category, it is well-known – but not obvious – that the unit of the first adjunction is invertible if and only if the counit of the second is; cf. [KL], Proposition 2.3.)

Thus, the comparison $i \colon M \to X^2$ is also an embedding. To prove that i is full, take a morphism in X^2, from iy to iy'; since i^- and i^+ are full, this morphism can be written as the commutative square below

$$
\begin{array}{ccc}
i^- y & \xrightarrow{\; iy \;} & i^+ y \\
{\scriptstyle i^- b'} \downarrow & & \downarrow {\scriptstyle i^+ b''} \\
i^- y' & \xrightarrow[\; iy' \;]{} & i^+ y'
\end{array}
\qquad i = \eta i^- = \varepsilon i^+. \tag{3.65}
$$

Applying p, and noting that the natural transformation pi is the identity, we deduce that $b' = b''$. Calling $b \colon y \to y'$ this morphism of M, the given square is $i(b) \colon iy \to iy'$. $\square$

3.4.9 Two structural pf-equivalences

(a) Every category X has a structural split pf-injection *into* its category of morphisms X^2, determined by the cocylinder structure of the latter (or, equivalently, by the structure of **2** as a reflexive graph in **Cat**)

$$e\colon X \rightleftarrows X^2 \colon \partial^-, \partial^+, \qquad (3.66)$$

$$
\begin{aligned}
&e(x) = 1_x \colon x \to x; \qquad \partial^\alpha(a\colon x^- \to x^+) = x^\alpha &&(\alpha = \pm), \\
&\varepsilon(a\colon x^- \to x^+) = (1, a)\colon 1_{x^-} \to a &&\text{(the counit)}, \\
&\eta(a\colon x^- \to x^+) = (a, 1)\colon a \to 1_{x^+} &&\text{(the unit)}, \\
&\partial = \mathrm{id}\colon X^2 \to X^2 &&\text{(the comparison)}.
\end{aligned}
$$

(b) Dually, there is a structural split pf-*projection* from $X \times \mathbf{2}$ *onto* X, determined by the cylinder structure of $X \times \mathbf{2}$

$$e\colon X \times \mathbf{2} \rightleftarrows X \colon \partial^-, \partial^+, \qquad (3.67)$$

$$
\begin{aligned}
&e(x, \alpha) = x; \qquad \partial^\alpha(x) = (x, \alpha) &&(\alpha = 0, 1), \\
&\varepsilon(x, \alpha) = (x, 0 \to \alpha)\colon (x, 0) \to (x, \alpha) &&\text{(the counit)}, \\
&\eta(x, \alpha) = (x, \alpha \to 1)\colon (x, \alpha) \to (x, 1) &&\text{(the unit)}, \\
&\partial = \mathrm{id}\colon X \times \mathbf{2} \to X \times \mathbf{2} &&\text{(the comparison)}.
\end{aligned}
$$

3.5 Injective and projective models of categories

Injective and projective models, defined in Section 3.5.1, will be the main tool of this section. A pf-presentation of a category, formed of a past and a future retract (Section 3.5.2), yields both an injective model (Theorem 3.5.3) and a projective one (Theorem 3.5.7), for the given category.

3.5.1 Main definitions

(a) Let $i\colon E \rightleftarrows X$ be a pf-embedding, i.e. a pf-equivalence where i is a full embedding (Section 3.4.4(a)). In this situation, we say that E is an *injective model* of X, and that X is *injectively modelled* by E.

Two categories will be said to be *injectively equivalent* if they can be linked by a finite chain of pf-embeddings, forward or backward. *Faithful* pf-injections (Section 3.4.1) give raise to *faithful injective models* and *faithfully injectively equivalent* categories.

(b) Similarly, a *surjective model* M of X is given by a pf-surjection $p\colon X \rightleftarrows M\colon r^\alpha$, i.e. a pf-equivalence where p is surjective on objects (Section 3.4.4(b)).

More particularly (and more interestingly), a *projective model* M of X is given by a pf-projection $p\colon X \rightleftarrows M\colon r^\alpha$, i.e. a pf-surjection where the associated functor $r\colon M \to X^2$ is a full embedding (Section 3.4.4(b)). Such a model will generally be seen as a full subcategory $r\colon M \to X^2$. The *projective equivalence* relation is generated by pf-projections. The *faithful* case is defined analogously.

In the rest of this section, *the faithful case will generally be inserted in square brackets.*

3.5.2 Pf-presentations

We now introduce another structure which combines past and future notions, and will be shown to give rise to an injective model (Theorem 3.5.3) and a projective one (Theorem 3.5.7).

A [faithful] *pf-presentation* of the category X will be a diagram consisting of a [faithful] past retract P and a [faithful] future retract F of X

$$P \underset{p^-}{\overset{i^-}{\rightleftarrows}} X \underset{i^+}{\overset{p^+}{\rightleftarrows}} F \tag{3.68}$$

$$\varepsilon\colon i^- p^- \to 1_X \qquad (p^- i^- = 1,\ p^- \varepsilon = 1,\ \varepsilon i^- = 1),$$
$$\eta\colon 1_X \to i^+ p^+ \qquad (p^+ i^+ = 1,\ p^+ \eta = 1,\ \eta i^+ = 1),$$

so that P is a full coreflective subcategory of X, while F is a full reflective subcategory.

We thus have two split adjunctions $i^- \dashv p^-$, $p^+ \dashv i^+$; and a composed one, from P to F, which no longer splits, with the following counit and unit

$$p^+ i^- \dashv p^- i^+,$$
$$p^+ \varepsilon i^+\colon p^+ i^- . p^- i^+ \to p^+ i^+ = 1_F, \qquad p^- \eta i^-\colon 1_P = p^- i^- \to p^- i^+ p^+ i^-.$$

3.5.3 Theorem and definition
(Pf-presentations and injective models)

Given a [faithful] pf-presentation of the category X (written as above, in (3.68)), let E be the full subcategory of X on $\mathrm{Ob}P \cup \mathrm{Ob}F$ and u its embedding in X.

(a) These data can be uniquely completed to a diagram with four commutative squares (adding the functors j^α, q^α of the lower row)

$$
\begin{array}{ccccc}
P & \underset{p^-}{\overset{i^-}{\rightrightarrows}} & X & \underset{i^+}{\overset{p^+}{\rightrightarrows}} & F \\[2pt]
\| & & \uparrow u & & \| \\[2pt]
P & \underset{q^-}{\overset{j^-}{\rightrightarrows}} & E & \underset{j^+}{\overset{q^+}{\rightrightarrows}} & F
\end{array}
\qquad\qquad
\begin{array}{c}
X \\[2pt]
r^-\,\Big\updownarrow\,r^+ \\[2pt]
E
\end{array}
\qquad\qquad (3.69)
$$

Moreover:

(b) there is a unique natural transformation $\varepsilon_E : j^- q^- \to 1_E$ such that $u\varepsilon_E = \varepsilon u$;

(c) there is a unique natural transformation $\eta_E : 1_E \to j^+ q^+$ such that $u\eta_E = \eta u$;

(d) these transformations make the lower row a [faithful] pf-presentation of E;

(e) letting $r^\alpha = j^\alpha p^\alpha : X \to E$ ($\alpha = \pm$), we get a [faithful] pf-embedding

$$
(u, r^-, r^+; \varepsilon_E, \varepsilon, \eta_E, \eta) \qquad\qquad u : E \to X,
$$

and E will be called the [faithful] injective model generated by the given [faithful] pf-presentation of X.

(f) The functors $ur^\alpha : X \to X$ are idempotents, with $ur^- \varepsilon = 1_{ur^-} = ur^- \varepsilon$ and $ur^+ \eta = 1_{ur^+} = ur^+ \eta$.

Proof

(a) We (must) take $j^+ : F \subset E$ (so that $uj^+ = i^+$) and $q^+ = p^+ u : E \to F$; dually $j^- : P \subset E$ and $q^- = p^- u : E \to P$.

Now, we prove (b) to (d), completing the lower row of diagram (3.69) to a pf-presentation of E, as stated. On the right-hand side, we already know that $q^+ j^+ = p^+ u j^+ = p^+ i^+ = 1_F$. Moreover, all the components of $\eta u : u \to i^+ p^+ u : E \to X$ belong to the (full) subcategory E, because both its functors take values there (since $i^+ p^+ u = uj^+ q^+$); there is thus a unique natural transformation $\eta_E : 1_E \to j^+ q^+$ such that $u\eta_E = \eta u$, and it is easy to verify that $\eta_E j^+ = 1$ and $q^+ \eta_E = 1$.

(e) We complete the pf-embedding by letting $r^\alpha = j^\alpha p^\alpha : X \to E$, and observe that:

$$
ur^+ = uj^+ p^+ = i^+ p^+, \qquad\qquad r^+ u = j^+ p^+ u = j^+ q^+.
$$

Therefore, we can take the natural transformation

$$\eta\colon 1_X \to i^+ p^+ = u r^+ ,$$

as main unit (cf. (3.57)) of the pf-embedding $u\colon E \;\rightleftarrows\; X : r^\alpha$; the secondary unit is η_E, by (c); similarly for counits. Finally, point (f) is a straightforward consequence of $i^\alpha p^\alpha = u r^\alpha$.

[The faithful case is proved in the same way. Point (d) requires a specific argument: we know that all the components of η are mono, whence the same holds for the components of $u\eta_E = \eta u$, and also for those of η_E, since u is faithful; dually for counits.] $\square$

3.5.4 Comments

Given a pf-presentation of the category X, with the same notation as in Theorem 3.5.3:

(a) composing the future equivalences $F \rightleftarrows X \rightleftarrows E$, one gets the pair
$j^+ : F \rightleftarrows E : q^+$;

(b) composing the future equivalences $F \rightleftarrows E \rightleftarrows X$, one gets the pair
$i^+ : F \rightleftarrows X : p^+$;

and symmetrically for the past retracts.

On the other hand, the future equivalence $u\colon E \rightleftarrows X : r^+$ *is not the composition* of the future equivalences $E \rightleftarrows F \rightleftarrows X$ (in general): the image of $i^+ q^+ : E \to X$ is F, instead of E.

3.5.5 Factorisation of adjunctions

We have already seen, in Theorem 3.3.5, that a future equivalence has a canonical factorisation into a future section followed by a future retraction. Similarly, we show now that an adjunction has a canonical factorisation into a *past section* (the embedding of a full coreflective subcategory) followed by a *future retraction* (the reflector onto a full reflective subcategory).

This factorisation is implicitly considered in Gray [Gy2] (see I,1.11–12). In the category of adjunctions, this factorisation is functorial (cf. [JT, GT]) and mono-epi, but we shall not need these facts.

Let $f\colon X \rightleftarrows Y : g$ be an adjunction, with unit $\eta\colon 1 \to gf$ and counit $\varepsilon\colon fg \to 1$. We shall factorise it through the following comma category, the *graph* of the adjunction

$$W = (X \downarrow g) = (f \downarrow Y), \tag{3.70}$$

where we *identify* an object $(x, y; u: x \to gy)$ of $(X \downarrow g)$ with the corresponding $(x, y; v: fx \to y)$ in $(f \downarrow Y)$. The factorisation is obvious:

$$X \underset{p^-}{\overset{i^-}{\rightleftarrows}} W \underset{i^+}{\overset{p^+}{\rightleftarrows}} Y \qquad i^- \dashv p^-, \qquad p^+ \dashv i^+, \tag{3.71}$$

$$i^-(x) = (x, fx; \eta x: x \to gfx) = (x, fx; 1_{fx}),$$
$$p^-(x, y; v: fx \to y) = x,$$
$$\varepsilon_W : i^- p^- \to 1_W,$$
$$\varepsilon_W(x, y; v: fx \to y) = (1_x, v): (x, fx; 1_{fx}) \to (x, y; v),$$
$$i^+(y) = (gy, y; 1_{gy}) = (gy, y; \varepsilon y: fgy \to y),$$
$$p^+(x, y; u: x \to gy) = y,$$
$$\eta_W : 1_W \to i^+ p^+,$$
$$\eta_W(x, y; u: x \to gy) = (u, 1_y): (x, y; u) \to (gy, y; 1_{gy}).$$

In fact, composing these split adjunctions we get back the original one:

$$p^+ i^-(x) = fx, \qquad p^- i^+(y) = gy,$$
$$(p^- \eta_W i^-)(x) = p^- \eta_W(x, fx; \eta x) = p^-(\eta x, 1y) = \eta x,$$
$$(p^+ \varepsilon_W i^+)(y) = p^+ \varepsilon_W(gy, y; \varepsilon y) = p^+(1_x, \varepsilon y) = \varepsilon y.$$

Functoriality can be easily checked, starting from a commutative square of adjunctions (whose rows are already factorised)

$$
\begin{array}{ccccc}
X & \underset{p^-}{\overset{i^-}{\rightleftarrows}} & W & \underset{i^+}{\overset{p^+}{\rightleftarrows}} & Y \\
h \Big\Updownarrow h' & & r \Big\Updownarrow r' & & k \Big\Updownarrow k' \\
X' & \underset{q^-}{\overset{j^-}{\rightleftarrows}} & W' & \underset{j^+}{\overset{q^+}{\rightleftarrows}} & Y'
\end{array}
$$

$$
\begin{array}{llll}
i^- \dashv p^-, & p^+ \dashv i^+, & f = p^+ i^- \dashv g = p^- i^+, \\
h \dashv h', & k \dashv k', & \\
j^- \dashv q^-, & q^+ \dashv j^+, & f' = q^+ j^- \dashv g' = q^- j^+.
\end{array}
$$

One defines the functors r, r' as follows

$r: W \to W', \qquad r': W' \to W,$
$r(x, y; v: fx \to y) = (hx, ky; kv: f'hx = kfx \to ky),$
$r'(x', y'; u': x' \to g'y') = (h'x', k'y'; h'u': h'x' \to h'g'y' = gk'y').$

and constructs an adjunction $r \dashv r'$ which gives commutative squares in the factorisation above.

3.5.6. Faithful adjunctions

We shall say that the adjunction $f \dashv g$ is *faithful* if the functors f, g are faithful, or – equivalently – if the components of ε are epi and the components of η are mono (Section A3.3). Obviously, faithful adjunctions compose.

Now, we can adapt the previous result, obtaining a similar factorisation into a *faithful past section* followed by a *faithful future retraction*.

We restrict W to its full subcategory W_0 (the *faithful graph*) of objects $(x, y; u: x \to gy) = (x, y; v: fx \to y)$ such that:

- $u: x \to gy$ is *mono* and the corresponding $v: fx \to y$ is *epi*.

Indeed, the functor i^- (resp. i^+) take values in W_0, because every ηx is mono and corresponds to 1_{fx} (resp. every 1_{gy} corresponds to εy, which is epi). Moreover, the restricted adjunctions are faithful, because the components of $\varepsilon_W (x, y; v) = (1_x, v)$ and $\eta_W (x, y; u) = (u, 1_y)$ on the objects of W_0 are, respectively, epi and mono.

3.5.7 Theorem and definition
(Pf-presentations and projective models)

(a) Given a pf-presentation of the category X (with notation as in (3.68)), there is an associated projective model M of X, constructed as follows

$$
\begin{array}{ccccc}
P \underset{p^-}{\overset{i^-}{\rightrightarrows}} X \underset{i^+}{\overset{p^+}{\leftleftarrows}} F & & X & & X \\
\| \quad\quad \downarrow f \quad\quad \| & & \uparrow\downarrow & & \uparrow\downarrow \\
P \underset{q^-}{\overset{j^-}{\rightrightarrows}} W \underset{j^+}{\overset{q^+}{\leftleftarrows}} F & & W & & M
\end{array}
\tag{3.72}
$$

The lower row is the canonical factorisation (see (3.71)) of the composed adjunction $P \rightleftarrows F$, through its graph, the category W, which (here) can be embedded as a full subcategory of X^2

$$
W = (P \downarrow p^- i^+) = (p^+ i^- \downarrow F) = (i^- \downarrow i^+) \subset X^2.
$$

Then, there exists a pf-equivalence $f: X \rightleftarrows W : r^\alpha$, with

$$
r^\alpha f = i^\alpha p^\alpha, \qquad j^\alpha = f i^\alpha,
\tag{3.73}
$$

which inherits the counit ε from the adjunction $i^- \dashv p^-$ and the unit η from $p^+ \dashv i^+$; its comparison $r: W \to X^2$ (cf. (3.52)) coincides with the embedding $(i^- \downarrow i^+) \subset X^2$.

Finally, replacing W with its full subcategory M of all objects of type fx (for $x \in X$) we have a projective model $p: X \rightleftarrows M : r^\alpha$. The adjunctions of the lower row can be restricted to M (since $j^\alpha = fi^\alpha$), so that P and F are also, canonically, a past and a future retract of M.

(b) If the given pf-presentation of X is faithful, so is the associated projective model $p: X \rightleftarrows M$, and M is a full subcategory of the faithful graph $W_0 \subset W$ (Section 3.5.6).

Proof (a) The comma category $W = (i^- \downarrow i^+)$ is a full subcategory of X^2, because both i^α are full embeddings; it has a canonical isomorphism with the 'graph' $(P \downarrow p^- i^+) = (p^+ i^- \downarrow F)$

$$(i^- \downarrow i^+) \to (P \downarrow p^- i^+), \qquad (P \downarrow p^- i^+) \to (i^- \downarrow i^+),$$
$$(x, y; w: i^- x \to i^+ y) \mapsto (x, y; p^- w: x \to p^- i^+ y),$$
$$(x, y; u: x \to p^- i^+ y) \mapsto (x, y; \varepsilon i^+ y . i^- u: i^- x \to i^+ y), \tag{3.74}$$
$$p^-(\varepsilon i^+ y . i^- u) = u, \qquad \varepsilon i^+ y . i^- p^- w = w . \varepsilon x = w . \varepsilon i^- p^- x = w.$$

We define the three functors $f: X \rightleftarrows W : r^\alpha$

$$f(x) = (p^- x, p^+ x; \ \eta x . \varepsilon x: i^- p^- x \to i^+ p^+ x),$$
$$r^-(x, y; \ w: i^- x \to i^+ y) = i^- x, \tag{3.75}$$
$$r^+(x, y; \ w: i^- x \to i^+ y) = i^+ y,$$

and observe that they satisfy the relations (3.73). Then, we complete the pf-equivalence with the following counits and units (ε and η are the 'original' ones, in the pf-presentation of X):

$$\varepsilon_X = \varepsilon: r^- f = i^- p^- \to 1_X, \qquad \eta_X = \eta: 1_X \to r^+ f = i^+ p^+,$$
$$\varepsilon_W: fr^- \to 1_W, \qquad \eta_W: 1_W \to fr^+,$$

$$\varepsilon_W(x, y; w: i^- x \to i^+ y) =$$
$$(1_x, p^+ w): (x, p^+ i^- x; \eta i^- x: x \to i^+ p^+ i^- x) \to (x, y; w: i^- x \to i^+ y),$$

$$\eta_W(x, y; w: i^- x \to i^+ y) =$$
$$(p^- w, 1_y): (x, y; w: i^- x \to i^+ y) \to (p^- i^+ y, y; \varepsilon i^+ y: i^- p^- i^+ y \to y).$$

The coherence conditions are easily verified. Moreover, the comparison functor $r: W \to X^2$ (coming from the natural transformation $r = \varepsilon_X r^+ . r^- \eta_W : r^- \to r^+$) coincides with the full embedding

$(i^- \downarrow i^+) \subset X^2$

$$r(x, y; w: i^- x \to i^+ y) = \varepsilon_X i^+ y.r^-(p^- w, 1y) = \varepsilon i^+ y.i^- p^- w = w,$$
$$r(a, b) = (i^- a, i^+ b).$$

The last assertion follows from Theorem 3.4.5 on the 'middle model', since $M \subset W \subset X^2$ is also full in the latter.

(b) Assume that the given pf-presentation is faithful. Since the faithful graph W_0 has been defined, in Section 3.5.6, as a full subcategory of $(P \downarrow p^- i^+) = (p^+ i^- \downarrow F)$, let us rewrite f via the identification (3.74)

$$
\begin{aligned}
f(x) &= (p^- x, p^+ x; \eta x.\varepsilon x: i^- p^- x \to i^+ p^+ x) \\
&= (p^- x, p^+ x; p^- \eta x: p^- x \to p^- i^+ p^+ x) \\
&= (p^- x, p^+ x; p^+ \varepsilon x: p^+ i^- p^- x \to p^+ x).
\end{aligned}
\tag{3.76}
$$

Now, $f(x) \in W_0$ because $p^- \eta x$ is mono (so is ηx, by hypothesis, and p^- is a right adjoint), while the corresponding $p^+ \varepsilon x$ is epi (dually). Moreover, restricting to M, the new units are componentwise mono

$$\eta_X(x) = \eta x, \qquad\qquad \eta_W(f(x)) = (p^- \eta x, 1p + x),$$

and the new counits are componentwise epi.

(Replacing W with W_0 in the proof of (a), above, we would arrive at the same result.) $\square$

3.5.8 *From injective to projective models*

In particular, a *split* injective model $i: E \;\rightleftarrows\; X : p^\alpha$ (Section 3.4.6) has an *associated* projective model (which need not be split, cf. Section 3.6.5). In fact, our structure gives a pf-presentation

$$
E \underset{p^-}{\overset{i}{\rightleftarrows}} X \underset{i}{\overset{p^+}{\rightleftarrows}} E \qquad\qquad \varepsilon: ip^- \to 1_X, \quad \eta: 1_X \to ip^+, \tag{3.77}
$$

Since $i: E \to X$ is full (Section 3.4.6), the category $W = (i \downarrow i) \subset X^2$ of the associated projective model can be identified with E^2, and the functor $f: X \to W$ (in (3.75)) with the comparison $p: p^- \to p^+$, viewed as a functor $p: X \to E^2$. The pf-equivalence between X and $W = E^2$ (in (3.73)) thus becomes

$$p: X \;\rightleftarrows\; E^2 : r^\alpha, \qquad p(x) = (p^- x, p^+ x; p^+ \varepsilon x.p^- \eta x),$$

and restricts to a pf-projection $p': X \;\rightleftarrows\; M$ with values in the full subcategory $M \subset E^2$ whose objects are the morphisms

$px: p^- x \to p^+ x$. (Example 3.6.5 will show that M can be a proper subcategory).

3.6 Minimal models of a category

In this section, pf-equivalences are used to analyse a category, via injective and projective models. The faithful case is considered at the end, in Section 3.6.7.

3.6.1 Ordinary skeleta

Let us briefly review the usual, non-directed notion of a skeleton in category theory (cf. [M3]). A category is said to be *skeletal* if it has a unique object in each class of isomorphic objects; two equivalent skeletal categories are necessarily isomorphic.

The *skeleton* of a category X is a skeletal category equivalent to the former, determined up to isomorphism of categories. It always exists: one can *choose* one object in each class of isomorphic objects (in X) and take their full subcategory X_0; then its embedding in X is faithful, full and essentially surjective on objects, and therefore an equivalence of categories (cf. Section A1.5). Two categories are equivalent if and only if their skeleta are isomorphic; therefore, *skeleta classify equivalence classes of categories*.

For our present analysis, it will be useful to note two facts. First, the skeleton of a category X can also be defined as a category E such that:

(a) E has an 'injective equivalence' into X;
(b) every injective equivalence $E' \to E$ is an isomorphism of categories;

where 'injective equivalence' denotes an equivalence of categories which is injective on objects (and, necessarily, on maps).

Second, the uniqueness of the skeleton of a category X can be expressed as follows: given two skeleta $i: E \to X$ and $j: E' \to X$

(c) there is a unique mapping $u: \mathrm{Ob}E \to \mathrm{Ob}E'$ such that, for every $z \in E$, $i(z) \cong ju(z)$ in X;

(d) for every choice of a family of isomorphisms $\lambda(z): i(z) \to ju(z)$, the mapping u has a unique extension to a functor $u: E \to E'$ making that family a natural transformation $\lambda: i \to ju$.

Thus, the injective equivalence $E \to X$ of a skeleton is determined up to a natural isomorphism, which is not unique.

3.6.2 Minimal models

(a) By definition, an injective model of the category X is given by a pf-embedding $i \colon E \rightleftharpoons X$ (Section 3.5.1). We say that E is a *minimal injective model* of X if:

(i) E is an injective model of every injective model E' of X;
(ii) every injective model E' of E is isomorphic to E.

We say that it is a *strongly minimal injective model* if it also satisfies the condition (i'), stronger than (i):

(i') E is an injective model of every category injectively equivalent to X (see Section 3.5.1).

Note that we are not requiring any consistency of the embeddings. Thus, the minimal injective model of a category X is determined up to isomorphism (when existing); *but the isomorphism itself is generally undetermined*, and the pf-embedding $E \to X$ will not even be determined up to isomorphism, as we will see in various examples (Sections 3.6.5 and 3.6.6).

Two categories having a common injective model are injectively equivalent. Moreover, *strongly minimal injective models classify injective equivalence* (when they exist): if the category X has a strongly minimal injective model E, then the category Y is injectively equivalent to X if and only if E is also an injective model of Y (in which case, it is also a strongly minimal injective model of the latter).

(b) Similarly, a *projective model* of X is given by a pf-projection $p \colon X \rightleftharpoons M$ (Section 3.5.1). We define a (strongly) minimal *projective model* of X as above, in the injective case.

We shall see that the two notions are different: a category with initial and terminal object is always projectively contractible, while it is injectively contractible if and only if it is pointed (Section 3.6.4). Other comparisons of these two kinds of models, after the hints already given above (Section 3.1.1), will be seen in Section 3.9.

(c) Let us begin by considering the plain case of a groupoid X. Every full subcategory E containing at least one object in each class of isomorphic objects is an injective model (since the embedding can be completed to an adjoint equivalence, which can be viewed as a past and a future equivalence). Therefore, the ordinary skeleton of a groupoid is its minimal injective model (and also its minimal projective model).

3.6.3 Lemma

Let $i\colon E \rightleftarrows X : r^\alpha$ be a pf-embedding (Section 3.4.4).

(a) The functor i preserves the initial and terminal object, while r^- preserves the initial one and r^+ the terminal one. All of them preserve the zero-object (if any).

(b) The category E has an initial (resp. terminal, zero) object if and only if X does.

Proof The first part of (a) follows from Lemma 3.3.8, as well as the fact that i preserves the zero-object. This also proves the 'only if' part of (b).

Suppose now that X has an initial object 0 and a terminal one, 1. Then $r^-(0)$ is initial and $r^+(1)$ is terminal in E; moreover $ir^-(0) \cong 0$ (because $ir^-(0)$ is initial in X) and $ir^+(1) \cong 1$, so that $0 \cong 1$ in X if and only if $r^-(0) \cong r^+(1)$ in E (again because i is full and faithful). □

3.6.4 Injective and projective contractibility

We say that a category X is *injectively* (resp. *projectively*) *contractible* if it is injectively (resp. projectively) equivalent to **1**.

Being injectively contractible is equivalent to each of the following conditions:

(a) X is pointed (i.e. it has a zero object);
(b) **1** is a (split) injective model of X;
(c) **1** is a strongly minimal injective model of X.

Indeed, if X is injectively equivalent to **1** then it is pointed (because of the previous lemma). If this is true, then we have functors $i\colon \mathbf{1} \rightleftarrows X : p$ with $p \dashv i \dashv p$, so that **1** is a (split) injective model of X; and, in this case, strong minimality is obvious. Finally, (c) trivially implies that X is injectively equivalent to **1**.

On the other hand, a category X with non-isomorphic initial and terminal object is injectively modelled by the ordinal $\mathbf{2} = \{0 \to 1\}$, with the obvious pf-embedding $i\colon \mathbf{2} \rightleftarrows X : r^\alpha$ (*not* split)

$$r^-(x) = 0; \qquad \varepsilon_E(z)\colon 0 \to z, \qquad \varepsilon(x)\colon 0 \to x,$$
$$r^+(x) = 1, \qquad \eta_E(z)\colon z \to 1, \qquad \eta(x)\colon x \to 1.$$

This is actually *the strongly minimal injective model* of X. Indeed, again by the previous lemma, every category injectively equivalent to X has an initial and terminal object which are not isomorphic, and is thus injectively modelled by **2**. Second, any injective model $E' \to \mathbf{2}$ is surjective on objects (and a full embedding), whence an isomorphism.

It is interesting to note that **2**, the *directed interval* of **Cat** (Section 3.3.0), is *not* injectively contractible.

On the other hand, on the projective side, the existence of the initial and terminal objects is necessary and sufficient to make a category X projectively contractible, via the split pf-projection $p \colon X \leftrightarrows \mathbf{1} \colon i^\alpha$, with $i^-(*) = 0$ and $i^+(*) = 1$.

In all these cases, the *faithful* notion of contractibility restricts the categories X to preorders (as in Section 3.3.7).

3.6.5 The model of the ordered line

(Here, all the categories will be ordered sets, so that all coherence conditions are automatically satisfied and all equivalences are faithful.) We want to model the ordered real line $\mathbf{r}$ as a category; note that $\mathbf{r}$ is the fundamental category of the ordered topological space $\uparrow\mathbf{R}$.

The full subcategory $\mathbf{z}$ of integers is a *split injective model* of $\mathbf{r}$, with $i \colon \mathbf{z} \subset \mathbf{r}$ and its adjoint retractions $(p^- i = \mathrm{id} = p^+ i)$

$$
\begin{aligned}
p^-(x) &= \max\{k \in \mathbf{z} \mid k \leqslant x\}, & \varepsilon x \colon ip^-(x) \leqslant x, & \quad (i \dashv p^-), \\
p^+(x) &= \min\{k \in \mathbf{z} \mid k \geqslant x\}, & \eta x \colon x \leqslant ip^+(x), & \quad (p^+ \dashv i),
\end{aligned}
\tag{3.78}
$$

consisting of the integral part $p^-(x) = [x]$ and of $p^+(x) = -[-x]$.

It is, in fact, a *minimal* injective model of $\mathbf{r}$. For every injective model $u \colon E \leftrightarrows \mathbf{r} \colon r^\alpha$, E is a subset of $\mathbf{r}$ with the induced preorder, necessarily *initial* in $\mathbf{r}$, i.e. unbounded below (since $ur^-(x) \leqslant x$), and final in $\mathbf{r}$, i.e. unbounded above; choosing an arbitrary order-preserving embedding $(x_k)_{k \in \mathbf{z}}$ of $\mathbf{z}$ into E, unbounded both ways, we have again a split pf-injection $\mathbf{z} \to E$ (with right and left adjoint constructed as above). Moreover, if E is an injective model of $\mathbf{z}$, then it is unbounded there and necessarily order-isomorphic to it.

Of course, $\mathbf{r}$ contains various minimal injective models, all isomorphic but not isomorphically embedded (since the only isomorphisms of the category $\mathbf{r}$ are the identities); e.g. $2\mathbf{z}$ (properly contained in $\mathbf{z}$) and $1/2 + \mathbf{z}$ (disjoint from $\mathbf{z}$).

By Section 3.5.8, the split pf-injection $\mathbf{z} \to \mathbf{r}$ has an associated pf-equivalence $p\colon \mathbf{r} \rightleftarrows \mathbf{z}^2 : r^\alpha$ with values in the order category of pairs of integers (k, k') with $k \leqslant k'$

$$p(x) = (p^- x, p^+ x), \qquad r^-(k, k') = k, \qquad r^+(k, k') = k',$$

which, essentially, sends a real number to the least interval $[k, k']$ with integral endpoints, containing it. Reducing the codomain of p to the full subcategory $\mathbf{z}' \subset \mathbf{z}^2$ of pairs (k, k') with $k \leqslant k' \leqslant k + 1$, we get the associated projective model $p'\colon \mathbf{r} \rightleftarrows \mathbf{z}'$, which is not split $(p'r^-(k, k') = (k, k))$.

It is interesting to note that there is no split pf-projection $p\colon \mathbf{r} \to \mathbf{z}$; in fact, the pre-images of integers would form a sequence of disjoint compact intervals $I_k = p^{-1}\{k\} = [i^-(k), i^+(k)]$, with I_k (strictly) preceding I_{k+1}; but such a sequence does not cover the line: it leaves gaps $]i^+(k), i^-(k+1)[$.

Injective models of trees will be considered later (Section 3.9.1).

3.6.6 The model of the directed circle

Consider now the fundamental category $\mathbf{c} = {\uparrow}\Pi_1{\uparrow}\mathbf{S}^1$ of the directed circle (cf. Section 3.2.7(d)), i.e. the subcategory of the fundamental groupoid $\Pi_1(\mathbf{S}^1)$ of the circle containing all points and the homotopy classes of those paths which move 'anticlockwise' in the oriented plane $\mathbf{R}^2$.

We prove now that the minimal injective model of $\mathbf{c}$ is its full subcategory $E = {\uparrow}\pi_1({\uparrow}\mathbf{S}^1, x)$ at a(ny) point x, which we identify with the additive monoid $\mathbf{N}$ of the natural numbers.

First, we show that the embedding $i\colon E \to \mathbf{c}$ has a left and a right adjoint, forming a split pf-injection

$$
\begin{aligned}
&i\colon E \to \mathbf{c}, && p^+ \dashv i \dashv p^-, \\
&\varepsilon\colon ip^- \to 1_{\mathbf{c}} \;\; (\textit{past counit}), && \eta\colon 1_{\mathbf{c}} \to ip^+ \;\; (\textit{future unit}).
\end{aligned}
\tag{3.79}
$$

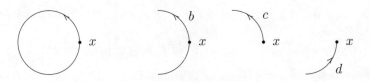

$$p^-[b] = 1, \;\; p^-[c] = 0, \;\; p^-[d] = 1, \qquad p^+[b] = 1, \;\; p^+[c] = 1, \;\; p^+[d] = 0.$$

Roughly speaking, both the functors $p^-, p^+ : c \to E$ count the number of times that a *directed* path a in $\uparrow\mathbf{S}^1$ crosses the point x, a number which only depends on the homotopy class $[a]$ in c, because of our restriction on paths. But the precise definition is different: $p^-[a]$ is the number of times that *a reaches x from below*, while $p^+[a]$ is the number of times that *a leaves x upwards* (the examples above show the difference). Then, the counit component $\varepsilon x' : x \to x'$ is the class of the 'least anticlockwise path' from x to x' (so that $p^-(\varepsilon x') = 0$ is indeed the identity of the monoid); and dually for $\eta x' : x' \to x$ (now, $p^+(\eta x') = 0$). The coherence properties (3.60) hold.

Now, if E' is an injective model of $\mathbf{c}$ (hence a full subcategory), the full subcategory of E' (and $\mathbf{c}$) on some point x' is pf-embedded in E' as above, and isomorphic to E; moreover, E – having just one object – is the unique injective model of itself.

Similarly, one can prove that the minimal injective model of the fundamental category of the directed torus $(\uparrow\mathbf{S}^1)^n$ is the fundamental monoid at any point, isomorphic to $\mathbf{N}^n$.

On the other hand, the projective model of $\mathbf{c}$ given by the split pf-injection $E \to \mathbf{c}$ is the full subcategory of the category E^2 on the two objects $0, 1 : x \to x$ (always identifying $E = \uparrow\pi_1(\uparrow\mathbf{S}^1, x) = \mathbf{N}$); which seems not to be of much interest.

Consider now the fundamental categories $\uparrow\Pi_1(C^+(\mathbf{S}^1))$ and $\uparrow\Pi_1(C^+(\uparrow\mathbf{S}^1))$, described in Section 3.2.8. One easily concludes that, in both cases, a minimal injective model is given by the full subcategory on two points, the terminal object v^+ and any other, $x \neq v^+$.

3.6.7 Minimal faithful models

The terminology of this section can be adapted to the faithful case (Section 3.5.1) in the obvious way, for the injective and the projective case. For instance, we say that E is a *minimal faithful injective model* of X if:

(i) E is a faithful injective model of every faithful injective model E' of X;

(ii) every faithful injective model E' of E is isomorphic to E.

If X is *balanced*, every faithful pf-embedding $i : E \to X$ is essentially surjective on objects (each component $\eta x : x \to ir^+x$ being an isomorphism), *whence an equivalence of categories*. Therefore, the minimal faithful injective model of X is simply its skeleton. (But note that the fundamental categories of the d-spaces which we are considering are often not balanced, cf. Section 3.9.)

A category can have a minimal injective model and a *different* minimal faithful injective model. For instance, the well-known category Δ^+ of finite ordinals (the site of augmented simplicial sets), being skeletal and balanced, is already a minimal *faithful* injective model (of itself), while its minimal injective model is **2** (Section 3.6.4); the same happens with the category of finite cardinals (and all mappings).

3.7 Future invariant properties

We now investigate various properties, of morphisms and objects, which are invariant under future (or past) equivalence. They arise from 'branching' or 'non-branching' properties, and will be used in the following sections to identify and construct minimal models of categories.

Most of the material of this section comes from [G14], but the present definition of the 'future regularity equivalence' $x \sim^+ x'$ (Definition 3.7.6) is finer, and many parts have been modified according to this. In the applications below, this modification gives better results in some cases (e.g. in Section 3.9.7), and the same results in most cases.

3.7.1 Future regularity

A morphism $a \colon x \to x'$ in X will be said to be V^+-*regular* if it satisfies condition (i), O^+-*regular* if it satisfies (ii), and *future regular* if it satisfies both:

(i) given $a' \colon x \to x''$, there is a commutative square $ha = ka'$ (V^+-regularity);

(ii) given $a_i \colon x' \to x''$ such that $a_1 a = a_2 a$, there is some h such that $ha_1 = ha_2$ (O^+-regularity),

$$
\begin{array}{ccc}
x & \xrightarrow{\ a\ } & x' \\
{\scriptstyle a'}\big\downarrow & & \big\downarrow{\scriptstyle h} \\
x'' & \dashrightarrow{\scriptstyle\ k\ } & \bullet
\end{array}
\qquad
x \xrightarrow{\ a\ } x' \underset{a_2}{\overset{a_1}{\rightrightarrows}} x'' \dashrightarrow{\scriptstyle h} \bullet
\qquad (3.80)
$$

We will see that these properties are closed under composition (Lemma 3.7.2). In a category with finite colimits or with a terminal object, all morphisms are future regular. In a preordered set, all arrows are O^+-regular, and future regularity coincides with V^+-regularity. (The relationship of these notions with filtered categories is dealt with in Section 3.7.4.)

On the other hand, we shall say that the map a is V^+-*branching* if it is not V^+-regular; that it is O^+-*branching* if it is not O^+-regular; that is a *future branching morphism* if it falls into at least one of the previous cases, i.e. if it is not future regular. In the category represented below, on the left, the morphism a is V^+-branching and O^+-regular, while on the right a is O^+-branching and V^+-regular

$$
\begin{array}{ccc}
\begin{array}{ccc}
x & \xrightarrow{\ a\ } & x' \\
\Big\downarrow{\scriptstyle a'} & & \\
x'' & &
\end{array}
&
\begin{array}{c}
x \xrightarrow{\ a\ } x' \\
{}_{b}\searrow \ \Big\downarrow\Big\downarrow {\scriptstyle a_1\ a_2} \\
x''
\end{array}
&
(b = a_1 a = a_2 a). \qquad (3.81)
\end{array}
$$

Dually, we have V^--*regular*, O^--*regular*, *past regular* morphisms and the corresponding *branching* morphisms.

3.7.2 Lemma (Future regular morphisms)

(a) V^+-regular, O^+-regular and future regular morphisms form (wide) subcategories, which contain all the isomorphisms.

(b) If a composite ba is V^+-regular, then the first map a is also.

(c) If a composite ba is O^+-regular, then the second map b is also.

(d) If ba is V^+-regular (resp. future regular) and a is O^+-regular, then b is V^+-regular (resp. future regular).

(Recall that a subcategory is said to be *wide* if it contains all the objects of the given category.)

Proof Take two consecutive morphisms in X, $a: x \to x'$ and $b: x' \to x''$.

First, let us consider the property of V^+-regularity. It is plainly consistent with composition. On the other hand, if ba is V^+-regular also a is: for every $a': x \to \bar{x}$ there is a commutative square $h(ba) = ka'$, which can be rewritten as $(hb).a = ka'$.

Second, let us consider O^+-regularity. If a and b are so (and composable), take two maps $b_i: x'' \to \bar{x}$ such that $b_1(ba) = b_2(ba)$; by O^+-regularity of a there is some h such that $h(b_1 b) = h(b_2 b)$; then, by O^+-regularity of b, there is some k such that $khb_1 = khb_2$. On the other hand, if the composite ba is O^+-regular, also b is: if $b_1 b = b_2 b$, then $b_1(ba) = b_2(ba)$ and there is some h such that $hb_1 = hb_2$.

Finally, for (d), it is sufficient to prove the first case, since the second will then follow from (c). Take a map $b' : x' \to \overline{x}$; since ba is V^+-regular, there are maps c, c' such that $c(ba) = c'(b'a)$, i.e. $(cb)a = (c'b')a$. But a is O^+-regular, hence there is some d such that $d(cb) = d(c'b')$

$$
\begin{array}{ccccc}
x & \xrightarrow{\ a\ } & x' & \xrightarrow{\ b\ } & x'' \\
& & \downarrow{\scriptstyle b'} & {\scriptstyle ?} & \downarrow{\scriptstyle c} \\
& & \overline{x} & \dashrightarrow[\;c'\;]{} \bullet \dashrightarrow[\;d\;]{} \bullet
\end{array}
$$

and the maps dc, dc' solve our problem. □

3.7.3 Theorem (Future equivalence and regular morphisms)

Consider a future equivalence $f \colon X \rightleftarrows Y \colon g$, with units $\varphi \colon 1 \to gf$, $\psi \colon 1 \to fg$.

(a) *All the components φx and ψy are future regular morphisms.*

(b) *The functors f and g preserve V^+-regular, O^+-regular and future regular morphisms.*

(c) *The functors f and g preserve V^+-branching, O^+-branching and future branching morphisms (i.e. reflect V^+-regular, O^+-regular and future regular morphisms).*

Proof The index i always takes values 1, 2.

(a) Take a component $\varphi x \colon x \to gfx$. Then, a map $a \colon x \to x'$ gives the left commutative diagram, showing that φx is V^+-regular

$$
\begin{array}{ccc}
x & \xrightarrow{\varphi x} & gfx \\
{\scriptstyle a}\downarrow & & \downarrow{\scriptstyle gfa} \\
x' & \xrightarrow{\varphi x'} & gfx'
\end{array}
\qquad
\begin{array}{ccccc}
x & \xrightarrow{\varphi x} & gfx & \xrightarrow{\varphi gfx} & gfgfx \\
& {\scriptstyle a}\searrow & \downarrow{\scriptstyle a_i} & \searrow{\scriptstyle gfa} & \downarrow{\scriptstyle gfa_i} \\
& & x' & \xrightarrow{\varphi x'} & gfx'
\end{array}
\tag{3.82}
$$

Moreover, given $a_i \colon gfx \to x'$ such that $a_i.\varphi x = a$, the right diagram shows that $\varphi x'$ coequalises a_1 and a_2: from $\varphi gf = gf\varphi$ we deduce $\varphi x'.a_i = gfa_i.\varphi gfx = gf(a_i.\varphi x) = gf(a)$.

(b) Suppose that $a \colon x \to x'$ is V^+-regular in X; we must prove that $fa \colon fx \to fx'$ is also, in Y. Given $b \colon fx \to y$, we can form the left commutative diagram in X, and then the right one, in Y

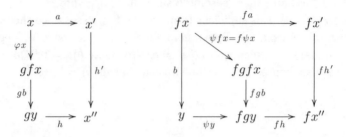

Second, suppose that $a: x \to x'$ is O^+-regular in X. Given two maps $b_i: fx' \to y$ such that $b_i.fa = b$, we have (on the left, below): $gb_i.\varphi x'.a = gb_i.gfa.\varphi x = gb.\varphi x$. Therefore, there exists an h in X such that the composite $h.gb_i.\varphi x'$ does not depend on i (see the left diagram below)

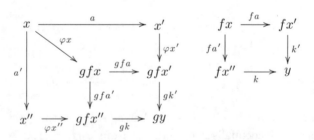

Then, in the right diagram above, the composite

$$fh.\psi y.b_i = fh.fgb_i.\psi fx' = f(h.gb_i.\varphi x'),$$

in Y, does not either depend on i.

(c) First, given $a: x \to x'$ in X, suppose that fa is V^+-regular (in Y); we must prove that a is also. Given $a': x \to x''$, we can form the right commutative diagram in Y, and then the left one, in X

Second, let us suppose that fa is O^+-regular in Y; given two maps $a_i: x' \to x''$ such that $a_i.a = a'$, there is some k such that $k.fa_i = k'$, in the right diagram; and then, on the left, $(gk.\varphi x'').a_i = g(k.fa_i).\varphi x' =$

$gk'.\varphi x'$ is independent of i

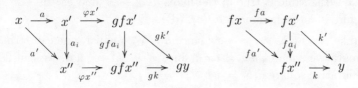

$$\square$$

3.7.4 Regular and branching points

We now consider properties *of points* of a category X, which will be proved to be future invariant (Theorem 3.7.7). (We have already seen a few, concerning maximal points, in Lemma 3.3.8.)

A point x will be said to be V^+-*regular* if it satisfies (i), O^+-*regular* if it satisfies (ii), *future regular* if it satisfies both:

- (i) every arrow starting from x is V^+-regular (equivalently, two arrows starting from x can always be completed to a commutative square);
- (ii) every arrow starting from x is O^+-regular (equivalently, given an arrow $a\colon x \to x'$ and two arrows $a_i\colon x' \to x''$ such that $a_1 a = a_2 a$, there exists an arrow h such that $ha_1 = ha_2$).

It is easy to verify that x is future regular in X if and only if the comma category $(x \downarrow X)$ of arrows starting from x is *filtered* ([M3], IX.1); but this will not be used here.

We shall say that x is a V^+-*branching point* in X if it is not V^+-regular (i.e. if there is some arrow starting from x which is V^+-branching); that x is an O^+-*branching point* if it is not O^+-regular; that x is a *future branching point* if it falls into at least one of the previous cases, i.e. if it is not future regular.

Dually, we have the notions of V^--, O^-- and *past regular* (resp. *branching*) *point* in X.

3.7.5 Lemma

Given a map $a\colon x \to x'$ in a category X

(a) if the point x is O^+-regular (resp. future regular) so is x';
(b) if the point x' and the map a are V^+-regular, also x is.

Proof (a) Take a map $b: x' \to x''$. Since x is O^+-regular (resp. future regular), the same is true of the maps $a: x \to x'$ and $ba: x \to x''$, and therefore of b (by Lemma 3.7.2(c), (d)).

(b) Take two maps $a_i: x \to x_i$; since a is V^+-regular, one can form two commutative squares $b_i a = c_i a_i$; since x' is V^+-regular, we can add a commutative square $d_1 b_1 = d_2 b_2$

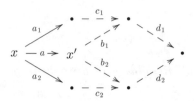

Finally, the maps $d_i c_i$ form a commutative square with the initial ones, a_i.   □

3.7.6 Definition (Future regularity equivalence)

For a category X, the *future regularity* equivalence relation $x \sim^+ x'$ in $\mathrm{Ob}X$ is defined as follows:

(i) there exists in X a zig-zag $x \to x_1 \leftarrow x_2 \ldots x_{n-1} \leftarrow x'$ of future regular maps;

(ii) the points x, x' are of the same V^+-type (regular or branching) and of the same O^+-type.

(In [G14], only (i) is considered.) The future regularity class of an object x will be written as $[x]^+$.

It will be useful to note that *each* of the following conditions implies $x \sim^+ x'$ (by the previous lemma):

(iii) x is future regular and there is a map $x \to x'$;

(iv) x is O^+-regular, x' is future regular and there is a future regular map $x \to x'$.

On the other hand, the category drawn in the second diagram of (3.81) shows a future regular map $b: x \to x''$ going from an O^+-branching point to a future regular one.

Note now that, in the fundamental category of the square annulus (Section 3.1.1), the starting point 0 is V^+-branching, but the choice between the different paths starting from it can be deferred, while at the point p the choice must be made. To distinguish these situations, we will say that a future branching point x is *effective* when every future

regular map $x \to x'$ *reaching a point in the same future regularity class*
(i.e. of the same V^+- and O^+-type) is a split monomorphism.

(One might expect to find here an *isomorphism* instead of a split
monomorphism, but the present formulation has various advantages:
e.g. it is future invariant, see Theorem 3.7.7, and works well in Theorem 3.8.2(d). For the fundamental categories of preordered spaces, studied in Section 3.9, the two conditions are equivalent, since there a split
monomorphism is always invertible.)

Dually, one has the *past regularity* equivalence relation $\sim^-$, with past
regularity classes $[x]^-$, and *effective past branching points*.

3.7.7 Theorem (Future equivalence and branching points)

The following properties of a point are future invariant *(i.e. preserved
by each functor of a future equivalence):*

(a) *being a V^+-regular, or an O^+-regular, or a future regular point;*
(b) *being a V^+-branching, or an O^+-branching, or a future branching
point, or an effective one (Definition 3.7.6).*

Proof Let $f \colon X \rightleftarrows Y \colon g$ be a future equivalence, with units $\varphi \colon \mathrm{id}X \to gf$ and $\psi \colon \mathrm{id}Y \to fg$.

(a) Let us take a point x which is V^+-regular in X, and prove that every
Y-arrow $b \colon fx \to y$ is V^+-regular. Indeed, the map $a = gb.\varphi x \colon x \to gy$ is
V^+-regular, whence also fa is (by Theorem 3.7.3); but $fa = fgb.f\varphi x = fgb.\psi fx = \psi y.b$, whence also the 'first map' b is V^+-regular (by
Lemma 3.7.2(b)).

We assume now that x is O^+-regular, and prove that every Y-map
$b \colon fx \to y$ is also. Now, the composite $gb.\varphi x \colon x \to gy$ is O^+-regular in
X, whence the 'second map' gb is O^+-regular (Lemma 3.7.2(c)), and b
itself is O^+-regular in Y (by the reflection property, Theorem 3.7.3(c)).

(b) We take a point x in X such that fx is V^+-regular (i.e. *not* V^+-
branching) and prove that also x is. For every $a \colon x \to x'$ in X, $fa \colon fx \to fx'$ is V^+-regular in Y; but then a is V^+-regular in X, by the reflection
property, Theorem 3.7.3(c). The same holds replacing the prefix V^+
with O^+.

Now, let x be an effective future branching point and $b \colon fx \to y$ a
future regular map between points of the same V^+- and O^+-type. Then,
we have already proved that also x and gy have the same V^+- and
O^+-type. Moreover, the morphism $a = gb.\varphi x \colon x \to gy$ is future regular

(by composition), whence a is a split mono and also fa is; but $fa = fgb.f\varphi x = fgb.\psi fx = \psi y.b$, whence also b is a split monomorphism. $\square$

3.7.8 Theorem (Future regularity equivalence)

A future equivalence $f \colon X \rightleftarrows Y \colon g$ induces a bijection between the quotients $(\mathrm{Ob}X)/\sim^+$ and $(\mathrm{Ob}Y)/\sim^+$ (of the set of objects, up to future regularity equivalence).

In other words:

(i) *the functors f and g preserve and reflect the future regularity equivalence relation $\sim^+$;*

(ii) *for every x in X and y in Y, $x \sim^+ gfx$ and $y \sim^+ fgy$.*

Proof Point (ii) and the preservation property in (i) follow from Theorems 3.7.3 and 3.7.7.

Therefore, we have induced mappings $\mathrm{Ob}X/\sim^+ \rightleftarrows \mathrm{Ob}Y/\sim^+$, which are inverses, by (ii). This implies the reflection property in (i). $\square$

3.8 Spectra and pf-equivalence of categories

We define now the *future* and the *past spectrum* of a category, and show that, when they exist, they are, respectively, its least full reflective and its least full coreflective subcategory. Their join forms the *pf-spectrum*, which is a strongly minimal injective model of the original category (Theorem 3.8.8) and classifies injective equivalence (Theorem 3.8.7). The pf-spectrum also yields a projective model (Section 3.8.5).

3.8.0 Least future retracts

By a *replete* subcategory of a category X we will mean a *full* subcategory which is closed (in X) under isomorphic copies of objects. If C is a full subcategory, its *replete closure* C' in X has the same skeleton.

Within full subcategories of X, we define the preorder of *essential inclusion* $C \prec D$ by the inclusion $C' \subset D'$ of their replete closures (which reduces to $C \subset D$, when X is skeletal – as will often be the case in our applications). We are interested in the *least full reflective subcategory*, or *least future retract* F of X, for this preorder. If it exists, its replete closure is strictly determined as the *least replete reflective subcategory* of X (with respect to inclusion); and a category Y is future equivalent to X if and only if F is also a future retract of Y (by Theorem 3.3.5). Similarly for full coreflective subcategories.

One could define the *future skeleton* of X as the skeleton of the least future retract of X. Rather than developing this notion, we shall study a stronger one, called the 'future spectrum', which will be easier to determine and yield the same results in the examples of Section 3.9.

The categories **r** and **c** have minimal future retracts, but do not have a least one (Sections 3.6.5 and 3.6.6).

The ordered set of (replete) reflective subcategories of a category was investigated in [Ke3] (see also its references). But these results, being based on the existence of limits in the original category, are of interest for the 'ordinary' categories of structured sets, rather than for the categories studied here.

3.8.1 Spectra

Recall that we have defined, in the set of objects $\mathrm{Ob}X$, the equivalence relation $x \sim^+ x'$ of future regularity, with equivalence classes $[x]^+$ (Definition 3.7.6).

A *future spectrum* $\mathrm{sp}^+(X)$ of the category X will be a subset of objects such that:

(sp^+.1) $\mathrm{sp}^+(X)$ contains precisely one object, written $\mathrm{sp}^+(x)$, in every future regularity class $[x]^+$;

(sp^+.2) for every $x \in X$ there is precisely one morphism $\eta x \colon x \to \mathrm{sp}^+(x)$ in X;

(sp^+.3) every morphism $a \colon x \to \mathrm{sp}^+(x')$ factorises as $a = h.\eta x$, for a unique $h \colon \mathrm{sp}^+(x) \to \mathrm{sp}^+(x')$.

The full subcategory of X on the set of objects $\mathrm{sp}^+(X)$, written $\mathrm{Sp}^+(X)$ (with a capital letter), will also be called the *future spectrum* of X. The second and third conditions above can be equivalently written as:

(sp^+.2$'$) for every $x \in X$, $\mathrm{sp}^+(x)$ is the terminal object of the full subcategory on $[x]^+$;

(sp^+.3$'$) for every $x \in X$, $\eta x \colon x \to \mathrm{sp}^+(x)$ is a universal arrow (Section A1.9) from the object x to the inclusion functor $i \colon \mathrm{Sp}^+(X) \to X$.

We shall prove that the future spectrum (when it exists) is the least future retract (Theorem 3.8.2), that it is determined up to a *canonical* isomorphism, and that the same is true of its embedding in the given category (Lemma 3.8.4). Therefore, the future spectrum is more strictly determined than the ordinary skeleton.

Dually we have the *past spectrum* $\mathrm{sp}^-(X)$ and its full subcategory $\mathrm{Sp}^-(X)$. The categories **r** and **c**, considered in Sections 3.6.5 and 3.6.6, do not have a future or past spectrum. Indeed, all their maps are future regular, and all objects form a unique future regularity class, which has no terminal object; and dually, their objects form a unique past regularity class, which has no initial object. It is also easy to see that a category has future spectrum **1** if and only if it is future equivalent to **1**, if and only if it has a terminal object (Proposition 3.3.6).

3.8.2 Theorem (Properties of the future spectrum)

Let $F = \mathrm{Sp}^+(X)$ be a future spectrum of the category X and $i: F \to X$ its inclusion.

(a) F is a future retract of X, with an essentially unique retraction p and unit η:

$$i: F \rightleftarrows X : p, \qquad \eta: 1_X \to ip: X \to X. \qquad (3.83)$$

(b) F is the least future retract of X, with respect to essential inclusion (of full subcategories, Section 3.8.0). It is a skeletal category, whose only endomorphisms are the identities.

The inclusion $i: F \to X$ preserves and reflects future regularity of maps and points.

(c) Replacing some objects of $\mathrm{sp}^+(X)$ with isomorphic copies, the new subset is still a future spectrum of X.

(d) Every point of $\mathrm{sp}^+(X)$ is either maximal *in X (Section 3.3.0) or an effective future branching* point *(Definition 3.7.6).*

Proof (a) The inclusion $i: \mathrm{Sp}^+(X) \to X$ has a left adjoint $p: X \to \mathrm{Sp}^+ X)$, with $ip = \mathrm{sp}^+$, because of $(\mathrm{sp}^+.3')$. Moreover, for $x_0 \in \mathrm{Sp}^+(X)$, the counit $\varepsilon x_0: x_0 = pi(x_0) \to x_0$ is necessarily the identity, by $(\mathrm{sp}^+.2)$. It follows that $\eta i = 1$ and $p\eta = 1$.

(b) Let $(j, q; \zeta): G \to X$ be a future retract of X; we want to prove that $\mathrm{Sp}^+(X)$ is contained in the replete closure of G. Take an object $x_0 \in \mathrm{Sp}^+(X)$; then $x = jqx_0 \sim^+ x_0$ (Theorem 3.7.8), whence $\mathrm{sp}^+(x) = x_0$ and the left composite below is the identity, by $(\mathrm{sp}^+.2)$:

$$\eta x.\zeta x_0: x_0 \to x \to x_0, \qquad e = \zeta x_0.\eta x: x \to x_0 \to x.$$

Now, for the right composite e, we have $e.\zeta x_0 = \zeta x_0 = 1x.\zeta x_0$; since the unit-component $\zeta x_0 \colon x_0 \to jqx_0$ is a universal arrow from x_0 to the full embedding $j \colon G \to X$ (see Section A3.1), it follows that $e = 1_x$.

Moreover, $\mathrm{Sp}^+(X)$ is skeletal by $(\mathrm{sp}^+.1)$ and has no endomorphisms, except the identities, by $(\mathrm{sp}^+.2)$. The last assertion follows from Theorems 3.7.3 and 3.7.7.

(c) Obvious. (This point is inserted for future convenience.)

(d) Let $x_0 \in \mathrm{sp}^+(X)$ be not maximal for the path preorder in X, and let us prove that x_0 is a future branching point in X. By hypothesis, there is some arrow $a \colon x_0 \to x$ in X with no arrows backwards; since x_0 is terminal in its future regularity class, these points cannot be equivalent under future regularity, and x_0 cannot be future regular (Definition 3.7.6(iii)).

Finally, as to effectiveness, let $b \colon x_0 \to x$ be a future regular map with $x \sim^+ x_0$; then $\eta x.b = \mathrm{id} x_0$, whence b is a split mono. $\qquad\square$

3.8.3 Lemma (Characterisation of future spectra)

The following conditions on a functor $i \colon F \to X$ are equivalent:

(a) the functor i is an embedding and $i(F)$ is a future spectrum of X;

(b) the category F has precisely one object in each future regularity class; the functor i is a future retract, i.e. it has a left adjoint $p \colon X \to F$ with $pi = 1_F$ as counit; moreover the unit-component $x \to ip(x)$ is the unique X-morphism with these endpoints;

(c) the category F has precisely one object in each future regularity class and only one endomorphism for each object; the functor i can be extended to a future equivalence $i \colon F \rightleftarrows X \colon p$, whose unit-component $x \to ip(x)$ is the unique X-morphism with these endpoints.

Note. The form (c) is appropriate to link future spectra and future equivalences, cf. Theorem 3.8.7(a).

Proof Identifying F with $\mathrm{Sp}^+(x)$ and i with the inclusion, the fact that (a) implies (b) and (c) has already been proved in Theorem 3.8.2(a). Then (c) implies (b): for every $x_0 \in F$, $x_0 \sim^+ pi(x_0)$ (by Theorem 3.7.8) whence $x_0 = pi(x_0)$ and the unit-component $x_0 \to pi(x_0)$ (of the future equivalence) is necessarily an identity. Finally, (b) implies (a): letting $\mathrm{sp}^+(x) = ip(x)$, the universal property of the unit of an adjunction gives $(\mathrm{sp}^+.3)$. $\qquad\square$

3.8.4 Lemma (Uniqueness of future spectra, I)

Let $i\colon F \to X$ and $j\colon G \to X$ be embeddings of future spectra of the category X.

(a) For every $x_0 \in F$ there is a unique $u(x_0)$ in G such that $i(x_0) \cong ju(x_0)$ in X. Furthermore, there is a unique morphism $\lambda x_0\colon i(x_0) \to ju(x_0)$ in X, and it is invertible.

(b) The mapping $u\colon \mathrm{Ob}F \to \mathrm{Ob}G$ so defined has a unique extension to a functor $u\colon F \to G$ making the family (λx_0) into a natural transformation $\lambda\colon i \to ju\colon F \to X$; the latter is invertible.

Note. A more complete uniqueness result will be given in Theorem 3.8.9.

Proof Obvious. $\qquad\qquad\qquad\qquad\qquad\qquad\qquad\qquad\qquad\qquad\qquad$ $\square$

3.8.5 Spectral presentations

The *spectral pf-presentation* of X (cf. Section 3.5.2) will be a diagram of functors and natural transformations satisfying the following conditions

$$P \underset{p^-}{\overset{i^-}{\rightleftarrows}} X \underset{i^+}{\overset{p^+}{\rightleftarrows}} F \tag{3.84}$$

$$\varepsilon\colon i^- p^- \to 1_X, \qquad p^- i^- = 1, \quad p^- \varepsilon = 1, \quad \varepsilon i^- = 1,$$
$$\eta\colon 1_X \to i^+ p^+, \qquad p^+ i^+ = 1, \quad p^+ \eta = 1, \quad \eta i^+ = 1,$$

(i) P is *the* past spectrum and F *the* future spectrum of X;

(ii) given $x \in \mathrm{Ob}P$ and $x' \in \mathrm{Ob}F$, if $x \cong x'$ in X then $x = x'$ (*linked choice condition*).

Such a presentation exists if and only if X has a past spectrum and a future one, since the linked-choice condition can always be realised replacing each object of P with its isomorphic copy in F, if any (Theorem 3.8.2(c)). The set of objects given by this linked choice will be called the *pf-spectrum* of X, or *spectral model*

$$\mathrm{sp}(X) = \mathrm{Ob}P \cup \mathrm{Ob}F = \mathrm{sp}^-(X) \cup \mathrm{sp}^+(X). \tag{3.85}$$

The full subcategory $\mathrm{Sp}(X)$ on these objects will also be called the *pf-spectrum* of X. We prove below that it is well determined, in the same form of future spectra (Lemma 3.8.4); and we shall prove that it is a strongly minimal injective model (Theorem 3.8.8); it is not a split injective model (Section 3.4.6), in general.

The projective model $X \to M$ which is associated to the spectral pf-presentation (as in Theorem 3.5.7) will be called the *spectral projective model* of X.

3.8.6 Theorem (Uniqueness of pf-spectra)

Two pf-spectra, $i\colon E \subset X$ and $j\colon E' \subset X$, are given.

(a) For every $x_0 \in E$ there is a unique $u(x_0)$ in E' such that $i(x_0) \cong ju(x_0)$ in X; and then there is a unique morphism $\lambda x_0 \colon i(x_0) \to ju(x_0)$ in X, which is invertible.

(b) The mapping $u\colon \mathrm{Ob}E \to \mathrm{Ob}E'$ so defined has a unique extension to a functor $u\colon E \to E'$ making the family (λx_0) into an (invertible) natural transformation $\lambda\colon i \to ju\colon E \to X$ (see the left diagram below)

$$
\begin{array}{ccc}
E \xrightarrow{\;i\;} X & \qquad P \xrightarrow{\;i^-\;} E \xleftarrow{\;i^+\;} F \\
\;u\downarrow\,\swarrow^{\lambda}\quad \big\| & \qquad u^-\downarrow \qquad u\downarrow \qquad \downarrow u^+ \\
E' \xrightarrow{\;j\;} X & \qquad P' \xrightarrow{\;j^-\;} E' \xleftarrow{\;j^+\;} F'
\end{array}
\qquad (3.86)
$$

(c) Let P, F be, respectively, the past and the future spectrum of X giving rise to E; and similarly P', F' for E'.

Their embeddings and the functor u give the commutative right diagram above, where u^- and u^+ are the isomorphisms resulting from the uniqueness of these directed spectra of X (Lemma 3.8.4).

Proof (a) We already know (by Lemma 3.8.4(a)) that, if the point x_0 belongs to P (resp. F), there is a unique $u^-(x_0)$ in P' (resp. $u^+(x_0)$ in F') such that $x_0 \cong u^\alpha(x_0)$ in X ($\alpha = \pm$); moreover, if x_0 is in $P \cap F$, then $u^-(x_0) \cong x_0 \cong u^+(x_0)$, whence $u^-(x_0) = u^+(x_0)$, because of the linked-choice condition in E'. We have thus a unique object $u(x_0)$ consistent with the right diagram above.

We also have (again by Lemma 3.8.4(a)) a unique map $\lambda x_0 \colon i(x_0) \to ju(x_0)$ in X, which is an isomorphism.

The points (b) and (c) follow now easily. $\qquad\qquad \square$

3.8.7 Theorem (Preservation of future spectra and pf-spectra)

(a) If $f\colon X \rightleftarrows Y \colon g$ is a future equivalence and $i\colon F \to X$ is the embedding of a future spectrum, then $fi\colon E \to Y$ is also.

(b) A pf-embedding preserves and reflects pf-spectra. More precisely, assuming that the category X has a pf-spectrum $i: E \subset X$, we have the following results (with $E_0 = \mathrm{Ob}E$):

(i) *given a pf-embedding $u: X \to Y$, the set of objects $u(E_0)$ is the pf-spectrum of Y;*

(ii) *given a pf-embedding $v: Y \to X$, the set of objects $v^{-1}(E_0)$ is the pf-spectrum of Y.*

(c) If the category X has a pf-spectrum E, then X is injectively equivalent to a category Y if and only if E is also a pf-spectrum of Y.

Proof (a) The units of the future equivalence will be written as $\varphi: 1 \to gf$ and $\psi: 1 \to fg$.

Let us use the characterisation Lemma 3.8.3(c) of embeddings of future spectra, taking into account the fact that future equivalences compose (Section 3.3.3). Let $p: X \to F$ be the retraction and $\eta: 1 \to ip$ the unit at F (Theorem 3.8.2(a)).

We know that F has one object in each future regularity class and only one endomorphism for every object. It remains to show that, for every $y \in Y$, the composed unit $\eta' y = f\eta gy.\psi y: y \to fipgy$ is the *unique* morphism between these points.

Let $x = ipgy \in i(F)$ and note that the composite $\eta gfx.\varphi x: x \to gfx \to ipgfx$ is the identity, because x and $ipgfx$ are equivalent up to future regularity, in $i(F)$. Take now any map $b: y \to fx$ in Y; by naturality of ψ we have a commutative (solid) diagram

$$(3.87)$$

where the lower row is the identity, because of the previous remark and because $\psi fx = f\varphi x$. Also the right triangle commutes, since $\eta gfx.gb: gy \to x$ must coincide with $\eta gy: gy \to ipgy = x$. Finally, we have the thesis: $b = f\eta gy.\psi y$.

(b) Point (i) follows from (a). As to point (ii), Y is past and future equivalent to X, whence it also has a past and a future spectrum, and therefore a pf-spectrum H_0, preserved by the pf-embedding $Y \to X$. Since the pf-spectrum of X is determined up to isomorphism, $v(H_0)$ coincides with E_0 up to isomorphic copies of objects. Since v is a full embedding, it follows that $v^{-1}(E_0)$ coincides with H_0 up to isomorphic copies of objects, and $v^{-1}(E_0)$ is also a pf-spectrum of Y.

(c) Is a straightforward consequence of (b). □

3.8.8 Theorem (Spectra and injective models)

Given a pf-presentation of the category X, let E be the injective model generated by this presentation, as defined in Theorem 3.5.3. Then:

(a) the pf-presentation of X is spectral if and only if the same holds for the pf-presentation of E, in (3.69);

(b) in this case, $E = \mathrm{sp}(X)$ is a strongly minimal injective model of X.

Proof (a) Follows immediately from Theorem 3.8.7(b).

(b) Assume that the given pf-presentation is spectral, and let us show that E is a strongly minimal injective model of X. Given a category Y injectively equivalent to X, we know (Theorem 3.8.7(c)) that E is an injective model of Y. Secondly, given an injective model $v\colon E' \to E$, we have to prove that v is surjective on objects, hence an isomorphism. Indeed, we have a composed pf-embedding $E' \to X$, therefore v must reach an isomorphic copy of every object of P and F, whence every object of E, by the linked-choice condition (Section 3.8.5). □

3.8.9 Theorem (Uniqueness of future spectra, II)

Two future spectra of the category X are given, as future retracts:

$$i\colon F \rightleftarrows X \colon p, \qquad j\colon G \rightleftarrows X \colon q$$
$$\eta\colon 1_X \to ip, \qquad \eta'\colon 1_X \to jq. \tag{3.88}$$

(a) We have: $qip = q$ and $q\eta = 1_q$; dually, $pjq = p$ and $p\eta' = 1_p$.

(b) There is a unique functor $u\colon F \to G$ such that $up = q$, namely $u = qi$, and it is an isomorphism.

(c) There is a unique natural transformation $\lambda\colon i \to ju\colon F \to X$, namely $\lambda = \eta'i$, and it is invertible

$$
\begin{array}{ccccc}
X & \xrightarrow{\ p\ } & F & \xrightarrow{\ i\ } & X \\[2pt]
\| & & {\scriptstyle u}\big\downarrow\nwarrow{\scriptstyle \lambda} & & \| \\[2pt]
X & \xrightarrow{\ q\ } & G & \xrightarrow{\ j\ } & X
\end{array}
$$

Note. This is a second statement on the uniqueness of future spectra, after Lemma 3.8.4. It is more complete than the first, being based on the whole structure of a future spectrum as a future retract; yet, it seems to be less useful than the first.

Proof (a) To prove that $qip = q$, let us begin by noting that this is true on every object $x \in X$, because $ip(x) \sim^+ x$ (by Theorem 3.7.8) and $qip(x) = q(x)$.

Now, the natural transformation $q\eta\colon q \to qip$ has general component $q\eta(x)\colon qx \to qipx = qx$; but there is a *unique* map from qx to itself, in the future spectrum G, namely the identity of qx. It follows that $qip = q$ is also true on maps, and $q\eta = 1_q$.

(b) Uniqueness is plain: $up = q$ implies $u = qi$. Existence follows from point (a): taking $u = qi\colon F \to G$, we have $up = qip = q$. Symmetrically, there is a unique functor $v\colon G \to F$ such that $vq = p$; and then u and v are inverses.

(c) We do have a natural transformation $\lambda = \eta'i\colon i \to jqi = ju$. Its component $\lambda(x_0)\colon i(x_0) \to jqi(x_0)$ is the unique X-morphism between such objects (because they are future regular equivalent and the second is in G). But there is also a unique X-morphism backwards $jqi(x_0) \to i(x_0)$, because $i(x_0)$ is in F; and their composites must be identities.   □

3.9 A gallery of spectra and models

After considering pf-spectra of preorders (Section 3.9.1), we will construct the pf-spectrum of the fundamental category of various ordered topological spaces, and of one *pre*ordered space (Section 3.9.5). All these pf-spectra yield *faithful* injective models, except in Section 3.9.8. We end with a few hints on applications (Section 3.9.9).

Speaking of branching points, the term 'effective' will generally be understood (cf. Definition 3.7.6), unless we want to stress this fact.

In this section, the arrows of a fundamental category are denoted by Greek letters $\alpha, \beta, \gamma, \ldots$

3.9.1 Future spectra of preorders

Let C be a preorder category. All morphisms in C are O^+-regular (Section 3.7.1), so that future regularity coincides with V^+-regularity and is always faithful. Explicitly, the arrow $x \prec x'$ is future regular in

C if, whenever $x \prec x''$ there exists some upper bound for x' and x'', i.e. some object $\bar{x}$ which follows both.

In this case, the existence (and choice) of a future spectrum $\mathrm{sp}^+(C)$ (which is necessarily faithful) amounts to these conditions:

(i) each future regularity class of objects $[x]^+$ has a maximum (determined up to the equivalence relation $\simeq$ of mutual precedence), and we choose one, called $\max[x]^+$ (of course, if C is ordered, the choice is determined);

(ii) if $x \prec x'$ in C, then $\max[x]^+ \prec \max[x']^+$.

Every finite tree C has a spectrum. Indeed, C is past contractible, with its root 0 as a past spectrum: $P = \{0\}$; its future spectrum F can be obtained omitting any point which has precisely one immediate successor. In the example below (ordered rightward), the points of the future spectrum are marked with a bigger bullet; the spectral injective model $E = P \cup F$ is shown on the right

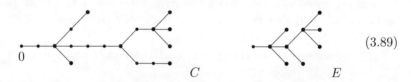

$$(3.89)$$

The associated projective model, the full subcategory $E' \subset E^2$ on the objects $0 \prec y$ ($y \in F$), is isomorphic to F (and not isomorphic to E, unless $0 \in F$).

3.9.2 Modelling an ordered space

In the sequel, we will *generally* consider *ordered* topological spaces X (equipped with the associated d-structure) with minimum (0) and maximum (1) and study the pf-spectrum of the fundamental category $C = \mathord{\uparrow}\Pi_1(X)$.

The latter inherits a privileged 'starting point' 0, which is a minimal point of C (but not necessarily the unique one, cf. Section 3.9.6) and a privileged 'ending point' 1, which is maximal in C. Furthermore, recall that C is skeletal (when X *is* ordered), so that the future spectrum – if it exists – is the least full (i.e. replete) reflective subcategory of C, and is strictly determined as a subset of C. Objects of C (i.e. points of X) will be denoted by letters $x, a, b, c, \ldots$; arrows of C (i.e. 'homotopy classes' of paths of X) by Greek letters $\alpha, \beta, \gamma, \ldots$

Consider, in the category p**Top**, the compact ordered space X in the left figure below: a subspace of the standard ordered square $\uparrow[0,1]^2$ obtained by taking out two *open* squares (marked with a cross)

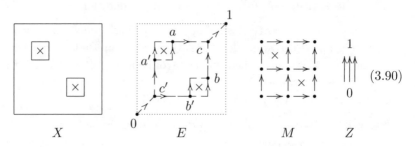

$$(3.90)$$

$$X \qquad E \qquad M \qquad Z$$

The fundamental category $C = \uparrow\Pi_1(X)$ is easy to determine (see Theorem 3.2.6 and Section 3.2.7).

We shall prove that its pf-spectrum is the full subcategory E, on eight vertices (where the two cells marked with a cross do not commute, while the central one does), that the associated projective model is M (see Section 3.9.3) and that the category Z is just a coarse model (of C, E and M).

First, we show that the category $C = \uparrow\Pi_1(X)$ has a past spectrum

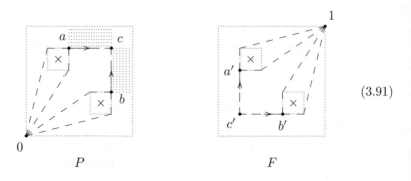

$$(3.91)$$

$$P \qquad F$$

In fact, there are four past regularity classes of objects, each having an initial object:

$$[c]^- = \{x \mid x \geqslant c\} \qquad \text{(unmarked)},$$
$$[a]^- = \{x \mid x \geqslant a\} \setminus [c]^-, \qquad \text{(marked with dots)},$$
$$[b]^- = \{x \mid x \geqslant b\} \setminus [c]^- \qquad \text{(marked with dots)},$$
$$[0]^- = X \setminus ([c]^- \cup [a]^- \cup [b]^-) \qquad \text{(unmarked)}.$$

Notice that a, b, c are effective V^--branching points, while 0 is the global minimum (for the path order), weakly initial in C.

These four points form the past spectrum $\mathrm{sp}^-(C) = \{0, a, b, c\}$, as is easily verified with the characterisation of Lemma 3.8.3(b): take the full subcategory $P \subset C$ on these objects (represented in the same picture), its embedding $i^- : P \subset C$ and the projection p^- sending each point $x \in C$ to the minimum of its past regularity class. Now $i^- \dashv p^-$, with a counit-component $\varepsilon(x) : i^- p^-(x) \to x$ which is uniquely determined in $\uparrow \Pi_1(X)$, since – within each of the four zones described above – there is at most one homotopy class of paths between two given points.

Symmetrically, we have the future spectrum: the full subcategory $F \subset C$ in the right figure above, on the following four objects (each of them being a maximum in its future regularity class):

- 1 (the global maximum of X, *weakly* terminal in C);

- a', b', c' (V^+-branching points).

The projection p^+ (left adjoint to $i^+ : F \subset C$) sends each point $x \in C$ to the maximum of its future regularity class (i.e. the lowest distinguished vertex $p^+(x) \geqslant x$); the unit-component $\eta(x) : x \to i^+ p^+(x)$ is, again, uniquely determined in $\uparrow \Pi_1(X)$.

Globally, we have constructed a spectral pf-presentation of C (Section 3.8.5); this generates the skeletal injective model E, as the full subcategory of C on $\mathrm{sp}(C) = \{0, a, b, c, a', b', c', 1\}$. The full subcategory $Z \subset E$ on the objects $0, 1$ is isomorphic to the past spectrum of F, as well as to the future spectrum of P, hence coarse equivalent (Section 3.3.3) to C and E.

Comments. The pf-spectrum E provides a category with the same past and future behaviour as C. This can be read as follows, in (3.90):

(a) the action begins at the 'starting point' 0, the minimum of X, from where we can only move to c';

(b) c' is an (effective) V^+-branching point, where we choose: either the upper/middle way or the lower/middle one;

(c) the first choice leads to a', a further V^+-branching point where we choose between the upper or the middle way; similarly, the second choice leads to the V^+-branching point b', where we choose between the lower or the middle way (which is the same as before);

(d) the routes of the first bifurcation considered in (c) join at a, those of the second at b (V$^-$-branching points);

(e) the two resulting routes come together at c (the last V$^-$-branching point);

(f) from where we can only move to the 'ending point' 1, the maximum of X.

The 'coarse model' Z only says that in C there are *three* homotopically distinct ways of going from 0 to 1, and loses relevant information on the branching structure of C.

The projective model is studied below.

3.9.3 The projective model

For the same category $C = \mathord{\uparrow}\Pi_1(X)$, the spectral projective model M, represented in the right figure below, is the full subcategory of C^2 on the nine arrows displayed in the left figure

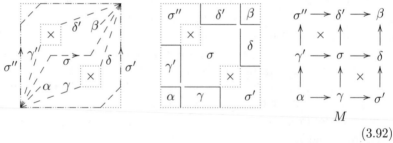

$$M \tag{3.92}$$

The projection $f(x) = (p^- x, p^+ x; p^- \eta x)$ (see (3.76)), from $X = \mathrm{Ob}C$ to $\mathrm{Ob}M \subset \mathrm{Mor}C$, thus has nine equivalence classes, analytically defined in (3.9.3) and 'sketched' in the middle figure above (the solid lines are meant to suggest that a certain boundary segment belongs to a certain region, as made precise below); in each of these regions, the morphism $f(x)$ is constant, and equal to $\alpha, \beta, \ldots$

$$
\begin{aligned}
f^{-1}(\alpha) &= [0, 1/5]^2, & \text{(closed in X),} \\
f^{-1}(\beta) &= [4/5, 1]^2 & \text{(closed in X),} \\
f^{-1}(\gamma) &= \,]1/5, 3/5] \times [0, 1/5], \\
f^{-1}(\gamma') &= [0, 1/5] \times \,]1/5, 3/5], \\
f^{-1}(\delta) &= [4/5, 1] \times [2/5, 4/5[,
\end{aligned}
\tag{3.93}
$$

$$
\begin{aligned}
f^{-1}(\delta') &= [2/5, 4/5[\times [4/5, 1], \\
f^{-1}(\sigma) &= X \cap \,]1/5, 4/5[^2 && \text{(open in X)}, \\
f^{-1}(\sigma') &= X \cap (\,]3/5, 1] \times [0, 2/5[) && \text{(open in X)}, \\
f^{-1}(\sigma'') &= X \cap ([0, 2/5[\times \,]3/5, 1]) && \text{(open in X)}.
\end{aligned}
$$

The interpretation of the projective model M is practically the same as above, in Section 3.9.2, with some differences:

(i) in M there is no distinction between the starting point and the first future branching point, nor between the ending point and the last past branching point;

(ii) the different paths produced by the obstructions are 'distinguished' in M by three new intermediate objects: $\sigma, \sigma', \sigma''$.

Note also that – here and in many cases – one *can* also embed M in C, by choosing a suitable point of a suitable path in each homotopy class $\alpha, \beta, ...$; but there is no canonical way of doing so.

In order to compare the injective model E and the projective model M, the examples below (in Section 3.9.4) will make clear that distinguishing 0 from c' (or c from 1) carries *some* information (like distinguishing the initial from the terminal object, in the injective model **2** of a non-pointed category having both, cf. Section 3.6.4). According to applications, one may decide whether this information is useful or redundant.

3.9.4 Variations

(a) Consider the previous ordered space X (Section 3.9.2) together with the spaces X' and X'', obtained by taking out, from the ordered square $\uparrow[0,1]^2$, two open squares placed in different positions, 'at' the boundary

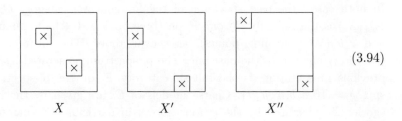

$$\begin{array}{ccc} X & X' & X'' \end{array} \tag{3.94}$$

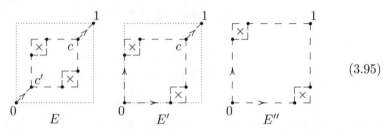

$$(3.95)$$

The pf-spectra E, E' and E'' distinguish these situations: in the second case the starting point 0 is an *effective* future branching point, and *we must make a choice from the very beginning* (either the upper/middle way or the middle/lower one); in the last case, this remains true and moreover the ending point is an *effective* past branching point. The projective models of these three spectra coincide (with the category M of Section 3.9.3).

(b) The following examples show similar situations, with a different injective (and projective) model. We start again from a (compact) ordered space $X_i \subset {\uparrow}[0,1]^2$, obtained by taking out two open squares

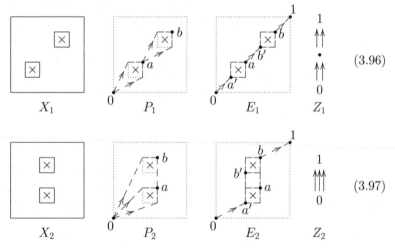

$$(3.96)$$

$$(3.97)$$

In both cases, the past spectrum of the fundamental category $C_i = {\uparrow}\Pi_1(X_i)$ is the full subcategory P_i on three objects: 0 (the minimum) and a, b (V^--branching points), as shown above. The future spectrum is symmetric to the past one. The pf-spectrum, generated by the previous presentation, is the full subcategory E_i on the pf-spectrum $\mathrm{sp}(C_i) = \{0, a, b, a', b', 1\}$. Coarse models of C_i are given by the categories Z_i generated by the graphs above; in particular, Z_1 has four arrows from 0 to 1.

The categories E_1, E_2 are not isomorphic, as more clearly shown below

$$b' \overset{\times}{\underset{}{\rightrightarrows}} b \rightarrow 1 \qquad\qquad b' \overset{\times}{\underset{}{\rightrightarrows}} b \rightarrow 1$$

$$0 \rightarrow a' \overset{\times}{\underset{}{\rightrightarrows}} a \qquad\qquad 0 \rightarrow a' \overset{\times}{\underset{}{\rightrightarrows}} a$$

$$E_1 \qquad\qquad\qquad\qquad E_2$$

3.9.5 A preordered space

The compact space $X \subset \mathbf{I}^2$ represented below (non-monometrically) is now equipped with the *preorder*: $(x,y) \prec (x',y')$ defined by the relation $y \leqslant y'$. Thus, all points having the same vertical coordinate are equivalent.

The fundamental category $C = {\uparrow}\Pi_1(X)$ is no longer skeletal. Let us choose $m = (1/2, 0)$ as a minimum of X (weakly initial in C) and $m' = (1/2, 1)$ as a maximum

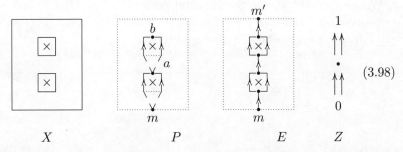

$$\text{(3.98)}$$

$$X \qquad\qquad P \qquad\qquad E \qquad Z$$

Now, the past spectrum of the fundamental category $C = {\uparrow}\Pi_1(X)$ is the full subcategory $P \subset C$ on three objects: m (a minimum), $a = (1/2, 2/5)$ and $b = (1/2, 4/5)$ (V^--branching points), as in the second figure above; of course, all of them can be equivalently moved, horizontally.

The future spectrum is symmetric to the past one: $a' = (1/2, 1/5)$, $b' = (1/2, 3/5)$ and m'. The pf-spectrum is the full subcategory E on these six points (or any equivalent sextuple). It is isomorphic to the pf-spectrum E_1 of (3.96).

3.9.6 The Swiss flag

Let us come back to ordered spaces. The following situation is often analysed as a basic one, in concurrency: the 'Swiss flag' $X \subset {\uparrow}[0,1]^2$.

See [FGR2, FRGH, GG, Go] for a description of 'the conflict of resources' which it depicts in the theory of concurrent systems, and [FRGH], p. 84, for an analysis of the fundamental category which leads to a 'category of components' similar to the projective model that we get below (in (3.101)):

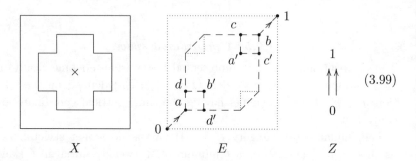

$$X \qquad E \qquad Z \qquad (3.99)$$

Working as above, the fundamental category $C = {\uparrow}\Pi_1(X)$ has an injective model E and a coarse model Z. In fact, the past spectrum is the full subcategory $P \subset C$ represented below:

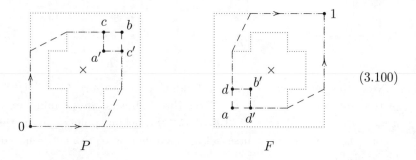

$$P \qquad F \qquad (3.100)$$

Its set of objects $\mathrm{sp}^-(C) = \{0, a', b, c, c'\}$ contains two minimal points $0, a'$ which are not comparable in the path-preorder of X and C, so that the starting point 0 is not a minimum for this preorder; the remaining points b, c, c' are V^--branching. Similarly, the future spectrum $\mathrm{Sp}^+(C)$ is the full subcategory $F \subset C$ in the right figure above, on the set of objects $\mathrm{sp}^+(C) = \{a, d, d', b', 1\}$.

The pf-spectrum of C is the full subcategory category E on $\mathrm{sp}(C) = \mathrm{sp}^-(C) \cup \mathrm{sp}^+(C)$.

The spectral projective model M is shown below, under the same conventions as in (3.92):

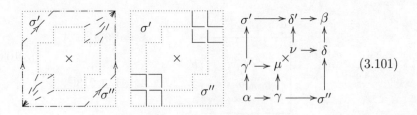

$$(3.101)$$

3.9.7 A three-dimensional case

Consider now the ordered compact space $X \subset \uparrow[0,1]^3$ represented below (the complement of the cube $]1/3, 2/3[^2 \times]2/3, 1]$ in $\uparrow \mathbf{I}^3$):

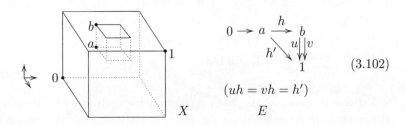

$$(3.102)$$

Then the category $C = \uparrow \Pi_1(X)$ is strongly past contractible: 0 is the initial object and past spectrum; it has future spectrum F formed of three points: 1 (the maximum, weakly terminal), a (an O^+-branching point), and b (a V^+-branching point). The pf-spectrum is the category E, embedded as the full subcategory on the objects $0, a, b, 1$.

(Here, the simpler definition of future regularity equivalence given in [G14] would yield a less neat analysis. In fact, since all the morphisms $0 \to x$ are future regular, all points of C *would be equivalent in that sense*, and C would have no future spectrum: F would just be a future retract of C, and E an injective model.)

3.9.8 Faithful and non-faithful spectra

The situation is very different for the ordered compact space $X \subset \uparrow[0,1]3$ of the figure below (taking out an open cube in the central position):

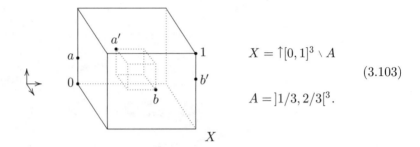

$$X = {\uparrow}[0,1]^3 \setminus A$$

$$(3.103)$$

$$A = \left]1/3, 2/3\right[^3.$$

The fundamental category $C = {\uparrow}\Pi_1(X)$ has an initial and a terminal object, 0 and 1. Therefore, C has pf-spectrum **2** (Section 3.6.4), which is *not* a faithful model: indeed, C is not a preorder, since the set $C(x, y)$ contains two arrows when (among other cases)

$$x < y, \qquad a < x < a', \qquad b < y < b', \qquad x_3,\, y_3 \in \left]1/3, 2/3\right[,$$

$$a = (0, 0, 1/3), \qquad a' = (1/3, 1/3, 2/3),$$
$$b = (2/3, 2/3, 1/3), \qquad b' = (1, 1, 2/3).$$

Various proposals have been suggested to analyse such situations, either by *a finer analysis of the fundamental category*, via categories of fractions [FRGH] and generalised quotients [GH, G18], or by *introducing and modelling the fundamental 2-category* [G15]. Yet the problem of finding a good solution still seems to be open.

3.9.9 Some hints at applications

Applications of directed algebraic topology to concurrency are well developed; the interested reader can begin from the references cited in the Introduction (Section 1) and see how the examples of this section can be interpreted in this domain. Here we want to hint at other possibilities, like the analysis of traffic networks, space-time models, directed images and biological systems.

(a) We begin by developing an example similar to that of the Introduction (Section 2). Consider the subspace $X \subset \mathbf{R} \times [-1, 1]$ obtained by taking out two open squares (marked with a cross)

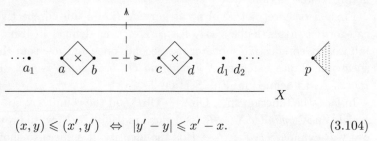

$$(x, y) \leqslant (x', y') \quad \Leftrightarrow \quad |y' - y| \leqslant x' - x. \tag{3.104}$$

It is equipped with the above order relation, whose 'cone of the future' at a point p is shown on the right.

First, this ordered space can be viewed as representing a stream with two islands; the stream moves rightward, with velocity v. The order expresses the fact that the observer can move, with respect to the stream, with an upper bound for scalar velocity, so that the composed velocity v'' can at most form an angle of $45°$ with the direction of the stream.

Secondly, one can view the coordinate x as time, the coordinate y as position in a *one-dimensional* physical medium and the order as the possibility of going from (x, y) to (x', y') with velocity $\leqslant 1$ (with respect to a 'rest frame', linked to the medium). The two forbidden squares are now *linear* obstacles in the medium, with a limited duration in time (first expanding and then contracting).

The fundamental category $\uparrow\Pi_1(X)$ reveals obstructions (islands, temporary obstacles,...). A minimal injective model E of the fundamental category is given by the full subcategory on the points marked above, in (3.104). E is generated by the following countable graph (under no conditions)

$$\dots a_2 \to a_1 \to a \rightrightarrows b \to c \rightrightarrows d \to d_1 \to d_2 \dots \tag{3.105}$$

The analysis is similar to that of (3.96). Moving the obstructions, one can get results similar to other previous cases (Section 3.9.2, etc.); the fundamental category will distinguish whether these obstructions occur one after the other (as above) or 'sensibly' at the same time (as in Section 3.9.2).

(b) Finally, we observe that the analysis of a category by *minimal past and future models*, as developed here, is closely related to notions recently introduced by A.C. Ehresmann [Eh], and used in a series of papers with J.P. Vanbremeersch for modelling biosystems, neural systems, etc. Clearly, such relationship arises from the common aim of studying non-reversible actions.

First, a past retract P of a category X (i.e. a full coreflective subcategory) is a particular case of a *corefract*, as defined in [Eh], 1.2: a full weakly coreflective subcategory. Second, one can show that the past spectrum P of a category X *having no O^--branchings* is necessarily a *root* of X, as defined in [Eh], Section 2.

In fact, the function $\mathrm{sp}^- \colon \mathrm{Ob}X \to \mathrm{Ob}P$ and the counit $\varepsilon x \colon \mathrm{sp}^-(x) \to x$ yield an X-*cylinder* $X \to P$, as defined in [Eh], 1.1. Now, to prove that every P-cylinder $(F, f) \colon P \to P$ is the identity, it suffices to show that $F(x) \sim^- x$, for all x in P, so that the map $fx \colon F(x) \to x$ must be the identity (because P is a past spectrum). First, fx is V^--regular in P, by definition of P-cylinder, and in X as well (the embedding $P \subset X$ being a past equivalence). Second, fx is O^--regular, by hypothesis. Finally, if x is past regular, so is $F(x)$, by Lemma 3.7.5(a); otherwise, x must be V^--branching, which implies that $F(x)$ is also, by Lemma 3.7.5(b).

In this section, all the examples of Sections 3.9.2–3.9.6 fall into this situation: *their past spectrum is a root and their future spectrum a co-root*. On the other hand, the second example of (3.81) – also present in Section 3.9.7 – shows a category having an O^+-branching; it is easy to see that its future spectrum is not a coroot. The categories $\mathbf{r}, \mathbf{c}$, whose minimal injective model is studied in Sections 3.6.5 and 3.6.6, have no root nor coroot, as well as no past nor future spectrum.

Part II
Higher directed homotopy theory

4

Settings for higher order homotopy

This chapter is a complete reworking of previous settings, which were mostly – not exclusively – aimed at the reversible case; they have been developed in various papers, in particular [G1, G3, G4].

Starting from the basic settings of Chapter 1, we arrive in Sections 4.1 and 4.2 at the notion of a 'symmetric dIP4-homotopical category' (Section 4.2.6), through various steps, called dI2 and dI3-category (or dI2 and dI3-homotopical category) and their duals, dP2-category, etc. (possibly symmetric).

Special care is given to single out the results which hold in the intermediate settings, and in particular do not depend on the transposition symmetry: we have already remarked that its presence has both advantages and drawbacks (Section 1.1.5).

Some basic examples are dealt with in Sections 4.3 and 4.4; many others will follow in Chapter 5. Thus, d**Top**, d**Top.** (pointed d-spaces) and **Cat** are symmetric dIP4-homotopical categories. On the other hand, the category of reflexive graphs is just dIP2-homotopical (Section 4.3.3), and the category **Cub** of cubical sets is just dIP1-homotopical, under two isomorphic structures for left and right homotopies (Section 4.3.4). Chain complexes on an additive category form a symmetric dIP4-homotopical category, which is regular and reversible; directed chain complexes have a regular dIP4-homotopical structure, which lifts the previous one but is no longer symmetric nor reversible (Section 4.4).

In the rest of this chapter we work out the general theory of dI2, dI3 and dI4-categories. In Section 4.5 we construct the homotopy 2-category $\mathrm{Ho}_2(\mathbf{A})$ and the fundamental category functor $\uparrow\Pi_1 : \mathbf{A} \to \mathbf{Cat}$, for a dI4-category $\mathbf{A}$. In Sections 4.6–4.8 we deal with higher properties of homotopy pushouts and cofibre sequences; we also examine the cone functor and the monad structure which it inherits from the cylinder.

We end in Section 4.9, studying the reversible case. This has peculiar properties, which will also be of interest for non-reversible structures, in the relative settings of Section 5.8. Under some additional hypotheses, a reversible dI4-homotopical category can be given a structure of 'cofibration category' in the sense of Baues (Theorem 4.9.6), which is a non-self-dual version of Quillen's model categories.

We will see in the next chapter (Section 5.8) how one can study the higher properties of directed homotopy in 'defective' cases, like cubical sets and directed chain complexes, by a relative setting consisting of a forgetful functor with values in a stronger framework.

4.1 Preserving homotopies and the transposition symmetry

We begin to study higher properties of directed homotopy, assuming that cylindrical colimits are preserved by the cylinder functor I (Section 4.1.2) and that a transposition symmetry is given (Section 4.1.4). In the presence of the latter, the cylinder functor acts on homotopies and the preservation of cylindrical colimits becomes equivalent to the preservation of h-pushouts.

4.1.0 The basic setting

Let us recall, from Chapter 1, that a dI1-category $\mathbf{A} = (\mathbf{A}, R, I, \partial^\alpha, e, r)$ comes equipped with a reversor (an involutive covariant endofunctor)

$$R \colon \mathbf{A} \to \mathbf{A}, \qquad (R(X) = X^{\mathrm{op}}, \quad R(f) = f^{\mathrm{op}}), \qquad (4.1)$$

and a cylinder endofunctor $I \colon \mathbf{A} \to \mathbf{A}$, with faces ∂^α, degeneracy e and reflection r

$$\partial^\alpha \colon 1 \underset{\longleftarrow}{\overset{\longrightarrow}{\rightrightarrows}} I \colon e, \qquad r \colon IR \to RI \qquad (\alpha = \pm). \qquad (4.2)$$

These data must satisfy the axioms

$$
\begin{aligned}
e\partial^\alpha &= 1 \colon \mathrm{id}\mathbf{A} \to \mathrm{id}\mathbf{A}, & RrR.r &= 1 \colon IR \to IR, \\
Re.r &= eR \colon IR \to R, & r.\partial^- R &= R\partial^+ \colon R \to RI.
\end{aligned}
\qquad (4.3)
$$

A homotopy $\varphi \colon f^- \to f^+ \colon X \to Y$ is defined as a map $\varphi \colon IX \to Y$ with $\varphi.\partial^\alpha X = f^\alpha$ (and the map can be written as $\hat\varphi$ to distinguish it from the homotopy which it represents). Whisker composition $k \circ \varphi \circ h$ is defined by $(k \circ \varphi \circ h)^\hat{} = k.\varphi.Ih$ (for $h \colon X' \to X$ and $k \colon Y \to Y'$, see Section 1.2.3).

Each map $f\colon X \to Y$ has a trivial endo-homotopy, $0_f\colon f \to f$, represented by $f.eX = eY.If\colon IX \to Y$. Each homotopy $\varphi\colon f \to g\colon X \to Y$ has a reflected homotopy

$$\varphi^{\mathrm{op}}\colon g^{\mathrm{op}} \to f^{\mathrm{op}}\colon X^{\mathrm{op}} \to Y^{\mathrm{op}}, \quad (\varphi^{\mathrm{op}})\hat{} = R(\hat{\varphi}).r\colon IRX \to RY,$$
$$(\varphi^{\mathrm{op}})^{\mathrm{op}} = \varphi, \quad (0_f)^{\mathrm{op}} = 0_{(f^{\mathrm{op}})}, \quad (k \circ \varphi \circ h)^{\mathrm{op}} = k^{\mathrm{op}} \circ \varphi^{\mathrm{op}} \circ h^{\mathrm{op}}. \tag{4.4}$$

We also recall, from Section 1.7.0, that a dI1-homotopical category is a dI1-category with all h-pushouts (Section 1.3.5) and a terminal object T.

The dual notions of dP1-category and dP1-homotopical category can be found in Section 1.2.2 and Section 1.8.2; the self-dual cases of dIP1-category and pointed dIP1-homotopical category are in Section 1.2.2 and Section 1.8.6.

4.1.1 Double homotopies and 2-homotopies

Let $\mathbf{A}$ be a general dI1-category. Double homotopies behave much as in d$\mathbf{Top}$, in Section 3.2.

The *second-order cylinder* I^2X has four (one-dimensional) faces, written

$$\partial_1^\alpha = I\partial^\alpha\colon IX \to I^2X, \qquad \partial_2^\alpha = \partial^\alpha I\colon IX \to I^2X. \tag{4.5}$$

A *double homotopy* is a map $\Phi\colon I^2X \to Y$; it has four faces, which are ordinary homotopies

$$\partial_1^\alpha(\Phi) = \Phi.\partial_1^\alpha = \Phi.I\partial^\alpha, \qquad \partial_2^\alpha(\Phi) = \Phi.\partial_2^\alpha = \Phi.\partial^\alpha I, \tag{4.6}$$

$$
\begin{array}{ccc}
f & \xrightarrow{\partial_2^-(\Phi)} & h \\
{\scriptstyle \partial_1^-(\Phi)}\downarrow & \Phi & \downarrow{\scriptstyle \partial_1^+(\Phi)} \\
k & \xrightarrow[\partial_2^+(\Phi)]{} & g
\end{array}
\qquad\qquad
\begin{array}{ccc}
\bullet & \xrightarrow{1} & \\
\downarrow{\scriptstyle 2} & & \\
& &
\end{array}
$$

Moreover, Φ has four vertices, the maps $\partial^-\partial_1^-(\Phi) = f = \partial^-\partial_2^-(\Phi)$, etc. Again, we *can* write $\hat{\Phi}\colon I^2X \to Y$ when we want to distinguish the *map* from the double homotopy which it represents.

Two 'horizontally' consecutive d-homotopies

$$\varphi\colon f^- \to f^+\colon X \to Y, \qquad\qquad \psi\colon g^- \to g^+\colon Y \to Z,$$

can be composed, to form a double homotopy $\psi \circ \varphi$

$$
\begin{array}{ccc}
h^- f^- & \xrightarrow{\ h^- \varphi\ } & h^- f^+ \\
{\scriptstyle \psi \circ f^-}\downarrow & \psi \circ \varphi & \downarrow{\scriptstyle \psi \circ f^+} \\
h^+ f^- & \xrightarrow[\ h^+ \varphi\ ]{} & h^+ f^+
\end{array}
\qquad
\begin{array}{c}
(\psi \circ \varphi)\hat{\ } = \hat{\psi}.(\hat{\varphi} \times \uparrow\mathbf{I}): \\[4pt]
I^2 X \ \to\ Z.
\end{array}
\tag{4.7}
$$

Together with the whisker composition, in Section 1.2.3, this is a particular instance of the cubical enrichment produced by the (co)cylinder functor, see (1.44).

A (directed) *2-homotopy* $\Phi \colon \varphi \to \psi \colon f \to g \colon X \to Y$ will be a double homotopy whose faces ∂_1^α are degenerate, while the faces ∂_2^α are φ, ψ (the symmetric choice becomes equivalent in the presence of a transposition, see Section 4.1.4)

$$
\begin{array}{ccc}
f & \xrightarrow{\ \varphi\ } & g \\
{\scriptstyle 0_f}\downarrow & \Phi & \downarrow{\scriptstyle 0_g} \\
f & \xrightarrow[\ \psi\ ]{} & g
\end{array}
\qquad
\begin{array}{ll}
\partial_2^-(\Phi) = \varphi, & \partial_2^+(\Phi) = \psi, \\[4pt]
\partial^- \varphi = f = \partial^- \psi, & \partial^+ \varphi = g = \partial^+ \psi, \\[4pt]
\partial_1^-(\Phi) = 0_f, & \partial_1^+(\Phi) = 0_g.
\end{array}
\tag{4.8}
$$

Using the natural transformation

$$
r_2 = rI.Ir \colon I^2 R \to RI^2, \qquad (\textit{double reflection of } I^2), \tag{4.9}
$$

a double homotopy $\Phi \colon I^2 X \to Y$ has a *double reflection*, which works on faces as below

$$
\Phi^{\mathrm{op}} \colon I^2(X^{\mathrm{op}}) \to Y^{\mathrm{op}}, \qquad (\Phi^{\mathrm{op}})\hat{\ } = R(\hat{\Phi}).r_2 \colon I^2 RX \to RY, \tag{4.10}
$$

$$
\begin{array}{ccc}
f & \xrightarrow{\ \varphi\ } & h \\
{\scriptstyle \sigma}\downarrow & \Phi & \downarrow{\scriptstyle \tau} \\
k & \xrightarrow[\ \psi\ ]{} & g
\end{array}
\qquad\qquad
\begin{array}{ccc}
g^{\mathrm{op}} & \xrightarrow{\ \psi^{\mathrm{op}}\ } & k^{\mathrm{op}} \\
{\scriptstyle \tau^{\mathrm{op}}}\downarrow & \Phi^{\mathrm{op}} & \downarrow{\scriptstyle \sigma^{\mathrm{op}}} \\
h^{\mathrm{op}} & \xrightarrow[\ \varphi^{\mathrm{op}}\ ]{} & f^{\mathrm{op}}
\end{array}
$$

(A simple reflection in *one* direction is only possible in the reversible case, see Section 4.9.1.) In particular, a 2-homotopy $\Phi \colon \varphi \to \psi \colon f \to g \colon X \to Y$ yields a 2-homotopy

$$
\Phi^{\mathrm{op}} \colon \psi^{\mathrm{op}} \to \varphi^{\mathrm{op}} \colon g^{\mathrm{op}} \to f^{\mathrm{op}} \colon X^{\mathrm{op}} \to Y^{\mathrm{op}}. \tag{4.11}
$$

4.1.2 The main preservation property

Let $\mathbf{A}$ be a dI1-homotopical category. We have seen, in Section 1.3.5, that the h-pushout of a span (f, g) in $\mathbf{A}$ can be described as the ordinary colimit of the left diagram below, called the *cylindrical colimit* of (f, g)

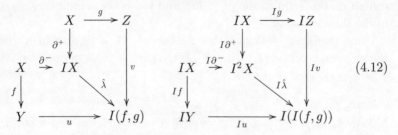

$$\begin{array}{ccc}
X \xrightarrow{\ g\ } Z & & IX \xrightarrow{\ Ig\ } IZ \\
\end{array} \qquad (4.12)$$

We say that the cylinder functor $I\colon \mathbf{A} \to \mathbf{A}$ *preserves cylindrical colimits* (as colimits) if, for every span (f, g) in $\mathbf{A}$, the right diagram above is also a colimit – a property already considered in Lemma 1.7.1 for strong dI1-functors.

Notice that the second diagram is not the *cylindrical* colimit of the maps (If, Ig): the latter would require the faces $\partial^\alpha(IX)$ of the object IX, instead of the faces $I\partial^\alpha(X)$ which occur above. Indeed, we will need the presence of a transposition to convert the faces $\partial_1^\alpha = I\partial^\alpha$ into the faces $\partial_2^\alpha = \partial^\alpha I$, and the colimit above into a cylindrical colimit (see Section 4.1.5).

4.1.3 Theorem (The higher property of h-pushouts)

Let $\mathbf{A}$ be a dI1-homotopical category and assume that the cylinder functor $I\colon \mathbf{A} \to \mathbf{A}$ preserves all cylindrical colimits (Section 4.1.2).

Then every h-pushout $A = I(f, g)$ also satisfies a two-dimensional universal property, concerned with two maps a, b, two homotopies σ, τ and a double homotopy Φ (Section 4.1.1) with the following boundaries

$$\begin{array}{ccccc}
 & & Y & & W \\
 & & & & \\
X & & A & \rightrightarrows & W \\
 & & & & \\
 & & Z & & W
\end{array}
\qquad
\begin{array}{ccc}
auf & \xrightarrow{\ a\lambda\ } & avg \\
\sigma f \downarrow & \Phi & \downarrow \tau g \\
buf & \xrightarrow{\ b\lambda\ } & bvg
\end{array}
\qquad (4.13)$$

$$a, b\colon A \to W, \qquad \sigma\colon au \to bu, \quad \tau\colon av \to bv, \quad \Phi\colon I^2 X \to W,$$

$$\partial_1^-(\Phi) = \Phi.(I\partial^- X) = \sigma \circ f, \qquad \partial_2^-(\Phi) = \Phi.(\partial^- IX) = a \circ \lambda, \ldots$$

Then there is some homotopy $\varphi\colon a \to b$ such that $\varphi \circ u = \sigma$, $\varphi \circ v = \tau$; and there is precisely one which also satisfies the condition $\varphi.I(\hat{\lambda}) = \Phi$.

Proof By hypothesis, the cylinder functor $I\colon \mathbf{A} \to \mathbf{A}$ preserves the colimit on the left diagram of (4.12), and yields the colimit on the right, in the same diagram.

Using the latter, we can factor the cocone $(W; \sigma.If, \Phi, \tau.Ig)$ through the universal cocone $(IA; Iu, I(\hat{\lambda}), Iv)$. There is thus precisely one map $\varphi\colon IA \to W$ such that

$$\varphi{\circ}u = \varphi.Iu = \sigma, \qquad \varphi{\circ}v = \varphi.Iv = \tau, \qquad \varphi.I(\hat{\lambda}) = \Phi.$$

Moreover, its lower face $\partial^- \varphi = \varphi.\partial^- A$ is a (and the upper one is b) because

$$\varphi.\partial^- A.u = \varphi.Iu.\partial^- X = \partial^- \sigma = au,$$

$$\varphi.\partial^- A.v = \varphi.Iv.\partial^- X = \partial^- \tau = av,$$

$$\varphi.\partial^- A.\hat{\lambda} = \varphi.I(\hat{\lambda}).\partial^- IX = \Phi.\partial^- (IX) = a{\circ}\lambda.$$

$\square$

4.1.4 Symmetric dI1-categories

A *symmetric dI1-category* $(\mathbf{A}, R, I, \partial^\alpha, e, r, s)$ is a dI1-category equipped with a natural transformation $s\colon I^2 \to I^2$, called *transposition*, which satisfies the conditions

$$ss = 1, \qquad Ie.s = eI, \qquad s.I\partial^\alpha = \partial^\alpha I, \qquad Rs.r_2 = r_2.sR, \qquad (4.14)$$

where $r_2 = rI.Ir\colon I^2 R \to RI^2$ is the double reflection of I^2 (cf. (4.9)).

This basic transposition generates higher transpositions

$$s_i = I^{n-1-1} s I^{i-1}\colon I^n \to I^n \qquad\qquad (\text{for } i = 1, ..., n-1),$$

and an action of the symmetric group S_n on the power I^n of the cylinder endofunctor. This action motivates the term 'symmetric', which in the following terminology about dI-, dP-, dIP-categories can always be replaced with 'permutable'. (See the discussion of symmetries in directed algebraic topology, in Section 1.1.5.)

Dually, a *symmetric dP1-category* $(\mathbf{A}, R, P, \partial^\alpha, e, r, s)$ is a dP1-category equipped with a *transposition* $s\colon P^2 \to P^2$ satisfying the conditions:

$$ss = 1, \qquad s.eP = Pe, \qquad \partial^\alpha P.s = P\partial^\alpha, \qquad r_2.sR = Rs.r_2, \qquad (4.15)$$

where $r_2 = Pr.rP\colon RP^2 \to P^2 R$ is now the double reflection of P^2.

A *symmetric dIP1-category* is a dIP1-category equipped with transpositions of the cylinder and cocylinder which are mates (Section A5.3) and satisfy the equivalent conditions (4.14) and (4.15).

A *symmetric dI1-homotopical category* is a symmetric dI1-category with a terminal object $\top$ and all cylindrical colimits, *preserved by I* (Section 4.1.2). By extending I to homotopies (Section 4.1.5), the last property will be equivalently expressed by saying that **A** has all h-pushouts, preserved by I.

Let **A** be a symmetric monoidal category with reversor; the unit object is E. We have seen, in Section 1.2.5, that a dI1-interval **I** has a structure consisting of four maps

$$\partial^\alpha : E \rightrightarrows \mathbf{I} : e, \qquad r : \mathbf{I} \to \mathbf{I}^{\mathrm{op}} \qquad (\alpha = \pm) \tag{4.16}$$

satisfying the conditions (1.48). This gives rise to a *symmetric monoidal dI1-structure* on the category **A** (Section 1.2.5)

$$
\begin{aligned}
I(X) = X \otimes \mathbf{I}, &\qquad \partial^\alpha X = X \otimes \partial^\alpha : X \to IX, \\
eX = X \otimes e : IX \to X, &\quad rX = X^{\mathrm{op}} \otimes r : IRX \to RIX.
\end{aligned}
\tag{4.17}
$$

It is now straightforward to verify that this structure is symmetric in the present sense, with transposition retrieved from the symmetry $s(X, Y) : X \otimes Y \to Y \otimes X$ of the tensor product

$$sX = X \otimes s(\mathbf{I}, \mathbf{I}) : X \otimes \mathbf{I} \otimes \mathbf{I} \to X \otimes \mathbf{I} \otimes \mathbf{I}. \tag{4.18}$$

4.1.5 Cylinder functor and homotopies

Let **A** be a *symmetric* dI1-category (Section 4.1.4). The existence of the transposition $s : I^2 \to I^2$ allows us to define I on homotopies.

In fact, $I : \mathbf{A} \to \mathbf{A}$ becomes a strong dI1-functor (Section 1.2.6), when equipped with the natural transformations $i = r^{-1} : RI \to IR$ and $s : I^2 \to I^2$ which make the following diagrams commute (because of the axioms on r, s, in (4.14))

$$
\begin{array}{ccccc}
I & \xrightarrow{\partial^\alpha I} & I^2 & \xrightarrow{eI} & I \\
& {\scriptstyle I\partial^\alpha}\searrow & \downarrow{\scriptstyle s} & \nearrow{\scriptstyle Ie} & \\
& & I^2 & &
\end{array}
\qquad
\begin{array}{ccccc}
IRI & \xrightarrow{Ii} & I^2R & \xrightarrow{sR} & I^2R \\
{\scriptstyle rI}\downarrow & & & & \downarrow{\scriptstyle Ir} \\
RI^2 & \xrightarrow{Rs} & RI^2 & \xrightarrow{iI} & IRI
\end{array}
$$

Therefore, as in (1.11) for **Top**, we define I on the homotopy $\varphi : f^- \to f^+ : X \to Y$, letting

$$
\begin{aligned}
(I\varphi)\hat{\ } &= I(\hat\varphi).sX : I^2 X \to IY, \\
(I(\hat\varphi).sX) . \partial^\alpha(IX) &= I(\hat\varphi).I(\partial^\alpha X) = If^\alpha.
\end{aligned}
\tag{4.19}
$$

As a consequence, I also preserves future and past homotopy equivalences.

Applying Lemma 1.7.1, it follows that, in a *symmetric* dI1-category (Section 4.1.4), the cylinder functor preserves an h-pushout if and only if it preserves the corresponding cylindrical colimit as a colimit (Section 4.1.2). This is always the case in a symmetric dI1-homotopical category (Section 4.1.4).

4.1.6 Theorem (The h-pushout functor on homotopies)

We already know that, if $\mathbf{A}$ is a dI1-homotopical category, the double mapping cylinder $I(f, g)$ gives a functor $\mathbf{A}^\vee \to \mathbf{A}$, where $\vee$ is the formal-span category (Section 1.3.7), as expressed in the left diagram below (with $I(f, g) = A$, $I(f', g') = A'$

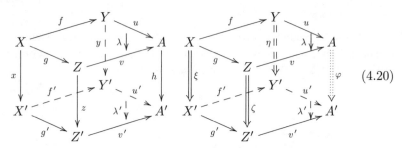

$$(4.20)$$

Now, if $\mathbf{A}$ is a symmetric dI1-homotopical category, the homotopy pushout carries coherent triples *of homotopies to homotopies.*

Precisely, let a coherent triple of homotopies (ξ, η, ζ) *be assigned, as in the right cube above (where the double arrow labelled ξ stands for $\xi \colon x \to x' \colon X \to X'$, and so on)*

$$(\xi, \eta, \zeta) \colon (x, y, z) \to (x', y', z') \colon (f, g) \to (f', g'),$$
$$\xi \colon x \to x', \qquad \eta \colon y \to y', \qquad \zeta \colon z \to z', \qquad (4.21)$$
$$f' \circ \xi = \eta \circ f, \qquad g' \circ \xi = \zeta \circ g,$$

Then, there is some homotopy $\varphi \colon h \to h' \colon I(f, g) \to I(f', g')$ which completes the cube in a coherent way

$$\varphi \colon h \to h', \qquad \varphi \circ u = u' \circ \eta, \qquad \varphi \circ v = v' \circ \zeta, \qquad (4.22)$$

and φ is uniquely determined if we also ask that $\varphi.I(\hat{\lambda}) = \lambda'.I(\hat{\xi}).sX$.

Proof Let us define $h = I(x, y, z)$ and $h' = I(x', y', z')$ as in Section 1.3.7. The thesis follows from the two-dimensional property of the h-pushout λ

of (f, g), with respect to the following double homotopy (Theorem 4.1.3)

$$\Phi = \lambda'.I(\hat{\xi}).sX : I^2 X \to I^2 X \to IX' \to A',$$

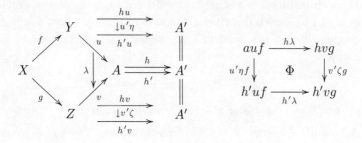

$$\partial_1^-(\Phi) = u'f'\circ\xi = u'\circ\eta\circ f, \qquad \partial_2^-(\Phi) = \lambda'\circ x = h\circ\lambda, ...$$

There is thus some homotopy $\varphi: h \to h'$ such that $\varphi\circ u = u'\circ\eta$, $\varphi\circ v = v'\circ\zeta$; and there is precisely one which also satisfies $\varphi.I(\hat{\lambda}) = \Phi$.

$\square$

4.1.7 Theorem
(Homotopy invariance of the cone and suspension functors)

Let $\mathbf{A}$ be a symmetric dI1-homotopical category (Section 4.1.4).

(a) We use the notation of Section 1.7.2 for the upper cone functor $C^+ : \mathbf{A} \to \mathbf{A}$, in the diagram below: u denotes the lower basis, v^+ the upper vertex and γ the structural homotopy. Then, for every homotopy $\varphi: f \to g: X \to Y$ there is some homotopy $\psi: C^+[f] \to C^+[g]: C^+X \to C^+Y$ such that

$$
\begin{array}{ccccc}
X & \xrightarrow{\ \mathrm{id}\ } & X & \xrightarrow[g]{\overset{f}{\downarrow\varphi}} & Y \\
\downarrow p & & \gamma \nearrow \downarrow u & & \downarrow u \\
\top & \xrightarrow[v^+]{} & C^+X & \xrightarrow{\downarrow\psi} & C^+Y
\end{array}
\qquad
\begin{array}{l}
\psi\circ uX = uY\circ\varphi, \\[2mm]
\psi\circ v^+X = 0.
\end{array}
\qquad (4.23)
$$

Moreover, ψ is uniquely determined if we also ask that $\psi.I(\gamma X) = \gamma Y.I(\hat{\varphi}).sX.$

As a consequence, the cone functors $C^\alpha : \mathbf{A} \to \mathbf{A}$ preserve future and past homotopy equivalences.

(b) Given a homotopy $\varphi: f \to g$, there is some homotopy $\psi: \Sigma f \to \Sigma g$ (and precisely one such that $\psi.I(\mathrm{ev}^X) = \mathrm{ev}^Y.(I(\hat{\varphi}).sX)$).

Again, the suspension functor Σ preserves future and past homotopy equivalences.

Proof Both results are a straightforward consequence of the previous theorem (4.1.6), which describes the action of the h-pushout functor on homotopies. In case (a), one lets Z be the terminal object (and modifies notation). $\qquad\qquad\square$

4.1.8 External transposition

Let $\mathbf{A}$ be a dl1-category. When the transposition $s\colon I^2 \to I^2$ is missing, one often has an *external transposition pair* $(S, s\colon SISI \to ISIS)$, as happens for cubical sets (see (1.134) and (1.155), for the path functor) and directed chain complexes (see Section 4.4.5).

This pair consists of an involutive endofunctor $S\colon \mathbf{A} \to \mathbf{A}$, the *transposer*, and of a natural transformation $s\colon SISI \to ISIS$, the *external transposition*, which satisfy the following axioms (again, s is invertible, with $s^{-1} = SsS$)

$$ RS = SR, \qquad\qquad SsS.s = 1, \qquad\qquad (4.24) $$

$$ I \xrightarrow{S\partial^\alpha SI} SISI \xrightarrow{SISe} SIS \qquad SISIR \xrightarrow{SISr} SISRI \xrightarrow{SrSI} SRISI $$

$$ ISISR \xrightarrow[ISrS]{} ISRIS \xrightarrow[rSIS]{} RISIS $$

This yields an associated *S-opposite cylinder SIS*, with structure: $S\partial^\alpha S, SeS, SrS$.

An *S-opposite homotopy* $\psi\colon f^- \to_S f^+\colon X \to Y$ is a map $\psi\colon SIS(X) \to Y$ with faces $\psi.S\partial^\alpha S = f^\alpha$. Each original homotopy $\varphi\colon f^- \to f^+\colon X \to Y$ defines an S-opposite homotopy $\psi\colon Sf^- \to_S Sf^+\colon SX \to SY$, as follows:

$$ \psi = S\varphi\colon SIS(SX) \to SY, $$
$$ S\varphi.(S\partial^\alpha S(SX)) = S(\varphi.\partial^\alpha X) = Sf^\alpha. \qquad (4.25) $$

This approach gives an effective framework for *left* and *right* homotopies in *non-symmetric* dl1-categories, like cubical sets and directed chain complexes (Section 4.4.5). But it is not clear if the external transposition can be of further help in studying homotopy theory.

4.2 A strong setting for directed homotopy

We arrive, through various intermediate steps, at the notion of a symmetric dIP4-homotopical category (Section 4.2.6).

4.2.1 Connections and transposition

A *dI2-category* $\mathbf{A} = (\mathbf{A}, R, I, \partial^\alpha, e, r, g^\alpha)$ is equipped with a reversor $R \colon \mathbf{A} \to \mathbf{A}$ (an involutive covariant endofunctor, as usual) and a cylinder endofunctor $I \colon \mathbf{A} \to \mathbf{A}$, with two faces ∂^α, a degeneracy e (as in the dI1-case) and two additional natural transformations, called the *lower* and *upper connection* g^-, g^+ (or higher degeneracies)

$$1 \underset{e}{\overset{\partial^\alpha}{\rightleftarrows}} I \overset{g^\alpha}{\longleftarrow} I^2 \qquad r \colon IR \to RI \qquad (\alpha = \pm). \tag{4.26}$$

This structure has to satisfy the following axioms (which *include* the axioms of dI1-categories)

$$
\begin{aligned}
e\partial^\alpha &= 1, & eg^\alpha &= e.Ie \ (= e.eI) & (\textit{degeneracy}), \\
g^\alpha.Ig^\alpha &= g^\alpha.g^\alpha I & & & (\textit{associativity}), \\
g^\alpha.I\partial^\alpha &= 1 = g^\alpha.\partial^\alpha I & & & (\textit{unit}), \\
g^\beta.I\partial^\alpha &= \partial^\alpha e = g^\beta.\partial^\alpha I & & & (\textit{absorbency; } \alpha \neq \beta), \\
RrR.r &= 1, & Re.r &= eR, \\
r.\partial^- R &= R\partial^+, & r.g^+ R &= Rg^-.r_2 & (\textit{reflection}).
\end{aligned}
\tag{4.27}
$$

Also here $r_2 = rI.Ir \colon I^2 R \to RI^2$ is the double reflection of I^2, defined in (4.9).

A dI2-category is *reversible* if R is the identity (so that I becomes an involutive diad on $\mathbf{A}$, see Section 1.1.8).

A *symmetric dI2-category* $\mathbf{A} = (\mathbf{A}, R, I, \partial^\alpha, e, r, g^\alpha, s)$ is a dI2-category equipped with a transposition $s \colon I^2 \to I^2$ satisfying the following conditions

$$
\begin{aligned}
s.s &= 1, & Ie.s &= eI, & s.I\partial^\alpha &= \partial^\alpha I, \\
Rs.r_2 &= r_2.sR, & & & g^\alpha.s &= g^\alpha,
\end{aligned}
\tag{4.28}
$$

where, after the conditions of symmetric dI1-categories (Section 4.1.4), we are requiring that the connections be invariant under s.

It will be useful to note that, in a dI2-category, the connections $g^\alpha X \colon I^2 X \to IX$ make the lower basis $\partial^- \colon X \to IX$ of the cylinder a past deformation retract (Section 1.3.1), and the upper basis

$\partial^+\colon X \to IX$ a future deformation retract, with the *same* retraction $e\colon IX \to X$

$$\partial^\alpha\colon X \rightrightarrows IX\colon e$$

$$
\begin{aligned}
e\partial^- &= \mathrm{id}X, & g^-&\colon \mathrm{id}(IX) \to \partial^- e\colon IX \to IX, & (4.29)\\
e\partial^+ &= \mathrm{id}X, & g^+&\colon \partial^+ e \to \mathrm{id}(IX)\colon IX \to IX.
\end{aligned}
$$

Furthermore, the maps $g^\alpha X\colon I^2 X \to X$ and the higher degeneracies $IeX, eIX\colon I^2 X \to IX$ can be viewed as double homotopies, with the following faces, where $\partial\colon \partial^- \to \partial^+\colon IX \to X$ is the structural homotopy, represented by 1_{IX} (cf. (1.82)) and $0 = \partial^\alpha e\colon \partial^\alpha \to \partial^\alpha$ is a trivial homotopy

$$
\begin{array}{cccccccc}
\partial^- \xrightarrow{\partial} \partial^+ & \quad & \partial^- \xrightarrow{0} \partial^- & \quad & \partial^- \xrightarrow{0} \partial^- & \quad & \partial^- \xrightarrow{\partial} \partial^+ \\[4pt]
\partial\Big\downarrow \; g^- \; \Big\downarrow 0 & & 0\Big\downarrow \; g^+ \; \Big\downarrow \partial & & \partial\Big\downarrow \; Ie \; \Big\downarrow \partial & & 0\Big\downarrow \; eI \; \Big\downarrow 0 \qquad (4.30)\\[4pt]
\partial^+ \xrightarrow[0]{} \partial^+ & & \partial^- \xrightarrow[\partial]{} \partial^+ & & \partial^+ \xrightarrow[0]{} \partial^+ & & \partial^- \xrightarrow[\partial]{} \partial^+
\end{array}
$$

For d-spaces, connections and transposition have already been defined (Section 3.1.3), in the same way as for topological spaces

$$
\begin{aligned}
g^\alpha\colon I^2 X \to IX, & \quad g^\alpha(x, t, t') = (x, g^\alpha(t, t')) & (\textit{connections}),\\
s\colon I^2 X \to I^2 X, & \quad s(x, t, t') = (x, t', t) & (\textit{transposition}),
\end{aligned}
\qquad (4.31)
$$

using the similar maps on the powers of the directed interval: $g^-(t, t') = \max(t, t')$, $g^+(t, t') = \min(t, t')$, $s(t, t') = (t', t)$.

We say that **A** is a (symmetric) dI2-*homotopical category* if:

(i) it is a (symmetric) dI2-category with terminal object $\top$;
(ii) it has all cylindrical colimits (Section 1.3.5), which are preserved by the functor I, as colimits.

We have seen that, *in the symmetric case*, condition (ii) amounts to the preservation of h-pushouts by the cylinder functor (Section 4.1.5).

Dually one defines (symmetric) *dP2-categories* and (symmetric) *dP2-homotopical categories*.

A *dIP2-category* can be equivalently defined as a dI2-category where the cylinder functor has a right adjoint, or a dP2-category where the path functor has a left adjoint.

4.2.2 Concatenation and dI3-categories

We introduce now a dI3-structure as a dI1-structure 'with concatenation'. The connections are *not* present here, but will be reinserted later, in a stronger structure (Section 4.2.5).

In a dI1-category, the *concatenation pushout* $J(X) = IX +_X IX$ of an object X, or *J-pushout*, is the pasting of two cylinders, one on top of the other

$$
\begin{array}{ccc}
X & \xrightarrow{\partial^+} & IX \\
{\scriptstyle \partial^-}\Big\downarrow & {\scriptstyle}\Big\downarrow{\scriptstyle c^-} & \\
IX & \xrightarrow[c^+]{} & JX
\end{array}
\tag{4.32}
$$

A dI3-*category*

$$
\mathbf{A} = (\mathbf{A}, R, I, \partial^\alpha, e, r, J, c),
$$

will be a dI1-category $(\mathbf{A}, R, I, \partial^\alpha, e, r)$ which has all concatenation pushouts $J(X) = IX +_X IX$. Moreover, these are *preserved by I* (as pushouts), and there is a natural transformation $c\colon I \to J$, called *concatenation*, which satisfies the axiom:

$$
c\partial^- = c^-\partial^-, \quad c\partial^+ = c^+\partial^+, \quad e^J.c = e, \quad r^J.cR = Rc.r. \tag{4.33}
$$

The natural transformations $e^J\colon J \to 1$ and $r^J\colon JR \to RJ$ which intervene here are induced by the degeneracy $e\colon I \to 1$ and the reflection $r\colon IR \to RI$, respectively. Namely, they are both defined by the universal property of a J-pushout, as follows:

$$
\begin{aligned}
e^J\colon JX \to X, & \qquad e^J.c^- = e^J.c^+ = e, \\
r^J\colon JRX \to RJX, & \qquad r^J.c^-R = Rc^+.r, \qquad r^J.c^+R = Rc^-.r.
\end{aligned}
\tag{4.34}
$$

It follows that:

$$
(RrJR).r^J = 1, \qquad\qquad Re^J.r^J = e^J R. \tag{4.35}
$$

Note that JX automatically exists if $\mathbf{A}$ has h-pushouts, since JX can be obtained as the h-pushout $I(1_X, \partial^+)$, or, equivalently, as $I(\partial^-, 1_X)$.

We already know that, in **Top** and d**Top**, one can take $JX = IX$ and $c = 1$, with c^- given by the 'first-half' embedding of the standard interval into itself, $t \mapsto t/2$, and c^+ by the 'second-half' embedding (Sections 1.1.1 and 3.1.3). On the other hand, in **Cat** and for chain complexes, JX is not isomorphic to IX (cf. Sections 4.3 and 4.4).

Coming back to the general situation, we have a functor $J\colon \mathbf{A} \to \mathbf{A}$ and two natural transformations $c^-, c^+\colon I \to J$, which give three faces $1 \to J$ (*lower, upper* and *middle face* of J)

$$
\partial^{--} = c^-\partial^-, \qquad \partial^{++} = c^+\partial^+, \qquad \partial^{\pm} = c^+\partial^- = c^-\partial^+. \tag{4.36}
$$

It will be useful to note that the functor J always preserves J-pushouts – a straightforward consequence of a general lemma of category theory: *pushouts preserve pushouts* (Lemma 4.2.9). Thus, $J^2 X$ can equivalently be expressed by the following pushouts

$$
\begin{array}{ccc}
JX & \xrightarrow{\partial^+ J} & IJX \\
\scriptstyle{\partial^- J} \downarrow & \llcorner \;-\; -\downarrow \scriptstyle{c^- J} & \\
IJX & \xrightarrow[c^+ J]{} & J^2 X
\end{array}
\qquad
\begin{array}{ccc}
JX & \xrightarrow{J\partial^+} & JIX \\
\scriptstyle{J\partial^-} \downarrow & \llcorner \;-\; -\downarrow \scriptstyle{Jc^-} & \\
JIX & \xrightarrow[Jc^+]{} & J^2 X
\end{array}
\qquad (4.37)
$$

We say that $\mathbf{A}$ is a dI3-*homotopical category* if:

(i) it is a dI3-category with terminal object $\top$;
(ii) it has all cylindrical colimits (Section 1.3.5), which are preserved by the functor I as colimits.

Dually one defines *dP3-categories* and *dP3-homotopical categories*

$$\mathbf{A} = (\mathbf{A}, R, P, \partial^\alpha, e, r, Q, c),$$

where $Q(Y) = PY \times_Y PY$ denotes the *concatenation pullback* or *Q-pullback* of the object Y

$$
\begin{array}{ccc}
QY & \xrightarrow{c^+} & PY \\
\scriptstyle{c^-} \downarrow & \urcorner \;-\; -\downarrow \scriptstyle{\partial^-} & \\
PY & \xrightarrow[\partial^+]{} & Y
\end{array}
\qquad (4.38)
$$

Q is called the functor of *pairs of consecutive paths*, and the transformation $c\colon Q \to P$ is called *path-concatenation*.

Following once more a general pattern, we can define a *dIP3-category* as a dI3-category where the functors I, J have right adjoints

$$I \dashv P, \qquad J \dashv Q. \qquad (4.39)$$

Indeed, once that they are equipped with all the natural transformations which are mate to the ones of the dI3-structure, the previous square diagram (4.38) is automatically a pullback, as a consequence of general facts on mates and (co)limits (Section A5.4).

4.2.3 Concatenating homotopies

In a dI3-category, the *concatenation* $\varphi + \psi\colon f \to h$, or *vertical composition* of consecutive homotopies $\varphi\colon f \to g$ and $\psi\colon g \to h$, is defined as represented by the map

$$
\begin{aligned}
(\varphi + \psi)\hat{} &= (\hat{\varphi} \vee \hat{\psi}).c \colon IX \to Y \\
(\hat{\varphi} \vee \hat{\psi}).c^- &= \hat{\varphi}, \qquad\qquad (\hat{\varphi} \vee \hat{\psi}).c^+ = \hat{\psi}.
\end{aligned}
\tag{4.40}
$$

where $\hat{\varphi} \vee \hat{\psi}$ denotes the obvious morphism defined on the pushout JX (as above, in the second line). By the previous axiom (4.33), this operation satisfies:

$$
\begin{aligned}
0_f + 0_f &= 0_f, \qquad\qquad (\varphi + \psi)^{\mathrm{op}} = \psi^{\mathrm{op}} + \varphi^{\mathrm{op}}, \\
k(\varphi + \psi)h &= k\varphi h + k\psi h.
\end{aligned}
\tag{4.41}
$$

In particular, c^- and c^+ are (represent) consecutive homotopies, with concatenation c

$$
\begin{aligned}
c^- &\colon \partial^{--} \to \partial^{\pm}, \qquad\qquad c^+ \colon \partial^{\pm} \to \partial^{++} \\
c^- + c^+ &= c \colon \partial^{--} \to \partial^{++} \colon X \to JX.
\end{aligned}
\tag{4.42}
$$

We say that $\mathbf{A}$ has a *regular concatenation*, or that it is a *regular* dI3-category, if the concatenation of homotopies behaves categorically (as it happens for chain complexes)

$$
(\varphi + \psi) + \chi = \varphi + (\psi + \chi), \qquad\qquad 0_f + \varphi = \varphi = \varphi + 0_g. \tag{4.43}
$$

Thus the dI3-category $\mathbf{A}$, equipped with homotopies as 2-cells, whisker composition and concatenation, becomes a sesquicategory (Section A5.1); but, with respect to this structure, it also has a reflection.

We say that $\mathbf{A}$ is *2-regular* if, moreover, these operations satisfy the *reduced interchange property* (as it happens in $\mathbf{Cat}$)

$$
X \xrightarrow[\ g\ ]{\overset{f}{\underset{\downarrow\varphi}{\longrightarrow}}} Y \xrightarrow[\ k\ ]{\overset{h}{\underset{\downarrow\psi}{\longrightarrow}}} Z \qquad\qquad \psi g . h\varphi = k\varphi . \psi f, \tag{4.44}
$$

which is equivalent to saying that our sesquicategory is actually a 2-category (Section A5.2).

In a dI3-category $\mathbf{A}$, the existence of a homotopy $f \to g$ yields a preorder relation $f \preceq_1 g$. If $\mathbf{A}$ is reversible (i.e. R is the identity), this relation coincides with the homotopy congruence $f \simeq_1 g$ which gives the homotopy category $\mathrm{Ho}_1(\mathbf{A}) = \mathbf{A}/\simeq_1$ (Section 1.3.3).

The concatenations of double homotopies will be studied in Section 4.5.

4.2.4 Symmetric dI3-categories

A *symmetric dI3-category*

$$\mathbf{A} = (\mathbf{A}, R, I, \partial^\alpha, e, r, s, J, c), \qquad Ic.s = s'.cI : I^2 \to IJ, \qquad (4.45)$$

is, at the same time, a symmetric dI1 and a dI3-category, where concatenation is consistent with transposition, as expressed in the right-hand equation above. Here, we are using a natural transformation

$$s' : JI \to IJ \qquad\qquad (IJ\text{-}transposition), \qquad (4.46)$$

which is defined componentwise (on the pushout $J(IX)$), by the (commutative) left diagram below

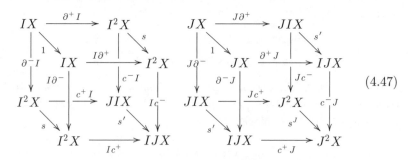

$$(4.47)$$

Thus, *in the presence of a transposition*, the condition that I preserve J-pushouts is equivalent to saying that this s' is an isomorphism.

On the other hand, the fact that J preserves J-pushouts (which is always true, see Section 4.2.2) implies that we have a natural isomorphism

$$s^J : J^2 X \to J^2 X \qquad\qquad (J\text{-}transposition), \qquad (4.48)$$

making the right diagram above commutative (since it is easy to verify that its upper and left square commute, on the injections $c^\alpha : IX \to JX$).

4.2.5 dI4-categories

A *dI4-category*

$$\mathbf{A} = (\mathbf{A}, R, I, \partial^\alpha, e, r, g^\alpha, J, c, z), \qquad (4.49)$$

is a dI2 and a dI3-category with additional structure: a natural transformation $z : I^2 \to I$, called *acceleration*, or *left-unit comparison*, which provides a 2-homotopy from the homotopy $0 + \partial : \partial^- \to \partial^+ : X \to IX$

to the homotopy $\partial\colon \partial^- \to \partial^+ \colon X \to IX$

$$
\begin{array}{ccc}
\partial^- & \xrightarrow{\;0+\partial\;} & \partial^+ \\
{\scriptstyle 0}\downarrow & z & \downarrow{\scriptstyle 0} \\
\partial^- & \xrightarrow{\;\partial\;} & \partial^+
\end{array}
\qquad
\begin{array}{l}
z.I\partial^- = \partial^- e, \qquad z.I\partial^+ = \partial^+ e, \\[4pt]
z.\partial^- I = \partial^- e + \partial, \qquad z.\partial^+ I = \partial.
\end{array}
\tag{4.50}
$$

Because of z, every homotopy $\varphi\colon f \to g$ has two *acceleration* 2-homotopies

$$
\begin{aligned}
\Theta'(\varphi) &= \varphi.zX\colon 0_f + \varphi \to \varphi, \\
\Theta''(\varphi) &= (\Theta'(\varphi^{\mathrm{op}}))^{\mathrm{op}}\colon \varphi \to \varphi + 0_g,
\end{aligned}
\tag{4.51}
$$

(but *not* the other way round: slowing down conflicts with direction). The second 2-homotopy is obtained by reflection of homotopies and double reflection of 2-homotopies (cf. (4.11)).

In d**Top**, the component zX can be defined as $X \times \zeta\colon I^2 X \to IX$, by means of the following map $\zeta\colon {\uparrow}\mathbf{I} \times {\uparrow}\mathbf{I} \to {\uparrow}\mathbf{I}$ (the affine homotopy from the path $f = 0 + \mathrm{id}\mathbf{I}\colon {\uparrow}\mathbf{I} \to {\uparrow}\mathbf{I}$ to $\mathrm{id}\mathbf{I}\colon {\uparrow}\mathbf{I} \to {\uparrow}\mathbf{I}$)

$$
\begin{array}{ccc}
\partial^- & \xrightarrow{\;0\;}\ \partial^-\ \xrightarrow{\;\partial\;} & \partial^+ \\
{\scriptstyle 0}\downarrow & \zeta & \downarrow{\scriptstyle 0} \\
\partial^- & \xrightarrow{\quad\partial\quad} & \partial^+
\end{array}
\qquad
\begin{array}{l}
\zeta(t,t') = (1 - t').f(t) + t'.t, \\[4pt]
f(t) = \max(0, 2t - 1).
\end{array}
\tag{4.52}
$$

If, in $\mathbf{A}$, homotopies have a *regular* concatenation (Section 4.2.3), the homotopy $0 + \partial\colon \partial^- \to \partial^+\colon X \to IX$ coincides with $\partial\colon \partial^- \to \partial^+$ and one can always take as z the trivial 2-homotopy $\partial \to \partial$, represented by $zX = eIX\colon I^2 X \to IX$. With this choice, we say that $\mathbf{A} = (\mathbf{A}, R, I, \partial^\alpha, e, r, g^\alpha, s, J, c)$ is a *regular* dI4-category (and a *2-regular* dI4-category if, moreover, $\mathbf{A}$ is a 2-category, as specified in Section 4.2.3).

A *symmetric dI4-category*

$$
\mathbf{A} = (\mathbf{A}, R, I, \partial^\alpha, e, r, g^\alpha, s, J, c, z)
$$

is a dI4-category equipped with a transposition consistent with the dI2- and dI3-structure.

Dually one defines a *dP4-category*

$$
\mathbf{A} = (\mathbf{A}, R, P, \partial^\alpha, e, r, g^\alpha, Q, c, z)
$$

and a *symmetric dP4-category* (which also has a transposition $s\colon P^2 \to P^2$). In both cases the structure contains the *concatenation pullback*, or Q-*pullback* (Section 4.2.2) with a *concatenation map* $c\colon Q \to P$ and an *acceleration* $z\colon P \to P^2$.

A *dIP4-category* is a dI4-category whose endofunctors I, J have right adjoints P, Q; then, these functors inherit a dP4-structure with the same reversor and with natural transformations which are mates to the transformations of I, J. Equivalently, a dIP4-category can also be defined as a dP4-category whose endofunctors P, Q have left adjoints I, J. The *symmetric* case is analogous.

4.2.6 Homotopical categories

We say that **A** is a (symmetric) *dI4-homotopical category* if:

(i) it is a (symmetric) dI4-category with terminal object $\top$;
(ii) it has all cylindrical colimits (Section 1.3.5), which are preserved by the functor I as colimits.

As we have already remarked in Section 4.2.2, condition (ii) implies, by itself, the existence of J-pushouts (and the fact that they are preserved by I).

Dually one defines a (symmetric) *dP4-homotopical category*.

Finally, **A** is a (symmetric) dIP4-homotopical category if:

(i′) it is a (symmetric) dIP4-category (4.2.5) with terminal object $\top$ and initial object $\bot$;
(ii′) it has all cylindrical colimits and cocylindrical limits (automatically preserved by I and P, respectively, because of the adjunction $I \dashv P$).

If, in this case, **A** is pointed (i.e. has a zero-object), we have further adjunctions (see (1.211)), with $\alpha = \pm$

$$
\begin{aligned}
C^\alpha \dashv E^\alpha &\qquad (\textit{cone-cocone}), \\
\Sigma \dashv \Omega &\qquad (\textit{suspension-loops}).
\end{aligned}
\qquad (4.53)
$$

The category d**Top** of d-spaces is a symmetric dIP4-homotopical category, which is complete and cocomplete.

In fact, it has all limits and colimits, and adjoint endofunctors $I \dashv P$, $J \dashv Q$. Furthermore, the cylinder has a symmetric dI4-structure, already considered above, step-by-step, which is transferred to the path functor along the adjunction (cf. Sections 1.2.2 and 4.2.2).

We also recall that the concatenation pushout $J(X) = IX +_X IX$ is realised as $JX = IX$ (cf. Sections 1.4.6 and 4.2.2); then $c^\alpha : IX \to JX$

is the 'first-half' or 'second-half' embedding of the standard interval into itself, and the concatenation map $c \colon IX \to JX$ is the identity.

Similarly the concatenation pullback $Q(Y) = PY \times_Y PY$ (cf. Section 3.1.4) is realised as $QY = PY$; then, $c^\alpha \colon QY \to PY$ restricts a path to its 'first-half' or 'second-half', and the concatenation map $c \colon QY \to PY$ is again the identity.

4.2.7 Functors and subcategories

Recall that a lax dI1-functor (Section 1.2.6) $H = (H, i, h) \colon \mathbf{A} \to \mathbf{X}$ is a functor H between dI1-categories, equipped with two natural transformations i, h, the comparisons, which satisfy the following conditions (so that i is invertible):

$$i \colon RH \to HR, \qquad h \colon IH \to HI, \qquad (RiR).i = 1_{RH}, \qquad (4.54)$$

$$
\begin{array}{ccc}
H \xrightarrow{\ \partial^\alpha H\ } IH \xrightarrow{\ eH\ } H & \qquad & IRH \xrightarrow{\ Ii\ } IHR \xrightarrow{\ hR\ } HIR \\
\quad {}_{H\partial^\alpha}\searrow \ \ {\downarrow}_h \ \ {}^{He}\nearrow & & {}_{rH}\downarrow \qquad\qquad\qquad\qquad \downarrow{}^{Hr} \\
HI & & RIH \xrightarrow[Rh]{} RHI \xrightarrow[iI]{} HRI
\end{array}
$$

Now, if $\mathbf{A}, \mathbf{X}$ are symmetric dI4-categories, we say that H is a *lax symmetric dI4-functor* if its comparisons are also consistent with the remaining structural transformations of $\mathbf{A}, \mathbf{X}$ (i.e. connections, transposition, concatenation and acceleration). Namely, the following diagrams must commute:

$$
\begin{array}{ccc}
I^2 H \xrightarrow{\ Ih\ } IHI \xrightarrow{\ hI\ } HI^2 & \qquad I^2 H \xrightarrow{\ Ih\ } IHI \xrightarrow{\ hI\ } HI^2 & \\
{}_{g^\alpha H}\downarrow \qquad\qquad\qquad \downarrow{}^{Hg^\alpha} & {}_{sH}\downarrow \qquad\qquad\qquad \downarrow{}^{Hs} & (4.55) \\
IH \xrightarrow[\quad h \quad]{} HI & \quad I^2 H \xrightarrow[Ih]{} IHI \xrightarrow[hI]{} HI^2 &
\end{array}
$$

$$
\begin{array}{ccc}
IH \xrightarrow{\ h\ } HI & \quad I^2 H \xrightarrow{\ Ih\ } IHI \xrightarrow{\ hI\ } HI^2 & \\
{}_{cH}\downarrow \qquad \downarrow{}^{Hc} & {}_{zH}\downarrow \qquad\qquad\qquad \downarrow{}^{Hz} & (4.56) \\
JH \xrightarrow[h']{} HJ & \quad IH \xrightarrow[\qquad h \qquad]{} HI &
\end{array}
$$

In the left diagram of (4.56), we have used the *J-comparison* $h' \colon JH \to HJ$ of H, which is defined componentwise, on the J-pushout

$J(HX)$ of **X**, by the following commutative diagram

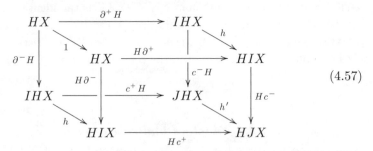

$$(4.57)$$

Dually, extending (1.59), one defines a *lax symmetric dP4-functor* $K = (K, i, k)\colon \mathbf{A} \to \mathbf{X}$ between symmetric dP4-categories, with natural transformations $i\colon KR \to RK$ and $k\colon KP \to PK$ coherent with the whole structure.

As in Section 1.2.6, a lax symmetric dI4-functor $(H, i, h)\colon \mathbf{A} \to \mathbf{X}$ between *symmetric dIP4*-categories becomes automatically a lax symmetric dP4-functor, with the inverse natural isomorphism $i^{-1}\colon HR \to RH$ and the comparison $k\colon HP \to PH$ which is mate to the comparison $h\colon IH \to HI$ (Section A5.3). Then, (H, i, h, k) is called a *lax symmetric dIP4-functor*.

The intermediate cases, dI2, dI3, dP2, dP3, dIP2, dIP3 (possibly symmetric), are dealt with in the same way. In all cases, we speak of *strong* (resp. *strict*) functors when all comparisons are invertible (resp. identities).

The notion of a dI1-subcategory was defined in Section 1.2.1. A *symmetric dI4-subcategory* $\mathbf{A}'$ of a dI4-category $\mathbf{A}$ is a subcategory closed in $\mathbf{A}$ with respect to the whole structure $(R, I, \partial^\alpha, e, r, g^\alpha, s, J, c, z)$, and amounts to an inclusion $\mathbf{A}' \to \mathbf{A}$ which is a strict symmetric dI4-functor. The notions of (symmetric) dI2, dI3, dP2, dP3, dP4, dIP2, dIP3, dIP4-*sub*category or *homotopical subcategory* are defined in the same way.

4.2.8 The structure of the directed interval

In the symmetric monoidal case, the structures considered above can be defined by the corresponding structures on a standard directed interval, as we have already seen in Section 1.2.5 for dI1 and dIP1-categories.

Let $\mathbf{A} = (\mathbf{A}, \otimes, E, s)$ be a symmetric monoidal category with reversor $R\colon \mathbf{A} \to \mathbf{A}$ (Section 1.2.5); again, we always omit the isomorphisms of the monoidal structure, except the symmetry.

A *symmetric dI2-interval* $\mathbf{I}$ in $\mathbf{A}$ comes equipped with *faces* (∂^α), *degeneracy* (e), *reflection* (r), *connections* (g^α) and the transposition $s = s(\mathbf{I}, \mathbf{I})$ obtained from the symmetry of the tensor product

$$E \underset{e}{\overset{\partial^\alpha}{\rightrightarrows}} \mathbf{I} \overset{g^\alpha}{\underset{}{\Leftarrow}} \mathbf{I} \otimes \mathbf{I} \qquad r \colon \mathbf{I} \to \mathbf{I}^{\mathrm{op}}, \quad s \colon \mathbf{I} \otimes \mathbf{I} \to \mathbf{I} \otimes \mathbf{I}. \tag{4.58}$$

These data must satisfy the 'same' axioms as the symmetric dI2-cylinder (Section 4.2.1), conveniently rewritten (much as in (1.32) for a closely related structure, the dioid): for instance, the associativity of the connections, which is $g^\alpha.Ig^\alpha = g^\alpha.g^\alpha I$ for the cylinder, here becomes $g^\alpha.g^\alpha \otimes \mathbf{I} = g^\alpha.\mathbf{I} \otimes g^\alpha$.

When $R = \mathrm{id}$, our structure is the same as a *symmetric involutive dioid*, as defined in Section 1.1.7. (In the general case, a symmetric dI2-interval could also be called a *symmetric dioid with reflection*.)

If the object $\mathbf{I}$ is exponentiable (Section A4.2) in $\mathbf{A}$, we have, accordingly, a *monoidal dIP2-structure*.

A *dI3-interval* $\mathbf{I}$ is a dI1-interval having a *standard concatenation pushout* $\mathbf{J}$, preserved by the tensor product, and a *concatenation map* c which satisfies conditions similar to (4.33)

$$
\begin{array}{ccc}
E & \overset{\partial^+}{\longrightarrow} & \mathbf{I} \\
{\scriptstyle \partial^-}\downarrow & {\scriptstyle c^-}\Big\downarrow & \\
\mathbf{I} & \underset{c^+}{\longrightarrow} & \mathbf{J}
\end{array}
\qquad\qquad c \colon \mathbf{I} \to \mathbf{J}. \tag{4.59}
$$

Finally, a *dI4-interval* has both the preceding structures, with an *acceleration* map $z \colon \mathbf{I}^2 \to \mathbf{I}$ so that the axioms (4.50), suitably rewritten, hold.

When the objects $\mathbf{I}$ and $\mathbf{J}$ are exponentiable in $\mathbf{A}$, we have, accordingly, a *monoidal dIP3-, or dIP4-structure*, with the right adjoint functors P, Q

$$I = - \otimes \mathbf{I} \dashv P = (-)^{\mathbf{I}}, \qquad\qquad J = - \otimes \mathbf{J} \dashv Q = (-)^{\mathbf{J}}. \tag{4.60}$$

When the tensor product is the categorical product and s is the ordinary transposition, all these structures are called *cartesian*, instead of monoidal.

If the tensor product is not symmetric, we get, as in Section 1.2.5, a left cylinder $\mathbf{I} \otimes X$ and a *right cylinder $X \otimes \mathbf{I}$* defining two dI2-, or dI3-, or dI4-structures. See the case of cubical sets, in Section 4.3.4.

4.2.9 Lemma (Pushouts preserve pushouts)

In a category $\mathbf{A}$, *we have a commutative diagram consisting of a span of commutative cubes*

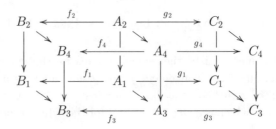

If the three squares of vertices A_i, B_i, C_i *are pushouts and all four pairs* (f_i, g_i) *of horizontal maps have a pushout* (k_i, h_i), *then the resulting new square* $D = (D_i)$ *is a pushout*

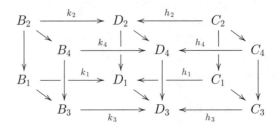

Formally, we are saying that, in the category $\mathbf{A}^{2\times 2}$ *of commutative squares, the pushout of a span of pushout-squares* $B \leftarrow A \rightarrow C$ *is a pushout-square, provided it exists and is* pointwise.

Proof Straightforward. □

4.3 Examples, I

We have seen that $\mathrm{d}\mathbf{Top}$ is a symmetric dIP4-homotopical category. We see now that the same is true of the category $\mathrm{d}\mathbf{Top}_{\bullet}$ of pointed d-spaces, and of $\mathbf{Cat}$.

On the other hand, the category of reflexive graphs is just dIP2-homotopical (Section 4.3.3), and the category $\mathbf{Cub}$ of cubical sets is just dIP1-homotopical, under two isomorphic structures (Section 4.3.4). We will see in Section 5.8 how one can study their higher properties of directed homotopy, by a relative setting using d-spaces.

4.3.1 Pointed spaces with distinguished paths

We know that d**Top** is a cartesian symmetric dIP4-homotopical category, which is complete and cocomplete (Section 4.2.6).

It is now easy to show that the category d**Top.** of pointed spaces with distinguished paths is a *pointed* symmetric *monoidal* dIP4-homotopical category, complete and cocomplete, with zero-object the (pointed) singleton $\{*\}$ and tensor product the smash product (1.122).

Extending what we have already seen, at the basic level (Section 1.5.5), the cocylinder structure of dTop. is that of dTop, enriched with the obvious base points. The cylinder structure is less simple, but can be deduced from the previous one, by adjunction.

In particular, let us recall that the pointed cocylinder is just the ordinary cocylinder, pointed at the constant loop at the base point

$$P\colon \mathrm{dTop.} \to \mathrm{dTop.}, \qquad P(Y, y_0) = (PY, \omega_0), \qquad (\omega_0 = 0_{y_0}), \quad (4.61)$$

while the cylinder functor

$$I\colon \mathrm{dTop.} \to \mathrm{dTop.}, \qquad I(X, x_0) = (IX/I\{x_0\}, [x_0, t]), \qquad (4.62)$$

is the quotient of the unpointed cylinder which collapses the fibre at the base-point, $I\{x_0\} = \{x_0\} \times \uparrow\mathbf{I}$.

The whole structure originates from the pointed directed interval $\uparrow\mathbf{I.} = (\uparrow\mathbf{I} + \{*\}, *)$ (Section 1.5.4), via smash product and exponentiation. Thus, $I(X, x_0) = (X, x_0) \wedge \uparrow\mathbf{I.}$.

4.3.2 Categories

We have already seen, in Section 1.2.2, that the category **Cat** of small categories has a cartesian dIP1-structure, with reversor $R(X) = X^{\mathrm{op}}$ and a directed interval consisting of the ordinal category $\uparrow\mathbf{i} = \mathbf{2} = \{0 \to 1\}$. In this structure, a path in X is an arrow, and a homotopy $\varphi\colon f \to g\colon X \to Y$ is a natural transformation. We show now that this structure is cartesian dIP4-homotopical.

First, it is well-known that **Cat** is complete, cocomplete and cartesian closed, with $[X, Y] = Y^X$ the category of functors $X \to Y$ and their natural transformations. The identity of the tensor product is the one-point ordinal category $\mathbf{1} = \{0\}$.

Now, the directed interval $\mathbf{2}$ is a symmetric dI4-interval (Section 4.2.8).

Its symmetric dI2-structure consists of the following functors (defined by their action on the objects)

$$1 \underset{e}{\overset{\partial^\alpha}{\rightrightarrows}} 2 \overset{g^\alpha}{\underset{}{\leftrightarrows}} 2^2 \qquad r: 2 \to 2^{\mathrm{op}}, \qquad s: 2^2 \to 2^2,$$

$$
\begin{aligned}
&\partial^\alpha(0) = \alpha, \qquad g^-(i,j) = \max(i,j), \quad g^+(i,j) = \min(i,j),\\
&s(i,j) = (j,i), \quad r(i) = 1 - i \qquad\qquad (\alpha, i, j = 0, 1).
\end{aligned}
\tag{4.63}
$$

Then, the *standard concatenation pushout* gives the ordinal **3**, which is equipped with the *standard concatenation map c*

$$
\begin{array}{ccc}
1 & \overset{\partial^+}{\longrightarrow} & 2 \\
{\scriptstyle \partial^-}\downarrow & {\scriptstyle --} \downarrow {\scriptstyle c^-} & \\
2 & \underset{c^+}{\longrightarrow} & 3
\end{array}
\qquad
\begin{aligned}
&c: 2 \to 3,\\
&c(0 \to 1) = 0 \to 2.
\end{aligned}
\tag{4.64}
$$

Altogether, we have a symmetric dIP4-structure, with cylinder $I(X) = X \times 2$, cocylinder $P(Y) = Y^2$, and the following concatenation pushout and pullback

$$
\begin{aligned}
J(X) = X \times 3, &\quad cX = X \times c: X \times 2 \to X \times 3,\\
Q(Y) = Y^3, &\quad cY = Y^c: Y^3 \to Y^2.
\end{aligned}
\tag{4.65}
$$

The effect of concatenation is simply the composition of consecutive arrows (i.e. paths). Thus, the concatenation is regular, and indeed 2-regular (Section 4.2.3). Finally, **Cat** is a regular cartesian dIP4-homotopical category (with trivial acceleration, see Section 4.2.5).

4.3.3 Reflexive graphs

Let us consider now the category **Gph** of (small) reflexive graphs, or 1-truncated cubical sets, or 1-truncated simplicial sets. To fix the notation, an object is a diagram in **Set**

$$X_0 \underset{e}{\overset{\partial^\alpha}{\rightrightarrows}} X_1 \qquad \partial^- e = 1 = \partial^+ e \qquad (\alpha = -, +),$$

consisting of a set of *vertices* X_0, a set of *arrows* (or *edges*) X_1, the *domain* and *codomain* mappings ∂^-, ∂^+, the *degeneracy* mapping e.

This category **Gph** will be equipped with the following symmetric monoidal closed structure. The *internal hom-functor* $[X, Y]$ is given by the reflexive graph consisting of morphisms of reflexive graphs $X \to Y$, with their transformations. The *tensor product* $X \otimes Y$ is the subgraph of $X \times Y$ containing all the objects $(x, y) \in X_0 \times Y_0$ and only those arrows $(u, v) \in X_1 \times Y_1$ such that either u or v is degenerated (an identity).

Consider now the ordinal $\uparrow\mathbf{i} = \mathbf{2} = \{0 \to 1\}$ as a reflexive graph, and a dI2-interval in $(\mathbf{Gph}, \otimes)$; the description is the same as above (in (4.63)), excepting the fact that the reflexive graph $\mathbf{2} \otimes \mathbf{2}$ has *four* non-degenerate arrows (and lacks the diagonal $(0, 0) \to (1, 1)$ of the category $\mathbf{2} \times \mathbf{2}$).

One obtains thus a symmetric monoidal dIP2-structure on **Gph**, with $IX = X \otimes \mathbf{2}$ and $PX = [\mathbf{2}, X]$. To assign a map $IX \to Y$ (or $X \to PY$) is here equivalent to give a transformation $\varphi \colon f^- \to f^+ \colon X \to Y$ between two morphisms $f^\alpha \colon X \to Y$ of reflexive graphs.

There is no concatenation of paths and homotopies.

4.3.4 Cubical sets

As we have seen in Section 1.6, directed homotopy in the category **Cub** is based on the directed interval $\uparrow\mathbf{i}$ (the cubical set freely generated by a 1-cube u), and on the classical tensor product. The latter is not symmetric (but induces the previous symmetric tensor product of reflexive graphs, by 1-truncation).

Thus, we have a left dIP1-structure $\mathbf{Cub}_L$ defined by the left cylinder functor $I(X) = \uparrow\mathbf{i} \otimes X$, and a right dIP1-structure $\mathbf{Cub}_R$ defined by the right cylinder $SIS(X) = X \otimes \uparrow\mathbf{i}$. The analytic description of left and right homotopies has been given in Section 1.6.5. The two structures are isomorphic, under the transposer $S \colon \mathbf{Cub}_L \to \mathbf{Cub}_R$, which reverses the order of faces, and we only need to consider one of them.

All the main enrichments of structure which we have considered in the two previous sections cannot be performed here. The structure $\mathbf{Cub}_L$ is not symmetric, but only has an external transposition (S, s) (see (1.155)). There are no connections, since any morphism $f \colon \uparrow\mathbf{i} \otimes \uparrow\mathbf{i} \to \uparrow\mathbf{i}$ must send the 2-cube $u \otimes u$ to a 2-cube of $\uparrow\mathbf{i}$, necessarily degenerate in direction 1 or 2, and therefore cannot satisfy both equations $f.\partial_1^\alpha = \mathrm{id} = f.\partial_2^\alpha$ (for a given $\alpha = \pm$). Finally, there is no concatenation of paths (i.e. edges), as in the 1-truncated case.

4.4 Examples, II. Chain complexes

Chain complexes on an additive category, with the usual homotopies, form a symmetric dIP4-homotopical category, which is regular and reversible. Directed chain complexes have a regular dIP4-homotopical structure, which lifts the previous one but is no longer symmetric nor reversible.

4.4.1 Chain complexes

Let $\mathbf{D}$ be an additive category (Section A4.6). Recall that $\mathrm{Ch}_{\bullet}\mathbf{D}$ denotes the category of its unbounded chain complexes $A = ((A_n), (\partial_n))$, indexed on $\mathbf{Z}$, with the usual morphisms (of degree zero). We show that it is a regular, reversible, symmetric dIP4-homotopical category. Of course, homotopies are the usual ones and the basic structure is classical, but the connections are less well known.

Some general remarks on duality will reduce computations. Writing X^*, f^* the object and arrow corresponding to X and f in the opposite category $\mathbf{D}^*$, the anti-isomorphism

$$\mathrm{Ch}_{\bullet}\mathbf{D} \to \mathrm{Ch}_{\bullet}(\mathbf{D}^*),$$
$$A = ((A_n), (\partial_n)) \mapsto A' = ((A^*_{-n}, (\partial^*_{-n+1})), \tag{4.66}$$

shows that $(\mathrm{Ch}_{\bullet}\mathbf{D})^*$ is again a category of chain complexes (on $\mathbf{D}^*$), and allows one to get the cylinder functor I of $\mathrm{Ch}_{\bullet}\mathbf{D}$ from the path functor P^* of $\mathrm{Ch}_{\bullet}(\mathbf{D}^*)$

$$I(A) = (P^*(A'))'.$$

The same holds for cones, suspensions, and h-pushouts. We also note that, for a small category $\mathbf{S}$, $(\mathrm{Ch}_{\bullet}\mathbf{D})^{\mathbf{S}} \cong \mathrm{Ch}_{\bullet}(\mathbf{D}^{\mathbf{S}})$ is still a category of chain complexes over an additive category. If $\mathbf{D}$ is complete, the same holds for sheaves on a small site $\mathbf{S}$: $\mathrm{Shv}(\mathbf{S}, \mathrm{Ch}_{\bullet}\mathbf{D}) \cong \mathrm{Ch}_{\bullet}(\mathrm{Shv}(\mathbf{S}, \mathbf{D}))$ (cf. Section 5.1.3).

A $\mathbf{D}$-map between finite biproducts $f: \oplus A_j \to \oplus B_i$ of components f_{ij} will be written 'on formal variables', as

$$f(x_1, ..., x_n) = (\textstyle\sum_j f_{1j}\, x_j, ..., \sum_j f_{mj}\, x_j). \tag{4.67}$$

This notation allows one to write computations as if $\mathbf{D}$ were a category of modules. It can be formally justified by letting $x_j = pr_j: \oplus A_j \to A_j$ be the j-th projection; then, $(x_1, ..., x_n)$, the morphism of components $x_1, ..., x_n$, is the identity of $\oplus A_j$, but also specifies the 'name of the variable', for each A_j.

4.4.2 The path functor

To fix notation, a homotopy in Ch.$\mathbf{D}$ is written as

$$\varphi\colon f \to g\colon A \to B, \qquad \varphi = (f, \varphi_\bullet, g), \qquad (4.68)$$

and satisfies

$$-f + g = \partial\varphi_\bullet + \varphi_\bullet\partial \qquad (-f_n + g_n = \partial_{n+1}\varphi_n + \varphi_{n-1}\partial_n). \qquad (4.69)$$

The sequence $\varphi_\bullet = (\varphi_n\colon A_n \to B_{n+1})_n$ is a map of graded objects, of degree 1, which will be called the *centre* of φ. We often write $\varphi(a)$ instead of $\varphi_n(a)$, where a denotes the variable of A_n.

Such homotopies are produced by a path endofunctor P

$$(PA)_n = A_n \oplus A_{n+1} \oplus A_n, \qquad \partial(a, h, b) = (\partial a, -a - \partial h + b, \partial b). \quad (4.70)$$

P has a *reversible* regular symmetric dP4-structure $(\partial^\alpha, e, r, g^\alpha, s, c)$, which can be written 'on variables', as specified above:

$$\begin{aligned}
\partial^\alpha\colon PA \to A, \qquad & \partial^\alpha(a^-, h, a^+) = a^\alpha && \textit{(faces)}, \\
e\colon A \to PA, \qquad & e(a) = (a, 0, a) && \textit{(degeneracy)}, \\
r\colon PA \to PA, \qquad & r(a, h, b) = (b, -h, a) && \textit{(reversion)}, \\
g^\alpha\colon PA \to P^2A, \qquad & \\
& g^-(a, h, b) = (a, h, b;\ h, 0, 0;\ b, 0, b), \\
& g^+(a, h, b) = (a, 0, a;\ 0, 0, h;\ a, h, b) && \textit{(connections)}, \\
s\colon P^2A \to P^2A, \qquad & s(a, h, b;\ u, z, v;\ c, k, d) \\
& = (a, u, c;\ h, -z, k;\ b, v, d) && \textit{(transposition)}, \\
c\colon PA \times_A PA \to PA, \qquad & \\
& c((a, h, c), (c, k, d)) = (a, h + k, d) && \textit{(concatenation)}.
\end{aligned} \qquad (4.71)$$

The second-order structure, where P^2A intervenes, will be analysed below (Section 4.4.3).

By duality (Section 4.4.1), homotopies are also represented by a cylinder endofunctor $I\colon$ Ch.$\mathbf{D} \to$ Ch.$\mathbf{D}$

$$(IA)_n = A_n \oplus A_{n-1} \oplus A_n, \qquad \partial(a, h, b) = (\partial a - h, -\partial h, \partial b + h). \quad (4.72)$$

The unit and counit of the adjunction $I \dashv P$ are computed as follows:

$$\eta\colon 1 \to PI, \qquad\qquad \varepsilon\colon IP \to 1, \qquad (4.73)$$

$$\eta_n\colon A_n \to (A_n \oplus A_{n-1} \oplus A_n) \oplus (A_{n+1} \oplus A_n \oplus A_{n+1}) \oplus (A_n \oplus A_{n-1} \oplus A_n),$$

$$\eta_n(a) = (a, 0, 0;\ 0, a, 0;\ 0, 0, a),$$

$$\varepsilon_n\colon (A_n \oplus A_{n+1} \oplus A_n) \oplus (A_{n-1} \oplus A_n \oplus A_{n-1}) \oplus (A_n \oplus A_{n+1} \oplus A_n) \to A_n,$$

$$\varepsilon_n(a, h, b;\ x, e, y;\ c, k, d) = a + e + d.$$

The structure of I can be obtained from the structure of P, in (4.71), by duality or by adjunction, equivalently. P and I respectively preserve the existing limits and colimits, and both preserve finite biproducts. All this gives rise to the usual trivial homotopies, whisker composition and concatenation of homotopies

$$
\begin{aligned}
&0_f = (f, 0, f), &&k\varphi h = (kfh, k\varphi_{\bullet}h, kgh), \\
&\varphi + \psi = (f, \varphi_{\bullet} + \psi_{\bullet}, h) &&(\text{for } \varphi\colon f \to g, \ \psi\colon g \to h).
\end{aligned}
\tag{4.74}
$$

Concatenation is obviously regular (but not 2-regular): strictly associative, with strict identities. Therefore, we do not have to introduce the acceleration 2-homotopy $z\colon P \to P^2$ (cf. Section 4.2.5). Notice, also, that the concatenation of homotopies $\varphi + \psi$ should not be confused with the sum of their representative morphisms in the abelian group $\mathrm{Ch}_{\bullet}\mathbf{D}(A, PB)$ (the relation between these operations will be considered in Section 4.8.6).

$\mathrm{Ch}_{\bullet}\mathbf{D}$ has thus a regular, reversible, symmetric dIP4 structure, which is actually dIP4-homotopical. In fact, given two morphisms $f\colon A \to C$ and $g\colon B \to C$, the standard homotopy pullback $P(f, g)$ is constructed making use of the biproducts of $\mathbf{D}$, as a simple extension of the construction of PA:

$$
\begin{aligned}
&(P(f, g))_n = A_n \oplus C_{n+1} \oplus B_n, \\
&\partial(a, c, b) = (\partial a, -fa - \partial c + gb, \partial b).
\end{aligned}
\tag{4.75}
$$

Homotopy pushouts also exist, by duality. Notice that $\mathrm{Ch}_{\bullet}\mathbf{D}$ *need not have all finite limits or colimits*: these exist if and only if they exist in $\mathbf{D}$ (cf. Section 1.3.5).

4.4.3 The higher structure

We describe now, in detail, the second-order structure of P, consisting of its connections and transposition.

The second-order path-object has the following components and differential

$$
(P^2 A)_n = (PA)_n \oplus (PA)_{n+1} \oplus (PA)_n =
\tag{4.76}
$$

$$
(A_n \oplus A_{n+1} \oplus A_n) \oplus (A_{n+1} \oplus A_{n+2} \oplus A_{n+1}) \oplus (A_n \oplus A_{n+1} \oplus A_n),
$$

$$
\partial(a, h, b; u, z, v; c, k, d) =
$$

$$
(\partial a, -a - \partial h + b, \partial b; -a - \partial u + c, \bar{z}, -b - \partial v + d; \partial c, -c - \partial k + d, \partial d)
$$

where $\bar{z} = -h + u + \partial z - v + k$.

It is convenient to represent the *variable* $\xi = (a, h, b; u, z, v; c, k, d)$ of $P^2 A$ (in the sense of Section 4.4.1) as a square diagram, so that its faces $\partial_1^\alpha = \partial^\alpha P$ and $\partial_2^\alpha = P \partial^\alpha$ show at the boundary of the square

$$
\begin{array}{ccc}
a & \xrightarrow{u} & c \\
h\downarrow & z & \downarrow k \\
b & \xrightarrow{v} & d
\end{array}
\qquad\qquad
\begin{array}{c}
\bullet \xrightarrow{1} \\
\downarrow 2
\end{array}
\tag{4.77}
$$

$$
\partial_1^-(\xi) = \partial^- P(\xi) = (a, h, b), \qquad \partial_1^+(\xi) = \partial^+ P(\xi) = (c, k, d),
$$
$$
\partial_2^-(\xi) = P\partial^-(\xi) = (a, u, c), \qquad \partial_2^+(\xi) = P\partial^+(\xi) = (b, v, d).
$$

The connections g^- and g^+ can thus be represented according to their geometrical meaning

$$
g^-(a, h, b) = (a, h, b; h, 0, 0; b, 0, b),
$$
$$
g^+(a, h, b) = (a, 0, a; 0, 0, h; a, h, b),
\tag{4.78}
$$

$$
\begin{array}{ccc}
a & \xrightarrow{h} & b \\
h\downarrow & 0 & \downarrow 0 \\
b & \xrightarrow{0} & b
\end{array}
\qquad\qquad
\begin{array}{ccc}
a & \xrightarrow{0} & a \\
0\downarrow & 0 & \downarrow h \\
a & \xrightarrow{h} & b
\end{array}
$$

Easy computations show that the connections satisfy 'their' axioms (whose duals are written in Section 4.2.1); for most of them (except coassociativity) this is evident from the above representation. We only write down the verification that g^- commutes with the differential (letting $h' = -a - \partial h + b$):

$$
g^-\partial(a, h, b) = g^-(\partial a, h', \partial b) = (\partial a, h', \partial b; h', 0, 0; \partial b, 0, \partial b),
$$
$$
\partial g^-(a, h, b) = \partial(a, h, b; h, 0, 0; b, 0, b)
$$
$$
= (\partial(a, h, b); -(a, h, b) - \partial(h, 0, 0) + (b, 0, b); \partial(b, 0, b))
$$
$$
= (\partial a, h', \partial b; -a - \partial h + b, -h + 0 + h, -b + b; \partial b,
$$
$$
-b + b, \partial b)
$$
$$
= (\partial a, h', \partial b; h', 0, 0; \partial b, 0, \partial b).
$$

Similarly, the transposition $s \colon P^2 A \to P^2 A$ comes from a symmetry with respect to the 'main diagonal', as in **Top**, together with a sign-change in the middle term

$$
s(a, h, b; u, z, v; c, k, d) = (a, u, c; h, -z, k; b, v, d).
\tag{4.79}
$$

$$
\begin{array}{ccc}
a & \xrightarrow{u} & c \\
h\downarrow & z & \downarrow k \\
b & \xrightarrow{v} & d
\end{array}
\qquad \mapsto \qquad
\begin{array}{ccc}
a & \xrightarrow{h} & b \\
u\downarrow & -z & \downarrow v \\
c & \xrightarrow{k} & d
\end{array}
$$

This representation makes evident that s satisfies 'its' axioms (Section 4.2.1, dualised): it is involutive, converts the horizontal faces into the vertical ones, makes the connections g^α commutative ($g^\alpha.s = g^\alpha$, since g^α in (4.78) is invariant under such symmetry and sign-change) and is consistent with the degeneracy e.

Finally, in order to prove that s commutes with the differential ∂ of $P^2 A$, recall that the latter is computed in (4.76). This formula shows that interchanging h/u, b/c, v/k, $z/-z$ in the original 9-tuple does yield in the final result the interchanging of the terms of place $2/4$, $3/7$, $6/8$ together with the sign-change in the middle term.

4.4.4 Positive chain complexes

The subcategory $Ch_+ \mathbf{D}$ of *positive* chain complexes (with $A_n = 0$ for $n < 0$) again has homotopies defined as above. It is a symmetric dI4-category, with a cylinder functor I' which is the restriction of the cylinder I for unbounded complexes, considered above

$$I' : Ch_+ \mathbf{D} \to Ch_+ \mathbf{D}, \qquad (I'A)_n = A_n \oplus A_{n-1} \oplus A_n. \qquad (4.80)$$

Assume now that the additive category $\mathbf{D}$ *has kernels* (and therefore all finite limits). Then I' has a right adjoint $P' : Ch_+ \mathbf{D} \to Ch_+ \mathbf{D}$, which differs from P in degree 0:

$$
\begin{aligned}
(P'A)_n &= A_n \oplus A_{n+1} \oplus A_n \qquad (n > 0), \\
(P'A)_0 &= \mathrm{Ker}\,(\partial_0^{PA} : A_0 \oplus A_1 \oplus A_0 \to A_0).
\end{aligned}
\qquad (4.81)
$$

Indeed, in this hypothesis (the existence of finite limits in $\mathbf{D}$), the embedding $U : Ch_+ \mathbf{D} \to Ch_\bullet \mathbf{D}$ has a reflector F and a coreflector G

$$
F : Ch_\bullet \mathbf{D} \to Ch_+ \mathbf{D}, \qquad (FA)_n =
\begin{cases}
A_n, & \text{for } n \geqslant 0, \\
0, & \text{for } n < 0.
\end{cases}
\qquad (4.82)
$$

$$
G : Ch_\bullet \mathbf{D} \to Ch_+ \mathbf{D}, \qquad (GA)_n =
\begin{cases}
A_n, & \text{for } n > 0, \\
\mathrm{Ker}(\partial_0), & \text{for } n = 0, \\
0, & \text{for } n < 0.
\end{cases}
\qquad (4.83)
$$

Therefore, the adjunctions $I \dashv P$ and $F \dashv U \dashv G$

$$
Ch_+ \mathbf{D} \; \underset{G}{\overset{U}{\rightleftarrows}} \; Ch_\bullet \mathbf{D} \; \underset{P}{\overset{I}{\rightleftarrows}} \; Ch_\bullet \mathbf{D} \; \underset{U}{\overset{F}{\rightleftarrows}} \; Ch_+ \mathbf{D}
\qquad (4.84)
$$

give a composed adjunction $I' \dashv P'$ (in $Ch_+ \mathbf{D}$), with $I' = FIU$, $P' = GPU$.

It follows that, for an additive category $\mathbf{D}$ with kernels, the category $Ch_+ \mathbf{D}$ of positive chain complexes is a symmetric dIP4-homotopical category. The embedding $U \colon Ch_+ \mathbf{D} \to Ch_\bullet \mathbf{D}$ is a strict symmetric dI4-functor and a *lax* symmetric dP4-functor; the comparison $k \colon UP' \to PU$ comes from the embedding $kA \colon UP'A \to PUA$ (obtained as in Section 4.2.7).

4.4.5 Directed chain complexes

Take now the category $dCh_+ \mathbf{Ab}$ of directed (positive) chain complexes of abelian groups, defined in Section 2.1.1 and recall that the differential is not assumed to preserve preorders.

Here, we prefer to work with the unrestricted case, $dCh_\bullet \mathbf{Ab}$. Let us begin by recalling that this category has all limits and colimits and is enriched on abelian monoids (Section 2.1.1).

The path and cylinder functor are defined as in $Ch_\bullet \mathbf{Ab}$ (Section 4.4.2)

$$
\begin{aligned}
(PA)_n &= A_n \oplus A_{n+1} \oplus A_n, \quad \partial(a, h, b) = (\partial a, -a - \partial h + b, \partial b), \\
(IA)_n &= A_n \oplus A_{n-1} \oplus A_n, \quad \partial(a, h, b) = (\partial a - h, -\partial h, \partial b + h),
\end{aligned}
\tag{4.85}
$$

equipping each component with the obvious preorder, which makes it a biproduct of preordered abelian groups. The adjunction $I \dashv P$ lifts to $dCh_\bullet \mathbf{Ab}$, since its unit and counit (computed in (4.73)) preserve these preorders.

The natural transformations $\partial^\alpha, e, g^\alpha, c$ defined in (4.71) for the path functor (and its powers) also *preserve* preorders and lift to $dCh_\bullet \mathbf{Ab}$, while this is not true of reversion and transposition (whose formulas make use of opposites). The corresponding natural transformations of the cylinder functor behave in the same way, of course.

In fact, the category $dCh_\bullet \mathbf{Ab}$ has a non-symmetric, non-reversible dIP4-homotopical structure, which lifts that of $Ch_\bullet \mathbf{Ab}$. This works with a *reversor* R which reverses the preorder of every component of *odd* degree (by a power of the involutive endofunctor $R(X) = X^{op}$ of $p\mathbf{Ab}$):

$$
RA = ((R^n A_n), (\partial_n)), \qquad Rf = (R^n f_n)_n. \tag{4.86}
$$

Now, the reversion of chain complexes is replaced with a *reflection* (or external reversion), which is defined by the same algebraic formula but operates between chain complexes with modified preorders (and preserves them):

$$(RPA)_n = (R^n A_n) \oplus (R^n A_{n+1}) \oplus (R^n A_n),$$
$$(PRA)_n = (R^n A_n) \oplus (R^{n+1} A_{n+1}) \oplus (R^n A_n), \qquad (4.87)$$
$$r \colon RPA \to PRA, \qquad r(a,h,b) = (b, -h, a).$$

The forgetful functor $dCh_\bullet \mathbf{Ab} \to Ch_\bullet \mathbf{Ab}$ is thus a strict dIP4-functor (Section 4.2.7).

The transposition of chain complexes can also be replaced with an *external transposition* having the same algebraic formula, but operating between chain complexes with modified preorders (much in the same way as already done for cubical sets, in Section 1.6.5).

To do this, we first define the *transposer S* which – in a directed chain complex – reverses the preorder of each component of degree n such that the integral part $[n/2]$ is odd (i.e. $n = 2, 3, 6, 7, 10, 11,...$)

$$SA = ((S_n A_n), (\partial_n)), \qquad Sf = (S_n f_n)_n \qquad (S_n = R^{[n/2]}),$$
$$SS = 1, \qquad\qquad RS = SR. \qquad (4.88)$$

The external transposition is now defined as the natural transformation

$$s \colon PSPS \to SPSP,$$
$$(PSPSA)_n = (A_n \oplus R^n A_{n+1} \oplus A_n) \oplus$$
$$\qquad (A_{n+1} \oplus R^{n+1} A_{n+2} \oplus A_{n+1}) \oplus (A_n \oplus R^n A_{n+1} \oplus A_n),$$
$$(SPSPA)_n = (A_n \oplus A_{n+1} \oplus A_n) \oplus \qquad\qquad (4.89)$$
$$\qquad (R^n A_{n+1} \oplus R^n A_{n+2} \oplus R^n A_{n+1}) \oplus (A_n \oplus A_{n+1} \oplus A_n),$$
$$s(a,h,b; u,z,v; c,h,d) = (a,u,c; h,-z,k; b,v,d).$$

To compute the components above, use the fact that

$$[n/2] + [(n+1)/2] = n,$$

whence $S_n S_{n+1} = R^n$ and $S_{n+1} S_{n+2} = R^{n+1}$.

4.4.6 Singular directed chains

The functor $\uparrow Ch_+ \colon d\mathbf{Top} \to dCh_+ \mathbf{Ab}$ (Section 2.1.2) becomes a lax dP1-functor (Section 1.2.6), via the natural transformations:

$$i \colon R \uparrow Ch_+(X) \to \uparrow Ch_+(RX), \qquad i(a) = (-1)^n a^{\mathrm{op}} . r^n,$$
$$k \colon \uparrow Ch_+(PX) \to P \uparrow Ch_+(X), \qquad k(b) = (\partial^- b, b', \partial^+ b), \qquad (4.90)$$

where

- a: $\uparrow\mathbf{I}^n \to X$ is a singular cube of X, and b: $\uparrow\mathbf{I}^n \to PX$ of PX;
- r^n: $\uparrow\mathbf{I}^n \to (\uparrow\mathbf{I}^n)^{\mathrm{op}}$ reverses all coordinates, sending (t_i) to $(1 - t_i)$;
- $b' = \mathrm{ev}.(b \times \uparrow\mathbf{I})$: $\uparrow\mathbf{I}^{n+1} \to X$ corresponds to b in the cylinder-path adjunction.

The coherence conditions are satisfied, i.e. the following diagrams commute (with $C_+ = \uparrow\mathrm{Ch}_+$):

$$C_+ \xrightarrow{C_+ e} C_+ P \xrightarrow{C_+ \partial^\alpha} C_+ \qquad RC_+ P \xrightarrow{iP} C_+ RP \xrightarrow{C_+ r} C_+ PR$$

For the first triangle, recall that a degenerate cube gives a null (normalised) chain. For the right-hand rectangle:

$$(kR.C_+ r.iP)(a: \uparrow\mathbf{I}^n \to PX)$$
$$= (-1)^n kR(rX.a^{\mathrm{op}}.r^n)$$
$$= (-1)^n ((\partial^+ a)^{\mathrm{op}}.r^n, \ \mathrm{ev}.((rX.a^{\mathrm{op}}.r^n) \times \uparrow\mathbf{I}), \ (\partial^- a)^{\mathrm{op}}.r^n),$$

$$(Pi.rC_+.Rk)(a: \uparrow\mathbf{I}^n \to PX)$$
$$= Pi(\partial^+ a, -a', \partial^- a)$$
$$= ((-1)^n (\partial^+ a)^{\mathrm{op}}.r^n, -(-1)^{n+1}\mathrm{ev}.(a \times \uparrow\mathbf{I})^{\mathrm{op}}.r^{n+1}, (-1)^n (\partial^- a)^{\mathrm{op}}.r^n)$$
$$= (-1)^n ((\partial^+ a)^{\mathrm{op}}.r^n, \ (\mathrm{ev}.(a \times \uparrow\mathbf{I}))^{\mathrm{op}}.r^{n+1}, \ (\partial^- a)^{\mathrm{op}}.r^n),$$

and the two results coincide, since

$$\mathrm{ev}.((rX.a^{\mathrm{op}}.r^n) \times \uparrow\mathbf{I}) \ = \ \mathrm{ev}.(a \times \uparrow\mathbf{I})^{\mathrm{op}}.r^{n+1} \colon \uparrow\mathbf{I}^{n+1} \to X^{\mathrm{op}}.$$

4.4.7 The monoidal structure

Let now K be a commutative unital ring and $\mathbf{D} = K\text{-}\mathbf{Mod}$ the category of K-modules, with the usual tensor product $\otimes = \otimes_K$ and internal hom-functor $\mathrm{Hom} = \mathrm{Hom}_K$. In this case the reversible symmetric dIP4-structures of $\mathrm{Ch}_\bullet\mathbf{D}$ and $\mathrm{Ch}_+\mathbf{D}$ are *monoidal*, i.e. produced by a reversible symmetric dIP4-interval (Section 4.2.8).

Indeed, the category $\mathrm{Ch}_\bullet\mathbf{D}$ of unbounded chain complexes (of K-modules) has a classical symmetric monoidal closed structure

([EK], p. 558)

$$(A \otimes B)_n = \oplus_p (A_p \otimes B_{n-p}),$$
$$\partial(a \otimes b) = (\partial a) \otimes b + (-1)^{|a|}.a \otimes (\partial b), \tag{4.91}$$

$$(\mathrm{Hom}(A,B))_n = \Pi_p \, \mathrm{Hom}(A_p, B_{n+p}),$$
$$(\partial f)x = \partial(fx) - (-1)^{|f|}.f(\partial x), \tag{4.92}$$

whose identity is the complex K, concentrated in degree zero. (Here, $|-|$ denotes the degree of an element.)

We obtain an interval-object $\mathbf{I}$ by setting $\mathbf{I} = I(K)$; it is a complex concentrated in degrees 0 and 1

$$\mathbf{I}_0 = K \oplus K, \qquad \mathbf{I}_1 = K, \qquad \partial_1(\lambda) = (-\lambda, \lambda), \tag{4.93}$$

and it is easy to verify that, in this case $(\mathbf{D} = K\text{-}\mathbf{Mod})$, the cylinder and path functor of $\mathrm{Ch}_\bullet \mathbf{D}$ (Section 4.4.2) are given by

$$I(A) \cong A \otimes \mathbf{I}, \qquad P(A) \cong \mathrm{Hom}(\mathbf{I}, A). \tag{4.94}$$

Further the object $\mathbf{I} = I(K)$ has a structure of a *reversible dI4-interval* in $(\mathrm{Ch}_\bullet \mathbf{D}, \otimes)$, coming from the reversible dI4-structure of I and the fact that $I^2(K) = I(I(K)) \cong \mathbf{I} \otimes \mathbf{I}$. Conversely, this structure on $\mathbf{I}$ defines the structure of the functors I, P, according to the general procedure for symmetric monoidal closed categories (Section 4.2.8).

The same argument applies to the category $\mathrm{Ch}_+(K\text{-}\mathbf{Mod})$ of positive chain complexes of modules, with the appropriate monoidal closed structure. The latter can be expressed with the reflector F and coreflector G (Section 4.4.4): the new tensor product is still computed as above, in (4.91), but the new hom has a different formula in degree zero

$$(\mathrm{Hom}_+(A,B))_0 =$$
$$\mathrm{Ker}(\partial_0 \colon (\Pi_p \, \mathrm{Hom}(A_p, B_p) \to \Pi_p \, \mathrm{Hom}(A_p, B_{p-1})). \tag{4.95}$$

4.5 Double homotopies and the fundamental category

We begin now the general theory of dI2, dI3 and dI4-categories. Double homotopies and 2-homotopies have been introduced in Section 4.1.1. We use them to construct the homotopy 2-category $\mathrm{Ho}_2(\mathbf{A})$ and the fundamental category functor $\uparrow \Pi_1 \colon \mathbf{A} \to \mathbf{Cat}$, for a dI4-category $\mathbf{A}$.

4.5.1 *Concatenation of double homotopies*

Let **A** be a dI3-category (Section 4.2.2), and recall that $c\colon I \to J$ denotes the concatenation map.

The *concatenation*, or *pasting*, of double homotopies *in direction* 1, is defined by means of the pushout $I(JX)$ and the map $Ic\colon I^2X \to IJX$

$$A +_1 B = (A \vee_1 B).Ic\colon I^2X \to IJX \to Y, \qquad (4.96)$$

where the double homotopies A, B are consecutive in direction 1 ($\partial_1^+ A = \partial_1^- B$), and of course $(A \vee_1 B).Ic^- = A$, $(A \vee_1 B).Ic^+ = B$.

Similarly, the *concatenation in direction* 2 is defined by the J-pushout $J(IX)$ and the map $cI\colon I^2X \to JIX$, for two double homotopies which are consecutive in direction 2

$$A +_2 C = (A \vee_2 C).cI\colon I^2X \to JIX \to Y \quad (\partial_2^+ A = \partial_2^- C). \qquad (4.97)$$

We shall see that, in the presence of a transposition, these operations can be obtained one from the other (Section 4.5.3). We also prove below (Theorem 4.5.2) that these operations satisfy a (strict) middle-four interchange property, which allows us to define the *matrix concatenation* of four double homotopies $A_{\alpha\beta}\colon I^2X \to Y$, with faces as below

$$\begin{bmatrix} A_{00} & A_{10} \\ A_{01} & A_{11} \end{bmatrix} = (A_{00} +_1 A_{10}) +_2 (A_{01} +_1 A_{11})$$
$$= (A_{00} +_2 A_{01}) +_1 (A_{10} +_2 A_{11}), \qquad (4.98)$$

Plainly, 2-homotopies (Section 4.1.1) are closed under concatenation in both directions (also because $0_f + 0_f = 0_f$, strictly).

The preorder $\varphi \preceq_2 \psi$ (i.e. there is a 2-homotopy $\varphi \to \psi$) spans an equivalence relation $\simeq_2$; two homotopies which satisfy the relation $\varphi \simeq_2 \psi$ are said to be 2-*homotopic*.

4.5.2 *Theorem (Double concatenations)*

Let **A** *be a dI3-category.*

(a) The two concatenation laws of double homotopies satisfy the middle-four interchange property (4.98).

(b) J^2X *(expressed by each of the two pushouts (4.37)) is also the colimit of the left diagram below*

$$
\begin{array}{ccc}
I^2X \xleftarrow{\partial_1^+} IX \xrightarrow{\partial_1^-} I^2X & \qquad I^2X \qquad\qquad I^2X \\
\partial_2^+ \uparrow \qquad\qquad \uparrow \partial_2^+ & \qquad c^{00} \searrow \qquad \swarrow c^{10} \\
IX \qquad\qquad IX & \qquad J^2X \\
\partial_2^- \downarrow \qquad\qquad \downarrow \partial_2^- & \qquad c^{01} \nearrow \qquad \nwarrow c^{11} \\
I^2X \xleftarrow{\partial_1^+} IX \xrightarrow{\partial_1^-} I^2X & \qquad I^2X \qquad\qquad I^2X
\end{array}
\qquad (4.99)
$$

with structural maps $c^{\alpha\beta} : I^2X \to J^2X$ *defined as:*

$$
c^{\alpha\beta} = c^\beta J.Ic^\alpha = Jc^\alpha.c^\beta I : I^2X \to J^2X \qquad (\alpha, \beta = 0, 1). \qquad (4.100)
$$

(As usual we use, according to convenience, the indices 0, 1 or $-$, $+$.)

(c) *The matrix concatenation (4.98) of all* $A_{\alpha\beta}$ *can be computed as* $Ac_2 : I^2X \to Y$, *where A is the pasting of all* $A_{\alpha\beta}$ *in the colimit (4.99) and* $c_2 : I^2 \to J^2$ *is defined below*

$$
A : J^2X \to Y, \qquad\qquad A.c^{\alpha\beta} = A_{\alpha\beta} \qquad (\alpha, \beta = 0, 1), \qquad (4.101)
$$

$$
c_2 = cJ.Ic = Jc.cI : I^2X \to J^2X \quad \text{(double-concatenation)}. \qquad (4.102)
$$

Proof Let us begin by considering the left diagram (4.99), and replace it with the left diagram below, which is commutative and equivalent to the former diagram, as far as their colimits are concerned

$$
\begin{array}{ccc}
I^2X \xleftarrow{I\partial^+} IX \xrightarrow{I\partial^-} I^2X & \qquad I^2X \xrightarrow{Ic^-} IJX \xleftarrow{Ic^+} I^2X \\
\partial^+ I \uparrow \quad \uparrow \partial^+ \quad \uparrow \partial^+ I \quad \partial^+ I \uparrow & \uparrow \partial^+ J \quad \uparrow \partial^+ I \\
IX \xleftarrow{\partial^+} X \xrightarrow{\partial^-} IX & \qquad IX \xrightarrow{c^-} JX \xleftarrow{c^+} IX \\
\partial^- I \downarrow \quad \downarrow \partial^- \quad \downarrow \partial^- I \quad \partial^- I \downarrow & \downarrow \partial^- J \quad \downarrow \partial^- I \\
I^2X \xleftarrow{I\partial^+} IX \xrightarrow{I\partial^-} I^2X & \qquad I^2X \xrightarrow{Ic^-} IJX \xleftarrow{Ic^+} I^2X
\end{array}
\qquad (4.103)
$$

Now, we form the right diagram above, replacing *each row* of the left diagram with its colimit (IJX for the upper and lower rows, because I preserves the J-pushout, by assumption). The colimit of the new central column is the J-pushout of JX, as in the left pushout (4.37), copied below

$$JX \xrightarrow{\partial^+ J} IJX$$

$$\partial^- J \downarrow \qquad - \quad -\downarrow c^- J$$

$$IJX \xrightarrow[c^+ J]{} J^2 X$$

All this proves that $J^2 X$ is the colimit of (4.99), with structural maps $c^{\alpha\beta} = c^\beta J.Ic^\alpha \colon I^2 X \to J^2 X$.

It follows that:

$$(A_{00} +_1 A_{10}) +_2 (A_{10} +_1 A_{11})$$
$$= ((A_{00} \vee_1 A_{10}).Ic \vee_2 (A_{10} \vee_1 A_{11}).Ic).cI$$
$$= ((A_{00} \vee_1 A_{10}) \vee_2 (A_{10} \vee_1 A_{11})).Jc.cI = Ac_2,$$

where, as in (4.101), $A \colon J^2 X \to Y$ is the global pasting of all $A_{\alpha\beta}$.

The fact that Ac_2 also coincides with the other expression, i.e.

$$(A_{00} +_2 A_{10}) +_1 (A_{10} +_2 A_{11}),$$

is proved in the symmetric way: take the colimit of the *columns* of the left diagram (4.103), then use the fact that J preserves J-pushouts and that $c^{\alpha\beta}$ can equivalently be written as $Jc^\alpha.c^\beta I$. $\square$

4.5.3 Transposition of concatenations

Let **A** be a symmetric dI3-category.

The 1-directed and 2-directed concatenations of double homotopies (Section 4.5.1) can be transformed one into the other, by means of the transposition s

$$A +_2 C = (As +_1 Cs).s. \qquad (4.104)$$

To prove this formula, we use the IJ-transposition $s' \colon JI \to IJ$ (Section 4.2.4), which makes the left square below commutative

$$
\begin{array}{ccccc}
I^2 X & \xrightarrow{cI} & JIX & \xrightarrow{A \vee_2 C} & Y \\
s \downarrow & & \downarrow s' & & \parallel \\
I^2 X & \xrightarrow[Ic]{} & IJX & \xrightarrow[As \vee_1 Cs]{} & Y
\end{array}
$$

Also the right square commutes, as is detected by the structural maps $c^\alpha I \colon IX \to JIX$ of the J-pushout on IX:

$$((As \vee_1 Cs)s').c^\alpha I = (As \vee_1 Cs).Ic^\alpha.s = As.s = A = (A \vee_2 C).c^\alpha I.$$

Finally, we have:

$$(As+_1Cs).s = (As\vee_1Cs).Ic.s = (As\vee_1Cs).s'.cI = (A\vee_2C).cI = A+_2C.$$

4.5.4 Folding

Let **A** be a dI4-category. A double homotopy $\Phi\colon I^2A \to X$ with faces $\sigma, \tau, \varphi, \psi$, as below, gives rise to a 2-homotopy Ψ (often called a *folding* of Φ), by pasting Φ with two double homotopies of connection (denoted by $\natural$)

$$\Psi\colon (0_f + \varphi) + \tau \to (\sigma + \psi) + 0_g\colon f \to g.$$

Combining this 2-homotopy with the accelerations (Section 4.2.5), we get that $\varphi+\tau \simeq_2 \sigma+\psi$. In the regular case we already have a 2-homotopy $\Psi\colon \varphi + \tau \to \sigma + \psi\colon f \to g$.

There is a 'weak' converse: given a 2-homotopy $\Psi\colon \varphi+\tau \to \sigma+\psi$, one can construct a double homotopy with faces 2-homotopic to $\sigma, \tau, \varphi, \psi$, by pasting Ψ with two double homotopies of connection and two degenerate ones.

4.5.5 The homotopy 2-category

Let **A** be a dI4-category. Recall that we write $\varphi \preceq_2 \psi$ to mean that there exists a 2-homotopy $\varphi \to \psi$ (Section 4.5.1), and $\simeq_2$ the equivalence relation, called 2-homotopy equivalence, spanned by this preorder $\preceq_2$.

We now want to construct a 2-category

$$\mathrm{Ho}_2(\mathbf{A}) = \mathbf{A}/\!\simeq_2 \qquad \text{(the *homotopy 2-category*)}, \qquad (4.105)$$

equipping the category **A** with 2-cells $[\varphi]\colon f \to g$, which are classes of homotopies up to 2-homotopy equivalence.

Following the presentation of a 2-category in Section A5.2 (as a sesquicategory satisfying the reduced interchange property) we define the concatenation of 2-cells and the whisker composition of maps and 2-cells, in the obvious way, which is easily seen to be legitimate

$$[\varphi] + [\psi] = [\varphi + \psi], \qquad k\circ[\varphi]\circ h = [k\circ\varphi\circ h]. \qquad (4.106)$$

Now, associativity of the whisker composition – in the appropriate sense – already holds at the level of homotopies (cf. (1.43)). We prove in the next theorem that the remaining properties are satisfied in the quotient $\mathbf{A}/\simeq_2$.

4.5.6 Theorem (Weak regularity of concatenation)

In a dI4-category $\mathbf{A}$, the concatenation of homotopies is associative and has identities up to the 2-homotopy equivalence relation $\simeq_2$. The reduced interchange property, between concatenation and whisker composition, also holds up to $\simeq_2$.

As a consequence, $\mathrm{Ho}_2(\mathbf{A}) = \mathbf{A}/\simeq_2$ is a 2-category.

Proof First, we already know (from (4.51)) that a homotopy $\varphi \colon f \to g$ has acceleration 2-homotopies $0_f + \varphi \to \varphi \to \varphi + 0_g$.

Moreover, given three consecutive homotopies $\varphi \colon f \to g$, $\chi \colon g \to h$, $\psi \colon h \to k$, we can link the ternary concatenations $\varphi + (\chi + \psi)$ and $(\varphi + \chi) + \psi$, by extending a procedure already used in the construction of the fundamental category of a d-space (Theorem 3.2.4).

First, we construct a 2-homotopy

$$B \colon (0 + \varphi) + (\chi + \psi) \to (\varphi + \chi) + (\psi + 0), \qquad (4.107)$$

pasting in a suitable order the following double homotopies of degeneracy and connection

Now, accelerations give 2-homotopies (which could only be pasted with the previous one using reversion)

$$A \colon (0+\varphi)+(\chi+\psi) \to \varphi+(\chi+\psi), \quad C \colon (\varphi+\chi)+\psi \to (\varphi+\chi)+(\psi+0).$$

Finally, as to the reduced interchange property, take two 'horizontally consecutive' homotopies φ, ψ

$$X \underset{g}{\overset{f}{\underset{\downarrow\varphi}{\longrightarrow}}} Y \underset{k}{\overset{h}{\underset{\downarrow\psi}{\longrightarrow}}} Z \qquad
\begin{array}{ccc}
hf & \overset{h\circ\varphi}{\longrightarrow} & hg \\
{\scriptstyle\psi\circ f}\downarrow & \psi\circ\varphi & \downarrow{\scriptstyle\psi\circ g} \\
kf & \underset{k\circ\varphi}{\longrightarrow} & kg
\end{array}$$

Then the double homotopy $\psi \circ \varphi = \psi.(I\varphi): I^2 X \to Z$ has the faces shown above, in diagram (4.7). By folding (Section 4.5.4), we get $\psi g.h\varphi \simeq_2 k\varphi.\psi f$, and therefore $[\psi]g.h[\varphi] = k[\varphi].[\psi]f$. $\qquad\square$

4.5.7 The fundamental category in the concrete case

Recall that a concrete dI1-category is a dI1-category $\mathbf{A}$ equipped with a reversive object E and a specified isomorphism $E \to E^{\mathrm{op}}$ (Section 1.2.4). E is called the standard point, or free point of $\mathbf{A}$. The associated forgetful functor is $U = |-| = \mathbf{A}(E, -): \mathbf{A} \to \mathbf{Set}$ (represented by E) and $UR \cong U$.

Again, for the sake of simplicity, we identify $E = E^{\mathrm{op}}$ and $UR = R$.

Then, the object $\mathbf{I} = I(E)$ is a dI1-interval in $\mathbf{A}$. A point of X is an element $x \in |X|$, i.e. a map $x: E \to X$, while a path in X is a map $a: \mathbf{I} \to X$, defined on $\mathbf{I} = I(E)$, with endpoints $x^\alpha = a\partial^\alpha: E \to X$. Every point $x: E \to \mathbf{A}$ has a trivial path $0_x = xe: \mathbf{I} \to X$.

We have defined, in Section 1.2.4, the fundamental graph $\uparrow\Gamma_1(X)$: its vertices are the points of X, its arrows $[a]: x^- \to x^+$ are the classes of paths a from x^- to x^+, up to the equivalence relation generated by homotopy with fixed endpoints.

Let now $\mathbf{A}$ be a *concrete dI4-category*, which means that it is a dI4-category made concrete as above (by assigning E). The fundamental reflexive graph $\uparrow\Gamma_1(X)$ will be called the *fundamental category* of X, and written $\uparrow\Pi_1(X)$, when equipped with the composition law induced by concatenation of consecutive paths, which we write additively

$$[a] + [b] = [a + b].$$

This operation is well defined, and gives indeed a category, as follows straightforwardly from the homotopy 2-category $\mathrm{Ho}_2(\mathbf{A}) = \mathbf{A}/\simeq_2$ (Section 4.5.5)

$$\uparrow\Pi_1(X) = \mathrm{Ho}_2(\mathbf{A})(E, X). \tag{4.108}$$

We thus have a functor $\uparrow\Pi_1\colon \mathbf{A} \to \mathbf{Cat}$ defined on a morphism $f\colon X \to Y$ by:

$$\uparrow\Pi_1(f) = \mathrm{Ho}_2(\mathbf{A})(E, f) = f_*\colon \uparrow\Pi_1(X) \to \uparrow\Pi_1(Y),$$
$$\uparrow\Pi_1(f)(x) = f\circ x, \qquad\qquad \uparrow\Pi_1(f)[a] = f_*[a] = [f\circ a]. \tag{4.109}$$

It is actually a (representable) 2-functor $\uparrow\Pi_1\colon \mathrm{Ho}_2(\mathbf{A}) \to \mathbf{Cat}$. A homotopy $\varphi\colon f \to g\colon X \to Y$ in $\mathbf{A}$ induces a natural transformation (a directed homotopy of categories, Section 1.1.6)

$$\uparrow\Pi_1(\varphi) = \mathrm{Ho}_2(\mathbf{A})(E, [\varphi]) = \varphi_*\colon f_* \to g_*\colon \uparrow\Pi_1 X \to \uparrow\Pi_1 Y,$$
$$\varphi_*(x) = [\varphi.Ix]\colon f\circ x \to g\circ x, \tag{4.110}$$

where $x\colon E \to X$ is a point of X and an object of $\uparrow\Pi_1(X)$, while $\varphi.Ix\colon \mathbf{I} \to Y$ is a path in Y, and its 2-homotopy class $[\varphi.Ix]$ is an arrow in $\uparrow\Pi_1(Y)$.

As a consequence, $\uparrow\Pi_1\colon \mathbf{A} \to \mathbf{Cat}$ preserves the homotopy preorder $f \preceq_1 g$, future homotopy equivalence and future deformation retracts (Section 1.3.1).

4.6 Higher properties of h-pushouts and cofibrations

We deal now with higher properties of homotopy pushouts, mostly in dI2 and dI4-categories.

The reversible case is deferred to Section 4.9.

4.6.1 Proposition (Special invariance)

(a) Let $\mathbf{A}$ be a dI1-category and consider an h-pushout $I(f, g) = A$ (see the diagram below, in the proof). If g is an isomorphism, then the 'opposite' morphism $u\colon Y \to A$ is a split monomorphism (the embedding of a retract).

(b) If, moreover, $\mathbf{A}$ is dI2-homotopical, then u is the embedding of a past deformation retract (Section 1.3.1). This extends a property of the lower face of a cylinder $\partial^-\colon X \to IX$, already remarked in (4.29).

Proof (a) Let $g' = g^{-1}\colon Z \to X$. We define a left inverse h of u, applying the first-order universal property of λ

$$h\colon A \to Y, \qquad\qquad hu = 1_Y,$$
$$hv = fg'\colon Z \to Y, \qquad\qquad h\lambda = 0_f\colon huf \to hvg.$$

(b) We can construct a homotopy $uh \to 1_A$, applying the higher property of λ (4.1.3) to the left diagram below

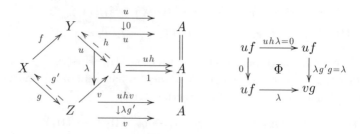

This works with the upper connection $\Phi = \lambda g^+ : I^2 X \to A$ shown in the right-hand part. $\qquad\square$

4.6.2 Pasting theorem for h-pushouts

Let **A** *be a dI4-homotopical category and let* λ, μ, ν *be standard homotopy pushouts*

$$
\begin{array}{ccccccc}
X & \xrightarrow{f} & Y & \xrightarrow{g} & Z & = & Z \\
\downarrow{x} & \lambda & \downarrow{y} & \mu & \downarrow{z} & & \downarrow{b} \\
A & \xrightarrow{u} & B & \xdashrightarrow{v} & C & \xrightarrow{\nu} & \\
\| & & & \searrow{w} & & & \\
A & & \xrightarrow{\qquad a \qquad} & & & D &
\end{array}
\tag{4.111}
$$

Then the canonical comparison $k : C \to D$ *(constructed in the pasting lemma 1.3.8) is a future homotopy equivalence (Section 1.3.1).*

More precisely, let us define $w : B \to D$, $k : C \to D$ *and* $k' : D \to C$ *by means of the universal property of* λ, μ, ν, *respectively (using also the concatenation of homotopies, for* k' *)*

$$
\begin{array}{cc}
w.u = a : A \to D, & w.y = bg : Y \to D, \\
\multicolumn{2}{c}{w \circ \lambda = \nu : ax \to bgf,} \\
k.v = w : B \to D, & k.z = b : Z \to D, \\
\multicolumn{2}{c}{k \circ \mu = 0 : wy \to bg,} \\
k'.a = vu : A \to C, & k'.b = z : Z \to C, \\
\multicolumn{2}{c}{k' \circ \nu = v \circ \lambda + \mu \circ f : vux \to zgf.}
\end{array}
\tag{4.112}
$$

Then k, k' *form a future homotopy equivalence, with homotopies* φ, ψ *which satisfy the following higher coherence relations (notice that we are* not *saying that $\psi \circ v = 0$)*:

$$\varphi \colon \mathrm{id}D \to kk', \qquad \varphi \circ a = 0_a, \qquad \varphi \circ b = 0_b,$$
$$\psi \colon \mathrm{id}C \to k'k, \qquad\qquad \psi \circ z = 0_z, \qquad\qquad (4.113)$$
$$\psi \circ vu = 0_{vu}, \qquad \psi \circ vy = 0 + \mu.$$

By reflection, if we reverse the direction of homotopies in diagram (4.111), *the pasting of the h-pushouts $I(f, x)$ and $I(g, y)$ is past homotopy equivalent to $I(gf, x)$.*

In other words, the comparison is a *future* homotopy equivalence when, as in diagram (4.111), the cells λ, μ can be pasted according to the 'order of construction' of the h-pushouts: first λ and then μ. We obtain a *past* homotopy equivalence in the other case.

Proof This result, which strengthens the pasting lemma 1.3.8, will be essential for the sequel, e.g. to prove the homotopical exactness of the cofibre sequence, in Theorem 4.7.5(c). (It is an enrichment of a similar, non-directed result in [G3], 3.4.)

First, the higher universal property of ν (in (4.13)) yields a homotopy $\varphi \colon \mathrm{id}D \to kk'$ (with $\varphi \circ a = 0_a$, $\varphi \circ b = 0_b$), provided by the *acceleration* double homotopy $\Theta'' \colon \nu \to \nu + 0$ (4.51)

$$
\begin{array}{ccc}
ax & \xrightarrow{\;\nu\;} & bgf \\
{\scriptstyle 0_a x}\big\downarrow & \Theta'' & \big\downarrow{\scriptstyle 0_b gf} \\
kk'ax & \xrightarrow[kk'\nu]{} & kk'bgf
\end{array}
\qquad
\begin{array}{l}
kk'a = kvu = a, \quad kk'b = kz = b, \\[4pt]
kk' \circ \nu = kv \circ \lambda + k \circ \mu \circ f \\[4pt]
\qquad = w \circ \lambda + 0 = \nu + 0.
\end{array}
\qquad (4.114)
$$

Now, we make use of the higher universal property of λ to link the maps $v, k'w \colon B \to C$. Consider the following three double homotopies (acceleration, degeneracy and upper connection), labelled $\sharp$

$$
\begin{array}{ccc}
vux & \xrightarrow{\;\;v\lambda\;\;} & vyf \\
{\scriptstyle 0x}\big\downarrow & \# & {\scriptstyle 0f}\big\downarrow \\
vux & \xrightarrow{v\lambda} vyf \xrightarrow{0} & vyf \\
{\scriptstyle 0x}\big\downarrow & \# \;\big\downarrow{\scriptstyle 0}\; \# \;{\scriptstyle \mu f}\big\downarrow \\
k'wux & \xrightarrow[v\lambda]{} vyf \xrightarrow[\mu f]{} & k'wyf
\end{array}
\Bigg\}{\scriptstyle (0+\mu)f}
\qquad
\begin{array}{l}
k'wu = ka = vu, \\[18pt]
k'w \circ \lambda = k' \circ \nu = v \circ \lambda + \mu \circ f.
\end{array}
$$

Their pasting yields a homotopy $\rho \colon v \to k'w$ such that $\rho \circ u = 0$, $\rho \circ y = 0 + \mu$. Finally, the higher property of μ gives rise to a homotopy $\psi \colon \mathrm{id}C \to k'k$ (such that $\psi \circ v = \rho$, $\psi \circ z = 0$), using a double homotopy which results from degeneracy and lower connection

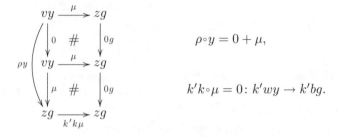

$$\rho \circ y = 0 + \mu,$$

$$k'k \circ \mu = 0 \colon k'wy \to k'bg.$$

$\square$

4.6.3 Cofibrations and fibrations

Let us come back to the basic setting, to give some definitions which are needed now: let **A** be a dI1-category *or* a dP1-category. (Actually, the cylinder or cocylinder functor is not used here: a dh1-category would suffice, see Section 1.2.9).

We say that a map $u \colon X \to A$ is an upper *cofibration* if (see the left diagram below), for every object W and every map $h \colon A \to W$, every homotopy $\psi \colon h' = hu \to k'$ can be 'extended' to a homotopy φ on A, so that $\varphi \circ u = \psi$ (and, in particular, $ku = k'$)

$$
\begin{array}{ccc}
X \xrightarrow[\quad k' \quad]{\overset{h'}{\underset{\downarrow\psi}{\longrightarrow}}} W & & X \xrightarrow[\quad k' \quad]{\overset{h'}{\underset{\downarrow\psi}{\longrightarrow}}} W \\
u \downarrow \quad\quad \| & & u \downarrow \quad\quad \| \\
A \xrightarrow[\quad k \quad]{\overset{h}{\underset{\downarrow\varphi}{\dashrightarrow}}} W & & A \xrightarrow[\quad k \quad]{\overset{h}{\underset{\downarrow\varphi}{\dashrightarrow}}} W
\end{array}
\qquad (4.115)
$$

It is easy to verify that upper cofibrations are closed under composition and contain all isomorphisms. The R-dual notion, shown in the right diagram above, will be called a *lower cofibration*: for every map $k \colon A \to W$ and homotopy $\psi \colon h' \to k' = ku$ there is some homotopy φ such that $\varphi \circ u = \psi$. A *bilateral cofibration* has to satisfy both conditions. (A motivation for the terms 'upper' and 'lower' comes from the faces of the cylinder, see Theorem 4.6.6.)

On the other hand, by categorical duality, the left diagram below shows the definition of an *upper fibration* $f \colon X \to B$: for every map h and homotopy $\psi \colon fh \to k'$ there is some homotopy φ which lifts ψ, in the sense that $f \circ \varphi = \psi$. The property of a *lower fibration* is shown on the right hand:

$$
\begin{array}{ccc}
W & \xrightarrow[\substack{k}]{\substack{h \\ \downarrow\varphi}} & X \\
\| & & \downarrow f \\
W & \xrightarrow[\substack{k'}]{\substack{h' \\ \downarrow\psi}} & B
\end{array}
\qquad
\begin{array}{ccc}
W & \xrightarrow[\substack{k}]{\substack{h \\ \downarrow\varphi}} & X \\
\| & & \downarrow f \\
W & \xrightarrow[\substack{k'}]{\substack{h' \\ \downarrow\psi}} & B
\end{array}
\qquad (4.116)
$$

4.6.4 Theorem (Pushouts of cofibrations as h-pushouts)

Let **A** *be a dI2-homotopical category. Let* $f\colon X \to Y$ *be an upper cofibration and* $g\colon X \to Z$ *any map, and assume that the ordinary pushout* V *of* f, g *exists and is preserved by the cylinder functor. (This last fact necessarily holds in the dIP2-homotopical case, when* I *is a left adjoint.)*

Then the obvious comparison map $k\colon A \to V$ *from the h-pushout* $A = I(f, g)$ *to the ordinary pushout* V, *defined by the three equations below, is a future homotopy equivalence*

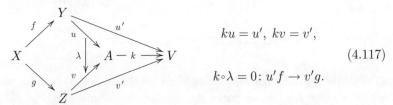

$$ku = u', \; kv = v',$$

$$(4.117)$$

$$k\circ\lambda = 0\colon u'f \to v'g.$$

In particular, taking $Z = \top$, *this shows that if* f *is an upper cofibration and the cokernel* $\mathrm{Cok}(f)$ *is preserved by the cylinder functor, the comparison map* $u\colon C^+ f \to \mathrm{Cok}(f)$ *is a future homotopy equivalence.*

Proof The extension property of f ensures that the homotopy $\lambda\colon uf \to vg$ can be extended to some homotopy $\varphi\colon u \to w\colon Y \to A$ such that $\varphi\circ f = \lambda$ and $wf = vg$, as in the left diagram below. This gives a new commutative square $wf = vg$, which we factorise in the right diagram through the pushout V, by a unique map $h\colon V \to A$ such that $hu' = w$, $hv' = v$

$$
\begin{array}{ccc}
X & \xrightarrow[\substack{vg}]{\substack{uf \\ \downarrow\lambda}} & A \\
\downarrow f & & \| \\
Y & \xrightarrow[\substack{w}]{\substack{u \\ \downarrow\varphi}} & A
\end{array}
\qquad
\begin{array}{ccc}
 & Y & \\
f\nearrow & \substack{u'\searrow} & \substack{w\searrow} \\
X & V - h \Rightarrow & A \\
g\searrow & \substack{v'\nearrow} & \substack{v\nearrow} \\
 & Z &
\end{array}
$$

We get a homotopy $1_A \to hk$, applying the higher property of λ (Section 4.1.3) to the left diagram below, and using a lower connection in the right-hand part

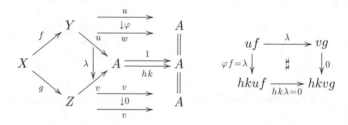

Finally, we want to construct a homotopy $1_V \to kh$, making use of the fact that the ordinary pushout V is preserved by I

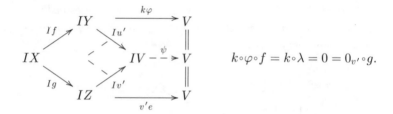

$$k \circ \varphi \circ f = k \circ \lambda = 0 = 0_{v'} \circ g.$$

Since the pair of homotopies $k \varkappa \varphi \colon u' \to kw$ and $0_{v'} \colon v' \to v'$ is coherent with the maps f, g (as computed above), there is one morphism $\psi \colon IV \to V$ such that $\psi \circ u' = k \circ \varphi$ and $\psi \circ v' = 0_{v'}$.

This morphism has the correct faces: embedding the original pushout $u'f = v'g$ into the pushout above by four lower faces ∂^- we get that $\psi \partial^- = 1_V$. Similarly, using four upper faces ∂^+ (and the equations $kw = khu'$, $v' = kv = khv'$) we get that $\psi \partial^+ = kh$. $\quad\square$

4.6.5 Extended acceleration

The rest of this section is devoted to construct the usual factorisation of a map via a cofibration. This can be obtained in a dI4-homotopical category *equipped with a stronger version of the acceleration* z (Section 4.2.5), which is still automatic in the regular case. In the sequel, all this part will only be used to establish a relationship with Baues 'cofibration categories', in Theorem 4.9.6.

If $\mathbf{A}$ is a dI4-category, an *extended acceleration* will be a natural transformation $w \colon I^2 \to J$ with the following faces (see Section 4.2.2 for the structure of J):

$$
\begin{array}{ccc}
\partial^{--} & \xrightarrow{\ c\ } & \partial^{++} \\
{\scriptstyle c^-}\downarrow & w & \downarrow{\scriptstyle 0} \\
\partial^{\pm} & \xrightarrow{\ c^+\ } & \partial^{++}
\end{array}
\qquad
\begin{array}{l}
w.I\partial^- = c^-, \quad w.I\partial^+ = \partial^{++}e, \\
\\
w.\partial^- I = c, \quad w.\partial^+ I = c^+.
\end{array}
\qquad (4.118)
$$

(Given w, one can construct $z\colon I^2 \to J$ by post-composing w with the natural transformation $(e\partial^-) \vee \mathrm{id}X\colon JX \to IX$.)

In this situation, every pair of consecutive homotopies $\varphi\colon f \to g$, $\psi\colon g \to h$ has two double homotopies, obtained from w and the map $\varphi \vee \psi\colon JX \to IX$ (on the pushout JX)

$$
\Theta'(\varphi,\psi) = (\varphi \vee \psi).wX, \qquad \Theta''(\varphi,\psi) = (\Theta'(\psi^{\mathrm{op}}, \varphi^{\mathrm{op}}))^{\mathrm{op}}, \qquad (4.119)
$$

$$
\begin{array}{ccc}
f & \xrightarrow{\ \varphi+\psi\ } & h \\
{\scriptstyle \varphi}\downarrow & \Theta'(\varphi,\psi) & \downarrow{\scriptstyle 0} \\
g & \xrightarrow{\ \psi\ } & h
\end{array}
\qquad\qquad
\begin{array}{ccc}
f & \xrightarrow{\ \varphi\ } & g \\
{\scriptstyle 0}\downarrow & \Theta''(\varphi,\psi) & \downarrow{\scriptstyle \psi} \\
f & \xrightarrow{\ \varphi+\psi\ } & h
\end{array}
$$

The category d**Top** can be given this additional structure: the component wX is defined as $X\times\omega\colon I^2X \to IX$, by means of the following map $\omega\colon \uparrow\!\mathbf{I}\times\uparrow\!\mathbf{I} \to \uparrow\!\mathbf{I} = \uparrow\!\mathbf{J}$ (the affine deformation from the path $\mathrm{id}\mathbf{I}\colon \uparrow\!\mathbf{I} \to \uparrow\!\mathbf{I}$ to the path $c^+\colon \uparrow\!\mathbf{I} \to \uparrow\!\mathbf{I}$)

$$
\begin{array}{ccc}
\partial^- & \xrightarrow{\ \partial\ } & \partial^+ \\
{\scriptstyle c^-}\downarrow & \omega & \downarrow{\scriptstyle 0} \\
\partial^{\pm} & \xrightarrow{\ c^+\ } & \partial^+
\end{array}
\qquad
\begin{array}{l}
\omega(t,t') = (1-t').t + t'.(t+1)/2, \\
\\
c^-(t) = t/2, \qquad c^+(t) = (t+1)/2.
\end{array}
$$

If $\mathbf{A}$ has a regular concatenation of homotopies (Section 4.2.3), the double homotopies (4.119) can be constructed using degeneracies and connections (and one can also construct w in the same way)

$$
\begin{array}{ccccc}
f & \xrightarrow{\ \varphi\ } & g & \xrightarrow{\ \psi\ } & h \\
{\scriptstyle \varphi}\downarrow & \sharp & \downarrow{\scriptstyle 0} & \sharp & \downarrow{\scriptstyle 0} \\
g & \xrightarrow{\ 0\ } & g & \xrightarrow{\ \psi\ } & h
\end{array}
\qquad
\begin{array}{ccccc}
f & \xrightarrow{\ \varphi\ } & g & \xrightarrow{\ 0\ } & g \\
{\scriptstyle 0}\downarrow & \sharp & \downarrow{\scriptstyle 0} & \sharp & \downarrow{\scriptstyle \psi} \\
f & \xrightarrow{\ \varphi\ } & g & \xrightarrow{\ \psi\ } & h
\end{array}
$$

Dually one defines dP4-categories *with extended acceleration*.

4.6.6 Theorem (h-pushouts and cofibrations)

(a) Let $\mathbf{A}$ be a dI4-category with extended acceleration (Section 4.6.5). In every h-pushout $A = I(f,g)$ (as in the diagram below), the first '*injection*' $u\colon Y \to A$ *is a lower cofibration, while the second* $v\colon Z \to A$ *is an upper cofibration. In particular, the lower face of a cylinder* $\partial^-\colon X \to IX$ *is a lower cofibration, and* $\partial^+\colon X \to IX$ *an upper one.*

(a*) Let **A** *be a dP4-category with extended acceleration. In every h-pullback, the first 'projection' is a lower fibration, while the second is an upper fibration. In particular, the lower face* $\partial^-\colon PY \to Y$ *is a lower fibration.*

Proof It is sufficient to verify the second statement of (a). Take a map $h\colon A \to W$ and a homotopy $\psi\colon h' = hv \to k'$. By the ordinary property of λ, there is precisely one map $k\colon A \to W$ such that

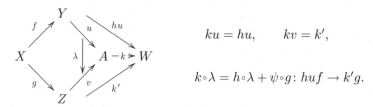

$$ku = hu, \qquad kv = k',$$

$$k\circ\lambda = h\circ\lambda + \psi\circ g\colon huf \to k'g.$$

The higher property of λ (Section 4.1.3) now yields a homotopy $\varphi\colon h \to k$ such that $\varphi\circ v = \psi$

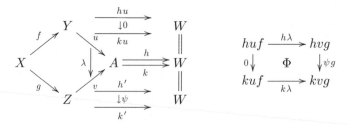

The double homotopy $\Phi = \Theta''(h\circ\lambda, \psi\circ g)$ of the right diagram above comes from the extended acceleration (4.119) and the relation $k\circ\lambda = h\circ\lambda + \psi\circ g$. □

4.6.7 Theorem (Factorisations via (co)fibrations)

(a) Let **A** *be a dI4-homotopical category with an extended acceleration (Section 4.6.5). Every map* $f\colon X \to Y$ *has two canonical factorisations, related by R-duality:*

$$f = hv\colon X \to I(f,1) \to Y$$
$$\textit{(upper cofibration -- past deformation retraction)},$$

(4.120)

$$f = ku \colon X \to I(1, f) \to Y$$
$$\text{(lower cofibration – future deformation retraction).} \tag{4.121}$$

(a*) Let **A** *be a dP4-homotopical category with extended acceleration. Every map* $f \colon X \to Y$ *has two canonical factorisations:*

$$f = vh \colon X \to P(f, 1) \to Y$$
$$\text{(embedding of a past deformation retract – upper fibration)},$$

$$f = uk \colon X \to P(1, f) \to Y$$
$$\text{(embedding of a future deformation retract – lower fibration)},$$

Proof It suffices to prove the first case of (a). In Proposition 4.6.1, take $g = 1_X$ and consider the h-pushout $A = I(f, 1)$, which yields the factorisation (4.120). We already know, from Proposition 4.6.1, that h is a past deformation retraction (and u the corresponding embedding). We also know, from Theorem 4.6.6(a), that the h-pushout 'injection' $v \colon X \to I(f, 1)$ is an upper cofibration. $\qquad\square$

4.7 Higher properties of cones and Puppe sequences

In a symmetric dI4-homotopical category, the cofibre sequence of a map has strong properties of 'homotopical exactness': it is homotopy equivalent to a sequence of iterated mapping cones. This will be used to simplify the exactness axiom of homology theories (Theorem 4.7.6).

The dI2-case is sufficient to obtain part of these results; this will be relevant in the relative settings of Section 5.8. This material essentially comes from [G3].

4.7.1 Theorem (The higher property of h-cokernels)

Let **A** *be a dI2-homotopical category. Take a map* $f \colon X \to Y$ *and its upper h-cokernel* $(C^+ f, u, v^+, \gamma \colon uf \to v^+ p)$. *Take also two maps* $a, b \colon C^+ f \to W$ *and a homotopy* $\sigma \colon au \to bu \colon Y \to W$, *as in the left diagram below*

$$\tag{4.122}$$

If the following three coherence conditions hold (the fourth is a consequence)

$$a\gamma = \sigma f, \quad b\gamma = 0_{buf}, \quad av^+ = bv^+ \quad (buf = bv^+p = av^+p), \qquad (4.123)$$

there is some homotopy $\varphi\colon a \to b$ extending σ on u (i.e. such that $\varphi u = \sigma$).

By reflection duality, let us consider the right diagram above, where

$$(C^- f, v^-, u, \gamma\colon v^- p \to uf)$$

is the lower h-cokernel of f, and assume that:

$$a\gamma = 0_{auf}, \quad b\gamma = \sigma f, \quad av^- = bv^- \quad (auf = av^- p = bv^- p). \qquad (4.124)$$

Then there is some homotopy $\varphi\colon a \to b$ extending σ on u.

Note. The construction of φ requires the connections: g^- in the first case, g^+ in the second.

Proof In the first case, we apply Theorem 4.1.3 to the homotopies $\sigma\colon au \to bu$ and $\tau = 0\colon av \to bv$, letting $\Phi\colon I^2 X \to W$ be defined as $\Phi = a\gamma.g^- = \sigma f.g^-$. The faces of Φ are indeed as required:

$$
\begin{array}{ccc}
auf & \xrightarrow{\ a\gamma\ } & avp \\
{\scriptstyle \sigma f}\big\downarrow & \Phi & \big\downarrow{\scriptstyle 0} \\
buf & \xrightarrow{\ 0\ } & bvp
\end{array}
$$

$\square$

4.7.2 The suspension functor

Let us recall, from Section 1.7.2, that in a dl1-homotopical category $\mathbf{A}$, the suspension ΣX is an upper and a lower cone, at the same time, with an upper and a lower vertex, v^+ and v^-

$$
\begin{array}{ccc}
X & \xrightarrow{\ p_X\ } & \top \\
{\scriptstyle p_X}\big\downarrow & {\scriptstyle ev^X} \ \nearrow\ \big\downarrow{\scriptstyle v^+} & \\
\top & \xrightarrow[\ v^-\ ]{} & \Sigma X
\end{array}
\qquad
\begin{aligned}
\Sigma X &= I(p_X, p_X) \\
&= C^+(p_X) = C^-(p_X), \qquad (4.125) \\
R.\Sigma &= \Sigma.R.
\end{aligned}
$$

The suspension is equipped with a homotopy (*suspension evaluation*)

$$\mathrm{ev}^X\colon v^- p_X \to v^+ p_X\colon X \to \Sigma X, \qquad (4.126)$$

which is universal for homotopies between constant maps.

As a particular case of the h-pushout functor (Section 1.3.7), the suspension Σ is an endofunctor of $\mathbf{A}$: given $f\colon X \to Y$, the suspended map $\Sigma f\colon \Sigma X \to \Sigma Y$ is the unique morphism which satisfies the conditions

$$\Sigma f.v^- = v^-, \qquad \Sigma f.v^+ = v^+, \qquad (\Sigma f)\circ\mathrm{ev}^X = \mathrm{ev}^Y \circ f. \qquad (4.127)$$

4.7.3 Lemma (The comparison map)

(a) In a dI4-homotopical category $\mathbf{A}$, the lower comparison map of $f\colon X \to Y$

$$k^-(f)\colon C^+u^- \to \Sigma X, \qquad u^- = \mathrm{hcok}^- f, \qquad (4.128)$$

defined in (1.185) is a future homotopy equivalence, while the upper comparison map (1.189) is a past homotopy equivalence:

$$k^+(f)\colon C^-u^+ \to \Sigma X, \qquad u^+ = \mathrm{hcok}^+ f. \qquad (4.129)$$

(a*) In a dP4-homotopical category $\mathbf{A}$, the lower comparison map $h^-(f)\colon \Omega Y \to E^+v^-$ of the lower fibre diagram of f (in (1.209), with $v^- = \mathrm{hker}^- f$) is a future homotopy equivalence, while the upper comparison map $h^+(f)\colon \Omega Y \to E^-v^+$ is a past homotopy equivalence (with $v^+ = \mathrm{hker}^+ f$).

Proof The first statement is a particular case of the pasting theorem for h-pushouts (Theorem 4.6.2), letting $A = Z = \top$ in (4.111). The others follow by reflection or categorical duality.   □

4.7.4 Lemma (The comparison square)

If $\mathbf{A}$ is dI2-homotopical, the lower *comparison square of f, defined in (1.186), is homotopically commutative. More precisely, there exists a homotopy*

$$\psi\colon d_2 \to \Sigma f.k_1\colon C^+u \to \Sigma Y, \qquad (4.130)$$

$$u = u_1 = \mathrm{hcok}^-(f), \qquad u_2 = \mathrm{hcok}^+(u), \qquad k_1 = k^-(f).$$

By reflection duality, the upper *comparison square of f has a homotopy in the opposite direction*

$$\psi: \Sigma f.k_1 \to d_2 : C^- u \to \Sigma Y, \tag{4.131}$$

$$
\begin{array}{ccccccccc}
X & \xrightarrow{f} & Y & \xrightarrow{u} & C^+ f & \xrightarrow{d} & \Sigma X & \xrightarrow{\Sigma f} & \Sigma Y \\
\| & & \| & & \| & & {\scriptstyle k_1}\uparrow \ {\scriptstyle \psi}\searrow & & \| \\
X & \xrightarrow{f} & Y & \xrightarrow{u_1} & C^+ f & \xrightarrow{u_2} & C^- u & \xrightarrow{d_2} & \Sigma Y
\end{array}
$$

$$u = u_1 = \mathrm{hcok}^+(f), \qquad u_2 = \mathrm{hcok}^-(u), \qquad k_1 = k^+(f).$$

Proof First, we use the higher property (Theorem 4.7.1) of the *lower* h-cokernel $\gamma: v^- p \to uf: X \to C^- f$ of f, to prove that the composition $\Sigma f.d$ is homotopically null.

More precisely, there is a homotopy φ which extends the cell $\sigma = \mathrm{ev}^Y: v^-.p_Y \to v^+ p_Y: Y \to \Sigma Y$ on u

$$\varphi: v^-.p(C^- f) \to \Sigma f.d: C^- f \to \Sigma Y, \qquad \varphi.u = \mathrm{ev}^Y,$$

$$
\begin{array}{ccccc}
X & \xrightarrow{f} & Y & \overset{v^- p}{\underset{v^+ p}{\rightrightarrows}}\ {\scriptstyle \downarrow \sigma} & \Sigma Y \\
{\scriptstyle p}\downarrow & {\scriptstyle \gamma} & \downarrow{\scriptstyle u} & & \| \\
\top & \xrightarrow{v^-} & C^- f & \overset{v^- p}{\underset{\Sigma f.d}{\rightrightarrows}} & \Sigma Y
\end{array}
$$

$$v^- Y.pC^- f.u = v^- Y.pY, \qquad \Sigma f.d.u = \Sigma f.v^+ X.pY = v^+ Y.pY.$$

In fact, the coherence conditions (4.124) follow from the definition of d (in (1.183)) and Σf (in (4.127)):

$$v^- Y.pC^- f.\gamma = 0,$$
$$\Sigma f.d.\gamma = \Sigma f.\mathrm{ev}^X = \mathrm{ev}^Y.f,$$
$$v^- Y.pY.v^- = v^- Y = \Sigma f.v^- X = \Sigma f.dv^-.$$

Now the same extension property of Theorem 4.7.1 for the *upper* h-cokernel $\gamma': u_2 u \to v^+ p$ of u allows one to extend this cell $\varphi: v^- p \to$

$\Sigma f.d$ on u_2, producing a cell

$$\psi \colon d_2 \to \Sigma f.k_1 \colon C^+ u \to \Sigma Y, \qquad \psi.u_2 = \varphi,$$

$$d_2 u_2 = v^-.pC^- f,$$

$$\Sigma f.k_1.u_2 = \Sigma f.d.$$

Here, the coherence conditions (4.123) follow from the definition of $k_1 = k^-(f)$ (in (1.185)) and $d_2 = d^+(u)$ (in (1.188))

$$d_2 \gamma' = \mathrm{ev}^Y = \varphi u,$$

$$\Sigma f.k_1.\gamma' = 0,$$

$$d_2.v^+ = v^+.pY = \Sigma f.v^+ X = \Sigma f.k_1.v^+.$$

One can notice that both connections g^α have been used in the proof, via Theorem 4.7.1. $\qquad\square$

4.7.5 Theorem (Higher properties of the cofibre diagram)

(a) If **A** *is dI2-homotopical, each elementary square of the expanded cofibre diagram of* f *(1.195) is either commutative up to a directed homotopy, in a suitable direction, or the image of such a square under a suitable power* Σ^n *of the suspension endofunctor*

$$(4.132)$$

Such squares, images of a power Σ^n, are marked with an arrow in paren-theses. Their arrow will stand for a homotopy in the stronger hypotheses below.

(b) If **A** *is a symmetric dI2-homotopical category, all these squares are commutative up to directed homotopies, in suitable directions (not consistent with vertical pasting, see the last column above). As a consequence, the lower cofibre diagram of f (1.193) is commutative up to the homotopy congruence $\simeq_1$ generated by the existence of a (directed) homotopy between two maps (cf. (1.74))*

$$
\begin{array}{ccccccccccc}
X & \xrightarrow{f} & Y & \xrightarrow{u} & C^-f & \xrightarrow{d} & \Sigma X & \xrightarrow{\Sigma f} & \Sigma Y & \xrightarrow{\Sigma u} & \Sigma C^-f & \xrightarrow{\Sigma d} & \Sigma^2 X ... \\
\| & & \| & & \| & & h_1 \uparrow & \simeq_1 & h_2 \uparrow & \simeq_1 & h_3 \uparrow & \simeq_1 & h_4 \uparrow \\
X & \xrightarrow{f} & Y & \xrightarrow{u_1} & C^-f & \xrightarrow{u_2} & C^+u_1 & \xrightarrow{u_3} & C^-u_2 & \xrightarrow{u_4} & C^+u_3 & \xrightarrow{u_5} & C^-u_4 ...
\end{array}
\tag{4.133}
$$

(c) If **A** *is a symmetric dI4-homotopical category, every vertical map of the earlier diagram is a composite of past homotopy equivalences and future homotopy equivalences. In particular, the first three vertical maps (i.e. h_1, h_2, h_3) are future homotopy equivalences.*

(d) If **A** *is a reversible symmetric dI4-homotopical category, the diagram is commutative up to homotopy and every vertical map of this diagram is a homotopy equivalence.*

Proof (a) Follows from Lemma 4.7.4. In fact, each elementary square of (4.132) is either the lower or upper comparison square of some map, or its image under a power of Σ.

(b) Follows from the fact that, in the presence of the transposition, the suspension preserves homotopies (Theorem 4.1.7); also because, in the cofibre diagram (4.133), each square is a finite vertical pasting of squares of the previous diagram.

(c) Follows from Lemma 4.7.3 and, again, the homotopy-preservation property of Σ in the symmetric case. Point (d) is a straightforward consequence of the previous one. $\square$

4.7.6 Theorem (Homology theories)

Let **A** *be a symmetric dI4-homotopical category. Suppose we have a sequence $(\uparrow H_n, h_n)$ satisfying the axioms (dhlt.0, 1, 2) (Section 2.6.2). Then the following 'reduced exactness condition' implies the full exactness axiom (dhlt.3):*

(dhlt.3a) *for every morphism* $f\colon X \to Y$ *in* **A** *and every* $n \in \mathbf{Z}$, *the following sequence is exact in* **pAb** *(with* $u = \mathrm{hcok}^-(f)\colon Y \to C^- f$)

$$\uparrow H_n X \xrightarrow{f_*} \uparrow H_n Y \xrightarrow{u_*} \uparrow H_n C^- f. \qquad (4.134)$$

Proof First, let us remark that the present condition (dhlt.3a) is invariant under R-duality, because the *upper* analogue of sequence (4.134), with $u = \mathrm{hcok}^+(f)\colon Y \to C^+ f$, amounts to the sequence (4.134) of the reflected map f^{op} (see (1.192)).

Now, consider the initial part of the lower cofibre diagram of f

$$
\begin{array}{ccccccccc}
X & \xrightarrow{f} & Y & \xrightarrow{u} & C^- f & \xrightarrow{d} & \Sigma X & \xrightarrow{\Sigma f} & \Sigma Y \\
\| & & \| & & \| & & k_1 \uparrow \xleftarrow{\;\psi\;} & & \uparrow k_2 \\
X & \xrightarrow[f]{} & Y & \xrightarrow[u_1]{} & C^- f & \xrightarrow[u_2]{} & C^+ u_1 & \xrightarrow[u_3]{} & C^- u_2
\end{array}
$$

We have proved, in the previous theorem, that it is commutative up to directed homotopy ψ and that its vertical arrows k_1, k_2 are future homotopy equivalences.

Now, applying $\uparrow H_n$, we get a commutative diagram, by (dhlt.1), whose vertical arrows are algebraic isomorphisms, because of the previous remark and by (dhlt.1). Moreover, in the lower row every map is a (lower or upper) h-cokernel of the preceding one. Therefore, by (dhlt.3a), this row is transformed by $\uparrow H_n$ into an exact sequence. Finally, the same holds for the upper row, which proves (dhlt.3): the following row is exact

$$\uparrow H_n X \xrightarrow{f_*} \uparrow H_n Y \xrightarrow{u_*} \uparrow H_n C^- f \xrightarrow{\delta_*} \uparrow H_n \Sigma X \xrightarrow{(\Sigma f)_*} \uparrow H_n \Sigma Y.$$

$\square$

4.8 The cone monad

In this section we study the (upper) cone functor and the monad structure (Section A4.4) which it inherits from the cylinder diad, in a dl2-category. Most of this material has been developed in [G1].

The upper cone functor will be written as $C = C^+$. It has a lower basis $u\colon X \to CX$, an upper vertex $v = v^+\colon \top \to CX$ and a structural homotopy $\gamma\colon IX \to CX$ (cf. (1.171)), with $\gamma.\partial^- X = u$ and $\gamma.\partial^+ X = v.pX\colon X \to \top \to CX$.

4.8.1 Theorem (The second order cone)

(a) Let $\mathbf{A}$ *be dI1-homotopical, with I-preserved cylindrical colimits.*

For every object X, *the following diagram commutes and the second order (upper) cone* $C^2 X = C^+ C^+ X$ *is the colimit of the following row, with structural maps* c_0, c_2, c_1

$$
IT \xleftarrow{\;Ip\;} IX \xrightarrow{\;I\partial^+\;} I^2 X \xleftarrow{\;\partial^+ I\;} IX \xrightarrow{\;e\;} X
$$

(4.135)

with downward maps c_0, c_2, c_1 to $C^2 X$.

$$
c_0 = \gamma C.Iv = Cv.\gamma, \quad c_2 = \gamma C.I\gamma = C\gamma.\gamma I, \quad c_1 = vC.pX.
$$

Moreover, c_1 *is determined by* c_0

$$
c_1 = c_0.\partial^+.pX,
$$

(4.136)

which suggests that it can be omitted. In fact, $C^2 X$ *can be equivalently described as the colimit of the solid diagram below, with structural maps* c_0, c_2

$$
\begin{array}{ccc}
 & IX & \\
\nearrow^{\partial^+ .pI} & & \searrow^{\partial^+ I} \\
IT \xleftarrow{\;Ip\;} IX & \xrightarrow{\;I\partial^+\;} & I^2 X \\
\searrow_{c_0} & & \nearrow_{c_2} \\
 & C^2 X &
\end{array}
$$

(4.137)

(b) If $\mathbf{A}$ *is pointed, with an I-preserved zero object,* $C^2 X$ *can be viewed as the colimit*

$$
0 \longleftarrow IX \xrightarrow{\;I\partial^+\;} I^2 X \xleftarrow{\;\partial^+ I\;} IX \longrightarrow 0
$$

(4.138)

with downward map c_2 to $C^2 X$.

This amounts to saying that the map $c_2 \colon I^2 X \to C^2 X$ *is the generalised coequaliser of the three maps* $I\partial^+, \partial^+ I, 0 \colon IX \to I^2 X$, *whence an epimorphism (which need not be true in the unpointed case, see Section 4.8.2).*

Proof It suffices to prove (a), since (b) is a straightforward consequence of the second description of $C^2 X$, in (4.137). Consider the left diagram below

$$
\begin{array}{ccc}
X \xrightarrow{\partial^+} IX \xrightarrow{e} X \\
\downarrow p \quad\quad \downarrow \gamma \quad\quad \downarrow p \\
\mathsf{T} \xrightarrow{v} CX \longrightarrow \mathsf{T} \\
\downarrow \partial^+ \quad \downarrow \partial^+ C \quad\quad \downarrow vC \\
IT \underset{Iv}{\rightrightarrows} ICX \underset{\gamma C}{\rightrightarrows} C^2 X
\end{array}
\qquad
\begin{array}{ccc}
IX \xrightarrow{e} X \\
\downarrow \partial^+ I \quad\quad \\
IX \xrightarrow{I\partial^+} I^2 X \quad\quad \downarrow c_1 \\
\downarrow Ip \quad\;\; \dashv\!\downarrow I\gamma \;\;\dashv \\
IT \underset{Iv}{\rightarrow} ICX \underset{\gamma C}{\rightarrow} C^2 X
\end{array}
\qquad (4.139)
$$

By definition of CX, the left upper square is a pushout. The right upper square is also (since their pasting is trivially a pushout). By definition of $C(CX)$, the right lower square is a pushout, and the right rectangle is also.

Now, its composed central column coincides with the composed central column of the right diagram above. Therefore, the right rectangle of the latter is also a pushout. But its left lower square is also (as the result of applying I to a cylindrical colimit). It follows that $C^2 X$ is indeed the colimit in (4.135), with structural maps as specified there. Equality (4.136) comes from the outer square of the left diagram (4.139), where $e\partial^+ = \mathrm{id}X$.

The second description of $C^2 X$ follows from the fact that the upper row of (4.135) and the solid diagram (4.137) have the 'same' cocones. More precisely, if (d_0, d_2, d_1) is a cocone for (4.135), then (d_0, d_2) is a cocone for (4.137), because:

$$
d_2.\partial^+ I = d_1 e = d_1 e.\partial^+ e = d_2.\partial^+ I.\partial^+ .e = d_2.I\partial^+ .\partial^+ .e
$$
$$
= d_0.IpX.\partial^+ e = d_0.\partial^+ .pX.e = d_0.\partial^+ .pIX.
$$

Conversely, if (d_0, d_2) is a cocone for (4.137), we define $d_1 = d_0.\partial^+ .pX$ and obtain a cocone (d_0, d_2, d_1) for (4.135):

$$
d_1 e = d_0.\partial^+ .pX.e = d_0.\partial^+ .pIX = d_2.\partial^+ I.
$$

$\square$

4.8.2 Examples

It will be useful to compute $C^2 X$ for topological spaces and pointed topological spaces, also in order to see clearly that the *un*pointed case is 'not symmetric' (has no transposition). One would work similarly in d**Top** and d**Top.**.

(a) In **Top**, let us begin by noting that $C^2\emptyset = C\{*\} = I$; and then $c_2\colon \emptyset \to C^2\emptyset$ is not surjective.

On the other hand, if $X \neq \emptyset$, $C^2 X$ is the quotient of $I^2 X$ which identifies the pairs of points (x, t_1, t_2), (x', u_1, u_2) where $(t_1 = u_1 = 1$ and $t_2 = u_2)$ or $(t_2 = u_2 = 1)$

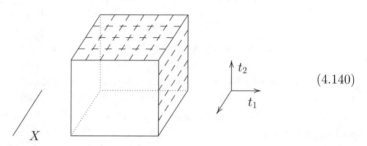

$$(4.140)$$

(The dashed lines suggest the classes of the equivalence relation on $I^2 X$. In particular, the top square $t_2 = 1$ is an equivalence class.)

One can note that the lower connection $g^-(x, s, t) = (x, \max(s, t))$ induces a map

$$g\colon C^2 X \to CX, \qquad g[x, s, t] = [x, \max(s, t)], \qquad (4.141)$$

which will be part of the monad structure of C (Theorem 4.8.3).

(b) In **Top$_\bullet$**, the second-order cone $C^2(X, x_0)$ is the quotient of the unpointed cylinder $I^2 X$ which identifies all points (x, t_1, t_2) where $x = x_0$, or $t_1 = 1$, or $t_2 = 1$. The operation g is defined as above. We now have a symmetric description, which is why there is an induced transposition which interchanges t_1 with t_2 (see Theorem 4.8.4).

4.8.3 Theorem (The cone monad)

Let $\mathbf{A}$ be dI2-homotopical.

The cylinder I gives rise to a monad (C, u, g) on the upper cone functor $C = C^+$. The unit is the lower basis $u = \gamma . \partial^- \colon 1 \to C$ and the multiplication $g\colon C^2 \to C$ is induced by $g^- \colon I^2 \to I$ (as made precise in the proof).

Furthermore, it is a based *monad, in the sense that there is a natural transformation $vX\colon \top \to CX$ (the upper vertex) which makes the following diagram commute*

$$(4.142)$$

It will also be useful to note that the following map is constant (i.e. factorises through the terminal object):

$$g.\gamma C.Iv = gc_0 = vp \colon I\top \to \top \to CX. \tag{4.143}$$

Proof Let us use the previous description of the second-order cone $C^2 X$ as a colimit (in (4.135)). The operation g^- gives a commutative diagram, because of the absorbency axiom on I

$$
\begin{array}{ccccccccc}
I\top & \xleftarrow{\,Ip\,} & IX & \xrightarrow{\,I\partial^+\,} & I^2X & \xleftarrow{\,\partial^+ I\,} & IX & \xrightarrow{\,e\,} & X \\
{\scriptstyle p}\downarrow & & {\scriptstyle e}\downarrow & & {\scriptstyle g^-}\downarrow & & {\scriptstyle e}\downarrow & & {\scriptstyle p}\downarrow \\
\top & \xleftarrow{} & X & \xrightarrow[\partial^+]{} & IX & \xleftarrow[\partial^+]{} & X & \xrightarrow{} & \top
\end{array}
$$

The colimit is our operation $g \colon C^2 X \to CX$, determined by the commutative diagram

$$
\begin{array}{ccccc}
I\top & \xrightarrow{\,c_0\,} & C^2X & \xleftarrow{\,c_2\,} & I^2X \\
{\scriptstyle p}\downarrow & & {\scriptstyle g}\downarrow & & {\scriptstyle g^-}\downarrow \\
\top & \xrightarrow[\,v\,]{} & CX & \xleftarrow[\,\gamma\,]{} & IX
\end{array}
$$

since the pair of maps c_0, c_2 is jointly epi, by (4.137).

It is now easy to see that (C, u, g) is indeed a monad, deducing the properties of (u, g) from the analogous ones of (∂^-, g^-). The properties of v in diagram (4.142) follow from the following computations (recall that $\gamma\top$ is an isomorphism, by Section 1.7.2(c), whence cancellable)

$$g.vC = g.c_0\partial^+ = vp\partial^+ = v,$$
$$g.Cv.\gamma\top = g.c_0 = v.p(I\top) = v.p(C\top).\gamma\top.$$

The last remark follows from the definition of c_0 (in Theorem 4.8.1) and the diagram above. $\qquad\square$

4.8.4 Theorem (The transposition of the cone)

Let $\mathbf{A}$ be a symmetric dI2-homotopical category, with an I-preserved zero object. Then the transposition $s \colon I^2 \to I^2$ induces an involutive transposition $s \colon C^2 \to C^2$, which interchanges the faces Cu and uC.

Proof Using the description of $C^2 X$ in (4.138), the transposition $s \colon I^2 \to I^2$ of the cylinder induces an involutive transformation $s \colon C^2 \to C^2$, defined by the commutative diagram on the right:

$$
\begin{array}{ccccccc}
0 \leftarrow IX & \xrightarrow{I\partial^+} & I^2 X & \xleftarrow{\partial^+ I} & IX \to 0 & & I^2 X \xrightarrow{c_2} C^2 X \\
\downarrow \quad 1\downarrow & & \downarrow s & & 1\downarrow \qquad \downarrow & & s\downarrow \qquad \downarrow s \\
0 \leftarrow IX & \xrightarrow[\partial^+ I]{} & I^2 X & \xleftarrow[I\partial^+]{} & IX \to 0 & & I^2 X \xrightarrow[c_2]{} C^2 X
\end{array}
$$

Finally, using the formula $c_2 = \gamma C.I\gamma$ of (Theorem 4.8.1) and recalling that $\gamma X \colon IX \to CX$ is always epi, in the pointed case, as remarked at the end of Section 1.7.2(b), we can prove that s interchanges the faces of C^2

$$
\begin{aligned}
(s.Cu).\gamma &= s.\gamma C.Iu = s.\gamma C.I\gamma.I\partial^- = s.c_2.I\partial^- = c_2.s.I\partial^- \\
&= c_2.\partial^- I = \gamma C.I\gamma.\partial^- I = \gamma C.\partial^- C.\gamma = uC.\gamma.
\end{aligned}
$$

$\square$

4.8.5 Proposition (Cones and contractibility)

Let **A** *be a dI2-homotopical category. Then every future cone $C^+ X$ is strongly future contractible.*

Proof By the previous monad structure on $C = C^+$, the basis $u^- C \colon CX \to C^2 X$ has a retraction $g \colon C^2 X \to CX$. By Lemma 1.7.3, it follows that CX is future contractible. Moreover, the contraction is given by the homotopy $g.\gamma C \colon I(CX) \to CX$, which on the vertex $v = v^+ \colon \top \to CX$ gives $g.\gamma C.Iv$, a constant map (cf. (4.143)). $\square$

4.8.6 Preadditive h-categories

In a category of chain complexes (on an additive category), homotopies are determined by null-homotopies $\nu \colon 0 \to a$, which can be represented by a cone functor $C(X)$ or by a cocone functor $E(X)$. Arbitrary homotopies $\varphi \colon f \to g$ are then defined as pairs (f, ν), for $\nu \colon 0 \to g - f$, and *can always be reversed*.

On the other hand, a category of *directed* chain complexes is only enriched on abelian monoids (see Section 2.1.1): there is no subtraction of morphisms, and *null-homotopies are not sufficient to define homotopies*.

Therefore, the additive case is of little interest for directed homotopy, and we will restrict ourselves here to sketching some basic definitions. This subsection will not be used in the sequel.

The additive notation for homotopy concatenation is not used here, as it would lead to ambiguity: degenerate homotopies are written as $e(f)$, reversed homotopies as φ^{op} and concatenated homotopies as $\varphi * \psi$.

A *preadditive h-category* will be a preadditive category $\mathbf{A}$ (Section A4.6) which is equipped with:

(a) abelian groups $\mathbf{A}_2(X, Y)$ of *homotopies* $\varphi \colon f \to g \colon X \to Y$ (with variable f, g);

(b) faces and degeneracy *homo*morphisms (where $\mathbf{A}_1(X, Y)$ denotes the abelian group of maps $X \to Y$)

$$\partial^\alpha \colon \mathbf{A}_2(X, Y) \rightrightarrows \mathbf{A}_1(X, Y) \colon e, \qquad \partial^\alpha e = 1, \qquad (4.144)$$

$$\begin{aligned}(\varphi \colon f \to g, \ \psi \colon f' \to g') &\Rightarrow (\varphi + \psi \colon f + f' \to g + g'), \\ (f, g \colon X \to Y) &\Rightarrow e(f) + e(g) = e(f + g);\end{aligned}$$

(c) a whisker composition of homotopies and maps, consistent with faces and degeneracy, which is trilinear and also satisfies the usual properties for associativity and identities:

$$k \circ (\varphi + \psi) \circ h = k \circ \varphi \circ h + k \circ \psi \circ h, \qquad \varphi \circ (h + h') = \varphi \circ h + \varphi \circ h',$$
$$(k + k') \circ \varphi = k \circ \varphi + k' \circ \varphi,$$
$$k'(k \circ \varphi \circ h)h' = (k'k) \circ \varphi \circ (hh') \qquad (\textit{associativity}),$$
$$1_Y \circ \varphi \circ 1_X = \varphi, \qquad k \circ e(f) \circ h = e(kfh) \qquad (\textit{identities}).$$

As a consequence, the degenerate homotopy $e(0)$ of the zero-map $A \to B$ is the zero-element of $\mathbf{A}_2(A, B)$.

It follows that $\mathbf{A}$ is a *reversible dh1-category* (Section 1.2.9), where $R = \mathrm{id}$ and the reversed homotopy φ^{op} comes forth from the algebraic opposite $-\varphi \colon -f \to -g$ (*but should not be confused with it*)

$$\varphi^{\mathrm{op}} = 0_f - \varphi + 0_g \colon g \to f \colon X \to Y.$$

A preadditive h-category also inherits a (regular) concatenation $*$ defined as follows, by means of the algebraic sum of homotopies in $\mathbf{A}_2(X, Y)$ (and again these two things should not be confused):

$$\varphi * \psi = \varphi - 0_g + \psi \colon f \to h \colon X \to Y \qquad (\varphi \colon f \to g, \ \psi \colon g \to h).$$

(We might say that $\mathbf{A}$ is a 'reversible dh3-category', according to a definition which has not been given, but can be easily abstracted from the regular dI3-case, defined in Section 4.2.3).

Since $\varphi = (\varphi - e(f)) + e(f)$, homotopies are determined by *null-homotopies* $\nu\colon 0 \to a\colon X \to Y$. In other words, a preadditive h-category can be equivalently described as a preadditive category **A** equipped with:

(a$'$) abelian groups $N_2(X,Y)$ of *null-homotopies* $\nu\colon 0 \to f\colon X \to Y$;

(b$'$) *upper face* homomorphisms $\partial^+\colon N_2(X,Y) \to \mathbf{A}_1(X,Y)$;

(c$'$) a whisker composition of null-homotopies and maps, consistent with upper face and degeneracy, which is trilinear and also satisfies the usual properties for associativity and identities.

Arbitrary homotopies $\varphi\colon f \to g$ are then defined as pairs (f,ν), for $\nu\colon 0 \to g - f$, and composed in the obvious way:

$$k(f,\nu)h = (kfh, k\nu h).$$

An *additive h-category* is further provided with a zero-object 0 and biproducts $A \oplus B$, always in the *two-dimensional* sense, i.e. satisfying the universal properties also for homotopies.

4.9 The reversible case

The reversible case has peculiar properties, which will also be of use in studying non-reversible structures equipped with a forgetful functor taking values in a reversible one (Section 5.8). We will end this section by establishing a relationship between reversible dI4-homotopical categories and Baues cofibration categories [Ba].

4.9.1 Reversing homotopies and 2-homotopies

Let **A** be a *reversible* dI3-category, which means that the reversor R is the identity.

Because of that, we already remarked (in Section 4.2.3) that the preorder relation $f \preccurlyeq_1 g$, i.e. the existence of a homotopy $f \to g$, coincides with the homotopy congruence $f \simeq_1 g$ which gives the homotopy category $\mathrm{Ho}_1(\mathbf{A}) = \mathbf{A}/\!\simeq_1$ (Section 1.3.3).

Furthermore, every double homotopy $\Phi\colon I^2 X \to Y$ can be reversed in both directions, by precomposing with $rI\colon I^2 X \to I^2 X$ or Ir

$$
\begin{array}{ccc}
\begin{array}{ccc}
f & \xrightarrow{\ \varphi\ } & h \\
{\scriptstyle \sigma}\downarrow & \Phi & \downarrow{\scriptstyle \tau} \\
k & \xrightarrow[\ \psi\ ]{} & g
\end{array}
&
\begin{array}{ccc}
k & \xrightarrow{\ \psi\ } & g \\
{\scriptstyle -\sigma}\downarrow & \Phi.rI & \downarrow{\scriptstyle -\tau} \\
f & \xrightarrow[\ \varphi\ ]{} & h
\end{array}
&
\begin{array}{ccc}
h & \xrightarrow{\ -\varphi\ } & f \\
{\scriptstyle \tau}\downarrow & \Phi.Ir & \downarrow{\scriptstyle \sigma} \\
g & \xrightarrow[\ -\varphi\ ]{} & k
\end{array}
\end{array}
\qquad (4.145)
$$

In particular, a 2-homotopy $\Phi\colon \varphi \to \psi$ yields a 2-homotopy $\Phi.rI\colon \psi \to \varphi$. Thus, the existence of a 2-homotopy $\varphi \to \psi$ is a symmetric relation, and coincides with the equivalence relation $\varphi \simeq_2 \psi$ (Section 4.5.1).

The procedures of (4.145) have no counterpart in the non-reversible case; but, applying both, we get a double reversion which corresponds to the double reflection of the non-reversible case (see (4.10)).

4.9.2 Theorem (The fundamental groupoid)

Let **A** *be a reversible dI4-category. Then the concatenation of homotopies is associative, has identities and inverses up to 2-homotopy.*

The 2-category $\mathrm{Ho}_2(\mathbf{A}) = \mathbf{A}/\simeq_2$ *(Section 4.5.5) has invertible cells. The fundamental category* $\uparrow\Pi_1(X)$ *(Section 4.5.7) becomes the fundamental groupoid, and should be preferably written as* $\Pi_1(X)$.

Proof After Theorem 4.5.6 and Section 4.5.7, it suffices to prove that, given a homotopy $\varphi\colon f \to g\colon X \to Y$, the reversed homotopy $\psi = -\varphi = \varphi r\colon g \to f$ is an inverse of φ, up to 2-homotopies

$$A\colon 0_f \to \varphi +_2 \psi\colon f \to f, \qquad B\colon 0_g \to \psi +_2 \varphi\colon g \to g.$$

These can be constructed as follows

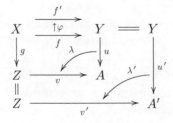

4.9.3 Theorem (Homotopy preservation)

Let **A** *be a reversible dI4-category and consider a homotopy* $\varphi\colon f \to f'\colon X \to Y$ *and two h-pushouts* $I(f,g) = A$, $I(f',g) = A'$

Then the comparison map $w: A \to A'$ *defined below is a homotopy equivalence*

$$wu = u', \quad wv = v', \qquad w \circ \lambda = u' \circ \varphi + \lambda': u'f' \to v'g.$$

Proof Since **A** is reversible, we also have a map $w': A' \to A$ constructed as above, from the reversed homotopy $\psi = -\varphi: f' \to f$

$$w'u' = u, \quad w'v' = v, \qquad w' \circ \lambda' = u \circ \psi + \lambda: uf \to vg.$$

Then $w'w \simeq \mathrm{id}: A \to A$, by the two-dimensional property of the h-pushout A, applied to the trivial homotopies of u, v

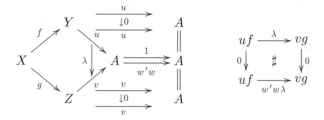

The existence of a 2-homotopy $\lambda \to w'w \circ \lambda$ is proved by the following relations, which come from Theorem 4.9.2, on the weak regularity of concatenation in a reversible dI4-category (including the fact that φ and ψ are inverses up to 2-homotopy)

$$w'w \circ \lambda = w'u' \circ \varphi + w' \circ \lambda' = u \circ \varphi + (u \circ \psi + \lambda) \simeq_2 u \circ (\varphi + \psi) + \lambda \simeq_2 \lambda.$$

$\square$

4.9.4 Invariance theorem

Let **A** *be a reversible dI4-homotopical category and consider an h-pushout* $I(f, g) = A$ *(as in the diagram below). If* g *is a homotopy equivalence, then the 'opposite' morphism* $g': Y \to A$ *is too.*

Proof We already know that, in more general hypotheses, the h-pushout of an isomorphism is a homotopy equivalence by Proposition 4.6.1(b). Now, in the present case, the map $g: X \to Z$ is only supposed to be a homotopy equivalence. We have to prove that g', the h-pushout of g along f, is a homotopy equivalence, or equivalently that its class $[g']$ in $\mathrm{Ho}_1(\mathbf{A})$ is an isomorphism.

Let $h\colon Z \to X$ be a homotopy inverse of g, with $hg \simeq_1 1_X$, $gh \simeq_1 1_Z$, and let us construct the following h-pushouts

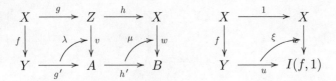

Now, by special invariance (Proposition 4.6.1), the h-pushout u of 1_X along f is a homotopy equivalence. Since $hg \simeq_1 1_X$, by homotopy preservation (Theorem 4.9.3), the h-pushout of hg along f is homotopy equivalent to u, hence a homotopy equivalence itself. By the pasting property (Theorem 4.6.2), also $h'g'$ is a homotopy equivalence, and $[h'g']$ is iso in $\mathrm{Ho}_1(\mathbf{A})$.

Working in the same way on the composition hg (and taking care to distinguish g' and g'')

$$
\begin{array}{ccccc}
Z & \xrightarrow{\ h\ } & X & \xrightarrow{\ g\ } & Z \\
{\scriptstyle v}\downarrow & \mu & \downarrow{\scriptstyle w} & \nu & \downarrow{\scriptstyle w'} \\
A & \xrightarrow{\ h'\ } & B & \xrightarrow{\ g''\ } & C
\end{array}
$$

one gets that $[g''h']$ is an isomorphism. Therefore $[h']$ is iso in $\mathrm{Ho}_1(\mathbf{A})$, and finally so is $[g']$. $\qquad\square$

4.9.5 Baues cofibration categories

We end this section by proving that, under an obvious condition on pushouts, a *reversible* dI4-homotopical category has a canonical structure of 'cofibration category', a non-self-dual version of Quillen's model categories defined in Baues' book on *Algebraic Homotopy* [Ba].

Let us recall that a *cofibration category*, in the sense of Baues, is a category $\mathbf{A}$ equipped with two classes of morphisms, called *cofibrations* and *weak equivalences*, which satisfy the following axioms.

(C1) *Composition.* All isomorphisms are cofibrations and weak equivalences. Cofibrations are closed under composition, while weak equivalences are closed under the 'two out of three' property: namely, if in a composite $h = gf$ two maps out of f, g, h are weak equivalences, then the third is also.

(C2) *Pushout.* A cofibration $f: X \to Y$ has a pushout (f', g') along any map $g: X \to Z$, and f' is a cofibration

$$
\begin{array}{ccc}
X & \xrightarrow{g} & Z \\
f \downarrow & \quad {-\,-} \quad & \downarrow f' \\
Y & \xrightarrow{g'} & A
\end{array}
\tag{4.146}
$$

Moreover:

(a) if g is a weak equivalence, so is g';

(b) if f is a weak equivalence, so is f'.

(Condition (b) is redundant, as proved in Baues' text, Lemma 1.4. Below, we will ignore it.)

(C3) *Factorisation.* Every map $f: X \to Y$ has a factorisation $f = f_2 f_1$, where f_1 is a cofibration and f_2 a weak equivalence.

(C4) *Fibrant models.* Every object X has a trivial cofibration $X \to RX$ with values in a fibrant object.

(A cofibration is said to be *trivial* if it is a weak equivalence. The object R is said to be *fibrant* if every trivial cofibration $i: R \to R'$ has a retraction $p: R' \to R$.)

4.9.6 Theorem

Let **A** *be a reversible dI4-homotopical category with extended acceleration (Section 4.6.5). We define* cofibrations *by the usual homotopy extension property (Section 4.6.3), and* weak equivalences *as homotopy equivalences (Section 1.3.1). Assume now that a cofibration $f: X \to Y$ has a pushout along any map, which is preserved by the cylinder functor I.*

Then **A** *is a* cofibration category, *in the sense recalled above, and every object is fibrant.*

Proof The properties of the composition axiom are obvious (and already stated in Section 4.6.3 and Section 1.3.3).

As to (C2), let us begin by proving that the map f' in diagram (4.146) is a cofibration. Take a map $h: A \to W$ and a homotopy $\psi: h' = hf' \to k'$. Then, the homotopy $\psi \circ g: (hg')f \to k'g$ has a lifting $\varphi: hg' \to w$, with $\varphi f = \psi \circ g$. Now, since I preserves the pushout square (4.146), there exists (precisely) one map $\chi\colon IA \to W$ such that $\chi g' = \varphi$ and $\chi f' = \psi$

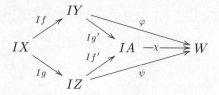

The property (a) is here a consequence of the invariance theorem 4.9.4: if g is a homotopy equivalence, so is g'. For the factorisation axiom, we have already proved, in Theorem 4.6.7, that every map has a canonical factorisation through its mapping cylinder $I(f, 1)$, formed of a cofibration followed by a deformation retraction, which is a homotopy equivalence.

Finally, every object is fibrant: take a trivial cofibration $f\colon X \to Y$, and a map $g\colon Y \to X$ forming a homotopy equivalence with f. Then, the homotopy $\psi\colon gf \to \mathrm{id}X$ can be lifted to a homotopy $\varphi\colon g \to p\colon Y \to X$, and $pf = \mathrm{id}X$

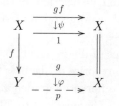

$\square$

5

Categories of functors and algebras, relative settings

The homotopical structures we are studying are 'categorically algebraic', in the sense that they are based on endofunctors and 'operations' on them (natural transformations between their powers), much in the same way as in the theory of monads.

This is why such a structure can generally be lifted from a ground category $\mathbf{A}$ to a categorical construction on the latter, yielding a second category $\mathbf{E}$ equipped with a forgetful functor $U\colon \mathbf{E} \to \mathbf{A}$, or with a family of functors $U_i\colon \mathbf{E} \to \mathbf{A}$.

We treat thus: categories of diagrams and sheaves (Section 5.1), slice categories (Section 5.2), categories of algebras for a monad (Sections 5.3 and 5.4) and categories of differential graded algebras (Sections 5.5–5.7).

Applying – for instance – these results to $\mathrm{d}\mathbf{Top}$, the symmetric dIP4-homotopical category of d-spaces, we obtain that any category of diagrams $\mathrm{d}\mathbf{Top}^{\mathbf{S}}$ and any slice category $\mathrm{d}\mathbf{Top}\backslash A$ or $\mathrm{d}\mathbf{Top}/B$ is a symmetric dIP4-homotopical category (Sections 5.1.5 and 5.2.6). Furthermore, any category of sheaves $\mathrm{Shv}(\mathbf{S}, \mathrm{d}\mathbf{Top})$, over any site $\mathbf{S}$, is a symmetric dP4-homotopical category (Section 5.1.5).

The same is also true of any category of algebras $\mathrm{d}\mathbf{Top}^{T}$, for every monad T on $\mathrm{d}\mathbf{Top}$ which is made consistent with the path functor, in a natural sense. This yields the homotopical structure of d-topological semigroups, or d-topological groups, or d-spaces equipped with an action of a fixed d-topological group, etc. (Section 5.4).

Similarly, the homotopy structure of $\mathbf{Cat}$ can be lifted to the category of strict monoidal categories (Section 5.4.6).

We end, in Section 5.8, by considering a relative setting, called a *relative dI-homotopical category* and based on a forgetful functor $U\colon \mathbf{A} \to \mathbf{B}$, where $\mathbf{A}$ is a dI1-homotopical category, $\mathbf{B}$ is a symmetric dI4-homotopical category and U is a dI1-homotopical functor. Now, we can prove higher properties of the fibre sequence in **A**, up to *relative equivalences*, i.e. maps of **A** which become past or future homotopy equivalences in **B**. This approach and the dual one allow us to treat 'weak' homotopy structures, like those of differential graded algebras (Section 5.8.4), directed chain complexes (Section 5.8.5), cubical sets (Section 5.8.6) and inequilogical spaces (Section 5.8.7).

5.1 Directed homotopy in categories of diagrams and sheaves

A d-homotopy structure based on a cylinder or cocylinder functor can be lifted from a category **A** to every category of diagrams $\mathbf{A}^\mathbf{S}$ defined on a small category; many different situations can be dealt with in this way (see Section 5.1.1), including group actions and equivariant homotopy.

For categories of *sheaves*, the lifting is fairly easy for the path functor (Theorem 5.1.4); on the other hand, to construct the cylinder functor requires stronger hypotheses, and a sheafification procedure to begin with; this will not be developed here.

The classical literature on such subjects is abundant. For homotopy in categories of diagrams and equivariant homotopy, we refer to Dror Farjoun [Dr], Brown and Loday [BL], Moerdijk and Svensson [MoS], and Cordier and Porter [CP]. For set-valued sheaves over a site, one can see the book by Mac Lane and Moerdijk [MM], while sheaves in general categories, but over a space, can be found in Gray [Gy1]. For Quillen structures in categories of sheaves, see Crans [Cr].

Most of the material of this section and the next comes from [G4].

5.1.1 Categories of diagrams

Let **S** be a small category and consider the category of diagrams $\mathbf{A}^\mathbf{S}$, i.e. functors $\mathbf{S} \to \mathbf{A}$ with their natural transformations. An object $X = ((X_i), (X_\iota))$ is thus a collection indexed over the objects i and the arrows $\iota \colon i \to j$ of **S**, which satisfies the functorial properties. A functor $\mathbf{S}^{\mathrm{op}} \to \mathbf{A}$ is also called an **A**-valued *presheaf* defined on **S**, and their category will be written as $\mathrm{Psh}(\mathbf{S}, \mathbf{A}) = \mathbf{A}^{\mathbf{S}^{\mathrm{op}}}$.

Such categories include, for instance:

- the cartesian power $\mathbf{A}^S$, for any set S (as a discrete category);
- the category $\mathbf{A}^\mathbf{2}$ of morphisms of **A**;
- the category $\mathbf{A}^\mathbf{z}$ of unbounded towers of **A** (where **z** is the order category of integers);

- the category $\mathbf{A}^G$ of actions in $\mathbf{A}$ of a fixed group, or monoid, G (as a one-object category);
- the categories of simplicial or cubical objects in $\mathbf{A}$;
- the category $\mathrm{Psh}(X, \mathbf{A}) = \mathbf{A}^{\mathbf{S}^{\mathrm{op}}}$ of $\mathbf{A}$-valued presheaves over a fixed topological space X (letting $\mathbf{S}$ be the category of open subsets of X, with their inclusion mappings).

We are interested in lifting the structure of $\mathbf{A}$ to $\mathbf{A}^{\mathbf{S}}$ along the (jointly faithful) family of evaluation functors $U_i \colon \mathbf{A}^{\mathbf{S}} \to \mathbf{A}$, $X \mapsto X_i$. Equivalently, we can consider *one* forgetful functor $U \colon \mathbf{A}^{\mathbf{S}} \to \mathbf{A}^{\mathrm{Ob}\mathbf{S}}$, of components U_i.

As a first step, if $\mathbf{A}$ is a dh1-category (Section 1.2.9), $\mathbf{A}^{\mathbf{S}}$ has a canonical dh1-structure, with the obvious reversor $R \colon \mathbf{A}^{\mathbf{S}} \to \mathbf{A}^{\mathbf{S}}$. A (*natural* or *equivariant*) homotopy $\varphi \colon f \to g \colon X \to Y$ in $\mathbf{A}^{\mathbf{S}}$ is defined to be a family of $\mathbf{A}$-homotopies $\varphi_i \colon f_i \to g_i \colon X_i \to Y_i$ ($i \in \mathrm{Ob}\mathbf{S}$) which is natural in the obvious sense provided by the whisker composition of maps and homotopies, in $\mathbf{A}$

$$Y_\iota \circ \varphi_i = \varphi_j \circ X_\iota \qquad (\iota \colon i \to j \text{ in } \mathbf{S}). \tag{5.1}$$

5.1.2 Theorem (Diagrams and homotopy)

If $\mathbf{A}$ is a dI1 (resp. dI2, dI3, dI4)-category or a dI1 (resp. dI2, dI3, dI4)-homotopical category, possibly symmetric, then the category of diagrams $\mathbf{A}^{\mathbf{S}}$ has a canonical structure of the same kind, by a pointwise lifting of the cylinder.

Similar results hold for the P- and IP-analogues.

Proof All the argument is obvious. For a dI1-category

$$\mathbf{A} = (\mathbf{A}.R, I, \partial^\alpha, e, r),$$

the dI1-structure of the category $\mathbf{A}^{\mathbf{S}}$ is obtained by post-composing a diagram $X \colon \mathbf{S} \to \mathbf{A}$ with the structure of $\mathbf{A}$:

$$\begin{aligned}
(RX)_i &= R(X_i), & (RX)_\iota &= R(X_\iota), \\
(IX)_i &= I(X_i), & (IX)_\iota &= I(X_\iota), \\
(\partial^\alpha X)_i &= \partial^\alpha X_i, & (eX)_i &= eX_i.
\end{aligned} \tag{5.2}$$

One works in the same way for the higher structures, up to reach the symmetric dI4-case $(R, I, \partial^\alpha, e, r, g^\alpha, s, J, c, z)$. Note that the concatenation pushout $J(X) = IX +_X IX$ exists and is pointwise calculated in $\mathbf{A}^{\mathbf{S}}$, which allows one to lift the concatenation $c \colon I \to J$.

In the dI1-homotopical case, the terminal object and all cylindrical colimits exist in $\mathbf{A}^\mathbf{S}$ and are pointwise calculated in $\mathbf{A}$. In the dI2-homotopical case, the preservation of cylindrical colimits by the cylinder functor of $\mathbf{A}$ automatically lifts to $\mathbf{A}^\mathbf{S}$.

The P-cases follow now by duality, and the IP-cases from the previous ones, taking into account that an endo-adjunction of $\mathbf{A}$ also lifts to $\mathbf{A}^\mathbf{S}$, in a canonical way.   □

5.1.3 Sheaves on a site

Let us recall the definition of sheaves on a site, from the text by Mac Lane and Moerdijk [MM]. Let $\mathbf{S}$ be a small category.

First, a *sieve* s of the object i in $\mathbf{S}$ is a right ideal of maps having codomain i, in the sense that: if $\iota \in s$, then $\iota\kappa \in s$, whenever the composition is defined.

Now, a *site* is a small category $\mathbf{S}$ equipped with a *Grothendieck topology* J; the latter assigns to every object i a set $J(i)$ of sieves of i (which are said to *cover* i), under three axioms which abstract the behaviour of (downwards closed) open coverings of open subsets, in a topological space ([MM], III.2, Def. 1). These axioms are:

(i) the maximal sieve t_i, consisting of all the arrows of codomain i, belongs to $J(i)$;

(ii) *(stability)* if $s \in J(i)$ and $\kappa\colon j \to i$ is in $\mathbf{S}$, then the sieve $\kappa^*(s) = \{\iota \mid \kappa\iota \in s\}$ belongs to $J(j)$;

(iii) *(transitivity)* if $s \in J(i)$ and r is sieve on i such that, for every arrow $\kappa\colon j \to i$ in s, $\kappa^*(r)$ belongs to $J(j)$, then $r \in J(i)$.

$\mathrm{Shv}(\mathbf{S}, \mathbf{A})$ is the full subcategory of $\mathrm{Psh}(\mathbf{S}, \mathbf{A})$ consisting of those presheaves $X = ((X_i), (\iota^*))$ which are *sheaves* (with respect to the Grothendieck topology J of $\mathbf{S}$), i.e. satisfy the following limit condition. For each object i and each sieve $s \in J(i)$, consider the (small) diagram $X|s$ in $\mathbf{A}$ having the following vertices and arrows

$$X_\iota = X_{\mathrm{Dom}\iota} \qquad (\iota \in s),$$

$$x_{\iota\kappa}\colon X_\iota \to X_{\iota\kappa} = \kappa^*\colon X_{\mathrm{Dom}\iota} \to X_{\mathrm{Dom}\kappa} \qquad (\iota \in s, \ \mathrm{Cod}\kappa = \mathrm{Dom}\iota).$$

Then X_i is required to be the limit in $\mathbf{A}$ of this diagram, with projections

$$\iota^*\colon X_i \to X_\iota = X_{\mathrm{Dom}\iota} \qquad (\iota \in s).$$

Note that the diagram $X|s$ is defined over the category $\mathrm{cat}(s)$, with objects $\iota \in s$, arrows

$$(\iota, \kappa) : \iota \to \iota\kappa \qquad (\text{for } \iota \in s,\ \mathrm{Cod}\kappa = \mathrm{Dom}\iota),$$

and composition $(\iota\kappa, \lambda).(\iota, \kappa) = (\iota, \kappa\lambda)$. Since $\mathbf{A}$ is not required to be complete, the sheaf condition cannot be expressed using products and equalisers; furthermore, the direct formulation above is often more manageable.

5.1.4 Theorem (Sheaves and cocylinder)

Let $\mathbf{A}$ be a dP1 (resp. dP2, dP3, dP4)-category, or a dP1 (resp. dP2, dP3, dP4)-homotopical category, possibly symmetric, and assume that the path functor P preserves all the existing limits (as it certainly does if it has a left adjoint).

Then the category $\mathrm{Shv}(\mathbf{S}, \mathbf{A})$ of sheaves of $\mathbf{A}$ over a small site $\mathbf{S}$ is a subcategory of the same kind (see Sections 1.2.2 and 4.2.7), in the category of presheaves $\mathrm{Psh}(\mathbf{S}, \mathbf{A}) = \mathbf{A}^{\mathbf{S}^{\mathrm{op}}}$.

Proof Let $\mathbf{A}$ be equipped with a (path) endofunctor P which preserves the existing limits. Then the path functor of presheaves (Theorem 5.1.2) $PX = ((PX_i), (P\iota^*))$ restricts trivially to sheaves.

After that, we have to show that the full subcategory of sheaves is closed in $\mathrm{Psh}(\mathbf{S}, \mathbf{A})$ under concatenation pullbacks or cocylindrical limits, whenever these are assumed to exist in $\mathbf{A}$ (and therefore also exist in $\mathrm{Psh}(\mathbf{S}, \mathbf{A})$). Since the existence of cocylindrical limits is equivalent to the existence of certain pullbacks, it suffices to prove that $\mathrm{Shv}(\mathbf{S}, \mathbf{A})$ is closed in $\mathrm{Psh}(\mathbf{S}, \mathbf{A})$ under the existing pullbacks.

Let X be the pullback of a cospan $A \to B \leftarrow C$ in $\mathrm{Psh}(\mathbf{S}, \mathbf{A})$, and assume that A, B, C are sheaves. Then X is also: the proof of the sheaf condition for X reduces to a straightforward diagram-chasing in the diagram below, for $\iota \in s$ and $\mathrm{Cod}\kappa = \mathrm{Dom}\iota$

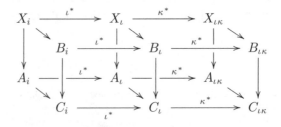

5.1.5 Topological examples

The categories $d\mathbf{Top}^\mathbf{S}$ and $(d\mathbf{Top}_\bullet)^\mathbf{S}$ of **S**-diagrams of d-spaces or pointed d-spaces are symmetric dIP4-homotopical.

This includes $d\mathbf{Top}^2$, $d\mathbf{Top}^z$, d-spaces with G-action (for a group or monoid G), presheaves of d-spaces on a fixed topological space and the pointed analogues (see Section 5.1.1).

Furthermore, for any site **S**, $\mathrm{Shv}(\mathbf{S}, d\mathbf{Top})$ and $\mathrm{Shv}(\mathbf{S}, d\mathbf{Top}_\bullet)$ are symmetric dP4-homotopical.

5.2 Directed homotopy in slice categories

We prove now that, under natural conditions, one can lift a d-homotopy structure from a ground category **A** to a slice category $\mathbf{A}\backslash A$ or $\mathbf{A}/B$, *under* or *over* a reversible object. In order to unify such arguments, we fix a map $u\colon A \to B$ of **A** and consider a category $\mathbf{A}(u)$ which extends the previous cases (Section 5.2.2).

Again, the literature in the classical case is much developed. For the homotopy theory of (strict or relaxed) slice categories of spaces see James [J1, J2], Baues [Ba], Hardie and Kamps [HK1, HK2, HK3], Hardie, Kamps and Porter [HKP]. Homotopy in a slice category $\mathbf{Top}/B$ is called *fibre-wise homotopy*.

5.2.1 Some remarks on pushouts

We say that a map $t\colon A \to B$ in the category **A** *has all pushouts* (or that **A** *has all t-pushouts*) if the pushout of t along an arbitrary map f exists

$$\begin{array}{ccc}
A & \xrightarrow{\ f\ } & \bullet \\
{\scriptstyle t}\downarrow & \ \dashrightarrow\ {\scriptstyle t'} & \downarrow \\
B & \longrightarrow & \bullet
\end{array}
\qquad
\begin{array}{ccccc}
A & \xrightarrow{\ f\ } & \bullet & \xrightarrow{\ g\ } & \bullet \\
{\scriptstyle t}\downarrow & \dashrightarrow\ {\scriptstyle t'} & \downarrow & \dashrightarrow\ {\scriptstyle t''} & \downarrow \\
B & \longrightarrow & \bullet & \longrightarrow & \bullet
\end{array}
\qquad (5.3)$$

This yields a map t' which again has all pushouts, as proved in the right diagram above: use the pushout of t along the composite gf, and 'factorise' it through the left-hand pushout; then the right-hand square is also a pushout, by a well-known lemma, easy to verify.

Therefore, if **A** has all t-pushouts and a functor $F\colon \mathbf{A} \to \mathbf{B}$ preserves them, it follows that F also preserves the pushouts of any map t' which is the pushout of t along some other map.

5.2.2 Bilateral slice categories and reversors

Let **A** be a category. The classical topological example of a slice category is **Top**$_{\bullet}$ = **Top**$\backslash\top$, the category of pointed spaces, or 'spaces under the point' $\top = \{*\}$: an object (X, a) is a map $a \colon \top \to X$ in **Top**; pointed maps preserve the base point. When studying the directed case d**Top**$_{\bullet}$ of pointed d-spaces, we begin by remarking that the reversor $R \colon$ d**Top** $\to$ d**Top** lifts to d**Top**$_{\bullet}$ letting $R(X, a) = (X^{\mathrm{op}}, a^{\mathrm{op}} \colon \top \to X^{\mathrm{op}})$.

Now, if **A** is a category and A an object therein, the slice category **A**$\backslash A$, of *objects under* A, has objects (X, a) consisting of a map $a \colon A \to X$ in **A**; a morphism $f \colon (X, a) \to (X', a')$ is given by a map $f \colon X \to X'$ in **A** such that $f \circ a = a'$, as in the left diagram below

$$
\begin{array}{ccc}
A & =\!\!=\!\!= & A \\
{\scriptstyle a}\downarrow & & \downarrow{\scriptstyle a'} \\
X & \xrightarrow{\ f\ } & X'
\end{array}
\qquad
\begin{array}{ccc}
X & \xrightarrow{\ f\ } & X' \\
{\scriptstyle b}\downarrow & & \downarrow{\scriptstyle b'} \\
B & =\!\!=\!\!= & B
\end{array}
\qquad\qquad (5.4)
$$

If **A** has a reversor $R \colon$ **A** $\to$ **A** (i.e. an involutive endofunctor) *and* the object A is invariant under it, we identify $R(A) = A$, for simplicity, and define the reversor $R \colon$ **A**$\backslash A \to$ **A**$\backslash A$ letting

$$
\begin{aligned}
R(X, a) &= (RX, Ra \colon A \to RX), \\
R(f \colon (X, a) \to (X', a')) &= R(f) \colon R(X, a) \to R(X', a').
\end{aligned}
\qquad (5.5)
$$

Dually, we have the slice category **A**$/B \cong ($**A**$^{*}\backslash B)^{*}$ of *objects over* B, whose morphisms make the right diagram above commutative (with a similar argument on reversors). We will unify these two types of slice categories, considering a more general, self-dual situation (as in [Ba], I.4.3).

Let a map $u \colon A \to B$ be fixed in **A**, and consider the *bilateral slice category along* u

$$
\mathbf{A}(u) = (\mathbf{A}\backslash A)/(B, u) = (\mathbf{A}/B)\backslash(A, u). \qquad (5.6)
$$

Explicitly, an object is a triple (X, a, b) with $ba = u$

$$
A \xrightarrow{\ a\ } X \xrightarrow{\ b\ } B
\qquad\qquad
\begin{array}{ccc}
A & \xrightarrow{\ a\ } X \xrightarrow{\ b\ } & B \\
\| & \downarrow{\scriptstyle f} & \| \\
A & \xrightarrow[\ a'\ ]{} X' \xrightarrow[\ b'\ ]{} & B
\end{array}
\qquad (5.7)
$$

and a map $f \colon (X, a, b) \to (X', a', b')$ is a morphism f of **A** which makes the right diagram above commute.

The category $\mathbf{A}(u)$ extends both cases of slice categories, since

$$\mathbf{A}\backslash A \cong \mathbf{A}(A \to \top), \qquad \mathbf{A}/B \cong \mathbf{A}(\bot \to B), \qquad (5.8)$$

provided $\mathbf{A}$ has initial and terminal object (which is no real restriction here, since such objects can always be formally added). On the other hand, $\mathbf{A}(u)$ has the following initial and terminal object (under no assumptions on $\mathbf{A}$)

$$\bot = (A, 1, u), \qquad \top = (B, u, 1). \qquad (5.9)$$

Notice that one could similarly consider the larger category of objects $(X, a: A \to X, b: X \to B)$ with A, B fixed and no assumption on the composite $ba: A \to B$. But the extension is only apparent, since the right diagram (5.7) shows that a morphism $f: (X, a, b) \to (X', a', b')$ can only exist when $ba = b'a'$; therefore, this larger category breaks into the sum of its connected components, the various $\mathbf{A}(u)$, for $u \in \mathbf{A}(A, B)$, each with its own initial and terminal object (5.9).

If $\mathbf{A}$ has a reversor $R: \mathbf{A} \to \mathbf{A}$, *and* one can identify $R(u) = u$, the reversor $R: \mathbf{A}(u) \to \mathbf{A}(u)$ is defined in the obvious way:

$$R(X, a, b) = (RX, Ra: A \to RX, Rb: RX \to B), \qquad (5.10)$$

It strictly commutes with the forgetful functor $U: \mathbf{A}(u) \to \mathbf{A}$, which sends (X, a, b) to X.

5.2.3 Theorem and Definition
(Lifting functors to slice categories)

Consider a bilateral slice category $\mathbf{A}(u)$, for $u: A \to B$ in $\mathbf{A}$.

(i) Let the pair (F, e) consist of an endofunctor $F: \mathbf{A} \to \mathbf{A}$ and a natural transformation $e: F \to 1$ (degeneracy) whose component $e_A: FA \to A$ on the object A has all pushouts (Section 5.2.1). Then F has a canonical lifting to an endofunctor $\mathbf{F}$ of $\mathbf{A}(u)$, defined by the following pushout

$$\mathbf{F}(X, a, b) = (F(X, a), a^F, b^F). \qquad (5.11)$$

The object $F(X, a)$ *and the map* $p(X, a)$ *do not depend on b. But the latter will also be written as* $p(X, a, b)$*, when useful.*

(ii) Every morphism $f \colon (F, e) \to (F', e')$ *of such pairs (consisting of a natural transformation* $f \colon F \to F'$ *such that* $e'f = e \colon F \to 1$*) lifts to a morphism* $(\mathbf{F}, e) \dashrightarrow (\mathbf{F}', c')$.

(iii) Given a pair (F, e) *as above, if* F *preserves all pushouts along* $e_A \colon FA \to A$*, then the composed endofunctor* $\mathbf{F}^2$ *is the canonical lifting of* F^2 *with respect to its associated degeneracy,*

$$e_2 = e.Fe = e.eF \colon F^2 \to 1.$$

Proof The first point is obvious, and (ii) comes from the following commutative cube, whose front and back faces are pushouts

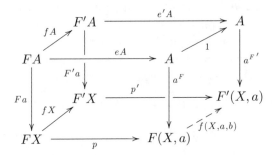

As to (iii), consider the following diagram, where we write $\mathbf{X} = (X, a, b)$, $Y = F(X, a)$ and $\mathbf{Y} = \mathbf{F}(X, a, b) = (Y, a^F, b^F)$

$$
\begin{array}{ccccccc}
F^2A & \xrightarrow{FeA} & FA & \xrightarrow{eA} & A & \!\!=\!\!=\!\! & A \\
{\scriptstyle F^2a}\downarrow & & \downarrow{\scriptstyle Fa^F} & & \downarrow{\scriptstyle \hat{a}} & & \\
F^2X & \xrightarrow{FpX} & FY & \xrightarrow{pY} & F(Y, a^F) & & {\scriptstyle u} \\
{\scriptstyle eFX}\downarrow & & \downarrow{\scriptstyle eFY} & & {\scriptstyle \hat{b}}\searrow & & \\
FX & \xrightarrow{pX} & Y & \xrightarrow{\hspace{2em} b^F \hspace{2em}} & & & B
\end{array}
$$

$\mathbf{F}^2(X, a, b)$ is a triple, whose object is computed as $F(F(X, a), a^F)$, in the right-hand pushout. The canonical lifting of F^2 on the object (X, a, b) also gives a triple, whose object is computed with the pasting of the two pushouts above, and coincides with the former object.

Both computations give the same map $\hat{a} \colon A \to F(Y, a^F)$. Finally, the lower left square commutes, by naturality of $e \colon F \to 1$ on the morphism

$p(X, a, b) \colon FX \to F(X, a)$. Henceforth, the map $\hat{b} \colon F(Y, a^F) \to B$ which satisfies the conditions for $\mathbf{F}^2(X, a, b)$:

$$\hat{b}.\hat{a} = u, \qquad \hat{b}.pY = b^F.eF(X, a),$$

also satisfies the similar conditions for the lifting of F^2, since:

$$\hat{b}.(p\mathbf{Y}.FpX) = b^F.p\mathbf{X}.eFX = b.eX.eFX = b.e_2X.$$

□

5.2.4 Theorem
(First order homotopy structure for slice categories)

Suppose we have a category $\mathbf{A}$ *equipped with a reversor* R *and an* $\mathbf{A}$-*morphism* $u \colon A \to B$ *such that* $Ru = u$. *The bilateral slice category* $\mathbf{A}(u)$ *is equipped with the lifted reversor* R, *as above (Section 5.2.2).*

(a) If $\mathbf{A}$ *is a dI1-category (resp. a dI1-homotopical category) and the degeneracy* $eA \colon IA \to A$ *of the domain of* u *has all pushouts in* $\mathbf{A}$ *(in the sense of Section 5.2.1), then the bilateral slice category* $\mathbf{A}(u)$ *has a canonical dI1-structure (resp. dI1-homotopical structure).*

Its cylinder functor $\mathbf{I}$ *is the canonical lifting of* (I, e), *constructed as follows (in the previous theorem)*

$$
\begin{array}{ccc}
IA \xrightarrow{\ eA\ } A = \!\!= \!\!= A & \quad & \mathbf{I}(X, a, b) \\
\end{array}
$$

$$\mathbf{I}(X, a, b) = (I(X, a), a^I, b^I). \tag{5.12}$$

The forgetful functor $U \colon \mathbf{A}(u) \to \mathbf{A}$ *is a lax dI1-functor (Section 1.2.6), with a comparison* $p \colon IU \to UI$ *whose general component is shown in the left square above.*

(a*) Dually, if $\mathbf{A}$ *is a dP1-category (resp. a dP1-homotopical category) and the degeneracy* $eB \colon B \to PB$ *of the codomain of* u *has all pullbacks, then* $\mathbf{A}(u)$ *is also dP1 (resp. dP1-homotopical), with a path functor* $\mathbf{P}$ *defined by the eB-pullback below (with* $wa^P = eX.a \colon A \to PX$ *).*

$$\mathbf{P}(X, a, b) = (P(X, b), a^P, b^P). \tag{5.13}$$

The forgetful functor $U: \mathbf{A}(u) \to \mathbf{A}$ is a lax dP1-functor (Section 1.2.6), with a comparison $w: UP \to PU$ whose general component is shown in the diagram above.

Proof We prove (a). The new cylinder functor $\mathbf{I}: \mathbf{A}(u) \to \mathbf{A}(u)$ is defined in (5.12). We have also defined the natural transformation

$$p: I U \to U\mathbf{I}, \qquad p(X,a,b): IX \to U\mathbf{I}(X,a,b),$$

and it will be useful to note that its components $p(X,a,b)$ have all pushouts in $\mathbf{A}$, by a remark in Section 5.2.1.

Faces and degeneracy of $\mathbf{I}$ are constructed in the dotted rectangles below (according to Theorem 5.2.3(ii))

$$\partial^\alpha = p(X,a,b).\partial^\alpha X: (X,a,b) \to \mathbf{I}(X,a,b),$$
$$e = e^I: \mathbf{I}(X,a,b) \to (X,a,b),$$

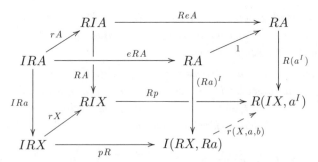

where $e^I.a^I = a$ and $e^I.p(X,a,b) = eX: IX \to X$.

Finally, the new reflection $r: \mathbf{IR} \to \mathbf{RI}$ is induced by the reflection of $\mathbf{A}$: in other words, the component $r(X,a,b)$ is defined on the pushout $I(RX, Ra)$ as the map which makes the following cube commute (recall that $RA = A$):

The axioms of dI1-category are easily verified, as well as the fact that (U, p) is a lax dI1-functor (taking into account the fact that $UR = RU$, as already noted in Section 5.2.2).

In the dI1-homotopical case, we have to prove the existence of the h-pushout $\mathbf{I}(f,g)$ of two maps, $f\colon (X,a,b) \to (X',a',b')$ and $g\colon (X,a,b) \to (X'',a'',b'')$, in $\mathbf{A}(u)$.

Let us begin by remarking that the lower row of the left diagram below must have a colimit in $\mathbf{A}$, say C

$$
\begin{array}{ccccccc}
X' \xleftarrow{f} X \xrightarrow{\partial^-} IX \xleftarrow{\partial^+} X \xrightarrow{g} X'' & & IX \xrightarrow{\lambda} I(f,g) \\
\end{array}
$$

$$
X' \xleftarrow{f} X \xrightarrow[\partial^-]{} I(X,a) \xleftarrow[\partial^+]{} X \xrightarrow[g]{} X'' \qquad I(X,a) \xrightarrow[\lambda']{} C
$$

In fact, we start from the colimit of the upper row, which is the h-pushout $I(f,g)$ in $\mathbf{A}$, with structural maps

$$
u\colon X' \to I(f,g), \qquad \lambda\colon IX \to I(f,g), \qquad v\colon X'' \to I(f,g).
$$

Since the map $p\colon IX \to I(X,a)$ has all pushouts, we form the right-hand pushout above. It is easy to verify that this yields the colimit C which we want, with structural maps

$$
p'u\colon X' \to C, \qquad \lambda'\colon I(X,a) \to C, \qquad p'v\colon X'' \to C.
$$

Finally, the h-pushout $\mathbf{I}(f,g)$ in $\mathbf{A}(u)$ is the triple $(C, \hat{a}, \hat{b})$, where the object C is equipped with the morphism

$$
\hat{a} = p'ua' = p'va''\colon A \to C,
$$

and the morphism $\hat{b}\colon C \to B$ defined on the colimit C of the lower row in the left diagram above, by the following three maps

$$
b'\colon X' \to B, \qquad b^I\colon I(X,a) \to B, \qquad b''\colon X'' \to B.
$$

$\square$

5.2.5 Theorem
(Higher homotopy structure for slice categories)

Consider again a bilateral slice category $\mathbf{A}(u)$, for $u\colon A \to B$ in $\mathbf{A}$, with $Ru = u$. If $\mathbf{A}$ is a symmetric dI4-category, or symmetric dI4-homotopical, so is $\mathbf{A}(u)$, provided that all eA-pushouts exist in $\mathbf{A}$ and are preserved by I.

Similar results hold for the non-symmetric case, and for the intermediate cases dI2, dI3 (for the latter, there is no need to assume that eA-pushouts are preserved by I).

Proof Let **A** be a symmetric dI4-category. Theorem 5.2.3 allows us to lift I and I^2, together with faces, degeneracies, reversion (as in Theorem 5.2.4), connections and transposition.

Furthermore, the endofunctor $JX = IX +_X IX$ of **A** lifts to the slice category, once we prove that the degeneracy $\bar{e} \colon JA \to A$ has pushout along any map $f \colon JA \to C$.

First form, in the diagram below, the eA-pushouts of the restrictions $f^\alpha = fc^\alpha$, producing two maps u^α which have all pushouts; then form their pushout D. It is easy to check that D is also the right-hand pushout below, via the maps $h = v^\alpha u^\alpha \colon C \to D$ and $h' = hfc^-\partial^+ = hfc^+\partial^- \colon A \to D$

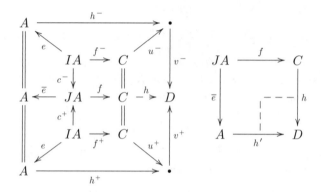

By Theorem 5.2.3(ii), the lifted functor **J** has transformations $c^-, c^+ \colon \mathbf{I} \to \mathbf{J}$; these still form the pushout of $\partial^-, \partial^+ \colon 1 \to \mathbf{I}$, because 'pushouts preserve pushouts' (Lemma 4.2.9). Finally, the concatenation $c \colon I \to J$ and the acceleration $z \colon I^2 \to I$ also lift to natural transformations $c \colon \mathbf{I} \to \mathbf{J}$ and $z \colon \mathbf{I}^2 \to \mathbf{I}$, which satisfy the axioms of Section 4.2.5.

The symmetric I4-homotopical case (Section 4.2.6) is now a trivial consequence of the previous ones. For the dI2 and dI3 cases, possibly symmetric, use the relevant part of the previous arguments.   □

5.2.6 Topological examples

We already know that the category d**Top.** of pointed d-spaces is a symmetric dIP4-homotopical category (Section 4.3.1). Since d**Top.** = d**Top**\⊤ is a slice category of the category of d-spaces, we can also obtain this fact from the same property of d**Top**, using the previous theorem.

Consistently with the constructions we have already used (in Section 1.5.5), the P-structure comes directly from that of d**Top** (cf. (5.13)) while the pointed cylinder $I(X, a) = (IX/I\{a\}, a^I)$ is formed by collapsing the subspace $I\{a\}$ in the non-pointed cylinder IX (see (5.12)).

More generally, if $u\colon A \to B$ is a map of d-spaces with $Ru = u$, the bilateral slice category d**Top**(u) is symmetric dIP4-homotopical, with cylinder and path functor as in Theorem 5.2.4. This includes the categories d**Top**$\backslash A$ of d-spaces under a reversible A (take $B = \top$) and d**Top**$/B$ of d-spaces over a reversible B (take $A = \bot$).

Notice that, for a *pointed* d-map $u\colon A \to B$ in d**Top**$_{\bullet}$, the slice category d**Top**$_{\bullet}(u)$ does not yield anything new. Indeed, if $|u|$ is the underlying map in d**Top**, the forgetful functor from d**Top**$_{\bullet}(u)$ to d**Top**$(|u|)$ is an isomorphism.

Preordered spaces give symmetric dIP4-homotopical categories of type p**Top**(u), while (ordinary) topological spaces give reversible symmetric dIP4-homotopical categories **Top**(u).

Finally, let us recall that a homotopy in a slice category **Top**$/B$ is called a *fibre-wise homotopy over B* (and the same terminology can be used for d-spaces). Indeed, given two fibre maps $f, g\colon (X, b) \to (Y, b')$, a map $\varphi\colon I(X, b) \to (Y, b')$ whose faces are f and g amounts to an ordinary homotopy $\varphi\colon f \to g\colon X \to Y$ such that, for all $x \in X$, the path $\varphi(\tau, x)$ ($\tau \in [0, 1]$) is contained in a b'-fibre of Y (namely, the fibre over $b(x) = b'f(x) = b'g(x)$).

5.3 Algebras for a monad and the path functor

We consider now a monad $T = (T, \eta, \mu)$ over a category **A** (Section A4.4). Then a homotopy structure for **A**, defined by a path endofunctor P, can be lifted to the category $\mathbf{A}^T$ of Eilenberg–Moore algebras over T, provided that the monad is made consistent with the path functor by a natural transformation $\lambda\colon TP \to PT$ satisfying some natural conditions.

The dual procedure, which will not be written down, allows one to lift a cylinder setting to a category of coalgebras over a comonad.

This section and the next are essentially an adaptation to the directed case of joint work with J. MacDonald [GMc]. We refer to Mac Lane's text [M3] for the basic theory of monads, their categories of algebras and the relationship with adjunctions (briefly recalled here in Sections A4.4 and A4.5).

5.3.1 Lifting functors to algebras

Let $T = (T, \eta, \mu)$ be a monad over the category $\mathbf{A}$. A T-algebra, or Eilenberg–Moore algebra for T, is a pair (X, t) consisting of an (*underlying*) object X and a morphism $t \colon TX \to X$ in $\mathbf{A}$, called the *algebraic structure*, under two (obvious) axioms of consistency with η, μ (see Section A4.5). A *morphism* of T-algebras, $f \colon (X, t) \to (X', t')$ is an $\mathbf{A}$-map $f \colon X \to X'$ such that $f.t = t'.Tf$.

As usual, $\mathbf{A}^T$ denotes the category of T-algebras, $U^T \colon \mathbf{A}^T \to \mathbf{A}$ the forgetful functor $U^T(X, t) = X$, and $F^T \colon \mathbf{A} \to \mathbf{A}^T$ its left adjoint, the *free-algebra functor* $F^T(X) = (TX, \mu X)$.

It is well known that the forgetful functor $U^T \colon \mathbf{A}^T \to \mathbf{A}$ *creates* [M3] the (existing) limits. For instance, given two morphisms of T-algebras $f_i \colon (X_i, t_i) \to (Y, u)$ $(i = 1, 2)$, if their underlying maps $f_i \colon X_i \to Y$ have a pullback X in $\mathbf{A}$, there is precisely one structure $t \colon TX \to X$ which makes the pullback-projections into T-morphisms (and is determined by the conditions $p_i.t = t_i.Tp_i$). Then (X, t) is the pullback of (f_1, f_2) in $\mathbf{A}^T$, with the same projections p_i

$$
\begin{array}{ccc}
(X, t) & \xrightarrow{\ p_1\ } & (X_1, t_1) \\
{\scriptstyle p_2}\downarrow\!- - - \,\downarrow & & \downarrow{\scriptstyle f_1} \qquad\qquad p_i.t = t_i.Tp_i \colon TX \to X_i. \qquad (5.14) \\
(X_2, t_2) & \xrightarrow{\ t_2\ } & (Y, u)
\end{array}
$$

We want now to consider an endomorphism $P \colon \mathbf{A} \to \mathbf{A}$, typically a path endofunctor, 'consistent' with the monad. But we prefer to deal, more generally, with two monads over two categories, as this distinction happens to make things clearer.

Given a second monad $S = (S, \eta', \mu')$ over $\mathbf{B}$ (with forgetful functor $U^S \colon \mathbf{B}^S \to \mathbf{B}$ and free-algebra functor F^S), a (lax) *morphism of monads* $(P, \lambda) \colon T \to S$ is a functor $P \colon \mathbf{A} \to \mathbf{B}$ equipped with a natural transformation $\lambda = \lambda^P \colon SP \to PT$, the *comparison*, satisfying

$$
\lambda.\eta' P = P\eta, \qquad\qquad \lambda.\mu' P = P\mu.\lambda_2,
$$
$$
(\lambda_2 = \lambda T.S\lambda \colon S^2 P \to PT^2). \qquad\qquad (5.15)
$$

$$
\begin{array}{ccc}
P & \xrightarrow{\ \eta' P\ } SP & \xleftarrow{\ \mu' P\ } S^2 P \\
& \searrow{\scriptstyle P\eta} \quad \downarrow{\scriptstyle \lambda} & \quad \downarrow{\scriptstyle \lambda_2} \\
& PT & \xleftarrow{\ P\mu\ } PT^2
\end{array}
$$

The morphism (P, λ) is said to be *strong* if λ is invertible, and *strict* if λ is an identity; the morphism will generally be written as P, leaving λ^P understood.

The composition $P'P$ with a morphism $P' \colon S \to R$ has the obvious comparison

$$\lambda^{P'P} = (P' \circ \lambda^P).(\lambda^{P'} \circ P) \colon R.P'P \to P'SP \to P'P.T. \qquad (5.16)$$

A 2-*cell* of these morphisms, or *natural transformation* $\varphi \colon P \to Q \colon T \to S$, is an ordinary natural transformation $\varphi \colon P \to Q \colon \mathbf{A} \to \mathbf{B}$ making the following square commute, for every object X

$$
\begin{array}{ccc}
SPX & \xrightarrow{\ S\varphi\ } & SQX \\
{\scriptstyle \lambda^P}\downarrow & & \downarrow{\scriptstyle \lambda^Q} \\
PTX & \xrightarrow[\ \varphi T\ ]{} & QTX
\end{array}
\qquad\qquad \varphi T.\lambda^P = \lambda Q.S\varphi. \qquad (5.17)
$$

A morphism of monads $P \colon T \to S$ has a canonical lifting $\overline{P} \colon \mathbf{A}^T \to \mathbf{B}^S$, with $U^S.\overline{P} = PU^T$

$$\overline{P}(X, t) = (PX, Pt.\lambda X), \qquad\qquad \overline{P}(f) = P(f), \qquad (5.18)$$

and every natural transformation $\varphi \colon P \to Q$ has a unique lifting $\overline{\varphi} \colon \overline{P} \to \overline{Q}$, with $U^S.\overline{\varphi} = \varphi U^T$, which will also be written as $\varphi \colon \overline{P} \to \overline{Q}$ since its components are 'the same' as those of φ

$$\overline{\varphi}(X, t) = \varphi X \colon \overline{P}(X, t) \to \overline{Q}(X, t).$$

For $T = S$, an endomorphism of monads $(P, \lambda) \colon T \to T$ will also be called a *T-functor*; it is an endofunctor $P \colon \mathbf{A} \to \mathbf{A}$ equipped with a natural transformation $\lambda = \lambda^P \colon TP \to PT$ such that

$$\lambda.\eta P = P\eta, \qquad \lambda.\mu P = P\mu.\lambda_2 \qquad (\lambda_2 = \lambda T.T\lambda \colon T^2 P \to PT^2).$$

Then, the lifting $\overline{P} \colon \mathbf{A}^T \to \mathbf{B}^T$ will also be written as P^T. Moreover P^2 is also a T-functor by means of the natural transformation

$$\lambda^{P^2} = (P \circ \lambda^P).(\lambda^P \circ P) \colon TP^2 \to P^2 T.$$

The lifting of P^2 coincides with the composite $(P^T)^2$.

In a setting where one can consider a 2-category **CAT** of large categories (see Section A1.1), a monad can be defined as a 2-functor $T \colon \mathbf{m} \to \mathbf{CAT}$, defined on the formal monad 2-category $\mathbf{m}$ (see Section 5.3.8). This yields naturally a 2-category **MON** of monads, lax morphisms and natural transformations; the lifting procedure respects the various compositions, forming a 2-functor $\mathbf{MON} \to \mathbf{CAT}$ which takes the monad T to its category of algebras $\mathbf{A}^T$. The lifting of adjunctions will be considered later (Theorem 5.3.7).

5.3.2 Remarks

(a) As noted in Johnstone [Jo] (Lemma 1, attributed to Appelgate's thesis), it is easy to see that, given two monads T and S as above and a mere functor $P \colon \mathbf{A} \to \mathbf{B}$, there is a bijective correspondence between liftings $\overline{P} \colon \mathbf{A}^T \to \mathbf{B}^S$ of P and natural transformations $\lambda \colon SP \to PT$ satisfying the conditions above (5.15), i.e. making P into a morphism of monads. We have already given this correspondence in one direction.

Conversely, if $\overline{P}$ is such a lifting, the transpose of $P\eta \colon P \to PU^T F^T = U^S \overline{P} F^T$ gives a natural transformation $\hat{\lambda} \colon F^S P \to \overline{P} F^T$, and $\lambda = U^S \hat{\lambda} \colon SP \to PT$ satisfies our conditions (5.15). The two procedures are inverses of each other.

(b) An oplax morphism of monads can be extended to the categories of Kleisli algebras [M3]. Details can be found in [GMc].

(c) If $(P, \lambda) \colon T \to T$ is an endomorphism of monads and P is part of a comonad (P, ε, δ) then λ is called a *bialgebra distributivity* in case $\varepsilon T.\lambda = T\varepsilon$ and $\delta T.\lambda = P\lambda.\lambda P.T\delta$, see MacDonald and Stone [MS]; a similar case, in which P is part of a monad, is dealt with in Beck [Be], p. 120.

5.3.3 Monads and path-functors

Let $\mathbf{A} = (\mathbf{A}, R, P, \partial^\alpha, e, r)$ be a dP1-category (Section 1.2.2, with $\alpha = \pm$)

$$\partial^\alpha \colon P \rightrightarrows 1 : e, \qquad r \colon RP \to PR \quad (R^2 = \mathrm{id}),$$
$$\partial^\alpha e = 1 \colon \mathrm{id}\mathbf{A} \to \mathrm{id}\mathbf{A}, \qquad RrR.r = 1 \colon RP \to RP, \qquad (5.19)$$
$$r.Re = eR \colon R \to RP, \qquad \partial^- R.r = R\partial^+ \colon RP \to R.$$

We also have a monad $T = (T, \eta, \mu)$ on the category $\mathbf{A}$, *strictly consistent* with the reversor R, in the sense that

$$RT = TR, \qquad R\eta = \eta R, \qquad R\mu = \mu R. \qquad (5.20)$$

This allows us to lift the reversor to T-algebras, letting $R \colon \mathbf{A}^T \to \mathbf{A}^T$ be defined as:

$$R(X, t) = (RX, Rt), \qquad Rt \colon T(RX) = RTX \to RX. \qquad (5.21)$$

By definition, the dP1-structure is *made consistent* with the monad T by a natural transformation $\lambda\colon TP \to PT$ which makes P into a T-functor and ∂^α, e, r into T-natural transformations (Section 5.3.1). In other words, the following equations hold, after (5.20)

$$\lambda.\eta P = P\eta, \quad \lambda.\mu P = P\mu.\lambda T.T\lambda,$$
$$\partial^\alpha T.\lambda = T\partial^\alpha, \qquad eT = \lambda.Te, \qquad rT.R\lambda = \lambda R.Tr. \tag{5.22}$$

Then the structure (5.19) lifts to a dP1-structure for algebras, yielding a path endofunctor $P^T = \overline{P}$

$$P^T \colon \mathbf{A}^T \to \mathbf{A}^T, \qquad\qquad P^T(X, t) = (PX, Pt.\lambda X), \tag{5.23}$$

whose faces, degeneracy and reflection will still be written $\partial^\alpha \colon P^T \to 1$, $e\colon 1 \to P^T, r\colon RP^T \to P^T R$.

The forgetful functor $U^T \colon \mathbf{A}^T \to \mathbf{A}$ extends obviously to homotopies, double homotopies and 2-homotopies, preserving faces

$$(U^T(\varphi))\hat{} = U^T(\hat{\varphi}), \qquad (UT(\Phi))\hat{} = U^T(\hat{\Phi}).$$

Recalling that the lifting of P^2 is $(P^T)^2$, the same lifting property holds for a symmetric dP2-structure $(R, P, \partial^\alpha, e, r, g^\alpha, s)$ *consistent* with T; this condition means that, moreover, the connections g^α and the transposition s have to satisfy

$$g^\alpha T.\lambda = P\lambda.\lambda P.Tg^\alpha, \qquad sT.P\lambda.\lambda P = P\lambda.\lambda P.Ts.$$

The same terminology can be used replacing the category of algebras with a category $\mathbf{C}$ *monadic* over $\mathbf{A}$ (Section A4.5).

5.3.4 The functor of consecutive pair of paths

Let $\mathbf{A}$ be a dP1-category consistent with the monad T, and let us assume that $\mathbf{A}$ has all concatenation pullbacks (Section 4.2.2)

$$\begin{array}{ccc} QX & \xrightarrow{\;c^+\;} & PX \\ {\scriptstyle c^-}\downarrow & \dashleftarrow\dashrightarrow & \downarrow{\scriptstyle \partial^-} \\ PX & \xrightarrow[\;\partial^+\;]{} & X \end{array} \tag{5.24}$$

Then the functor $Q\colon \mathbf{A} \to \mathbf{A}$ has a canonical T-structure $\lambda^Q \colon TQ \to QT$, which makes the following diagram commute (the inner square is the Q-pullback of TX)

$$
\begin{array}{ccc}
TQX & \xrightarrow{\;Tc^+\;} & TPX \\
\end{array}
$$

In fact, the coherence conditions with η, μ (5.15)

$$
\lambda^Q.\eta Q = Q\eta \colon QX \to QTX,
$$
$$
\lambda^Q.\mu Q = Q\mu.\lambda^Q T.T\lambda^Q \colon T^2QX \to QTX,
$$

follow from the analogues for P, by composing with the projections $c^\alpha T$ of QTX; for instance

$$
c^\alpha T.(\lambda^Q.\eta Q) = \lambda.Tc^\alpha.\eta Q = \lambda.\eta P.c^\alpha = P\eta.c^\alpha = c^\alpha T.(Q\eta).
$$

Now, the lifted functor $Q^T(X,t) = (QX, Qt.\lambda QX)$ is the concatenation pullback for algebras, with projections $c^\alpha \colon Q^T \to P^T$ which lift the original ones

$$
\begin{array}{ccc}
Q^T(X,t) & \xrightarrow{\;c^+\;} & P^T(X,t) \\
{\scriptstyle c^-}\downarrow & & \downarrow{\scriptstyle \partial^-} \\
P^T(X,t) & \xrightarrow[\;\partial^+\;]{} & (X,t)
\end{array}
$$

In fact, the structure $Qt.\lambda^Q X$ is precisely the one created by U^T over the concatenation pullback in $\mathbf{A}$ (see (5.14)):

$$
c^\alpha.(Qt.\lambda QX) = Pt.c^\alpha T.\lambda QX = Pt.\lambda X.Tc^\alpha.
$$

5.3.5 The remaining second-order structure

In the same way, we say that a symmetric dP4-category structure

$$
(R, P, \partial^\alpha, e, r, g^\alpha, s, Q, c, z)
$$

over $\mathbf{A}$ is *consistent* with the monad T, if P and all the listed natural transformations are consistent; recall that the consistency of Q is automatic (Section 5.3.4). Then, $\mathbf{A}^T$ is a symmetric dP4-category.

Homotopy pullbacks can be viewed as cocylindrical limits (Section 1.8.1), and are thus created (Section 5.3.1) by the forgetful functor $U^T \colon \mathbf{A}^T \to \mathbf{A}$ (when they exist in $\mathbf{A}$). Finally, we get the following results.

5.3.6 Theorem (Lifting the path functor to algebras)

Let $\mathbf{A}$ be a category with a monad T.

(a) A structure of (symmetric) dP4-category or dP4-homotopical category over $\mathbf{A}$, made consistent with T by a natural transformation $\lambda \colon TP \to PT$ (as specified above, in Sections 5.3.3–5.3.5) can always be lifted to a structure of the same type over the category of algebras, with cocylinder

$$P^T \colon \mathbf{A}^T \to \mathbf{A}^T, \qquad P^T(X, t) = (PX, Pt.\lambda). \tag{5.25}$$

The weaker cases, from dP1 to dP3, work similarly.

(b) If $\mathbf{A}$ is a (symmetric) dP4-homotopical category consistent with T, P^T has a left adjoint I^T and $\mathbf{A}^T$ has all cylindrical colimits, then $\mathbf{A}^T$ is a (symmetric) dIP4-homotopical category.

(c) If $\mathbf{A}$ is a dP1-category consistent with T, $\mathbf{A}^T$ has coequalisers and P has a left adjoint I, then P^T has a left adjoint I^T, which extends I, in the sense that $I^T F^T \cong F^T I$.

Note. Generally, I^T is not a lifting of I; see for instance the case of topological semigroups in Section 5.4.2.

Proof Points (a) and (b) summarise the results of this section, including some previous ones on dIP4-homotopical categories (Section 4.2). Point (c) is a particular case of the following well-known theorem. $\qquad \square$

5.3.7 Theorem (Lifting adjunctions to algebras)

Consider a morphism of monads $P = (P, \lambda) \colon T \to S$ over the categories $\mathbf{A}, \mathbf{B}$ (Section 5.3.1). Assume that the underlying functor $P \colon \mathbf{A} \to \mathbf{B}$ has a left adjoint $I \colon \mathbf{B} \to \mathbf{A}$ and that the category of algebras $\mathbf{A}^T$ has coequalisers.

Then, the lifted functor

$$\mathbf{P} \colon \mathbf{A}^T \to \mathbf{B}^S, \qquad \mathbf{P}(X, t) = (PX, Pt.\lambda X),$$

has a left adjoint $\mathbf{I}\colon \mathbf{B}^S \to \mathbf{A}^T$, *which extends* I *with respect to the free-algebra functors* $(\mathbf{I}F^S \cong F^T I)$.

In particular, if $P\colon \mathbf{A} \to \mathbf{A}$ *is the identity,* $\lambda\colon S \to T$ *is a morphism of monads over* $\mathbf{A}$, *i.e.* $\lambda.\eta^S = \eta^T$ *and* $\lambda.\mu^S = \mu^T.\lambda_2$. *(This case can also be found in Beck [Be], p. 119.)*

Proof The proof is outlined in [Jo], Theorem 2, where one can also find references to various earlier versions. Here we give a detailed argument along the same lines.

First, let us note that, if $\mathbf{P}$ has a left adjoint $\mathbf{I}$, then $\mathbf{I}F^S$ and $F^T I$ are both left adjoints to $U^S\mathbf{P} = PU^T$, hence canonically isomorphic.

Recall that $\lambda = \lambda^P\colon SP \to PT$ satisfies the equations $\lambda.\eta'P = P\eta$, $\lambda.\mu'P = P\mu.\lambda T.S\lambda$. Write the unit and counit of $I \dashv P$ as $u\colon 1_\mathbf{B} \to PI$ and $v\colon IP \to 1_\mathbf{A}$, and the free-algebra adjunctions of $T = (T, \eta, \mu)$ and $S = (S, \eta', \mu')$ as

$$F^T\colon \mathbf{A} \rightleftarrows \mathbf{A}^T\colon U^T,$$
$$U^T(X, t) = X, \qquad F^T(X) = (TX, \mu X),$$
$$\eta\colon 1 \to T = U^T F^T, \qquad \varepsilon\colon F^T U^T \to 1;$$
$$U^T \varepsilon(X, t) = t\colon TX \to X;$$
$$\tag{5.26}$$

$$F^S\colon \mathbf{B} \rightleftarrows \mathbf{B}^S\colon U^S,$$
$$\eta'\colon 1 \to S = U^S F^S, \qquad \varepsilon'\colon F^S U^S \to 1.$$
$$\tag{5.27}$$

To construct $\mathbf{I}$, note that, because of the adjunction $I \dashv P$, the functor I inherits a natural transformation $\lambda^*\colon IS \to TI$, the mate of $\lambda\colon SP \to PT$ (Section A5.3)

$$\lambda^* = vTI.I\lambda I.ISu\colon IS \to ISPI \to IPTI \to TI, \tag{5.28}$$

and let us record the fact that

$$P\lambda^*.uS = PvTI.PI(\lambda I.Su).uS = PvTI.uPTI.\lambda I.Su = \lambda I.Su. \tag{5.29}$$

Now, λ^* makes I into an oplax morphism of monads (which would just allow us to extend I to free algebras, according to Section 5.3.2(b)). However, since every algebra is a coequaliser of free ones, and left adjoints preserve the existing colimits, we get the value of the functor $\mathbf{I}$ over the S-algebra $(Y, s\colon SY \to Y)$ as the coequaliser in $\mathbf{A}^T$ of the following two maps of free T-algebras

$$F^T ISU^S(Y,s) \xrightarrow[\overline{\lambda}]{\overset{FTIUS\varepsilon'}{\longrightarrow}} F^T IU^S(Y,s) \xrightarrow{p} \mathbf{I}(Y,s) \qquad (5.30)$$

$F^T IUS\varepsilon'(Y,s) = F^T Is$:
$$\qquad F^T IS(Y) = (TISY, \mu ISY) \to (TIY, \mu IY) = F^T I(Y),$$
$\overline{\lambda} = \varepsilon F^T IUS.F^T \lambda^* US$:
$$\qquad F^T IS(Y) \to F^T TI(Y) = F^T U^T.F^T I(Y) \to F^T I(Y).$$

Its structure $\overline{s}\colon TU^T\mathbf{I}(Y,s) \to U^T\mathbf{I}(Y,s)$ makes p an S-morphism

$$
\begin{array}{ccc}
TI(Y) & \xrightarrow{\;U^T p\;} & U^T\mathbf{I}(Y,s) \\[4pt]
{\scriptstyle \mu IU^T}\Big\uparrow & & \Big\uparrow{\scriptstyle \overline{s}} \\[4pt]
T^2 I(Y) & \xrightarrow[\;TU^T p\;]{} & TU^T\mathbf{I}(Y,s)
\end{array}
\qquad (5.31)
$$

The functor $\mathbf{I}$ is well defined, letting its value on morphisms be the induced map on coequalisers. The coequaliser-maps form a natural transformation $p\colon F^T IU^S \to \mathbf{I}$.

We next go on to define a unit and a counit for the new adjunction. For the unit $\mathbf{u}\colon 1 \to \mathbf{PI}$, we want its underlying transformation $U^S\mathbf{u}\colon U^S \to U^S\mathbf{PI} = PU^T\mathbf{I}$ to be the composite

$$U^S\mathbf{u} = PU^T p.P\eta IU^S.\mathbf{u}U^S :$$
$$\qquad U^S \to PIU^S \to PU^T F^T IU^S \to PU^T\mathbf{I}. \qquad (5.32)$$

In fact, for an S-algebra $(Y,s\colon SY \to Y)$, we do get a $\mathbf{B}^S$-morphism with values in $\mathbf{PI}(Y,s)$ (whose structure is $P\overline{s}.\lambda U^T\mathbf{I}(Y,s)$, by (5.18))

$$U^S\mathbf{u}(Y,s).s = PU^T p(Y,s).(P\eta I.u)Y.s$$
$$= PU^T p(Y,s).PU^T F^T Is.(P\eta I.u)SY$$
$$= PU^T p(Y,s).P\mu IY.PU^T F^T \lambda^* Y.P\eta ISY.uSY \quad \text{(by (5.30))},$$
$$= PU^T p(Y,s).P\mu IY.P\eta TIY.P\lambda^* Y.uSY$$
$$= PU^T p(Y,s).\lambda IY.SuY \qquad\qquad \text{(by (5.29))},$$

$$P\overline{s}.\lambda U^T\mathbf{I}(Y,s).SU^S\mathbf{u}(Y,s)$$
$$= P\overline{s}.\lambda U^T\mathbf{I}(Y,s).SP(U^T p(Y,s).\eta IY).SuY$$
$$= P\overline{s}.PT(U^T p(Y,s).\eta IY).\lambda IY.SuY$$
$$= PU^T p(Y,s).P\mu IY.PT\eta IY.\lambda IY.SuY$$
$$= PU^T p(Y,s).\lambda IY.SuY. \qquad\qquad \text{(by (5.31))}.$$

For the counit $\mathbf{v}\colon \mathbf{IP} \to 1$, we require that

$$\mathbf{v}.p\mathbf{P} = \varepsilon.F^T v U^T : F^T I U^S \mathbf{P} = F^T I P U^T \to F^T U^T \to 1. \qquad (5.33)$$

The solution exists (and is unique), provided we show that the following natural transformations $f, g\colon F^T I U^S F^S U^S \mathbf{P} \to 1$ coincide

$$f = \varepsilon.F^T v U^T.F^T I U^S \varepsilon' \mathbf{P},$$
$$g = \varepsilon.F^T v U^T.\varepsilon F^T I U^S \mathbf{P}.F^T \lambda^* U^S \mathbf{P}.$$

Applying these to an object (X, t) of $\mathbf{A}^T$ we get a pair of morphisms of $\mathbf{A}^T$; it is sufficient to show that the underlying $\mathbf{A}$-morphisms, $f_0 = U^T f(X, t)$ and $g_0 = U^T g(X, t)$ are equal

$$f_0 = t.TvX.TIU^S \varepsilon'(PX, Pt.\lambda X) = t.TvX.TIPt.TI\lambda X$$
$$= t.Tt.TvTX.TI\lambda X = t.\mu X.TvTX.TI\lambda X,$$

$$g_0 = U^T g(X, t) = t.TvX.U^T \varepsilon F^T IPX.U^T F^T \lambda^* PX$$
$$= t.TvX.\mu IPX.TvTIPX.TI\lambda IPX.TISuPX \qquad \text{(by (5.28))},$$
$$= t.TvX.(\mu.TvT.TI\lambda)IPX.TISuPX$$
$$= t.(\mu.TvT.TI\lambda)X.TISPvX.TISuPX = t.\mu X.TvTX.TI\lambda X.$$

Finally, to verify the triangular identities for $\mathbf{u}, \mathbf{v}$, it is sufficient to show that $U^S(\mathbf{Pv}.\mathbf{uP}) = 1$ and $(\mathbf{vI}.\mathbf{Iu}).p = p$; this follows from (5.32), (5.33) and the triangular identities of u, v. $\qquad \square$

5.3.8 Formal remarks

The morphisms and 2-cells of monads that we have considered above (Section 5.3.1) are 'natural', as soon as we consider the 2-category $\mathbf{m}$ (the *formal-monad* 2-category) generated by one object $*$, one arrow $t\colon * \to *$, and two cells $e\colon 1 \to t$, $m\colon t^2 \to t$ subject to the relations

$$m.et = 1 = m.te, \qquad m.mt = m.tm. \qquad (5.34)$$

Notice that (t, e, m) can be viewed as a *monad on the object $*$ of the 2-category* $\mathbf{m}$; see Kelly and Street [KS] for further information regarding monads on objects of a 2-category.

Now, in a convenient set-theoretical setting, a monad (T, η, μ) on the category $\mathbf{A}$ amounts to a strict 2-functor $\mathbf{T}\colon \mathbf{m} \to \mathbf{CAT}$, with $\mathbf{A} = \mathbf{T}(*)$, $T = \mathbf{T}(t)$, $\eta = \mathbf{T}(e)$.

Given a second monad S over $\mathbf{B}$, a *lax natural transformation of 2-functors* $\mathbf{T} \to \mathbf{S}\colon \mathbf{m} \to \mathbf{CAT}$ amounts precisely to a functor $P\colon \mathbf{A} \to \mathbf{B}$

(corresponding to the object $*$) and a natural transformation $\lambda\colon SP \to PT$ (corresponding to the generating arrow $t\colon * \to *$ of $\mathbf{m}$) satisfying our conditions (5.15), i.e. a lax morphism $(P, \lambda)\colon T \to S$.

Similarly, a *modification* $\varphi\colon P \to Q\colon \mathbf{T} \to \mathbf{S}$ of lax natural transformations amounts to an ordinary natural transformation $\varphi\colon P \to Q\colon \mathbf{A} \to \mathbf{B}$ (corresponding to the object $*$) satisfying the appropriate condition, (5.17).

Equivalently, one could also view a monad as a *lax functor* $\mathbf{T}\colon \mathbf{1} \to \mathbf{CAT}$, consisting of a category $\mathbf{A} = \mathbf{T}(*)$, an endofunctor $T = \mathbf{T}(\mathrm{id}*)$ and two natural transformations

$$\eta\colon 1_{\mathbf{T}(*)} \to \mathbf{T}(\mathrm{id}*), \qquad \mu\colon \mathbf{T}(\mathrm{id}*).\mathbf{T}(\mathrm{id}*) \to \mathbf{T}(\mathrm{id} * .\mathrm{id}*) \qquad (5.35)$$

under conditions coinciding with the axioms of monads. Lax natural transformations of such lax functors, and their modifications, would give the same notions as above.

5.4 Applications to d-spaces and small categories

Applying the theory developed in the previous section, the homotopy structure of d-spaces can be lifted to various categories of algebras on d**Top**, like d-topological semigroups and groups. Equivariant homotopy, for d-spaces equipped with an action of a fixed d-topological group, is considered in Section 5.4.4.

The homotopy structure of **Cat** can be lifted to strict monoidal categories (Section 5.4.6). Slice categories under (or over) an object can also be recovered as categories of algebras (or coalgebras), see Section 5.4.7.

5.4.1 Directed topological semigroups and monoids

Let us start from the category d**Top** of d-spaces, which is complete and cocomplete, and forget for the moment its homotopy structure.

Let us define a *d-topological semigroup* X as an internal semigroup in d**Top**: X is a d-space equipped with an associative multiplication $X \times X \to X$ which is a morphism of d**Top**. Equivalently, the d-space X is a topological semigroup and its (continuous) multiplication preserves d-paths, in the sense that for every pair of d-paths $a, b\colon \uparrow\mathbf{I} \to X$, the following path is also directed

$$c\colon \uparrow\mathbf{I} \to X, \qquad c(\tau) = a(\tau) \cdot b(\tau). \qquad (5.36)$$

With the obvious morphisms (of d-spaces and semigroups, at the same time) we have the category Sgr-d**Top** of d-topological semigroups. We prove now that this category is monadic over d**Top**, and then (in Section 5.4.2) that the symmetric dP4-homotopical structure lifts to it, and is actually dIP4-homotopical.

The forgetful functor U: Sgr-d**Top** → **Top** has a left adjoint F, which is constructed as the free semigroup on a set

$$FX = (\Sigma_{n>0} X^n, *),$$
$$(x_1, ..., x_p) * (x_{p+1}, ..., x_n) = (x_1, ..., x_n), \tag{5.37}$$
$$\eta\colon X \subset UFX, \quad \varepsilon\colon FUA \to A, \quad \varepsilon(x_1, ..., x_n) = x_1 \cdot ... \cdot x_n.$$

Here, FX is the free semigroup over the underlying set $|X|$, endowed with the sum of the product d-structures X^n, i.e. the sum of the product topologies (the finest topology making all the embeddings $X^n \subset FX$ continuous), where a path $a\colon \uparrow\mathbf{I} \to X^n \subset FX$ is distinguished if this is true in X^n.

FX is a d-topological semigroup, since the juxtaposition $*\colon X^p \times X^q \to X^{p+q}$ is an isomorphism of d-spaces, and every cartesian product in d**Top** distributes over arbitrary sums (whence $TX \times TX = \Sigma_{p,q>0} X^p \times X^q$). It follows easily that FX is indeed the free d-topological semigroup over the space X.

The adjunction gives rise to the *free-semigroup* monad over d**Top**

$$T = UF\colon \textbf{Top} \to \textbf{Top}, \qquad TX = \Sigma_{n>0} X^n,$$
$$\mu = U\varepsilon F\colon T^2 \to T, \tag{5.38}$$
$$\mu((x_{11}, ..., x_{1p_1}), ..., (x_{n1}, ..., x_{np_n})) = (x_{11}, ..., x_{np_n}).$$

A T-algebra (X, t) is 'the same' as a topological semigroup $(X, \cdot)$ with multiplication $\cdot = t_2\colon X^2 \to X$; a map of T-algebras is a d-continuous homomorphism. We identify $\textbf{Top}^T = $ Sgr-d**Top**.

Similarly, the category Mon-d**Top** of *d-topological monoids* is monadic over d**Top**. One uses now the *free-monoid* monad on d**Top**, with $TX = \Sigma_{n\geqslant 0} X^n$.

5.4.2 The homotopy structure of directed topological semigroups

Now, d**Top** has a symmetric dIP4-homotopical structure (Section 4.2.6), based on the adjunction $I \dashv P$ of the cylinder and path endofunctors.

The cocylinder $P: \mathrm{d}\mathbf{Top} \to \mathrm{d}\mathbf{Top}$ preserves powers (as a right adjoint) and also sums. It is a strong T-functor, as proved by the following relations (for $a, a_i, a_{ij} \in PX$)

$$\lambda: TP \to PT, \qquad \lambda X: \Sigma_{n>0} (PX)^n \cong P(\Sigma_{n>0} X^n),$$
$$(a_1, ..., a_n) \mapsto \langle a_1, ..., a_n \rangle: [0,1] \to X^n,$$

$$\lambda.\eta P(a) = a = P\eta(a),$$
$$\lambda.\mu P((a_{11}, ..., a_{1p_1}), ..., (a_{n1}, ..., a_{np_n})) = \lambda(a_{11}, ..., a_{np_n})$$
$$= \langle a_{11}, ..., a_{np_n} \rangle = P\mu(\langle\langle a_{11}, ..., a_{1p_1}\rangle, ..., \langle a_{n1}, ..., a_{np_n}\rangle\rangle)$$
$$= P\mu.\lambda T.T\lambda((a_{11}, ..., a_{1p_1}), ..., (a_{n1}, ..., a_{np_n})).$$

P can thus be canonically lifted to topological semigroups

$$P^T : \mathbf{Top}^T \to \mathbf{Top}^T, \qquad P^T(X, t) = (PX, Pt.\lambda), \qquad (5.39)$$

which simply means that $P^T(X, \cdot)$ is the path d-space $PX = X^{\uparrow \mathbf{I}}$ with the *pointwise multiplication* $(a \bullet a')(\tau) = a(\tau) \cdot a'(\tau)$.

We prove now that all the operations of the symmetric dP4-structure of $\mathrm{d}\mathbf{Top}$ are T-transformations.

Leaving apart, for the moment, $c: Q \to P$, each of the remaining natural transformations $\partial^\alpha, e, g^\alpha, r, s, z$ is defined by pre-composition with some continuous increasing function between powers of the unit interval

$$f_0: [0,1]^q \to [0,1]^p,$$
$$fX: P^p X \to P^q X, \qquad (a: [0,1]^p \to X) \mapsto (af_0: [0,1]^q \to X).$$

Moreover, the natural transformation making P^q a T-functor is

$$\lambda^{P^q} = P^{q-1}\lambda \circ ... \circ P\lambda^{P^{q-2}} \circ \lambda^{P^{q-1}} : T.P^q \to P^q.T,$$
$$\lambda^{P^q} X: \Sigma_{n>0} (P^q X)^n \cong P^q(\Sigma_{n>0} X^n),$$
$$(a_1, ..., a_n) \mapsto \langle a_1, ..., a_n \rangle: [0,1]^q \to X^n.$$

Now, the consistency property of the natural transformation f defined by the mapping f_0, namely $fT.\lambda^P p = \lambda^P q.Tf$, is an easy consequence

$$fT.\lambda^{P^q}(a_1, ..., a_n) = fT\langle a_1, ..., a_n \rangle$$
$$= \langle a_1 f_0, ..., a_n f_0 \rangle = \lambda^{P^q}(a_1 f_0, ..., a_n f_0) \qquad (5.40)$$
$$= \lambda^{P^q}.Tf(a_1, ..., a_n).$$

Finally, recall our choice of $Q = P$ for the concatenation pullback in $\mathrm{d}\mathbf{Top}$ (Section 4.2.6), with $c^\alpha: Q \to P$ given by the first-half or second-half embedding $c_0^\alpha: [0,1] \to [0,1]$, and concatenation map $c = 1$.

Then the lifting of c^α to d-topological semigroups makes $P^T A$ into the concatenation pullback of A; finally, $c = \mathrm{id}$ obviously lifts.

By Theorems 5.3.6 and 5.3.7, we have thus proved that the category $\mathbf{Top}^T = \mathrm{Sgr\text{-}d}\mathbf{Top}$ of d-topological semigroups is a symmetric dIP4-homotopical category, with path functor P^T. The cylinder functor $I^T \dashv P^T$ can be directly calculated as

$$I^T : \mathrm{Sgr\text{-}d}\mathbf{Top} \to \mathrm{Sgr\text{-}d}\mathbf{Top}, \qquad I^T(X, \cdot) = (FIX)/R, \qquad (5.41)$$

where $IX = {\uparrow}[0,1]{\times}X$ is the cylinder of d-spaces and R is the congruence of semigroups over $F(IX)$ spanned by the following relation, based on the multiplication of X

$$(\tau, x) * (\tau, y) \; R_0 \; (\tau, x \cdot y) \qquad (\tau \in [0,1]; x, y \in X).$$

It follows that, for every instant $\tau \in [0, 1]$, the mapping

$$u_\tau : (X, \cdot) \to I^T(X, \cdot), \qquad u_\tau(x) = [\tau, x]$$

is a d-continuous homomorphism. The unit of the adjunction is

$$u : (X, \cdot) \to P^T I^T(X, \cdot) = P((FIX)/R, *), \quad u(x) \colon \tau \mapsto u_\tau(x) = [\tau, x].$$

Similarly, the category $\mathrm{Mon\text{-}d}\mathbf{Top}$ of d-topological monoids is symmetric dIP4-homotopical.

5.4.3 Directed topological groups

As we have already remarked (in Section 1.4.4), the definition of a d-topological group requires some care: we must allow the 'inversion' to reverse directions, much in the same way as in an ordered group.

Thus, a *d-topological group* X will be a d-space equipped with a group structure which consists of morphisms of d$\mathbf{Top}$:

$$
\begin{aligned}
X \times X &\to X, & (x, y) &\mapsto x \cdot y, \\
\{*\} &\to X, & * &\mapsto e, & (5.42) \\
X &\to X^{\mathrm{op}}, & x &\mapsto x^{-1}.
\end{aligned}
$$

Equivalently, X is both a d-space and a topological group, its multiplication preserves d-paths (as in Section 5.4.1), and every d-path $a \colon {\uparrow}\mathbf{I} \to X$ gives a d-path $\tau \mapsto (a(\tau))^{-1}$ in X^{op}. The category Gp-d$\mathbf{Top}$ of d-topological groups is a full subcategory of Sgr-d$\mathbf{Top}$ and Mon-d$\mathbf{Top}$.

Now, we can follow a procedure similar to the previous one, for topological semigroups. The *free d-topological group* FX on a d-space X can be constructed as a quotient of the free d-topological monoid on the d-space $X + X^{\mathrm{op}}$

$$FX = (\Sigma_{n \geqslant 0} (X + X^{\mathrm{op}})^n)/R.$$

The quotient is taken modulo the (usual) congruence of monoids R generated by the relation R_0

$$(x * x^{\mathrm{op}})\ R_0\ e, \qquad\qquad (x^{\mathrm{op}} * x)\ R_0\ e,$$

where x, x^{op} denote any two 'corresponding' elements of X and X^{op}, while e is the identity of the free monoid, i.e. the empty word (the unique element of $(X + X^{\mathrm{op}})^0$).

But it is simpler to lift the path functor (together with its operations)

$$P^T : \text{Gp-d}\mathbf{Top} \to \text{Gp-d}\mathbf{Top},$$

by letting $P^T(X, \cdot)$ be the path d-space $PX = X^{\uparrow \mathbf{I}}$ with pointwise multiplication (as above, in (5.39)).

The monad procedure gives the same result (as we have noted, in general, in Section 5.3.2(a)). Indeed, the *free-group monad* $T = UF \colon \mathrm{d}\mathbf{Top} \to \mathrm{d}\mathbf{Top}$ allows us to identify $\mathrm{d}\mathbf{Top}^T = \text{Gp-d}\mathbf{Top}$. Now, there is a unique d-homomorphism $\lambda X \colon TPX \to PTX$ such that $\lambda X.\eta PX = P(\eta X) \colon PX \to PTX$; it provides a natural transformation $\lambda \colon TP \to PT$, which also satisfies the 'multiplicative' condition $\lambda.\mu P = P\mu.\lambda T.T\lambda$.

Thus, $\mathrm{d}\mathbf{Top}^T = \text{Gp-d}\mathbf{Top}$ is a symmetric dIP4-homotopical category, with path functor P^T (Theorem 5.3.6). Its left adjoint cylinder functor I^T can be directly calculated as above (in (5.41)), using now a *group-congruence* R.

5.4.4 Equivariant directed homotopy

Let G be a d-topological group (Section 5.4.3) in additive notation and G-d$\mathbf{Top}$ the category of G-*d-spaces*, i.e. d-spaces X equipped with a right action

$$X \times G \to X, \qquad (x, g) \mapsto x + g, \qquad\qquad (5.43)$$

which is a morphism of d-spaces satisfying the usual conditions:

$$x + 0 = x, \quad (x + g) + g' = x + (g + g') \qquad (x \in X;\ g, g' \in G). \quad (5.44)$$

(If G is just a topological group, we apply this notion with respect to the *discrete* d-structure on G, so that, besides the continuity of the mapping (5.43) and the algebraic axioms (5.44), we are just requiring that each operator $(-) + g \colon X \to X$ is a d-map. If G is a discrete group, we go back to a situation which has already been studied in Section 5.1, as a functor $G \to \mathrm{d}\mathbf{Top}$ defined on the associated one-object category.) The forgetful functor $U \colon G\text{-}\mathrm{d}\mathbf{Top} \to \mathrm{d}\mathbf{Top}$ has left adjoint

$$F(X) = X \times G \qquad (\eta \colon 1 \to UF, \ \varepsilon \colon FU \to 1),$$

where $X \times G$ is the product of d-spaces, with action $(x, g) + g' = (x, g + g')$. This yields a monad over $\mathrm{d}\mathbf{Top}$

$$\begin{aligned}
T = UF \colon \mathrm{d}\mathbf{Top} \to \mathrm{d}\mathbf{Top}, \qquad & TX = X \times G, \\
\eta X \colon X \to X \times G, \qquad & x \mapsto (x, 0), \\
\mu = U\varepsilon F \colon T^2 \to T, \qquad & \mu X \colon X \times G \times G \to X \times G, \\
\mu X(x, g, g') = (x, g + g'). &
\end{aligned} \qquad (5.45)$$

A G-d-space X is the same as a T-algebra $(X, t \colon X \times G \to X)$. The cocylinder $P \colon \mathrm{d}\mathbf{Top} \to \mathrm{d}\mathbf{Top}$ becomes a T-functor, using as follows the degenerate-path embedding $e \colon G \to PG$

$$\lambda \colon TP \to PT,$$
$$\lambda X = PX \times eG \colon PX \times G \to P(X \times G) = (PX) \times (PG),$$
$$\lambda X(a, g) = \langle a, e(g) \rangle \colon \uparrow\!\mathbf{I} \to X \times G,$$
$$\lambda.\eta P(a) = \lambda(a, 0) = \langle a, e(0) \rangle = P\eta(a),$$
$$\begin{aligned}
\lambda.\mu P(a, g, g') &= \langle a, e(g + g') \rangle = P\mu \langle a, e(g), e(g') \rangle \\
&= P\mu.\lambda T.T\lambda(a, g, g').
\end{aligned}$$

P can thus be canonically lifted to G-d-spaces,

$$P^T \colon G\text{-}\mathrm{d}\mathbf{Top} \to G\text{-}\mathrm{d}\mathbf{Top}, \qquad P^T(X, t) = (PX, Pt.\lambda),$$

which means that $P^T(X, t)$ is the path d-space $PX = X^{\uparrow\mathbf{I}}$ with the pointwise action of G

$$(a + g)(\tau) = a(\tau) + g.$$

A homotopy $\varphi \colon (X, t) \to P^T(Y, u)$ is thus an *equivariant d-homotopy*, i.e. a homotopy $\varphi \colon X \to PY$ of d-spaces such that

$$\varphi(x, \tau) + g = \varphi(x + g, \tau) \qquad (x \in X, \ \tau \in [0, 1], \ g \in G).$$

To show the coherence of the symmetric dP4-structure with λ, we can now go on as for d-topological semigroups, replacing (5.40) with the following equation

$$fT.\lambda^{P^q}(a, g) = fT\langle a, e_p(g) \rangle = \langle af_0, e_q(g) \rangle = \lambda^{P^q}.Tf(a, g),$$

where the natural transformation $f \colon P^p \to Pq$ is induced by a continuous increasing function $f_0 \colon [0,1]^q \to [0,1]^p$.

The category $d\mathbf{Top}^T = G\text{-}d\mathbf{Top}$, with path functor P^T, is thus a symmetric dIP4-homotopical category (Theorem 5.3.6). The cylinder functor I^T (obtained from Theorem 5.3.7 or directly computed as left adjoint to P^T) can be expressed as

$$I^T \colon G\text{-}d\mathbf{Top} \to G\text{-}d\mathbf{Top}, \qquad I^T(X) = (FIX)/R.$$

Here $IX = X \times {\uparrow}\mathbf{I}$ is the cylinder of d-spaces and R is the congruence of G-sets over $F(IX)$ spanned by the following relation, based on the G-action over X

$$(x, \tau, g)\, R_0\, (x + g, \tau, 0) \qquad (x \in X,\ \tau \in [0,1],\ g \in G).$$

5.4.5 Algebras for pointed d-spaces

The (pointed) category $d\mathbf{Top}_\bullet$ of pointed d-spaces is symmetric dIP4-homotopical (Section 4.3.1). Recall that the P-structure comes directly from that of $d\mathbf{Top}$, adding to the original path-space PX the constant loop at the base-point

$$P(X, x) = (PX, x^P), \qquad x^P = e_X.x \colon \{*\} \to PX.$$

($d\mathbf{Top}_\bullet$ itself can be seen as a category of algebras over $d\mathbf{Top}$, for a monad consistent with P, see Section 5.4.7; but we are not interested in this fact here.)

A monoid or a group in $d\mathbf{Top}_\bullet$ is the same as in $\mathbf{Top}$, which we have already considered. But a semigroup (X, x) in $d\mathbf{Top}_\bullet$ is a *d-topological semigroup X with an assigned idempotent element x*. Sgr-$d\mathbf{Top}_\bullet$ is the category of algebras of the free-semigroup monad over $\mathbf{Top}_\bullet$

$$T \colon d\mathbf{Top}_\bullet \to d\mathbf{Top}_\bullet, \qquad T(X, x) = \textstyle\sum_{n>0} (X, x)^n,$$

where the powers and the sum belong now to the category $d\mathbf{Top}_\bullet$ (recall that the sum of pointed spaces has the base-points identified).

The path functor $P \colon d\mathbf{Top}_\bullet \to d\mathbf{Top}_\bullet$ does not preserve sums; it is a *non-strong* T-functor, via the natural transformation

$$\lambda \colon TP \to PT, \qquad \lambda X \colon \textstyle\sum_{n>0} (P(X, x))^n \to P(\textstyle\sum_{n>0} (X, x)^n),$$
$$\lambda X(a_1, ..., a_n) = \langle a_1, ..., a_n \rangle \colon {\uparrow}\mathbf{I} \to X^n,$$

which is plainly consistent with the identifications in our sums of pointed spaces.

One shows now, as above, that Sgr-d**Top.** is a symmetric dIP4-homotopical category, with path-functor equipped with the pointwise multiplication of paths

$$P^T(X, x, \cdot) = (PX, x^P, \bullet).$$

5.4.6 Strict monoidal categories

Recall that the category **Cat** of small categories and functors is regular symmetric dIP4-homotopical (Section 4.3.2), *with homotopies given by natural transformations.* The structure is based on the interval-object $\mathbf{2} = \{0 \to 1\}$, and $PX = X^{\mathbf{2}}$ is the category of morphisms of X.

The category Mon**Cat** of (small) *strict* monoidal categories and strict monoidal functors can be studied along the same lines as d-topological monoids, in Section 5.4.2: the symmetric dP4-homotopical structure of **Cat** is consistent with the monad, and lifts to a symmetric dIP4-homotopical structure for Mon**Cat**.

First, the category Mon**Cat** is made monadic over **Cat** by the forgetful functor U and its left adjoint F

$$F \colon \mathbf{Cat} \rightleftarrows \mathrm{Mon}\mathbf{Cat} \colon U, \qquad \eta \colon 1 \to UF, \qquad \varepsilon \colon FU \to 1,$$

$$FX = (\Sigma_{n \geqslant 0} X^n, \otimes), \quad (x_1, ..., x_p) \otimes (x_{p+1}, ..., x_n) = (x_1, ..., x_n),$$

$$\eta \colon X \subset UFX, \quad \varepsilon \colon FUA \to A, \quad \varepsilon(x_1, ..., x_n) = x_1 \otimes ... \otimes x_n.$$

FX, the free strict-monoidal category over X, is the sum of the power categories X^n.

Again, the cocylinder $P \colon \mathbf{Cat} \to \mathbf{Cat}$, $PX = X^{\mathbf{2}}$ preserves powers (as a right adjoint) and also sums; it is a strong T-functor (same calculations as in Section 5.4.2), by identifying an n-tuple of isomorphisms in X with an isomorphism of X^n

$$\lambda \colon TP \to PT, \qquad \lambda X \colon \Sigma_{n>0} (PX)^n \cong P(\Sigma_{n>0} X^n),$$

$$\lambda X(a_1, ..., a_n) = \langle a_1, ..., a_n \rangle \colon \mathbf{2} \to X^n.$$

In order to get *monoidal categories* in the usual relaxed sense, one should consider *homotopy coherent algebras* in **Cat**, satisfying the axioms of T-algebra up to specified, coherent and reversible, homotopies.

5.4.7 Objects under A as algebras

Let us assume that the category **A** *has finite sums.* Then, the slice category $\mathbf{A} \backslash A$ of *objects under* A can be viewed as a category of algebras over **A**.

In fact, the (obvious) forgetful functor $U \colon \mathbf{A} \backslash A \to \mathbf{A}$ has the following left adjoint $F \colon \mathbf{A} \to \mathbf{A} \backslash A$, with unit $\eta \colon 1 \to UF$ and counit $\varepsilon \colon FU \to 1$

$$FX = (X + A, j \colon A \subset X + A), \qquad \eta X \colon X \subset X + A,$$
$$\varepsilon(X, t) \colon (X + A, j) \to (X, t), \qquad \varepsilon(x) = x, \qquad \varepsilon(a) = t(a).$$

This yields a monad over $\mathbf{A}$ (where in_j denotes an injection into a categorical sum)

$$T = UF \colon \mathbf{A} \to \mathbf{A}, \qquad TX = X + A,$$
$$\eta X \colon X \subset X + A, \qquad \mu = U\varepsilon F \colon T^2 \to T,$$
$$(\mu X \colon X + A + A \to X + A, \quad \mu.\mathrm{in}_1 = \mathrm{in}_1, \quad \mu.\mathrm{in}_2 = \mu.\mathrm{in}_3 = \mathrm{in}_2).$$

One easily sees that a T-algebra $(X, \tau \colon X + A \to X)$ has just to satisfy $\tau.\eta = \mathrm{id}X$ (the consistency of τ with μ being trivially satisfied) and reduces to an arbitrary object under A, $(X, t \colon A \to X)$, letting $t = \tau.j$. We thus identify $\mathbf{A}^T = \mathbf{A} \backslash A$.

Finally, if $\mathbf{A}$ has a path functor, this description of $\mathbf{A} \backslash A$ as $\mathbf{A}^T$ yields the same results on homotopy as those presented in Section 5.2. A drawback of the present approach is the hypothesis of finite sums, which is not assumed in the previous one.

Dually, *if the category* $\mathbf{A}$ *has finite products*, a slice category $\mathbf{A}/B$ of *objects over* B can be viewed as a category of coalgebras over $\mathbf{A}$, finding again the results for the lifting of the cylinder functor developed in Section 5.2.

5.5 The path functor of differential graded algebras

We now investigate the category $\mathbf{Dga}$ of *differential graded algebras* over a fixed commutative unital ring K. Three sections are devoted to construct a *non-reversible* symmetric dIP2-homotopical structure on this category, along the lines of [G3].

First-order properties are developed in the present section, including the fibre sequence of a morphism. Its study takes advantage of the forgetful functor with values in the category $\mathbf{Dgm} = \mathrm{Ch}^+(K\text{-}\mathbf{Mod})$ of cochain complexes of K-modules, and of the resulting 'relative equivalences' (Section 5.5.2).

Higher homotopy properties of $\mathbf{Dga}$ are dealt with in Section 5.6; in the next, we show how this structure arises from a suitable (nonstandard) structure of $\mathbf{Dgm}$, lifted to its internal semigroups.

In a cochain complex, the degree of an element x is often written as $|x|$, and the corresponding 'sign' as $\varepsilon = (-1)^{|x|}$.

5.5.0 Terminology for the homotopy theory of algebras

First, we must say something about terminology. Homotopy limits of algebras have been named in *two* opposite ways: for instance, a polynomial ring $K[t]$ has been called either the *cylinder* of the ring K, or its *cocylinder (path algebra)*.

The first terminology is *geometric*, viewing rings as representing spaces (and ignoring the fact that this representation is *contravariant*). The second is *structural*, consistent with the fact that the polynomial ring $K[t]$ has faces $\partial^{\pm} \colon K[t] \to K$, which evaluate a polynomial at 0 or 1 and represent the initial or terminal point of a *path* in the affine line K.

The geometric terminology can be found, for instance, in Blackadar [Bl] and Murphy [Mr]; the structural one in Karoubi and Villamayor [KV] and Munkholm [Mn]. Gersten [Ge1, Ge2] uses the structural terms, referred to Karoubi and Villamayor, but says in a note of [Ge1]: '*I believe that a more appropriate term for EA and ΩA* [the cocone and loop algebras] *would have been cone and suspension*'.

Since our approach is based on structural properties of homotopies, *we shall always use the structural terminology*, naming homotopy limits and colimits after their universal properties, in the same way as a *product* of algebras is always called by its structural name, corresponding to the universal property that it satisfies rather then to the *sum* of spaces that it can be thought to represent.

5.5.1 Generalities

Consider the category K-**Dga**, or **Dga**, of (positive) *cochain algebras, or differential graded (associative) algebras*, or dg-algebras for short, over the (commutative, unital) ring K. Such algebras are not assumed to have a unit (see Section 5.6.9 for the unital case).

An object $A = ((A^n), (\partial^n))$ is a positive cochain complex of K-modules (indexed over $\mathbf{Z}$, with $A^n = 0$ for $n < 0$), equipped with a multiplication of graded K-algebras consistent with the differential $\partial^n \colon A^n \to A^{n+1}$

$$\partial(x.y) = \partial x.y + (-1)^{|x|}x.\partial y. \tag{5.46}$$

Dga has a zero object, the zero-algebra 0, and all limits, which are computed componentwise.

The forgetful functor with values in the category of *dg-modules* (or positive cochain complexes of K-modules) will be written as

$$U = |-|: \mathbf{Dga} \to \mathbf{Dgm} \qquad (\mathbf{Dgm} = \mathrm{Ch}^+(K\text{-}\mathbf{Mod})).$$

5.5.2 *Homotopy and relative homotopy*

A *homotopy* in **Dga** is a homotopy of dg-modules which respects the multiplicative structure as follows

$$\varphi: f \to g: A \to B, \qquad \varphi = (f, g, (\varphi^n: A^n \to B^{n-1})),$$
$$-f + g = \varphi^{n+1} \partial^n + \partial^{n-1} \varphi^n, \qquad (5.47)$$
$$\varphi(x.y) = \varphi x.gy + (-1)^{|x|}.fx.\varphi y.$$

The sequence $\varphi^\bullet = (\varphi^n)$ is a map of graded objects, of degree -1, which will be called the *centre* of φ (as in Section 4.4.2); we often write $\varphi(x)$ instead of $\varphi^n(x)$, for $x \in A^n$. These *multiplicative* homotopies cannot be reversed, but reflected (as we shall see), and cannot be concatenated.

We say that two dga-homomorphisms $f, g: A \to B$ are *relatively homotopic*, or *linearly homotopic*, and we write $f \simeq_U g$, if there is a homotopy of dg-modules linking them, possibly not consistent with the product. Or, in other words, if the forgetful functor $|-|: \mathbf{Dga} \to \mathbf{Dgm}$ carries f and g to homotopic maps; this relation is a congruence of categories in **Dga**. In the same way, we say that a homomorphism $f: A \to B$ is a *relative equivalence* if $|f|$ is a homotopy equivalence, i.e. if there is some map $g: |B| \to |A|$ of dg-modules such that $g.|f| \simeq 1$, $|f|.g \simeq 1$ in **Dgm**.

We are going to show, here and in the next section, that **Dga** has a path endofunctor P, representing homotopies, which makes it into a dIP2-homotopical category.

5.5.3 *The path functor*

The path endofunctor P is defined as follows, enriching with a multiplication the path functor of cochain complexes (we let $\varepsilon = (-1)^{|a|}$):

$$(PA)^n = A^n \oplus A^{n-1} \oplus A^n,$$
$$(a, h, b).(c, k, d) = (ac, hd + \varepsilon ak, bd), \qquad (5.48)$$
$$\partial(a, h, b) = (\partial a, -a - \partial h + b, \partial b).$$

We only write down the main verifications, for the associativity of the product and the multiplicativity of the differential (with $\varepsilon = (-1)^{|a|}$, $\varepsilon' = (-1)^{|a'|}$)

$$((a,h,b).(a',h',b')).(a'',h'',b'') = (aa',hb'+\varepsilon ah',bb').(a'',h'',b'')$$
$$= (aa'a'',(hb'+\varepsilon ah')b''+\varepsilon\varepsilon'(aa')h'',bb'b'')$$
$$= (aa'a'',hb'b''+\varepsilon ah'b''+\varepsilon\varepsilon'aa'h'',bb'b''),$$

$$(a,h,b).((a',h',b').(a'',h'',b'')) = (a,h,b).(a'a'',h'b''+\varepsilon'a'h'',b'b'')$$
$$= (aa'a'',h(b'b'')+\varepsilon a(h'b''+\varepsilon'a'h''),bb'b'')$$
$$= (aa'a'',hb'b''+\varepsilon ah'b''+\varepsilon\varepsilon'aa'h'',bb'b'').$$

$$\partial((a,h,b).(a',h',b')) = \partial(aa',hb'+\varepsilon ah',bb')$$
$$= (\partial(aa'),-aa'-\partial(hb'+\varepsilon ah')+bb',\partial(bb'))$$
$$= (\partial(aa'),-aa'-\partial h.b'+\varepsilon h.\partial b'-\varepsilon\partial a.h'-a.\partial h'+bb',\partial(bb')),$$

$$\partial(a,h,b).(a',h',b')+\varepsilon(a,h,b).\partial(a',h',b')$$
$$= (\partial a,-a-\partial h+b,\partial b).(a',h',b') +$$
$$\qquad + \varepsilon.(a,h,b).(\partial a',-a'-\partial h'+b',\partial b')$$
$$= ((\partial a).a',(-a-\partial h+b).b'-\varepsilon(\partial a).h',(\partial b).b') +$$
$$\qquad + \varepsilon(a.\partial a',h.\partial b'+\varepsilon a.(-a'-\partial h'+b'),b.\partial b')$$
$$= (\partial(aa'),-ab'-\partial h.b'+bb'-\varepsilon\partial a.h' +$$
$$\qquad + \varepsilon h.\partial b'-aa'-a.\partial h'+ab',\partial(bb')).$$

The basic structure of P is defined as for cochain complexes:

$$\partial^-: P \to 1, \quad \partial^-(a,h,b) = a \qquad \textit{(lower face)},$$
$$\partial^+: P \to 1, \quad \partial^+(a,h,b) = b \qquad \textit{(upper face)}, \qquad (5.49)$$
$$e: 1 \to P, \quad e(a) = (a,0,a) \qquad \textit{(degeneracy)},$$

but reversion has to be replaced with a reflection (Section 5.5.4).

With this structure, P yields the homotopies we have considered above, in (5.47). Whisker composition (of homotopies and maps) and trivial homotopies work as for cochain complexes

$$k\varphi h = (kfh,(k\varphi^n h),kgh), \qquad 0_f = (f,0,f).$$

It is easy to see that the path endofunctor $P\colon \mathbf{Dga} \to \mathbf{Dga}$ preserves products and equalisers, hence all limits. We will see later that it has a left adjoint (Section 5.6.7).

5.5.4 Reflection

Dga becomes a complete dP1-homotopical category, with the following reflection pair (R, r).

First, the reversor $R \colon \mathbf{Dga} \to \mathbf{Dga}$ is defined by the *opposite* algebra $RA = A^{\mathrm{op}}$, with the same structure of graded module, skew-opposite differential (written ∂^*) and reversed product (written $a * b$)

$$\partial^*(a) = (-1)^{|a|}\partial a, \qquad a * b = b.a,$$

$$\begin{aligned}
\partial^*(a * b) = \partial^*(b.a) &= \varepsilon\eta.\partial(ba) \\
&= \varepsilon\eta(\partial b.a + \eta b.\partial a) = \varepsilon\eta a * (\partial b) + \varepsilon(\partial a) * b \\
&= (\partial^* a) * b + \varepsilon a * (\partial^* b) \qquad (\varepsilon = (-1)^{|a|}, \eta = (-1)^{|b|}).
\end{aligned} \tag{5.50}$$

There is now a *reflection* $r \colon RP \to PR$

$$r \colon (PA)^{\mathrm{op}} \to P(A^{\mathrm{op}}), \quad r(a, h, b) = (b, \varepsilon h, a) \ (\varepsilon = (-1)^{|a|}). \tag{5.51}$$

Indeed, r is consistent with the differential and product, since (for $\varepsilon = (-1)^{|a|}$, $\eta = (-1)^{|c|}$)

$$\begin{aligned}
r\partial^*(a, h, b) &= \varepsilon r(\partial a, -a - \partial h + b, \partial b) \\
&= \varepsilon(\partial b, \varepsilon a + \varepsilon.\partial h - \varepsilon b, \partial a), \\
\partial^* r(a, h, b) &= \partial^*(b, \varepsilon h, a) = (\partial^*(b), -b - \partial^*(\varepsilon h) + a, \partial^*(a)) \\
&= (\varepsilon.\partial b, -b + \partial h + a, \varepsilon.\partial a), \\
r((c, k, d) * (a, h, b)) &= r(ac, hd + \varepsilon ak, bd)) = (bd, \varepsilon\eta(hd + \varepsilon ak), ac) \\
&= (d * b, \varepsilon\eta(d * h + \varepsilon k * a), c * a), \\
r(c, k, d) * r(a, h, b) &= (d, \eta k, c) * (b, \varepsilon h, a) \\
&= (d * b, \eta k * a + \eta d * \varepsilon h, c * a).
\end{aligned}$$

Finally, r satisfies the axioms (1.39)

$$\begin{aligned}
RrR.r(a, h, b) &= RrR(b, \varepsilon h, a) = (a, h, b), \\
r.Re(a) &= r(a, 0, a) = (a, 0, a) = eR(a), \\
\partial^- R.r(a, h, b) &= \partial^- R(b, \varepsilon h, a) = b = R\partial^+(a, h, b).
\end{aligned}$$

5.5.5 Homotopy pullbacks

The homotopy pullback $X = P(f, g)$ of a cospan (f, g) in **Dga**

$$\tag{5.52}$$

exists, and can be constructed as a cocylindrical limit (cf. (1.200)), starting from the path object PB, constructed above (Section 5.5.3). It thus consists of the homotopy pullback X of cochain complexes, enriched with *the* multiplication consistent with p, q and ξ (writing $\varepsilon = (-1)^{|a|}$):

$$X^n = A^n \oplus B^{n-1} \oplus C^n,$$
$$(a, b, c).(a', b', c') = (aa', b.gc' + \varepsilon fa.b', cc'),$$
$$\partial(a, b, c) = (\partial a, -fa + gc - \partial b, \partial c),$$
$$p(a, b, c) = a, \qquad q(a, b, c) = c, \qquad \xi(a, b, c) = b.$$

(5.53)

In particular, the upper h-kernel $E^+ f$ of the morphism $f \colon A \to B$ is the homotopy pullback of f and 0, as in the left diagram below

$$
\begin{array}{ccc}
E^+ f \longrightarrow 0 & \qquad & E^- f \xrightarrow{\ u\ } A \\
{\scriptstyle u}\downarrow \ \ {\scriptstyle \xi}\nearrow \quad \downarrow & & \downarrow \ \ {\scriptstyle \xi}\nearrow \quad \downarrow {\scriptstyle f} \\
A \xrightarrow{\ f\ } B & & 0 \longrightarrow B
\end{array}
$$

(5.54)

$E^+ f$ is given by the following formulas, where $a \in A^n$, $\varepsilon = (-1)^n$

$$(E^+ f)^n = A^n \oplus B^{n-1},$$
$$(a, b).(a', b') = (aa', \varepsilon fa.b'), \qquad \partial(a, b) = (\partial a, -fa - \partial b),$$
$$u(a, b) = a, \qquad \xi(a, b) = b.$$

Symmetrically, the lower h-kernel $E^- f$, displayed in the right diagram above, is computed as:

$$(E^- f)^n = B^{n-1} \oplus A^n,$$
$$(b, a).(b', a') = (b.fa', aa'), \qquad \partial(b, a) = (fa - \partial b, \partial a),$$
$$u(b, a) = a, \qquad \xi(b, a) = b.$$

5.5.6 The fibre sequence

The *lower cocone* algebra of a dg-algebra A is

$$E^- A = E^-(\mathrm{id} \colon A \to A), \qquad (E^+ A)^n = A^{n-1} \oplus A^n.$$

(5.55)

The *loop-algebra*

$$\Omega A = E^+(0 \to A) = E^-(0 \to A)$$

is the shifted chain complex $\Omega|A|$, with a null component in degree 0 and null multiplication

$$(\Omega A)^n = A^{n-1}, \qquad \mathrm{ev}_A : 0 \to 0 : \Omega A \to A,$$

$$a.a' = 0, \qquad \partial^n_{\Omega A}(a) = -\partial^{n-1} a, \qquad \mathrm{ev}_A(a) = a. \tag{5.56}$$

The lower fibre sequence (Section 1.8.5) of the morphism $f : A \to B$ is:

$$\cdots \xrightarrow{\Omega d} \Omega E^- f \xrightarrow{\Omega v} \Omega A \xrightarrow{\Omega f} \Omega B \xrightarrow{d} E^- f \xrightarrow{v} A \xrightarrow{f} B \tag{5.57}$$

$$d(b) = (b, 0), \qquad\qquad v(b, a) = a.$$

5.5.7 *Forgetting multiplication*

When useful, the path-functors of dg-algebras and dg-modules will be written as P_a and P_m, respectively.

We have seen that the forgetful functor $U = |-| : \mathbf{Dga} \to \mathbf{Dgm}$ preserves h-pullbacks (Section 5.5.5). It is actually a dP1-homotopical functor (i.e. a strong dP1-functor which preserves the initial object and h-pullbacks; see Section 1.8.2), when equipped with the following natural isomorphism i and the following identity

$$i : UR \to RU = U, \qquad iA : |A^{\mathrm{op}}| \to |A|, \qquad i^n = (-\mathrm{id})^{[n/2]},$$

$$UP_a = P_m U, \qquad |P_a A| = P_m |A|, \tag{5.58}$$

$$A^0 \xrightarrow{\partial^0} A^1 \xrightarrow{-\partial^1} A^2 \xrightarrow{\partial^2} A^3 \xrightarrow{-\partial^3} A^4 \xrightarrow{\partial^4} \cdots \qquad\qquad |A^{\mathrm{op}}|$$

$$\begin{array}{cccccc} \downarrow 1 & \downarrow 1 & \downarrow -1 & \downarrow -1 & \downarrow 1 & \qquad \downarrow i \end{array} \tag{5.59}$$

$$A^0 \xrightarrow{\partial^0} A^1 \xrightarrow{\partial^1} A^2 \xrightarrow{\partial^2} A^3 \xrightarrow{\partial^3} A^4 \xrightarrow{\partial^4} \cdots \qquad\qquad |A|$$

The only non-trivial fact is the consistency of i with reflection (see (1.59)), which here reduces to the commutativity of the square:

$$\begin{array}{ccc} URP_a & \xrightarrow{iP} RUP_a = RP_m U \\ {\scriptstyle Ur}\downarrow & \qquad\qquad \downarrow{\scriptstyle rU} \\ UP_a R = P_m UR & \xrightarrow{Pi} P_m RU \end{array} \tag{5.60}$$

In fact, if $(a, h, b) \in (PA)^n$ and $\varepsilon = (-1)^n$, $\eta = (-1)^{[n/2]}$, $\eta' = (-1)^{[(n-1)/2]}$, we have

$$rU.iP(a, h, b) = \eta.rU(a, h, b) = (\eta b, -\eta h, \eta a),$$

$$Pi.Ur(a, h, b) = Pi(b, \varepsilon h, a) = (\eta b, \varepsilon \eta' h, \eta a).$$

To show that these two results coincide, start from the (obvious) relation $[n/2] + [(n-1)/2] = n - 1$, which can be rewritten as $[n/2] + 1 = n - [(n-1)/2]$; thus, $-\eta = \varepsilon\eta'$.

5.5.8 Fibre diagrams of dg-algebras

The forgetful functor $U = |-|\colon \mathbf{Dga} \to \mathbf{Dgm}$, being dP1-homotopical, preserves the whole structure based on the path functor, in particular h-kernels, cocone and loop objects, fibre sequences and fibre diagrams (Section 1.8). We can thus make use of the stronger properties of $\mathbf{Dgm}$ – a reversible, symmetric dP4-homotopical category – which have been developed in Chapter 4.

In particular, let us examine the lower fibre diagram of a morphism $f\colon A \to B$ of dg-algebras (see (1.209)), where the $\sharp$-marked squares need not commute

$$
\begin{array}{ccccccccccccc}
\dots\ \Omega E^- f & \xrightarrow{\Omega v} & \Omega X & \xrightarrow{\Omega f} & \Omega Y & \xrightarrow{d} & E^- f & \xrightarrow{v} & X & \xrightarrow{f} & Y \\
{\scriptstyle k_3}\downarrow & {\scriptstyle \sharp} & {\scriptstyle k_2}\downarrow & {\scriptstyle \sharp} & {\scriptstyle k_1}\downarrow & & \| & & \| & & \| \\
\dots\ E^+ v_3 & \underset{v_4}{\dashrightarrow} & E^- v_2 & \underset{v_3}{\to} & E^+ v & \underset{v_2}{\to} & E^- f & \underset{v_1}{\to} & X & \underset{f}{\to} & Y
\end{array}
\tag{5.61}
$$

U transforms this diagram into the (lower) fibre diagram of Uf in $\mathbf{Dgm}$. By Theorem 4.7.5 (dualised), the transformed diagram is commutative up to homotopy and its vertical arrows Uk_i are homotopy equivalences.

5.5.9 Cohomology of dg-algebras

The usual cohomology of cochain algebras yields a covariant functor with values in the category of (positive, associative) graded K-algebras

$$
H^*\colon K\text{-}\mathbf{Dga} \to K\text{-}\mathbf{Gra}.
\tag{5.62}
$$

This can be viewed as a contravariant cohomology theory on the opposite dI1-homotopical category $(K\text{-}\mathbf{Dga})^*$, whose suspension Σ is the loop endofunctor of $K\text{-}\mathbf{Dga}$.

5.6 Higher structure and cylinder of dg-algebras

We complete the structure of $\mathbf{Dga}$ to a symmetric dIP2-homotopical category. After introducing connections and transposition for the path functor, we show that the latter is monoidal, $P(A) \cong \mathbf{P} \otimes A$, with respect to the tensor product of dg-algebras and the dP2-interval $\mathbf{P} = P(K)$.

This object is coexponentiable and gives rise to the cylinder endofunctor $I \dashv P$.

5.6.1 Second order paths

For an arbitrary element ξ of the second-order path object

$$\xi = (a, h, b;\ u, z, v;\ c, k, d) \in (P^2 A)^n, \tag{5.63}$$

$$(P^2 A)^n = (PA)^n \oplus (PA)^{n-1} \oplus (PA)^n =$$
$$(A^n \oplus A^{n-1} \oplus A^n) \oplus (A^{n-1} \oplus A^{n-2} \oplus A^{n-1}) \oplus (A^n \oplus A^{n-1} \oplus A^n),$$

we follow the same convention already used above for chain complexes (4.4.3), representing ξ as a square diagram, with its faces at the boundary of the square

$$
\begin{array}{ccc}
a & \xrightarrow{\ u\ } & c \\
h\downarrow & z & \downarrow k \\
b & \xrightarrow{\ v\ } & d
\end{array}
\qquad\qquad
\begin{array}{c}
\bullet \xrightarrow{\ 1\ } \\
\quad\downarrow 2 \\
\quad
\end{array}
$$

$$\partial_1^-(\xi) = \partial^- P(\xi) = (a, h, b), \qquad \partial_1^+(\xi) = \partial^+ P(\xi) = (c, k, d),$$
$$\partial_2^-(\xi) = P\partial^-(\xi) = (a, u, c), \qquad \partial_2^+(\xi) = P\partial^+(\xi) = (b, v, d).$$

Let us recall that the differential of $P^2 A$ is (cf. Section 4.4.3):

$$\partial(a, h, b;\ u, z, v;\ c, k, d) =$$
$$(\partial a, -a - \partial h + b, \partial b; -a - \partial u + c, \bar{z}, -b - \partial v + d; \partial c, -c - \partial k + d, \partial d)$$

where $\bar{z} = -h + u + \partial z - v + k$.

The product of $P^2 A$ arises from the product of PA (Section 5.5.3) and is (for $\varepsilon = (-1)^{|a|}$):

$$(a, h, b; u, z, v; c, k, d) \cdot (a', h', b'; u', z', v'; c', k', d') =$$
$$= ((a, h, b).(a', h', b');\ (u, z, v).(c', k', d') +$$
$$+ \varepsilon(a, h, b).(u', z', v');\ (c, k, d).(c', k', d')) \tag{5.64}$$
$$= (aa', hb' + \varepsilon ah', bb';\ uc' + \varepsilon au', zd' - \varepsilon uk' + \varepsilon hv' + az',$$
$$vd' + \varepsilon bv';\ cc', kd' + \varepsilon ck', dd').$$

5.6.2 The connections

The maps g^- and g^+ are defined as for chain complexes (in (4.78))

$$g^-(a, h, b) = (a, h, b; h, 0, 0; b, 0, 0),$$
$$g^+(a, h, b) = (a, 0, a; 0, 0, h; a, h, b), \qquad (5.65)$$

$$
\begin{array}{ccc}
a & \xrightarrow{\ h\ } & b \\
{\scriptstyle h}\big\downarrow & 0 & \big\downarrow{\scriptstyle 0} \\
b & \xrightarrow{\ 0\ } & b
\end{array}
\qquad\qquad
\begin{array}{ccc}
a & \xrightarrow{\ 0\ } & a \\
{\scriptstyle 0}\big\downarrow & 0 & \big\downarrow{\scriptstyle h} \\
a & \xrightarrow{\ h\ } & b
\end{array}
$$

We only have to verify that they preserve multiplication. For g^-, writing $\varepsilon = (-1)^{|a|}$, we have:

$$g^-((a, h, b).(a', h', b')) = g^-(aa', hb' + \varepsilon ah', bb')$$
$$= (aa', hb' + \varepsilon ah', bb'; hb' + \varepsilon ah', 0, 0; bb', 0, bb'),$$

$$g^-(a, h, b).g^-(a', h', b')$$
$$= (a, h, b; h, 0, 0; b, 0, 0).(a', h', b'; h', 0, 0; b', 0, b')$$
$$= ((a, h, b).(a', h', b'); (h, 0, 0).(b', 0, b') +$$
$$\qquad + \varepsilon(a, h, b).(h', 0, 0); (b, 0, b).(b', 0, b'))$$
$$= ((aa', hb' + \varepsilon ah', bb'); (hb', 0, 0) + \varepsilon(ah', 0, 0); (bb', 0, bb')).$$

5.6.3 The transposition

Again, the transposition $s\colon P^2 A \to P^2 A$ is defined as for chain complexes (in (4.79)), by a symmetry with respect to the 'main diagonal' (a, z, d) and a sign-change in the middle term

$$s(a, h, b; u, z, v; c, k, d) = (a, u, c; h, -z, k; b, v, d), \qquad (5.66)$$

$$
\begin{array}{ccc}
a & \xrightarrow{\ u\ } & c \\
{\scriptstyle h}\big\downarrow & z & \big\downarrow{\scriptstyle k} \\
b & \xrightarrow{\ v\ } & d
\end{array}
\qquad \longmapsto \qquad
\begin{array}{ccc}
a & \xrightarrow{\ h\ } & b \\
{\scriptstyle u}\big\downarrow & -z & \big\downarrow{\scriptstyle v} \\
c & \xrightarrow{\ k\ } & d
\end{array}
$$

To verify that s preserves the multiplication of $P^2 A$, computed in (5.64)

$$(a, h, b; u, z, v; c, k, d) . (a', h', b'; u', z', v'; c', k', d') =$$
$$= (aa', hb' + \varepsilon ah', bb'; \ uc' + \varepsilon au', zd' - \varepsilon uk' + \varepsilon hv' + az', \quad (5.67)$$
$$vd' + \varepsilon bv'; \ cc', kd' + \varepsilon ck', dd'),$$

it suffices to remark that the interchange h/u, b/c, v/k, $z/-z$, h'/u', ..., $z'/-z'$ in the two given 9-tuples has a similar effect on their product.

To show that the axioms of symmetric dP2-categories are satisfied, we have only to verify the consistency of the transposition with the reflection pair, since the latter is different from that of cochain complexes.

In other words, according to Section 4.1.4, we have to verify that the double reflection $r_2 = Pr.rP \colon RP^2 \to P^2R$ commutes with the transposition s, or more precisely that $r_2.Rs = sR.r_2$. But r_2 acts as below, and the commutation property is obvious:

$$r_2(a, h, b; u, z, v; c, k, d) = (d, \varepsilon k, c; \varepsilon v, z, \varepsilon u; b, \varepsilon h, a), \tag{5.68}$$

$$
\begin{array}{ccc}
a & \xrightarrow{\ u\ } & c \\
{\scriptstyle h}\downarrow & {\scriptstyle z} & \downarrow{\scriptstyle k} \\
b & \xrightarrow[\ v\ ]{} & d
\end{array}
\qquad \longmapsto \qquad
\begin{array}{ccc}
d & \xrightarrow{\ \varepsilon v\ } & b \\
{\scriptstyle \varepsilon k}\downarrow & {\scriptstyle z} & \downarrow{\scriptstyle \varepsilon h} \\
c & \xrightarrow[\ \varepsilon u\ ]{} & a
\end{array}
$$

Since all limits exist and P preserves them, **Dga** is a symmetric dP2-homotopical category. As a consequence, we get the homotopy-preservation property of the endofunctors P, E^α and Ω (the dual statements, for a symmetric dI1-homotopical category, can be found in Section 4.1.5 and Theorem 4.1.7).

5.6.4 *Tensor product*

The tensor product of dg-algebras is given by the tensor product of cochain complexes of K-modules, enriched with the following multiplicative structure

$$
\begin{aligned}
(A \otimes B)^n &= \bigoplus_{p+q=n} (A^p \otimes_K B^q), \\
\partial(a \otimes b) &= \partial a \otimes b + (-1)^{|a|}.a \otimes \partial b, \\
(a \otimes b).(a' \otimes b') &= (-1)^{|b|.|a'|}.aa' \otimes bb'.
\end{aligned}
\tag{5.69}
$$

Its identity is the dg-algebra K (in degree zero). The tensor product is symmetric, under the isomorphism

$$s(A, B) \colon A \otimes B \to B \otimes A, \qquad a \otimes b \mapsto (-1)^{|a|.|b|}.b \otimes a. \tag{5.70}$$

As for K-algebras, one can note that, restricting to the full subcategory of *commutative unital* dg-algebras, $A \otimes B$ is the categorical sum of A and B, with injections

$$
\begin{aligned}
i \colon A \to A \otimes B, &\qquad i(a) = a \otimes 1_B, \\
j \colon B \to A \otimes B, &\qquad j(b) = 1_A \otimes b.
\end{aligned}
$$

This symmetric monoidal structure is not co-closed. A characterisation of the coexponentiable objects A (such that $A \otimes -$ has a left adjoint) for algebras, graded algebras and cochain algebras can be found in [N1, N2].

5.6.5 The co-interval

The path functor P is *monoidal*, i.e. it is produced by a symmetric *dP2-interval* $\mathbf{P}$, as $P(A) = \mathbf{P} \otimes A$. Of course, the notion of dP2-interval is dual to dI2-interval, defined in Section 4.2.8; less specifically, we will also use the term 'co-interval'.

Indeed, let us consider the path-object $\mathbf{P} = P(K)$ of the monoidal unit, and write e_1, e_2 (resp. e) the free generators on K of the component $\mathbf{P}^0$ (resp. $\mathbf{P}^1$)

$$
\begin{aligned}
\mathbf{P} = P(K) = (K^2 \to K \to 0 \to 0 \to \ldots), \\
\partial^0(e_1) = -e, \qquad \partial^0(e_2) = e, \\
e_i.e_i = e_i, \qquad e_i.e_j = 0 \qquad (i \neq j), \\
e_1.e = e = e.e_2, \qquad e_2.e = 0 = e.e_1.
\end{aligned}
\tag{5.71}
$$

For every dg-algebra A, we identify $P(A)$ with $\mathbf{P} \otimes A$, via the natural isomorphism

$$
\begin{aligned}
\mathbf{i}_A : P(A) \to \mathbf{P} \otimes A, \qquad A^n \oplus A^{n-1} \oplus A^n \to (\mathbf{P} \otimes A)^n, \\
\mathbf{i}(a, h, b) = e_1 \otimes a + e_2 \otimes b + e \otimes h,
\end{aligned}
\tag{5.72}
$$

$$
\begin{aligned}
\mathbf{i}((a,h,b).(c,k,d)) &= \mathbf{i}(ac, hd + \varepsilon ak, bd) \\
&= e_1 \otimes (ac) + e_2 \otimes (bd) + e \otimes (hd + \varepsilon ak), \\
\mathbf{i}(a,h,b).\mathbf{i}(c,k,d) & \\
&= (e_1 \otimes a + e_2 \otimes b + e \otimes h).(e_1 \otimes c + e_2 \otimes d + e \otimes k) \\
&= e_1 \otimes (ac) + e \otimes (\varepsilon ak) + e_2 \otimes (bd) + e \otimes (hd), \\
\mathbf{i}(\partial(a,h,b)) &= \mathbf{i}(\partial a, -a - \partial h + b, \partial b) \\
&= e_1 \otimes \partial a + e_2 \otimes \partial b + e \otimes (-a - \partial h + b), \\
\partial(\mathbf{i}(a,h,b)) &= \partial(e_1 \otimes a + e_2 \otimes b + e \otimes h) \\
&= -e \otimes a + e_1 \otimes \partial a + e \otimes b + e_2 \otimes \partial b - e \otimes \partial h.
\end{aligned}
$$

The object $\mathbf{P}$ thus obtains the structure of a symmetric dP2-interval in $(\mathbf{Dga}, \otimes)$, from the symmetric dP2-structure of the functor P

$$
K \underset{e}{\overset{\partial^\alpha}{\rightleftarrows}} \mathbf{P} \overset{g^\alpha}{\rightrightarrows} P(\mathbf{P}) = \mathbf{P} \otimes \mathbf{P},
\tag{5.73}
$$

$$
r : \mathbf{P}^{\mathrm{op}} \to \mathbf{P}, \qquad\qquad s = s(\mathbf{P}, \mathbf{P}) : \mathbf{P} \otimes \mathbf{P} \to \mathbf{P} \otimes \mathbf{P}.
$$

Conversely, some standard calculations show that this co-dioid gives back the symmetric dP2-structure of P, by applying to its structural maps the functor $- \otimes A$.

5.6.6 The tensor algebra of a dg-module

The forgetful functor $U = |-|$ from dg-algebras to dg-modules has a left adjoint F. This carries a dg-module X to the tensor dg-algebra FX, defined as a categorical sum of tensor powers of cochain complexes

$$F: \mathbf{Dgm} \rightleftarrows \mathbf{Dga} : U, \qquad \eta : 1 \to UF, \qquad \vartheta : FU \to 1,$$

$$FX = (\Sigma_{n>0} X^{\otimes n}, \otimes),$$

$$(x_1 \otimes ... \otimes x_p) \otimes (x_{p+1} \otimes ... \otimes x_n) = x_1 \otimes ... \otimes x_n;$$

$$\eta : X \subset UFX; \quad \vartheta : FUA \to A, \ \vartheta(x_1 \otimes ... \otimes x_n) = x_1 \cdot ... \cdot x_n. \tag{5.74}$$

The n-component and differential of FX are:

$$(FX)^n = \bigoplus_{|p|=n} X^{p_1} \otimes X^{p_2} \otimes ... \otimes X^{p_r}$$

$$(|p| = p_1 + ... + p_r),$$

$$\partial(x_1 \otimes ... \otimes x_r) = \Sigma_{i=1,...,r} (-1)^{q_i} x_1 \otimes ... \otimes \partial x_i \otimes ... \otimes x_r \tag{5.75}$$

$$(q_i = \deg x_1 + ... + \deg x_{i-1}).$$

5.6.7 The cylinder functor

The path functor P has a left adjoint I, and the symmetric dP2-structure of P, transferred along the adjunction $I \dashv P$, yields a symmetric dIP2-homotopical structure. The dg-algebra $I(A)$ can be constructed as follows:

$$I(A) = F(I(|A|))/J_A. \tag{5.76}$$

First, we form the cylinder over the underlying dg-module $|A|$

$$(I|A|)^n = A^n \oplus A^{n+1} \oplus A^n,$$

$$\partial^I(a, h, b) = (\partial a - h, -\partial h, \partial b + h) \qquad (n > 0),$$

$$(I|A|)^0 = \mathrm{Cok}(\partial^{-1} : A^0 \to A^0 \oplus A^1 \oplus A^0),$$

$$\partial^{-1}(h) = (-h, -\partial h, h),$$

and then the tensor dg-algebra $F(I|A|)$ over the latter (Section 5.6.6). Finally, we use the product of A to quotient $F(I|A|)$ modulo the bilateral K-ideal J_A of the tensor algebra, spanned by the elements of the following types

$$(a, 0, 0) \otimes (b, 0, 0) - (a.b, 0, 0), \qquad (0, 0, a) \otimes (0, 0, b) - (0, 0, a.b),$$
$$(0, a, 0) \otimes (0, 0, b) + (-1)^{|a|}.(a, 0, 0) \otimes (0, b, 0) - (0, a.b, 0). \tag{5.77}$$

The unit η of the adjunction is

$$\eta A \colon A \to PIA, \qquad \eta(a) = ([a, 0, 0], [0, a, 0], [0, 0, a]),$$

$$\begin{aligned}
\eta(a).\eta(b) &= ([a, 0, 0], [0, a, 0], [0, 0, a]).([b, 0, 0], [0, b, 0], [0, 0, b]) \\
&= ([a, 0, 0] \otimes [b, 0, 0], [0, a, 0] \otimes [0, 0, b] + \\
&\quad + (-1)^{|a|}[a, 0, 0] \otimes [0, b, 0], [0, 0, a] \otimes [0, 0, b]) \\
&= ([ab, 0, 0], [0, ab, 0], [0, 0, ab]) = \eta(a.b), \\
\partial^P(\eta(a)) &= \partial^P([a, 0, 0], [0, a, 0], [0, 0, a]) \\
&= (\partial I[a, 0, 0], -[a, 0, 0] + [0, 0, a] - \partial I[0, a, 0], \partial I[0, 0, a]) \\
&= ([\partial a, 0, 0], -[a, 0, 0] + [0, 0, a] - [-a, -\partial a, a], [0, 0, \partial a]) \\
&= ([\partial a, 0, 0], [0, \partial a, 0], [0, 0, \partial a]) = \eta(\partial a).
\end{aligned}$$

The counit $\zeta A \colon IPA \to A$ is obtained as follows. Take first the counit for dg-modules, $\omega|A| \colon IP|A| \to |A|$

$$\omega|A| \colon (A^n \oplus A^{n-1} \oplus A^n) \oplus (A^{n+1} \oplus A^n \oplus A^{n+1}) \oplus (A^n \oplus A^{n-1} \oplus A^n) \to A^n,$$
$$\omega(a, h, b; \, x, e, y; \, c, k, d) = a + e + d,$$

and recall that $P|A| = |PA|$. Extend ω to the tensor algebra, $\overline{\omega} \colon FI|PA| \to A$, and define ζ as the induced morphism, which exists because $\overline{\omega}$ annihilates over the generators (5.77) of the ideal J_{PA}

$$\begin{aligned}
\overline{\omega}&(((a, h, b), 0, 0) \otimes ((c, k, d), 0, 0) - ((a, h, b).(c, k, d), 0, 0)) \\
&= \omega((a, h, b), 0, 0).\omega((c, k, d), 0, 0) - \omega((a, h, b).(c, k, d), 0, 0)) \\
&= a.c - \omega((ac, hd + \varepsilon ak, bd), 0, 0) = ac - ac = 0, \\
\overline{\omega}&((0, (a, h, b), 0) \otimes (0, 0, (c, k, d)) + \\
&\quad + \varepsilon((a, h, b), 0, 0) \otimes (0, (c, k, d), 0) - (0, (a, h, b).(c, k, d), 0)) \\
&= \omega(0, (a, h, b), 0).\omega(0, 0, (c, k, d) + \\
&\quad + \varepsilon\omega((a, h, b), 0, 0).\omega(0, (c, k, d), 0) - \omega(0, (a, h, b).(c, k, d), 0) \\
&= h.d + \varepsilon a.k - \omega(0, (ac, hd + \varepsilon ak, bd), 0) = 0.
\end{aligned}$$

5.6.8 Free dg-algebras

In the literature, the cylinder IA is usually considered for dg-algebras which are free in some sense (cf. [Mn, AL, Ba]). Indeed, if A is *free over dg-modules*, i.e. a tensor algebra FX for some dg-module X, there is a simple description of IA

$$I(FX) = F(IX), \qquad (5.78)$$

which follows from the relation $|PA| = P|A|$ (Section 5.5.3) and the adjunctions

$$\mathbf{Dgm} \xrightarrow{I_m} \mathbf{Dgm} \xrightarrow{F} \mathbf{Dga} \xrightarrow{I_a} \mathbf{Dga} \qquad (5.79)$$

$$I_a F \dashv U P_a, \qquad FI_m \dashv P_m U, \qquad U P_a = P_m U.$$

5.6.9 Unital dg-algebras

Finally, we briefly consider the category **DGA** of *unital* dg-algebras, always on the commutative, unital ring K.

This category is not pointed: the terminal object is still the null algebra, but the initial object is the unit K of the tensor product, with

$$j_A : K \to A, \qquad \lambda \mapsto \lambda.1_A,$$

(the element $\lambda.1_A$ is necessarily a cocycle, because $\partial(1_A) = \partial(1_A.1_A) = \partial(1_A) + \partial(1_A)$).

DGA is again a dIP2-homotopical category, with the 'same' construction of the path functor and of homotopy pullbacks, but a slightly more complicated description of h-kernels and loop-algebras. For instance, ΩA is no longer trivial in degree zero

$$(\Omega A)^n = K^n \oplus A^{n-1} \oplus K^n,$$

$$\partial^0_{\Omega A}(\lambda, \mu) = -\lambda_A + \mu_A, \qquad \partial^n_{\Omega A} = -\partial^{n-1}_A \qquad (n > 0),$$

$$K \oplus K \xrightarrow{\partial^0} A^0 \xrightarrow{\partial^1} A^1 \xrightarrow{\partial^2} \cdots$$

Its multiplication is:

$$(\lambda, a, \mu).(\lambda', a', \mu') = (\lambda\lambda', \mu'a + \lambda a', \mu\mu').$$

The result is null, unless one factor at least is of degree zero; indeed, an element $(\lambda, a, \mu) \in (\Omega A)^n$ has $a = 0$ in degree 0, and $\lambda = \mu = 0$ in degree > 0.

Here, the forgetful functor to cochain complexes must be replaced with

$$V : \mathbf{DGA} \to \mathbf{Dgm} \backslash K, \qquad V(A) = (|A|, j_A : |K| \to |A|),$$

which *preserves the initial object*, and is dP2-homotopical.

Augmented unital dg-algebras $(A, \zeta \colon A \to K)$ are the *co-pointed* objects of **DGA**. They form a slice category **DGA**\$\backslash K$, which is equivalent to the previous category **Dga** (without unit assumption) by well-known constructions: sending an augmented unital dg-algebra (A, ζ) to $A^- = \mathrm{Ker}(\zeta)$ and adding a unit to a dg-algebra B in the usual way, $B^+ = B \oplus K$.

Loosely speaking, **DGA** is an algebraic dual-counterpart of topological spaces, and **Dga** (or **DGA**\$\backslash K$) of *pointed* topological spaces. But notice that their homotopy structures are non-reversible.

5.7 Cochain algebras as internal semigroups in cochain complexes

We construct now a new symmetric dP4-structure on the category **Dgm** $= \mathrm{Ch}^+(K\text{-}\mathbf{Mod})$. This structure is *not reversible*, but is isomorphic to the standard (reversible) one, studied in Section 4.4, and could be said to be 'weakly reversible'.

Only the symmetric dP2-structure which it contains can be lifted to algebras, producing the symmetric dP2-structure we have already considered: all the rest, including the previous isomorphism and the concatenation of homotopies, is not consistent with the monad which gives rise to dg-algebras – it is only so up to homotopy. The last point makes evident the interest of studying a homotopy relaxation of algebras, like the strongly homotopy associative cochain algebras introduced by Stasheff (see [Sf, G5]).

Again, K is a fixed commutative unital ring.

5.7.1 The standard reversion

The category $\mathbf{A} = \mathrm{Ch}^+ \mathbf{D}$ of (positive) cochain complexes over an additive category $\mathbf{D}$ has a canonical reversible, symmetric dP4-structure

$$\mathbf{A}_0 = (\mathbf{A}, R_0, P, \partial^\alpha, e, \mathbf{r}, g^\alpha, s, Q, c, z), \qquad R_0 = \mathrm{id}\mathbf{A},$$

studied in Section 4.4 (in the case of chain complexes). Here, its reversion (4.71) will be written as:

$$\mathbf{r} \colon PA \to PA, \qquad \mathbf{r}(a, h, b) = (b, -h, a). \tag{5.80}$$

Notice that we are still using the notation based on 'formal variables' (cf. (4.67)), which allows us to compute on direct sums as if $\mathbf{D}$ were a category of modules.

Now, if A is a cochain K-algebra and PA is endowed with the product defined in Section 5.5.3

$$(a, h, b).(c, k, d) = (ac, hd + \varepsilon ak, bd) \qquad (\varepsilon = (-1)^{|a|}), \qquad (5.81)$$

the cochain morphism (5.80) does not respect the product:

$$\mathbf{r}(a, h, b).\mathbf{r}(c, k, d) = (b, -h, a).(d, -k, c) = (bd, -hc - \varepsilon bk, ac),$$
$$\mathbf{r}(ac, hd + \varepsilon ak, bd) = (bd, -hd - \varepsilon ak, ac).$$

We introduce therefore a new dP4-structure on $\mathrm{Ch}^+\mathbf{D}$, replacing the pair $(R_0, \mathbf{r})$ with a non-reversible pair (R, r), so that, when $\mathbf{D} = K\text{-}\mathbf{Mod}$, the restricted symmetric dP2-structure $(R, P, \partial^\alpha, e, r, g^\alpha, s)$ lifts to internal semigroups. The new structure on cochain complexes is isomorphic to the standard one; and, of course, is defined 'as' for cochain algebras (in Section 5.5), forgetting about multiplication but replacing $K\text{-}\mathbf{Mod}$ with an arbitrary additive category.

5.7.2 The skew structure of cochain complexes

The involutive endofunctor $R\colon \mathbf{A} \to \mathbf{A}$ $(\mathbf{A} = \mathrm{Ch}^+\mathbf{D})$ is based on the *opposite* chain complex $RA = A^{\mathrm{op}}$, having the same structure of graded module and a skew-opposite differential, written ∂^*

$$\partial^*(a) = (-1)^{|a|}\,(\partial a). \qquad (5.82)$$

This reversor is isomorphic to the identity, by a natural transformation already considered above, in a slightly different form (cf. (5.58))

$$i\colon R \to \mathrm{id}\mathbf{A}, \qquad i^n = (-1)^{[n/2]}, \qquad (5.83)$$

whose inverse is $Ri = iR\colon \mathrm{id}\mathbf{A} \to R$.

There is now a reflection $r\colon RP \to PR$ for cochain complexes

$$r\colon (PA)^{\mathrm{op}} \to P(A^{\mathrm{op}}), \qquad r(a, h, b) = (b, (-1)^{|a|}h, a), \qquad (5.84)$$

which comes from the original reversion $\mathbf{r}\colon P \to P$ and the natural isomorphism i, in the sense that r makes the following diagram commute (as in (5.60))

$$
\begin{array}{ccc}
RP & \xrightarrow{\ iP\ } & P \\
{\scriptstyle r}\downarrow & & \downarrow{\scriptstyle \mathbf{r}} \\
PR & \xrightarrow[\ Pi\ ]{} & P
\end{array}
$$

The fact that r is consistent with the differential follows from the diagram above (and has also been checked directly, in Section 5.5.4).

Finally, the next proposition says that $\mathbf{A}$ has a new symmetric dP4-structure

$$\mathbf{A}_1 = (\mathbf{A}, R, P, \partial^\alpha, e, r, g^\alpha, s, Q, c, z),$$

strongly isomorphic to the original one (called $\mathbf{A}_0$ in Section 5.7.1).

The new structure only differs from the original (reversible) one by replacing the reversion pair $(\mathrm{id}\mathbf{A}, \mathbf{r})$ with the reflection pair (R, r). The fact that the new structure also agrees with the cylinder is automatic, since all the natural transformations of the cylinder functor can be deduced from the adjunction $I \dashv P$.

5.7.3 Proposition

Let $\mathbf{A}$ be equipped with a symmetric dP4-structure

$$\mathbf{A}_0 = (\mathbf{A}, R_0, P, \partial^\alpha, e, r_0, g^\alpha, s, Q, c, z).$$

Let $R_1\colon \mathbf{A} \to \mathbf{A}$ be another involutive (covariant) endofunctor and $i\colon R_1 \to R_0$ a natural transformation such that

$$(R_1 i R_0).i = \mathrm{id}R_0.$$

Then i has inverse $i^{-1} = R_1 i R_0$. The natural transformation $r_1\colon R_1 P \to PR_1$ which makes the following diagram commute

$$\begin{array}{ccc} R_1 P & \xrightarrow{\;iP\;} & R_0 P \\ {\scriptstyle r_1}\downarrow & & \downarrow{\scriptstyle r_0} \\ PR_1 & \xrightarrow[\;Pi\;]{} & PR_0 \end{array}$$

gives a new symmetric dP4-structure $\mathbf{A}_1$ on $\mathbf{A}$ where the reflection pair (R_0, r_0) is replaced with (R_1, r_1). Furthermore, we have a strong symmetric dP4-functor (Section 4.2.7)

$$(\mathrm{id}\mathbf{A}, i, \mathrm{id}P)\colon \mathbf{A}_0 \to \mathbf{A}_1,$$

which is invertible, with inverse $(\mathrm{id}\mathbf{A}, i^{-1}, \mathrm{id}P)$.

Proof This is a straightforward consequence of the assumptions. □

5.7.4 The monad of cochain algebras

Dga is monadic over **Dgm**, via the monad $T = UF$ given by the forgetful functor $U\colon \mathbf{Dga} \to \mathbf{Dgm}$ and its left adjoint F, described in

Section 5.6.6

$$T = UF \colon \mathbf{Dgm} \to \mathbf{Dgm}, \qquad TX = \Sigma_{n \geqslant 0} \, X^{\otimes n}, \qquad (5.85)$$

$$\eta \colon X \subset UFX, \qquad \mu = U\vartheta F \colon T^2 \to T,$$
$$\mu((x_{11} \otimes \ldots \otimes x_{1p_1}) \otimes \ldots \otimes (x_{n1} \otimes \ldots \otimes x_{np_n})) = x_{11} \otimes \ldots \otimes x_{np_n}.$$

One can now verify that the *non-reversible* symmetric dP2-structure $(R, P, \partial^\alpha, e, r, g^\alpha, s)$ of $\mathbf{Dgm}$ is made consistent with the monad by the natural transformation

$$\lambda \colon TP \to PT, \qquad \lambda X \colon \Sigma_{n>0} \, (PX)^{\otimes n} \to P(\Sigma_{n>0} \, X^{\otimes n}),$$

$$(x_1, z_1, y_1) \otimes \ldots \otimes (x_n, z_n, y_n) \mapsto$$
$$(x_1 \otimes \ldots \otimes x_n, \Sigma_i (\overline{x}_1 \otimes \ldots \otimes \overline{x}_{i-1} \otimes z_i \otimes y_{i+1} \otimes \ldots \otimes y_n), y_1 \otimes \ldots \otimes y_n),$$

where $\overline{x} = (-1)^{|x|}.x$.

In fact, $P^T(X, t) = (PX, Pt.\lambda)$ gives the path-module PX, with the multiplication we have been using above:

$$(x_1, z_1, y_1).(x_2, z_2, y_2) = (x_1.x_2, \ \overline{x}_1.z_2 + z_1.y_2, \ y_1.y_2).$$

One concludes again that the category $\mathbf{Dga} = \mathbf{Dgm}^T$ is symmetric dP2-homotopical, with path functor $\mathbf{P} = P^T$.

5.8 Relative settings based on forgetful functors

The forgetful functor $U \colon \mathbf{Dga} \to \mathbf{Dgm}$ has been used in Section 5.5 to prove higher properties of the fibre sequence of a morphism of differential graded algebras, up to relative equivalences, i.e. maps of $\mathbf{Dga}$ which become homotopy equivalences in the strong domain of cochain complexes.

We now abstract this procedure, which can also be used for other 'weak' homotopy structures, like directed chain complexes (Section 5.8.5), cubical sets (Section 5.8.6) and inequilogical spaces (Section 5.8.7).

5.8.1 The main definitions

A *relative dI-homotopical category* will be a dI1-homotopical category $\mathbf{A}$ equipped with a dI1-homotopical functor $U = (U, i, h)$ (Section 1.7.0)

$$U \colon \mathbf{A} \to \mathbf{B} \qquad (i \colon RU \to UR, \quad h \colon IU \to UI), \qquad (5.86)$$

with values in a symmetric dI4-homotopical category $\mathbf{B}$. Recall that U is a strong dI1-functor (Section 1.2.6) which preserves the terminal object and h-pushouts. It is *not* assumed to be faithful.

A *U-equivalence*, or *relative equivalence*, in $\mathbf{A}$ will be any map f of $\mathbf{A}$ such that $U(f)$ is a past or future homotopy equivalence of $\mathbf{B}$. If $\mathbf{B}$ is reversible (i.e. its reversor is the identity), our condition simply means that $U(f)$ is a homotopy equivalence of $\mathbf{B}$; *in this case*, relative equivalences of $\mathbf{A}$ necessarily satisfy the 'two out of three' property (as homotopy equivalences of $\mathbf{B}$ do), and it may be useful to write $f \simeq_U g$ in $\mathbf{A}$ when $U(f) \simeq U(g)$ in $\mathbf{B}$ (as already done for dg-algebras, in Section 5.5.2).

Given a map $f \colon X \to Y$ in $\mathbf{A}$, the functor U preserves its lower cofibre sequence (1.190) and its lower cofibre diagram (1.192), as well as the upper analogues. By Theorem 4.7.5, a cofibre diagram in $\mathbf{A}$ satisfies the following properties:

(i) its vertical arrows are U-equivalences;
(ii) its image in $\mathbf{B}$ is a diagram commutative up to homotopy.

Dually, a *relative dP-homotopical category* will be a dP1-homotopical category $\mathbf{A}$ equipped with a dP1-homotopical functor $U \colon \mathbf{A} \to \mathbf{B}$, with values in a symmetric dP4-homotopical category.

Finally, a *relative dIP-homotopical category* is a dIP1-homotopical category $\mathbf{A}$ equipped with a dIP1-homotopical functor $U \colon \mathbf{A} \to \mathbf{B}$, with values in a symmetric dIP4-homotopical category.

5.8.2 Lemma

Let $\mathbf{A}$ be a dI2-category and $H \colon \mathbf{A} \to C$ a functor with values in an arbitrary category. Then condition (i) implies condition (ii):

(i) every future equivalence in $\mathbf{A}$ is carried by H to an isomorphism of C;
(ii) every pair of maps $f^-, f^+ \colon X \to Y$ in $\mathbf{A}$ such that there exists a homotopy $f^- \to f^+$ is identified by H.

Proof By hypothesis, there exists a map $\varphi \colon IX \to Y$ in $\mathbf{B}$ such that $\varphi.\partial^\alpha = f^\alpha$. Because of the existence of connections in $\mathbf{A}$, both faces $\partial^\alpha \colon X \to IX$ are embeddings of (past or future) deformation retracts, with the *same* retraction $e \colon IX \to X$ (in (4.29)). Therefore, e is a future equivalence and $H(e)$ is an algebraic isomorphism (of abelian groups).

From the relation $H(e).H(\partial^-) = H(\mathrm{id}X) = H(e).H(\partial^+)$, cancelling the isomorphism $H(e)$, we deduce that $H(\partial^-) = H(\partial^+)$ and finally $H(f^-) = H(f^+)$. $\qquad\qquad\square$

5.8.3 Theorem (Homology theories in the relative setting)

Let $U\colon \mathbf{A} \to \mathbf{B}$ be a relative dI-homotopical category.

Suppose we have a sequence of functors $\uparrow H_n$ and natural transformations h_n

$$\uparrow H_n\colon \mathbf{A} \to \mathrm{p}\mathbf{Ab}, \qquad h_n\colon \uparrow H_n \to \uparrow H_{n+1}\Sigma \qquad (n \in \mathbf{Z}). \qquad (5.87)$$

which satisfy the following axiom of homology theories (Section 2.6.2)

(dhlt.2) (algebraic stability) every component $h_n X\colon \uparrow H_n X \to \uparrow H_{n+1}\Sigma X$ is an algebraic isomorphism (of abelian groups).

Then the following conditions imply that these data also satisfy the homotopy invariance axiom (dhlt.1) and the exactness axiom (dhlt.3), so that they form a (reduced) theory of directed homology on $\mathbf{A}$, as defined in Section 2.6.2:

(i) every U-equivalence $f\colon X \to Y$ in $\mathbf{A}$ is taken by each functor $\uparrow H_n\colon \mathbf{A} \to \mathrm{p}\mathbf{Ab}$ to an algebraic isomorphism;

(ii) every pair of maps $f^-, f^+\colon X \to Y$ in $\mathbf{A}$ such that there exists a homotopy $Uf^- \to Uf^+$ in $\mathbf{B}$ is identified by each functor $\uparrow H_n$;

(iii) for every morphism $f\colon X \to Y$ in $\mathbf{A}$ and every $n \in \mathbf{Z}$, the following sequence is exact (in $\mathrm{p}\mathbf{Ab}$), with $u = \mathrm{hcok}^-(f)\colon Y \to C^- f$

$$\uparrow H_n(X) \xrightarrow{\ f_*\ } \uparrow H_n(Y) \xrightarrow{\ u_*\ } \uparrow H_n(C^- f). \qquad (5.88)$$

Moreover, if $\mathbf{A}$ is a dI2-homotopical category (in its own right), condition (ii) can be omitted.

Proof The homotopy invariance axiom (dhlt.1) is a trivial consequence of (ii). To prove the exactness axiom (dhlt.3) we operate as in Theorem 4.7.6. Consider the initial part of the lower cofibre diagram of f in $\mathbf{A}$, and recall that the right-hand square need not commute

$$
\begin{array}{ccccccccc}
X & \xrightarrow{\ f\ } & Y & \xrightarrow{\ u\ } & C^- f & \xrightarrow{\ d\ } & \Sigma X & \xrightarrow{\ \Sigma f\ } & \Sigma Y \\
\Big\| & & \Big\| & & \Big\| & & \Big\uparrow{\scriptstyle k_1} & \natural & \Big\uparrow{\scriptstyle k_2} \\
X & \xrightarrow[\ f\ ]{} & Y & \xrightarrow[\ u_1\ ]{} & C^- f & \xrightarrow[\ u_2\ ]{} & C^+ u_1 & \xrightarrow[\ u_3\ ]{} & C^- u_2
\end{array}
$$

This diagram is transformed by U into the initial part of the lower cofibre diagram of Uf in **B**, which is a symmetric dI4-homotopical category. By Theorem 4.7.6, the transformed diagram is commutative up to directed homotopy and its vertical arrows Uk_1, Uk_2 are future homotopy equivalences.

Now, applying $\uparrow H_n$ to the diagram above, we get a commutative diagram, by (ii), whose vertical arrows are algebraic isomorphisms, by (i). Moreover, in the lower row of the diagram every map is a (lower or upper) h-cokernel of the preceding one. Therefore, by (iii), this row is transformed by $\uparrow H_n$ into an exact sequence. Finally, the same holds for the upper row, which proves (dhlt.3): the following row is exact

$$\uparrow H_n(X) \xrightarrow{f_*} \uparrow H_n(Y) \xrightarrow{u_*} \uparrow H_n(C^-f) \xrightarrow{d_*} \uparrow H_n(\Sigma X) \xrightarrow{(\Sigma f)_*} \uparrow H_n(\Sigma Y).$$

As to the last assertion, if **A** is a dI2-homotopical it follows from Lemma 4.7.4 that one can insert a homotopy in the right-hand square of the diagram above. Then, applying Lemma 5.8.2 to the functor

$$H = W.\uparrow H_n : \mathbf{A} \to \mathrm{pAb} \to \mathbf{Ab},$$

(where W forgets preorders) condition (i) is sufficient to conclude that H takes the diagram above to a commutative diagram, and therefore $\uparrow H_n$ does also. $\qquad\square$

5.8.4 Differential graded algebras

We end this section with some examples of relative settings. We have seen, in Sections 5.5–5.6, that the category **Dga** of cochain algebras over the commutative unital ring K has a symmetric dIP2-homotopical structure.

Moreover, the forgetful functor to cochain complexes of K-modules, which forgets the multiplicative structure

$$U : \mathbf{Dga} \to \mathbf{Dgm} = \mathrm{Ch}^+(K\text{-}\mathbf{Mod}), \qquad (5.89)$$

is a dP2-homotopical functor (and a lax dI2-functor, with a comparison $I|A| \to |IA|$, cf. Section 1.2.6), with values in a reversible, symmetric dIP4-homotopical category.

It follows that **Dga**, equipped with the forgetful functor U, is a relative dP-homotopical category (Section 5.8.1). Its relative equivalences are those morphisms of dg-algebras which are homotopy equivalences of cochain complexes (forgetting multiplication, for their quasi-inverse and the relevant homotopies). Relative equivalences satisfy the 'two out of three' property.

5.8.5 Directed chain complexes

We have seen, in Section 4.4.5, that the category dCh$_\bullet$**Ab** of (unrestricted) directed chain complexes of abelian groups has a non-symmetric dIP4-homotopical structure.

Let us equip it with the forgetful functor

$$U \colon \mathrm{dCh}_\bullet\mathbf{Ab} \to \mathrm{Ch}_\bullet\mathbf{Ab}, \tag{5.90}$$

which forgets preorders; it is a dIP4-homotopical functor, with values in a reversible, symmetric dIP4-homotopical category.

Therefore, dCh$_\bullet$**Ab** is a relative dIP-homotopical category. Here, a relative equivalence is any morphism of directed chain complexes which is a homotopy equivalence of chain complexes (forgetting preorders). Again, relative equivalences satisfy the 'two out of three' property.

5.8.6 Cubical sets

For the category **Cub** of cubical sets, let us consider, for instance, the *left* dIP1-homotopical structure **Cub**$_L$ (Section 1.6.5), with cylinder $IX = {\uparrow}\mathbf{i}{\otimes}X$. There are no connections, no transposition and no concatenation. While considering the directed geometric realisation (Sections 1.6.7 and 1.7.0)

$$U = {\uparrow}\mathcal{R} \colon \mathbf{Cub}_L \to \mathrm{d}\mathbf{Top}, \tag{5.91}$$

we have already seen that it is a dI1-homotopical functor. We thus have a relative dI-homotopical category **Cub**$_L$.

One can also use a more drastic forgetful functor, the classical geometric realisation $V = \mathcal{R} \colon \mathbf{Cub}_L \to \mathbf{Top}$, with values in a reversible domain. This has advantages, since V-equivalences satisfy the 'two out of three' property, and drawbacks, since V misses information which we may prefer to keep, e.g. with respect to the applications of Chapter 2.

5.8.7 Inequilogical spaces

We have seen, in Section 1.9.1, that the category p**Eql** of *inequilogical spaces* is a cartesian closed dIP1-category, based on the directed interval ${\uparrow}\mathbf{I} = {\uparrow}[0,1]$.

Moreover, it has all limits and colimits ([G11], Theorem 1.5); in particular, a product ΠX_i is the product of the (preordered) supports $X_i^\#$, equipped with the product of all equivalence relation. p**Eql** becomes a dIP2-homotopical category, with the same structure on $\uparrow\mathbf{I}$ as in p**Top** (Section 3.1.3). But paths and homotopies cannot be concatenated, unless we extend maps and homotopies using the 'local' ones ([G11], 2.3).

Here, it will be simpler to consider the forgetful functor

$$U \colon \mathrm{p}\mathbf{Eql} \to \mathrm{p}\mathbf{Top}, \qquad X = (X^\#, \sim_X) \mapsto |X| = X^\# / \sim_X, \qquad (5.92)$$

where $X^\# / \sim_X$ has the induced topology and the induced preorder.

This is a dI2-homotopical functor with values in a symmetric dIP4-homotopical category. First, U preserves all colimits, since it has a right adjoint, $D'(Y) = (Y, =)$. Second, it preserves the cylinder functor: using the fact that $I = - \times \uparrow\mathbf{I}\colon \mathrm{p}\mathbf{Top} \to \mathrm{p}\mathbf{Top}$ is a left adjoint and preserves colimits, we have:

$$UI(X) = U(X^\# \times \uparrow\mathbf{I}, \ \sim_X \times \mathrm{id}\mathbf{I}) = (X^\# / \sim_X) \times \uparrow\mathbf{I} = IU(X). \qquad (5.93)$$

Therefore, $(\mathrm{p}\mathbf{Eql}, U)$ is a relative dI-homotopical category.

6

Elements of weighted algebraic topology

As we have seen, directed algebraic topology studies 'directed spaces' with 'directed algebraic structures' produced by homotopy or homology functors: on the one hand the fundamental *category* (and possibly its higher dimensional versions), and on the other *preordered* homology groups. Its general aim is *modelling non-reversible phenomena*.

We now sketch an enrichment of this subject: we replace the truth-valued approach of directed algebraic topology (where a path is licit or not) with a measure of costs, taking values in the interval $[0, \infty]$ of extended (weakly) positive real numbers. The general aim is, now, *measuring the cost of (possibly non-reversible) phenomena*.

Weighted algebraic topology will study 'weighted spaces', like (generalised) metric spaces, with 'weighted' algebraic structures, like the fundamental *weighted* (or normed) category, defined here, and the *weighted* homology groups, developed in [G10] for weighted (or normed) cubical sets.

Lawvere's generalised metric spaces, endowed with a possibly non-symmetric distance taking values in $[0, \infty]$ (already considered in Section 1.9.6), are a basic setting where weighted algebraic topology can be developed (see Section 6.1). This approach is based on the *standard generalised metric interval* $\delta \mathbf{I}$, with distance $\delta(x, y) = y - x$, if $x \leqslant y$, and $\delta(x, y) = \infty$ otherwise; the resulting cylinder functor $I(X) = X \otimes \delta \mathbf{I}$ has the l_1-type metric (Section 6.2). We define the fundamental weighted category $w\Pi_1(X)$ of a generalised metric spaces, and begin its study (Sections 6.3 and 6.4).

We work with *elementary* and *extended* homotopies, which are given by 1-Lipschitz maps (i.e. weak contractions) and Lipschitz maps, respectively (see Section 6.2). This gives a relative dI-homotopical category $U: \delta\mathbf{Mtr} \subset \delta_\infty\mathbf{Mtr}$. Thus, elementary homotopies are used to define the main homotopical constructions, namely cylinder, cone, suspension and – dually – cocylinder, cocone, loop-object (in the pointed case); and then to obtain the (co)fibration sequence of a map. But extended homotopies can be concatenated, and are essential to obtain the higher order properties of such sequences. Moreover, in the fundamental weighted category, an arrow is a class of *extended* paths, up to *extended* homotopy with fixed endpoints.

We also introduce, in Sections 6.5 and 6.6, the more flexible setting of *w-spaces*, or *spaces with weighted paths*, which can be viewed as a weighted version of d-spaces. This allows us to take on, in Section 6.7, the study of the irrational rotation C*-algebras A_ϑ, obtaining a more precise analogy than that which we have conducted with the cubical set $C_\vartheta = (\square\!\uparrow\!\mathbf{R})/G_\vartheta$ or the corresponding c-set (Section 2.5). Here, we start from the *standard w-line* $w\mathbf{R}$, which assigns a finite weight $w(a) = a(1) - a(0)$ to each *increasing* path $a: \mathbf{I} \to \mathbf{R}$, and $w(a) = \infty$ otherwise (Section 6.5.5). Now, the quotient w-space $W_\vartheta = (w\mathbf{R})/G_\vartheta$ has a non-trivial fundamental weighted monoid (at any point), isomorphic to the additive monoid $G_\vartheta^+ = G_\vartheta \cap \mathbf{R}^+$ with the natural weight $w(x) = x$. We prove in Theorems 6.7.3 and 6.7.4 that the *irrational rotation w-space* W_ϑ has the same classification up to isometric isomorphism (resp. Lipschitz isomorphism) as the C*-algebra A_ϑ up to isomorphism (resp. strong Morita equivalence).

We end with some hints about a possible formal setting for weighted algebraic topology (Section 6.8).

This chapter is an adaptation of two papers, [G19, G20], to the approach developed in the previous part. A previous paper [G10] already contains some of these concepts, based on 'normed cubical sets' (which would be called here *weighted* cubical sets).

6.1 Generalised metric spaces

Lawvere's generalised metric spaces [Lw1] have already been briefly considered in Section 1.9.6 and will be used below as a first setting for weighted algebraic topology.

After considering some of their basic properties, we develop some new points, like the standard models (Section 6.1.5), the *reflective* symmetric distance (Section 6.1.6), the length of paths (Definition 6.1.8), the associated *symmetric topology* and d-structure (Section 6.1.9).

6.1.1 Real weights

The basic ingredient is the strict symmetric monoidal closed category of extended positive real numbers, introduced by Lawvere [Lw1], which we write $\mathbf{w}^+$ (the original notation is $\mathbf{R}$). It has objects $\lambda \in [0, \infty]$, morphisms $\lambda \geqslant \mu$, and tensor product $\lambda + \mu$ (with $\lambda + \infty = \infty$, for all λ).

As a complete lattice, this category has all limits and colimits, which amount to products and sums

$$
\begin{aligned}
\text{product:} \quad &\sup(\lambda_i) = \vee \lambda_i, \quad \text{terminal object:} \quad 0, \\
\text{sum:} \quad &\inf(\lambda_i) = \wedge \lambda_i, \quad \text{initial object:} \quad \infty.
\end{aligned}
\tag{6.1}
$$

The internal hom is given by *truncated subtraction*, which will be written as a difference:

$$
\lambda + \mu \geqslant \nu \quad \Leftrightarrow \quad \lambda \geqslant hom^+(\mu, \nu) = \nu - \mu.
\tag{6.2}
$$

In other words, as in [Lw1], we write $\nu - \mu$ for $\max(0, \nu - \mu)$. Notice that the 'undetermined form' $\infty - \infty$ is assigned a precise value, namely 0; indeed, $0 + \infty \geqslant \infty$ implies $0 \geqslant \infty - \infty$.

Let $\mathbf{v}$ denote the full subcategory of $\mathbf{w}^+$ on the objects $0, \infty$; in this subcategory, the cartesian product $\max(\lambda, \mu)$ coincides with the tensor product $\lambda + \mu$. Thus, the following embedding of the Boolean algebra $\mathbf{2} = (\{0, 1\}, \leqslant)$ of *truth-values* (which is covariant with respect to the *given* orders)

$$
M : \mathbf{2} \to \mathbf{w}^+, \qquad M(0) = \infty, \quad M(1) = 0,
\tag{6.3}
$$

is strict monoidal with respect to the cartesian product in $\mathbf{2}$ (the operation min) and the cartesian *or* tensor product of $\mathbf{w}^+$. Moreover, M has left and right adjoint

$$
P \dashv M \dashv Q, \qquad P(\lambda) = 1 \Leftrightarrow \lambda < \infty, \quad Q(\lambda) = 1 \Leftrightarrow \lambda = 0.
\tag{6.4}
$$

A function $w \colon A \to [0, \infty]$ defined on a set (or a class) equipped with a partial operation $a * b$, will be said to be *(sub)additive* if $w(a * b) \leqslant w(a) + w(b)$ whenever $a * b$ is defined; and *strictly additive*, or *linear*, if $w(a * b) = w(a) + w(b)$ (again, when $a * b$ is defined). *The main property being the former*, the prefix 'sub' will generally be omitted: for instance, an 'additively weighted' category will have an 'additive' weight function on morphisms (Section 6.3.1), in the first sense.

Occasionally, we shall also use the same category $\mathbf{w} = ([0, \infty], \geqslant)$ equipped with the strict symmetric monoidal closed structure $\mathbf{w}^\bullet$ defined by the *multiplicative* tensor product $\lambda.\mu$. Here we (must) take $\lambda.\infty = \infty$ for all λ, including 0 (since tensoring with *any* element λ must preserve ∞, which is a colimit: the initial object for the 'direction' $\lambda \geqslant \mu$). Again, a *multiplicative* function with values in $\mathbf{w}^\bullet$ will mean a sub-multiplicative one.

6.1.2 Directed metrics

Now, as already recalled in Section 1.9.6, a generalised metric space X in the sense of Lawvere [Lw1], called here a *directed metric space* or δ-*metric space*, is a set X equipped with a δ-*metric* $\delta\colon X \times X \to [0,\infty]$, satisfying *two* of the classical axioms

$$\delta(x,x) = 0, \qquad \delta(x,y) + \delta(y,z) \geqslant \delta(x,z). \tag{6.5}$$

This amounts to a category enriched over the symmetric monoidal closed category $\mathbf{w}^+$ considered above, with $\delta(x,y) = X(x,y)$ the hom-object in $[0,\infty]$. (If the value ∞ is forbidden, δ is often called a *quasi-pseudo-metric*, cf. [Ky]; but including it has crucial advantages, e.g. with respect to limits and colimits.)

$\delta\mathbf{Mtr}$ denotes the category of such δ-metric spaces, with (weak) *contractions* $f\colon X \to Y$, satisfying $\delta(x,x') \geqslant \delta(f(x), f(x'))$ for all $x, x' \in X$; these mappings will also be called *1-Lipschitz maps* or δ-*maps*. Isomorphisms in this category are isometric, and will be called *isometric isomorphisms* or *1-Lipschitz isomorphisms*. Limits and colimits exist and are calculated as in **Set**, with the δ-metric specified in Section 1.9.6.

The reversor endofunctor $R\colon \delta\mathbf{Mtr} \to \delta\mathbf{Mtr}$ is based on the *opposite* δ-metric space $R(X) = X^{\mathrm{op}}$, with the opposite δ-metric, $\delta^{\mathrm{op}}(x,y) = \delta(y,x)$.

A *symmetric* δ-metric, with $\delta = \delta^{\mathrm{op}}$, is the same as an *écart* in Bourbaki [Bk]. In this case, the object X will be called a δ-metric space *with symmetric δ-metric*, even if the analogy with d-spaces would rather require us to speak of a 'reversible' δ-metric space. Unfortunately, clashes of terminology coming from different frameworks cannot be entirely avoided. More generally, according to a general definition (Section 1.2.1), a δ-metric space is *reversive* if it is isometrically isomorphic to its opposite.

The notation $X \leqslant X'$ means that these δ-metric spaces have the same underlying set and $\delta_X \leqslant \delta_{X'}$, or equivalently that the identity of the underlying set is a δ-map $X' \to X$.

A δ-metric space has a canonical preorder (to be used later)

$$x \prec_\infty x' \quad \text{if} \quad \delta(x, x') < \infty. \tag{6.6}$$

A preordered set is the same as a δ-metric space with a truth-valued metric $\delta \colon X \times X \to \mathbf{v}$, which takes values in $\{0, \infty\}$, so that $x \prec x' \Leftrightarrow \delta(x, x') = 0$. The canonical preorder thus gives the left adjoint to this embedding of the category of preordered sets in $\delta\mathbf{Mtr}$.

A δ-metric space X has a past topology and a future topology, defined in Section 1.9.6. *We will not use these constructions*, but a 'symmetric' topology enriched with the previous preorder, or with a d-structure (see Section 6.1.9).

6.1.3 Lipschitz maps

We now introduce the wider category $\delta_\infty \mathbf{Mtr}$ of δ-metric spaces and *Lipschitz maps*, also called δ_∞-*maps*.

An arbitrary mapping $f \colon X \to Y$ between δ-metric spaces has a *Lipschitz weight* $||f|| \in [0, \infty]$, defined as:

$$\min\{\lambda \in [0, \infty] \mid \forall x, x' \in X, \ \delta(f(x), f(x')) \leqslant \lambda.\delta(x, x')\}, \tag{6.7}$$

and f is a weak contraction, or 1-Lipschitz, if and only if $||f|| \leqslant 1$.

More generally, we say that f is *Lipschitz*, or a *Lipschitz map*, when $||f||$ is finite. A *Lipschitz isomorphism* will be an isomorphism of this wider category $\delta_\infty \mathbf{Mtr}$. The latter is *finitely* complete and cocomplete, and the inclusion $\delta\mathbf{Mtr} \subset \delta_\infty \mathbf{Mtr}$ preserves finite limits and colimits; but, now, the δ-metric of a (co)limit is only determined up to Lipschitz isomorphism.

The category $\delta_\infty \mathbf{Mtr}$ is *multiplicatively weighted* by the Lipschitz weight. By this we mean that every morphism f is assigned a weight $||f|| \in [0, \infty]$ so that:

$$||gf|| \leqslant ||f||.||g||, \qquad ||\mathrm{id}X|| \leqslant 1. \tag{6.8}$$

(These two axioms imply that the weight of an identity can only be 1 or 0.) The subcategory $\delta\mathbf{Mtr}$, characterised by the condition $||f|| \leqslant 1$, inherits the same weight.

If X is a δ-metric space and $\lambda \in [0, \infty[$, we will write λX for the same set equipped with the δ-metric $\lambda.\delta_X$. (Recall that $\lambda.\infty = \infty$, for all λ, cf. Section 6.1.1.) Thus, a δ_∞-map $f \colon X \to Y$ with $||f|| \leqslant \lambda$ is the same as a δ-map $\lambda X \to Y$. More generally, as in [Lw1], one can define λX where $\lambda \colon [0, \infty] \to [0, \infty]$ is any increasing mapping with

$\lambda(0) = 0$ and $\lambda(\mu + \nu) \leqslant \lambda(\mu) + \lambda(\nu)$ (i.e. a lax monoidal functor $\mathbf{w}^+ \to \mathbf{w}^+$). For instance, the square-root mapping gives the δ-metric space $\sqrt{X}$.

6.1.4 Tensor product

The category $\delta\mathbf{Mtr}$ has a 'natural' symmetric monoidal closed structure ([Lw1], p. 153). The tensor product $X \otimes Y$ is the cartesian product of the underlying sets, with the l_1-type δ-metric (instead of the l_∞-type δ-metric of the categorical product)

$$\delta((x, y), (x', y')) = \delta(x, x') + \delta(y, y'). \tag{6.9}$$

It solves the usual universal problem, with respect to mappings which are 1-Lipschitz in each variable. The exponential Z^Y is the set of 1-Lipschitz maps $Y \to Z$ equipped with the δ-metric *of uniform convergence*

$$\delta(h, k) = \sup_y \delta_Z(h(y), k(y)) \qquad\qquad (y \in Y),$$
$$= \sup_{yy'} (\delta_Z(h(y), k(y')) - \delta_Y(y, y')) \quad (y, y' \in Y). \tag{6.10}$$

The proof of the adjunction is standard (and can be deduced from the proof of Theorem 6.5.7). The cartesian and tensor products satisfy the following inequalities

$$X \times Y \leqslant X \otimes Y \leqslant 2.(X \times Y). \tag{6.11}$$

In $\delta_\infty \mathbf{Mtr}$, these products are isomorphic and denote isomorphic functors (in two variables). But, again, we will distinguish such objects (and functors): the notation $X \times Y$ (resp. $X \otimes Y$) will always denote the *realisation* of the cartesian product given by the δ-metric of l_∞-type (resp. l_1-type).

6.1.5 Standard models

The line $\mathbf{R}$ and the standard interval $\mathbf{I}$ have the euclidean metric $|x - y|$. Then, $\mathbf{R}^n$ and $\mathbf{I}^n$ have the *product metric*, $\sup_i |x_i - y_i|$, while $\mathbf{R}^{\otimes n}$ and $\mathbf{I}^{\otimes n}$ have the *tensor product metric*, $\sum_i |x_i - y_i|$.

But we are more interested in the following *non-symmetric* δ-metrics. The *standard δ-line* $\delta\mathbf{R}$ has the δ-metric

$$\delta(x, y) = y - x, \quad \text{if } x \leqslant y, \qquad \delta(x, y) = \infty, \quad \text{otherwise.} \tag{6.12}$$

Its associated preorder is the natural order $x \leqslant y$ (cf. Section 6.1.2).

The *standard δ-interval* $\delta\mathbf{I} = \delta[0, 1]$ has the subspace structure of the δ-line. This also provides the cartesian powers $\delta\mathbf{R}^n$, $\delta\mathbf{I}^n$ and the tensor powers $\delta\mathbf{R}^{\otimes n}$, $\delta\mathbf{I}^{\otimes n}$. These δ-metric spaces have a non-symmetric δ-metric (for $n > 0$), but are reversive; in particular, the canonical reflecting isomorphism

$$r \colon \delta\mathbf{I} \to (\delta\mathbf{I})^{\mathrm{op}}, \qquad\qquad r(t) = 1 - t, \qquad\qquad (6.13)$$

will play a role, in *reflecting* paths and homotopies (in the opposite space).

The *standard δ-circle* $\delta\mathbf{S}^1$ will be the coequaliser in $\delta\mathbf{Mtr}$ of each of the following two pairs of maps (equivalently)

$$\partial^-, \partial^+ \colon \{*\} \rightrightarrows \delta\mathbf{I}, \qquad \partial^-(*) = 0, \quad \partial^+(*) = 1,$$

$$\mathrm{id}, f \colon \delta\mathbf{R} \rightrightarrows \delta\mathbf{R}, \qquad f(x) = x + 1.$$

The 'standard realisation' of the first coequaliser is the quotient $(\delta\mathbf{I})/\partial\mathbf{I}$, which identifies the endpoints; $\delta(x, y)$ takes values in $[0, 1[$, and can be viewed as measuring the length of the 'counterclockwise arc' from x to y, with respect to the whole circle. The structure $2\pi.\delta\mathbf{S}^1$ is also of interest: now, arcs are measured in radians.

More generally, the *n-dimensional δ-sphere* will be the quotient of the monoidal δ-cube $\delta\mathbf{I}^{\otimes n}$ modulo its (ordinary) boundary $\partial\mathbf{I}^n$,

$$\delta\mathbf{S}^n = (\delta\mathbf{I}^{\otimes n})/(\partial\mathbf{I}^n) \qquad\qquad (n > 0). \qquad\qquad (6.14)$$

Thus, for $\delta\mathbf{S}^2$, $\delta(x', x'')$ and $\delta(x'', x')$ are respectively the length of the dashed and the dotted path, below

On the other hand, $\delta\mathbf{S}^0 = \{-1, 1\}$ will be given the *discrete* δ-metric, infinite out of the diagonal (so that every mapping from this space to any other is a contraction). All δ-spheres are reversive.

6.1.6 The symmetric case

The full subcategory $\mathbf{Mtr} \subset \delta\mathbf{Mtr}$ of δ-metric spaces with a symmetric δ-metric (Section 6.1.2) is reflective and coreflective in $\delta\mathbf{Mtr}$; the inclusion preserves limits and colimits.

The coreflector, right adjoint to the embedding, is the well-known procedure of symmetrisation $d(x, x') = \max(\delta(x, x'), \delta(x', x))$, based on the least symmetric δ-metric $d \geqslant \delta$. It will not be used here, since (for instance) it transforms the δ-metric of $\delta\mathbf{R}$ into the discrete δ-metric – infinite out of the diagonal.

But we shall frequently use the *reflector*, based on the greatest symmetric δ-metric $!\delta \leqslant \delta$, which will be called the *symmetrised δ-metric* of δ

$$!: \delta\mathbf{Mtr} \to \mathbf{Mtr}, \qquad !(X, \delta) = (X, !\delta),$$

$$!\delta(x, x') = \inf_{\mathbf{x}} \left(\sum_j \left(\delta(x_{j-1}, x_j) \wedge \delta(x_j, x_{j-1}) \right) \right) \qquad (6.15)$$

$$(\mathbf{x} = (x_0, ..., x_p), \qquad x_0 = x, \qquad x_p = x'),$$

The associated topology will be called the *symmetric topology* of the δ-metric space X; it is the one we are interested in.

This operation carries the δ-metric of $\delta\mathbf{R}$ to the euclidean metric. On $\delta\mathbf{R}^n$ the reflector gives a δ-metric $!(\delta\mathbf{R}^n)$ with ε-disc as in the second figure below, the convex hull of $[-\varepsilon, 0]^n \cup [0, \varepsilon]^n$

$$(!\delta\mathbf{R})^2 \qquad !(\delta\mathbf{R}^2) \qquad \begin{array}{c} !(\delta\mathbf{R}^{\otimes 2}) \\ = (!\delta\mathbf{R})^{\otimes 2} \end{array} \qquad 2.(!\delta\mathbf{R})^2 \qquad (6.16)$$

while on the tensor powers $\delta\mathbf{R}^{\otimes n}$ it gives precisely the l_∞-metric $(!\delta\mathbf{R})^{\otimes n}$, with ε-disc as above, in the third figure. All these δ-metrics are Lipschitz-equivalent (i.e. give Lipschitz isomorphic objects), as follows from comparing the discs above, and from the following more general result.

6.1.7 Proposition (Symmetrisation and products)

Given a finite family of δ-metric spaces $X_1, ..., X_n$ we have the following inequalities (or equalities) for the δ-metrics obtained by symmetrisation, product and tensor product (on the cartesian product of the underlying sets $|X_1| \times ... \times |X_n|$)

$$\Pi(!X_i) \leqslant !(\Pi X_i) \leqslant !(\otimes X_i) = \otimes(!X_i) \leqslant n.\Pi(!X_i). \qquad (6.17)$$

Therefore all these symmetric δ-metrics are Lipschitz-equivalent and induce the same topology.

Proof An element of $\Pi_i|X_i|$ is written as $\mathbf{x} = (x_i)$. Recall the notation $\delta_X(x, x') = X(x, x')$, which comes from viewing a δ-metric space as an enriched category (Section 6.1.2). The only non-standard point of the argument is the 'backward' inequality for the tensor product, proved in (c).

(a) First, to compare $\Pi(!X_i)$ and $!(\Pi X_i)$, note that

$$\Pi(!X_i)(\mathbf{x}, \mathbf{y}) = \sup_i (!X_i)(x_i, y_i) \leqslant \sup_i X_i(x_i, y_i) = (\Pi X_i)(\mathbf{x}, \mathbf{y});$$

since $\Pi(!X_i)$ is symmetric, it follows that $\Pi(!X_i) \leqslant !(\Pi X_i)$.

(b) The second inequality, $!(\Pi X_i) \leqslant !(\otimes X_i)$, is a straightforward consequence of $\Pi X_i \leqslant \otimes X_i$.

(c) We now prove that $!(\otimes X_i) = \otimes(!X_i)$. The δ-metric of the latter is:

$$\otimes(!X_i)(\mathbf{x}, \mathbf{y}) = \Sigma_i(!X_i)(x_i, y_i) \leqslant \Sigma_i X_i(x_i, y_i) = (\otimes X_i)(\mathbf{x}, \mathbf{y}).$$

Since $\otimes(!X_i)$ is symmetric, we have $\otimes(!X_i) \leqslant !(\otimes X_i)$. The opposite inequality is more subtle: take a sequence of $n+1$ points $\mathbf{z}^j$, which varies from $\mathbf{x}$ to $\mathbf{y}$, by changing one coordinate at a time

$$\mathbf{z}^j = (y_1, ..., y_j, x_{j+1}, ..., x_n), \qquad \mathbf{z}^0 = \mathbf{x}, \quad \mathbf{z}^n = \mathbf{y} \qquad (j = 0, ..., n).$$

and apply the triangular inequality:

$$!(\otimes X_i)(\mathbf{x}, \mathbf{y}) \leqslant \Sigma_j !(\otimes X_i)(\mathbf{z}^{j-1}, \mathbf{z}^j).$$

Now, $\mathbf{z}^{j-1}$ and $\mathbf{z}^j$ only differ at the j-th coordinate (x_j or y_j, respectively); restricting the domain of the 'inf' in the right term above to those sequences in $!(\otimes X_i)$ where only the j-th coordinate changes, we get the δ-metric $!X_j$, and the inequality:

$$\Sigma_j !(\otimes X_i)(\mathbf{z}^{j-1}, \mathbf{z}^j) \leqslant \Sigma_j (!X_j)(x_j, y_j) = \otimes (!X_i)(\mathbf{x}, \mathbf{y}).$$

(d) Finally, the last inequality in (6.17) is obvious. $\square$

6.1.8 Definition and Proposition (The length of paths)

Let X be a δ-metric space and $a \colon \mathbf{I} \to X$ a mapping of sets. We define its span *$\mathrm{spn}(a)$ and its* length *$L(a)$ with the following functions, taking values in $[0, \infty]$ (***t*** *stands for a finite strictly increasing sequence (t_i) of points in* $\mathbf{I}$)

$$\mathrm{spn}(a) = \sup_{\mathbf{t}} \delta(a(t_0), a(t_1)) \qquad\qquad (0 \leqslant t_0 < t_1 \leqslant 1),$$
$$L_{\mathbf{t}}(a) = \Sigma_j \delta(a(t_{j-1}), a(t_j)) \qquad (0 = t_0 < t_1 < ... < t_p = 1), \quad (6.18)$$
$$L(a) = \sup_{\mathbf{t}} L_{\mathbf{t}}(a).$$

These functions satisfy the following properties, where $||a||$ is the Lipschitz weight (Section 6.1.3), 0_x is the constant path at a point x, $a + b$ denotes a concatenation of consecutive paths, and Y is another δ-metric space:

(a) $\mathrm{spn}(0_x) = L(0_x) = 0$;

(b) $\mathrm{spn}(a + b) \leqslant \mathrm{spn}(a) + \mathrm{spn}(b)$, $\qquad L(a + b) = L(a) + L(b)$;

(c) $\mathrm{spn}(a\rho) \leqslant \mathrm{spn}(a)$, $\qquad L(a\rho) \leqslant L(a)$
    (for every weakly increasing map $\rho\colon \mathbf{I} \to \mathbf{I}$);

(d) $\mathrm{spn}(a\rho) = \mathrm{spn}(a)$, $\qquad L(a\rho) = L(a)$
    (for every increasing homeomorphism $\rho\colon \mathbf{I} \to \mathbf{I}$);

(e) $\mathrm{spn}(a, b) = \mathrm{spn}(a) + \mathrm{spn}(b)$, $\qquad L(a, b) = L(a) + L(b)$
    (for all paths $(a, b)\colon \mathbf{I} \to X \otimes Y$);

(f) $\mathrm{spn}(a) \leqslant L(a) \leqslant ||a||$;

(g) *L is the least function on 'set-theoretical paths' which is strictly additive for concatenation, invariant for reparametrisation on increasing homeomorphisms $\mathbf{I} \to \mathbf{I}$ and satisfies $L \geqslant \mathrm{spn}$;*

(h) $\mathrm{spn}(f \circ a) \leqslant ||f||.\mathrm{spn}(a)$, $\qquad L(f \circ a) \leqslant ||f||.L(a)$
    (for all δ_∞-maps $f\colon X \to Y$);

(i) *for a mapping $a\colon \mathbf{I} \to \delta\mathbf{R}^{\otimes n}$, we have:*
    $$L(a) = \mathrm{spn}(a) = \Sigma_i(a_i(1) - a_i(0)) \text{ if } a \text{ is (weakly) increasing,}$$
    and $L(a) = \infty$ otherwise.

Finally, note that the length $L(a)$ can be finite even when a is not Lipschitz, i.e. $||a|| = \infty$.

Proof The properties of the span being obvious, we only verify those of the length. Note that, if the partition $\mathbf{t}'$ is finer than $\mathbf{t}$, then $L_{\mathbf{t}}(a) \leqslant L_{\mathbf{t}'}(a)$, because of the triangular property of δ-metrics.

Point (a) is obvious. For (b), the inequality $L(a + b) \leqslant L(a) + L(b)$ follows easily from the previous remark: given a partition $\mathbf{t}$ for $c = a + b$, call $\mathbf{t}'$ its refinement by introducing the point $t = 1/2$ (if missing); thus $L_{\mathbf{t}}(c) \leqslant L_{\mathbf{t}'}(c)$ and the latter term can be split into two summands $\leqslant L(a) + L(b)$. For the other inequality, it is sufficient to note that a partition for a and one for b yield a partition of $[0, 2]$, which can be scaled down to the standard interval.

Point (c) is obvious, since $L(a\rho)$ is computed on the partitions of $\rho[0, 1] \subset [0, 1]$; (d) is a consequence. For (e), the inequality $L(a, b) \leqslant L(a) + L(b)$ is obvious, and the other follows again from the first remark: given a partition for a and one for b, by using a common refinement for both we get higher values.

Finally, points (f)–(h) are plain; (i) is obvious for $n = 1$, and follows from (e) for higher n. For the last remark, taking $X = \delta\mathbf{R}$, the square-root map $f\colon \mathbf{I} \to \mathbf{R}$ is not Lipschitz but has a finite length, as any increasing path: $L(f) = f(1) - f(0) = 1$. □

6.1.9 The associated topology and direction

A δ-metric space X will be equipped with the *symmetric topology*, defined by the symmetric δ-metric $!\delta$ (Section 6.1.6). In other words, we are composing the functor $!\colon \delta\mathbf{Mtr} \to \mathbf{Mtr}$ with the ordinary forgetful functor $\mathbf{Mtr} \to \mathbf{Top}$. Furthermore, we will keep a trace of the 'directed' information of the original δ in two main ways, based on the previous settings for directed algebraic topology.

The simplest, if a rather poor way, is by the *associated preorder* $x \prec_\infty y$, defined by $\delta(x, y) < \infty$ (see (6.6)). We have thus a forgetful functor with values in the category p$\mathbf{Top}$ of preordered spaces and continuous preorder-preserving mappings

$$\mathbf{p}\colon \delta\mathbf{Mtr} \to \mathrm{p}\mathbf{Top}, \qquad (\delta_\infty \mathbf{Mtr} \to \mathrm{p}\mathbf{Top}), \qquad (6.19)$$

which preserves finite products, since the symmetrising functor $!\colon \delta\mathbf{Mtr} \to \mathbf{Mtr}$ preserves finite products up to Lipschitz isomorphism (6.1.7), and $\delta(x, y) < \infty$ if and only if this holds on all components.

Thus $\mathbf{p}(\delta\mathbf{R}^n) = \mathbf{p}(\delta\mathbf{R}^{\otimes n}) = \uparrow\mathbf{R}^n$ is the euclidean n-dimensional space with the product order. Similarly, $\mathbf{p}(\delta\mathbf{I}^n) = \mathbf{p}(\delta\mathbf{I}^{\otimes n}) = \uparrow\mathbf{I}^n$. But a preorder is a poor way of describing direction, which does not allow for non-reversible loops. For instance, $\mathbf{p}(\delta\mathbf{S}^1)$ gets the indiscrete preorder and misses any information of direction.

A more accurate way of keeping the 'directed' information of the original δ is using d-spaces. We now have a forgetful functor

$$\mathbf{d}\colon \delta\mathbf{Mtr} \to \mathrm{d}\mathbf{Top}, \qquad (\delta_\infty \mathbf{Mtr} \to \mathrm{d}\mathbf{Top}), \qquad (6.20)$$

which equips a δ-metric space X with the associated symmetric topology *and* the d-structure where a (continuous) path $a\colon \mathbf{I} \to X$ is distinguished if and only if it is *L-feasible*, i.e. it has a finite length $L(a)$ (see (6.18)).

The axioms of d-spaces are satisfied (by Proposition 6.1.8): distinguished paths contain all the constant ones, are closed under concatenation and partial reparametrisation by weakly increasing maps $\mathbf{I} \to \mathbf{I}$. And, of course, a Lipschitz map $f\colon X \to Y$ of δ-metric spaces preserves feasible paths. It will be important to note that this functor takes the tensor (or cartesian) product of $\delta_\infty \mathbf{Mtr}$ (or $\delta \mathbf{Mtr}$) to the cartesian product of d-spaces (where a path is distinguished if and only if its two components are).

Now, in $d\mathbf{Top}$, $\mathbf{d}(\delta \mathbf{S}^1) = \uparrow\mathbf{S}^1 = \uparrow\mathbf{I}/\partial\mathbf{I}$ is the standard directed circle (Section 1.4.3), where paths are only allowed to turn in a given direction. In fact, it will be sufficient to consider continuous mapping $a\colon \mathbf{I} \to \mathbf{I}$ its image in the circle $a'\colon \mathbf{I} \to \mathbf{I}/\partial\mathbf{I}$ and $\mathbf{H}$. If a is increasing, $L(a') = L(a) = a(1) - a(0)$ is a 'finite' number L. Otherwise, it is easy to see that $L(a') = \infty$.

The functor $\mathbf{d}$ need not preserve quotients: for instance, $\mathbf{d}(\delta\mathbf{R}) = \uparrow\mathbf{R}$ is the standard directed line, with the increasing paths as distinguished ones; the quotient d-space $\uparrow\mathbf{R}/G_\vartheta$, modulo the action of the dense subgroup $G_\vartheta = \mathbf{Z} + \vartheta\mathbf{Z}$ (for an irrational number ϑ), is a non-trivial object, with the homology of the two-dimensional torus (Section 2.5.2), while $(\delta\mathbf{R})/G_\vartheta$ has a trivial δ-metric, always zero. A finer notion of 'weighted space', studied in Sections 6.5–6.7, will be able to express such phenomena within *weighted* algebraic topology (not just the *directed* one).

Note that the forgetful functor $d\mathbf{Top} \to p\mathbf{Top}$ provided by the path-preorder $x \preceq x'$ (there is a distinguished path from x to x'), applied to a d-space of type $\mathbf{d}X$, gives a *finer* preorder than $\mathbf{p}X$, generally more interesting than the latter (two points at a finite distance may be disconnected, or not linked by a feasible path.)

6.2 Elementary and extended homotopies

The standard δ-interval $\delta\mathbf{I}$ generates a cylinder endofunctor, which yields *elementary* homotopies in $\delta\mathbf{Mtr}$, and *extended* homotopies in $\delta_\infty\mathbf{Mtr}$. We need both.

6.2.1 Elementary and extended paths

Let X be a δ-metric space. An *elementary path* (resp. an *extended path*, or *Lipschitz path*) in X will be a 1-Lipschitz (resp. a Lipschitz) map $a\colon \delta\mathbf{I} \to X$. Applying the forgetful functors that we have seen in the previous section (cf. (6.20)), we always have an associated path $\mathbf{d}a\colon \uparrow\mathbf{I} \to \mathbf{d}X$.

Thus, a set-theoretical mapping $a\colon \mathbf{I} \to X$ is an elementary path if and only if $||a|| \leqslant 1$, for the Lipschitz weight (6.7)

$$||a|| = \min\{\lambda \in [0, \infty] \mid \forall t \leqslant t',\ \delta(a(t), a(t')) \leqslant \lambda.(t' - t)\}, \qquad (6.21)$$

and is an extended path if and only if $||a|| < \infty$. Elementary paths cannot be concatenated, because – loosely speaking – this procedure doubles the velocity, whose least upper bound is the Lipschitz weight. Recall that the (finite) length $L(a) \leqslant ||a||$ has been defined in Definition 6.1.8, and that a continuous path in X, of finite length, need not be Lipschitz.

The *reflected* (elementary or extended) path is obtained in the obvious way

$$a^{\mathrm{op}} = ar \colon \delta\mathbf{I} \to X^{\mathrm{op}}, \qquad r(t) = 1 - t, \tag{6.22}$$

A *reversible* extended path is a mapping $a \colon \mathbf{I} \to X$ such that both a and a^{op} are extended paths $\delta\mathbf{I} \to X$. This amounts to a Lipschitz map $a \colon !\delta\mathbf{I} \to X$, with respect to the ordinary metric $|t - t'|$ of the euclidean interval.

6.2.2 The elementary cylinder

The symmetric monoidal closed category $\delta\mathbf{Mtr}$ has a symmetric monoidal dIP2-homotopical structure, which arises from the δ-interval $\delta\mathbf{I}$.

Indeed, the latter is a monoidal symmetric dIP2-interval in $\delta\mathbf{Mtr}$, with the usual structural maps (for $\alpha = \pm$)

$$\{*\} \mathrel{\mathop{\rightrightarrows}^{\partial^\alpha}_{e}} \delta\mathbf{I} \mathrel{\mathop{\leftleftarrows}^{g^\alpha}} \delta\mathbf{I}^{\otimes 2} \qquad r \colon \delta\mathbf{I} \to \delta\mathbf{I}^{\mathrm{op}}, \quad s \colon \delta\mathbf{I}^{\otimes 2} \to \delta\mathbf{I}^{\otimes 2}. \tag{6.23}$$

$$\partial^\alpha(*) = \alpha, \qquad g^-(t,t') = \max(t,t'), \qquad g^+(t,t') = \min(t,t'),$$
$$r(t) = 1 - t, \qquad\qquad s(t,t') = (t',t).$$

We thus have the *elementary cylinder* endofunctor

$$I \colon \delta\mathbf{Mtr} \to \delta\mathbf{Mtr}, \qquad I(-) = - \otimes \delta\mathbf{I}, \tag{6.24}$$

with natural transformations, denoted, as always, by the same symbols and names

$$1 \mathrel{\mathop{\rightrightarrows}^{\partial^\alpha}_{e}} I \mathrel{\mathop{\leftleftarrows}^{g^\alpha}} I^2 \qquad r \colon IR \to RI, \qquad s \colon I^2 \to I^2. \tag{6.25}$$

Its right adjoint, the *elementary-path functor*, or *elementary cocylinder*, is

$$P \colon \delta\mathbf{Mtr} \to \delta\mathbf{Mtr}, \qquad P(Y) = Y^{\delta\mathbf{I}}. \tag{6.26}$$

The δ-metric space $Y^{\delta I}$ is the set of elementary paths $\delta\mathbf{Mtr}(\delta I, Y)$ with the δ-metric of uniform convergence (6.10). The lattice structure of δI in $d\mathbf{Top}$ gives – contravariantly – a dual structure on P.

The name 'elementary paths' is meant to suggest that such paths cannot be concatenated as such, as we consider now.

6.2.3 The Lipschitz cylinder

The category $\delta_\infty \mathbf{Mtr}$ has a symmetric dI4-homotopical structure, based on the *Lipschitz cylinder* endofunctor

$$I\colon \delta_\infty \mathbf{Mtr} \to \delta_\infty \mathbf{Mtr}, \qquad I(-) = - \otimes \delta I. \qquad (6.27)$$

After the basic part, which is the same as above, we have to consider concatenation. Given two consecutive Lipschitz paths $a, b\colon \delta I \to X$, with $a(1) = b(0)$, the usual construction gives a concatenated path

$$a + b\colon \delta I \to X, \qquad ||a + b|| \leqslant 2.\max(||a||, ||b||), \qquad (6.28)$$

(which need not be elementary, when a and b are). As usual, this can be dealt with by a concatenation pushout.

In the category $\delta\mathbf{Mtr}$, pasting two copies of the standard δ-interval, one after the other, can be realised as $\delta[0, 2] \subset \delta\mathbf{R}$, or (isometrically) as $2.\delta I$ (with the double δ-metric)

$$
\begin{array}{ccc}
\{*\} & \xrightarrow{\partial^+} & \delta I \\
\partial^- \downarrow & {}^{-}\!\!\dashv\!\!\downarrow c^- & \\
\delta I & \xrightarrow{c^+} & 2.\delta I
\end{array}
\qquad
\begin{array}{l}
c^-(t) = t/2, \\[2mm]
c^+(t) = (t+1)/2.
\end{array}
\qquad (6.29)
$$

Of course, this is of no help to concatenate *elementary* paths. *But, in the category $\delta_\infty \mathbf{Mtr}$, this pushout can be realised as the Lipschitz-isomorphic object δI. This yields the left diagram below, which will be called the standard concatenation pushout in $\delta_\infty \mathbf{Mtr}$ (it lives in $\delta\mathbf{Mtr}$, but is not a pushout there)*

$$
\begin{array}{ccc}
\{*\} & \xrightarrow{\partial^+} & \delta I \\
\partial^- \downarrow & {}^{-}\!\!\dashv\!\!\downarrow c^- & \\
\delta I & \xrightarrow{c^+} & \delta I
\end{array}
\qquad
\begin{array}{ccc}
X & \xrightarrow{\partial^+} & X \otimes \delta I \\
\partial^- \downarrow & {}^{-}\!\!\dashv\!\!\downarrow c^- & \\
X \otimes \delta I & \xrightarrow{c^+} & X \otimes \delta I
\end{array}
\qquad (6.30)
$$

$$c^-(x, t) = (x, t/2), \qquad c^+(x, t) = (x, (t+1)/2).$$

This pushout is preserved by any functor $X \otimes -$, yielding the right-hand pushout above (or, equivalently, by $X \times -$). In fact, $X \otimes -\colon \delta\mathbf{Mtr} \to \delta\mathbf{Mtr}$ preserves the pushout (6.29), as a left adjoint; and the embedding $\delta\mathbf{Mtr} \to \delta_\infty \mathbf{Mtr}$ preserves finite colimits (Section 6.1.3).

Now we complete the dI4-homotopical structure of $\delta_\infty \mathbf{Mtr}$, as for d-spaces: we take $J = I$, $c = \mathrm{id}I$ and define the acceleration $\zeta\colon I^2 \to I$ by $zX = X \otimes \zeta$, where $\zeta\colon \delta\mathbf{I} \otimes \delta\mathbf{I} \to \delta\mathbf{I}$ is defined in (4.52) (and is Lipschitz).

The embedding $U\colon \delta\mathbf{Mtr} \to \delta_\infty \mathbf{Mtr}$ is a dI2-homotopical functor. It gives to $\delta\mathbf{Mtr}$ the structure of a relative dI-homotopical category (Section 5.8.1), with relative equivalences consisting of the δ-maps which are homotopy equivalences in $\delta_\infty \mathbf{Mtr}$.

6.2.4 Homotopies

An *elementary homotopy* $\varphi\colon f \to g\colon X \to Y$ is defined as a δ-map $\varphi\colon IX = X \otimes \delta\mathbf{I} \to Y$ whose two faces are f and g, respectively: $\partial^-(\varphi) = \varphi\partial^- = f$, $\partial^+(\varphi) = \varphi\partial^+ = g$. In particular, an elementary path is a homotopy between two points, $a\colon x \to x'\colon \{*\} \to X$.

More generally, an *extended homotopy*, or *Lipschitz homotopy*, is a Lipschitz map $\varphi\colon X \otimes \delta\mathbf{I} \to Y$; an extended path is an extended homotopy between two points, $a\colon x \to x'\colon \{*\} \to X$. (Note that a Lipschitz map defined on the singleton is always a δ-map.)

An extended homotopy has a Lipschitz weight $||\varphi||$, which is $\leqslant 1$ if and only if φ is elementary.

Reflected homotopies (elementary or extended) live in the opposite 'spaces' (as for paths, in (6.22))

$$\varphi^{\mathrm{op}}\colon Rg \to Rf\colon RX \to RY, \qquad \varphi^{\mathrm{op}} = R\varphi.rX : I(RX) \to RY. \quad (6.31)$$

In both cases, elementary and extended, the main operations given by the cylinder functor (for $\varphi\colon f \to g\colon X \to Y$; $h\colon X' \to X$; $k\colon Y \to Y'$; $\psi\colon g \to h\colon X \to Y$) are:

(a) *whisker composition* of (elementary or extended) maps and homotopies

$$k \circ \varphi \circ h\colon kfh \to kgh \qquad (k \circ \varphi \circ h = k.\varphi.Ih\colon IX' \to Y');$$

(b) *trivial homotopies*: $\qquad 0_f\colon f \to f \qquad (0_f = fe\colon IX \to Y);$

and satisfy the axioms of dh1-categories (Section 1.2.9), for associativity, identities and reflection.

In $\delta_\infty \mathbf{Mtr}$ consecutive homotopies can be pasted via the *concatenation pushout* of the cylinder functor (the right-hand diagram in (6.30)). The concatenation $\varphi + \psi$ of two consecutive homotopies $(\partial^+ \varphi = \partial^- \psi)$ is thus computed as usual:

$$(\varphi + \psi)(x, t) = \begin{cases} \varphi(x, 2t), & \text{for } 0 \leqslant t \leqslant 1/2, \\ \psi(x, 2t - 1), & \text{for } 1/2 \leqslant t \leqslant 1. \end{cases} \qquad (6.32)$$

As always in directed algebraic topology, *homotopy equivalence is a complex notion*, which has to be considered not only for 'spaces' but also for their algebraic counterpart – weighted categories. This will be briefly considered in Section 6.4.

Extended double homotopies and 2-homotopies are based on the *second-order cylinder* $I^2 X = X \otimes \delta \mathbf{I}^2$, and treated as in any dI4-category: see Sections 4.1.1 and 4.5.

All the homotopy constructions of Chapter 1 can now be performed, in $\delta \mathbf{Mtr}$ and (consistently) in $\delta_\infty \mathbf{Mtr}$: homotopy pushouts and pullbacks; mapping cones, suspension and cofibration sequences; homotopy fibres, loop-objects and fibration sequences (in the pointed case). The higher properties of this machinery, as studied in Chapter 4, need concatenation, and work in $\delta_\infty \mathbf{Mtr}$, but can be partially extended to $\delta \mathbf{Mtr}$ in the relative sense of Section 5.8.

6.3 The fundamental weighted category

After sketching the theory of weighted categories and their elementary and extended homotopies, we will define the fundamental weighted category of a δ-metric space. Non-obvious computations are based on a van Kampen-type theorem (Theorem 6.3.8).

Small categories are generally written in additive notation.

6.3.1 Weighted categories

An *additively weighted category* will be a category X where every morphism a is equipped with a *weight*, or *cost*, $w(a) \in [0, \infty]$ (or $w_X(a)$), so that two obvious axioms are satisfied, for identities and composition (written in additive notation):

$(\mathrm{w}^+ \mathrm{cat}.0)$ $w(0_x) = 0$, for all objects x of X;

$(\mathrm{w}^+ \mathrm{cat}.1)$ $w(a + b) \leqslant w(a) + w(b)$, for all consecutive arrows a, b.

This was called a 'normed category' by Lawvere (see [Lw1]), but we also have to consider *multiplicatively* weighted categories, as δ_∞ **Mtr** (Section 6.1.3). We will generally let the term 'additive' be understood, and specify the term 'multiplicative' when it is the case.

We also speak of a w^+-*category*, for short. The weight is said to be *linear*, or *strictly additive*, if $w(a + b) = w(a) + w(b)$ for all composites.

A w^+-*functor* $f\colon X \to Y$, or *1-Lipschitz functor*, is a functor between such categories which satisfies the condition $w(f(a)) \leqslant w(a)$, for all morphisms a of X. We write w**Cat**, or w^+**Cat**, the category of (small) w^+-categories and such functors. (Here we do not use the multiplicative analogue w$^\bullet$**Cat**, for which one can see [G19].)

The *opposite weighted category* X^{op} is the opposite category with the 'same' weight.

As for δ-metric spaces, we also need the bigger category w_∞**Cat** of weighted categories and *Lipschitz functors* $f\colon X \to Y$, or w_∞-*functors*, i.e. the functors between weighted categories having a finite Lipschitz weight

$$||f|| = \min\{\lambda \in [0, \infty] \mid \forall a \in \mathrm{Mor}(X),\ w_Y(f(a)) \leqslant \lambda.w_X(a)\}. \quad (6.33)$$

(Lipschitz natural transformations will be defined in Section 6.3.4.) With this weight, the category w_∞**Cat** is *multiplicatively weighted* (cf. Section 6.1.3). This is also true of w**Cat**, which is the subcategory of all the functors f such that $||f|| \leqslant 1$.

A *weighted monoid* is a small weighted category on one (formal) object. We thus have the full subcategories w**Mon** and w_∞**Mon** of w**Cat** and w_∞**Cat**, respectively.

Weighted categories can be viewed as categories enriched over the symmetric monoidal closed category w^+**Set** of *weighted sets*: an object is a set X equipped with a mapping $w\colon X \to \mathbf{w}^+$; a morphism is a mapping $f\colon X \to Y$ between weighted sets, such that $w(f(x)) \leqslant w(x)$ for all $x \in X$. The tensor product $X \otimes Y$ is the cartesian product of the underlying sets, with $w(x, y) = w(x) + w(y)$ (see [G19]).

6.3.2 Proposition (The monoidal closed structure)

The category w**Cat** *has a symmetric monoidal closed structure, with tensor product* $X \otimes Y$ *consisting of the cartesian product of the underlying categories, equipped with the* l_1-*weight on a map* $(a, b)\colon (x, y) \to (x', y')$

$$w_\otimes(a, b) = w(a) + w(b). \quad (6.34)$$

The internal hom Z^Y is the category of 1-Lipschitz functors $h\colon Y \to Z$ and all *natural transformations* $\varphi\colon h \to k$ between such functors, with the (plainly subadditive) weight:

$$W(\varphi) = \sup_y w_Z(\varphi(y)), \qquad (y \in \mathrm{Ob}Y). \qquad (6.35)$$

Proof First, a 1-Lipschitz functor $f\colon X \otimes Y \to Z$ defines a functor $g\colon X \to Z^Y$, sending an object x to the 1-Lipschitz functor

$$g(x) = f(x, -)\colon Y \to Z,$$
$$w_Z(g(x)(b)) = w_Z(f(0_x, b)) \leqslant w_\otimes(0_x, b) = w(b),$$

and the X-morphism $a\colon x \to x'$ to the natural transformation $g(a) = f(a, -)\colon g(x) \to g(x')$. The functor g itself is a contraction:

$$W(g(a)) = \sup_y w_Z(g(a)(y)) = \sup_y w_Z(f(a, 0_y))$$
$$\leqslant \sup_y w_\otimes(a, 0_y) = w(a).$$

Conversely, given a 1-Lipschitz functor $g\colon X \to Z^Y$, we define the functor $f\colon X \otimes Y \to Z$ in the usual, obvious way, and verify that it is 1-Lipschitz, on a map $(a, b)\colon (x, y) \to (x', y')$

$$w_Z(f(a, b)) = w_Z(g(a)(y) + g(x')(b)) \leqslant w_Z(g(a)(y)) + w_Z(g(x')(b))$$
$$\leqslant w(a) + w(b).$$

The last inequality above comes from:

$$w_Z(g(a)(y)) \leqslant \sup_{y'} w_Z(g(a)(y')) = W(g(a)) \leqslant w(a).$$

$\square$

6.3.3 The elementary cylinder of weighted categories

Directed homotopy in w**Cat** is a non-trivial enrichment of what we have already seen in **Cat** (Section 1.1.6).

The *directed interval* w**2** is now the usual order category $\mathbf{2} = \{0 \to 1\}$ on two objects, enriched with the weight $w(0 \to 1) = 1$ on the unique non-trivial arrow. Taking into account the symmetric monoidal closed structure of w**Cat** considered above, w**2** is a monoidal dIP2-interval, and yields a symmetric monoidal dIP2-homotopical structure, made concrete by the monoidal unit **1**.

We thus have the *elementary cylinder* endofunctor of weighted categories

$$I\colon \text{w}\mathbf{Cat} \to \text{w}\mathbf{Cat}, \qquad I(-) = - \otimes \text{w}\mathbf{2}. \qquad (6.36)$$

An *elementary path* in the weighted category Y is a map $\mathbf{w2} \to Y$, and amounts to an *elementary* arrow $b\colon y \to y'$ (of weight $\leqslant 1$).

The right adjoint to I, called the *elementary path functor*, or elementary cocylinder, is

$$P\colon \mathbf{wCat} \to \mathbf{wCat}, \qquad P(Y) = Y^{\mathbf{w2}}, \qquad (6.37)$$

where $Y^{\mathbf{w2}}$ is the category of elementary arrows of Y and their 'unbounded' commutative squares in Y, with weight

$$W(b, b') = \max(w(b), w(b')).$$

Therefore, an *elementary homotopy*, or *elementary natural transformation* $\varphi\colon f \to g\colon X \to Y$, is the same as a natural transformation between 1-Lipschitz functors, *which satisfies* $w(\varphi(x)) \leqslant 1$ for all $x \in X$.

Such homotopies cannot be concatenated, since their vertical composition need not be elementary. An *elementary isomorphism* of 1-Lipschitz functors will be an elementary natural transformation having an inverse in the same domain; this amounts to an invertible natural transformation φ such that $\max(w(\varphi(x)), w(\varphi^{-1}(x))) \leqslant 1$, for all points x.

6.3.4 *The Lipschitz cylinder of weighted categories*

On the other hand, the category $\mathrm{w}_\infty \mathbf{Cat}$ has a regular symmetric dI4-homotopical structure, based on the same directed interval and the resultant *Lipschitz cylinder* endofunctor (which here is 'cartesian')

$$I\colon \mathrm{w}_\infty \mathbf{Cat} \to \mathrm{w}_\infty \mathbf{Cat}, \qquad I(-) = - \otimes \mathbf{w2}. \qquad (6.38)$$

Now, an *extended path* in the weighted category X is a Lipschitz functor $a\colon \mathbf{w2} \to X$, and amounts to a *feasible* arrow $a\colon x \to x'$, i.e. an arrow *with a finite weight* $w(a)$; the latter coincides with the Lipschitz weight $\|a\|$, as a functor on $\mathbf{w2}$.

The *standard concatenation pushout* gives the ordinal $\mathbf{w3}$ (cf. Section 4.3.2 for $\mathbf{Cat}$), equipped with the linear weight resulting from the pasting

$$
\begin{array}{ccc}
\mathbf{1} & \xrightarrow{\ \partial^+\ } & \mathbf{w2} \\
{\scriptstyle \partial^-}\big\downarrow & \ \ \big\downarrow{\scriptstyle c^-} & \\
\mathbf{w2} & \xrightarrow[\ c^+\ ]{} & \mathbf{w3}
\end{array}
\qquad
\begin{array}{l}
w(0 \to 1) = w(1 \to 2) = 1, \\[4mm]
\\[2mm]
w(0 \to 2) = 2,
\end{array}
\qquad (6.39)
$$

$$c\colon \mathbf{w2} \to \mathbf{w3}, \qquad c(0 \to 1) = (0 \to 2) \qquad \text{(concatenation map)}.$$

The concatenation map c is the same as in **Cat**; notice that its weight is 2. Concatenation of extended paths amounts to composition in X, and is thus strictly associative, with strict identities.

This pushout is preserved by any functor $X \otimes -$, yielding the concatenation pushout JX of the Lipschitz cylinder and the transformation $c\colon I \to J$, which complete the regular symmetric dI4-homotopical structure of $\mathrm{w}_\infty\mathbf{Cat}$.

A *Lipschitz homotopy*, or *Lipschitz natural transformation* $\varphi\colon f \to g\colon X \to Y$, is a Lipschitz functor $\varphi\colon X \otimes \mathrm{w}\mathbf{2} \to Y$. In other words, it is an ordinary natural transformation, viewed as a functor $\varphi\colon X \times \mathbf{2} \to Y$ whose Lipschitz weight $||\varphi||$ is finite. We prove below that:

$$||\varphi|| = \max(||f||, ||g||, |\varphi|),$$

where we call $|\varphi|$ the *reduced weight* of f

$$|\varphi| = \sup_x w_Y(\varphi(x))$$
$$= \min\{\lambda \in [0, \infty] \mid \forall x \in X, \; w_Y(\varphi(x)) \leqslant \lambda\}. \qquad (6.40)$$

Equivalently, φ is a natural transformation of *Lipschitz functors* which has a *finite reduced weight* $|\varphi|$. The concatenation of such natural transformations, computed with the J-pushout, is by vertical composition.

The symbol $\mathrm{w}_\infty\mathbf{Cat}$ will also denote the *2-category* of weighted categories, Lipschitz functors and Lipschitz natural transformations.

The embedding $U\colon \mathrm{w}\mathbf{Cat} \to \mathrm{w}_\infty\mathbf{Cat}$ is a dI2-homotopical functor. It gives to $\mathrm{w}\mathbf{Cat}$ the structure of a relative dI-homotopical category (Section 5.8.1), with relative equivalences consisting of the 1-Lipschitz functors which are Lipschitz equivalences in $\mathrm{w}_\infty\mathbf{Cat}$.

6.3.5 Lemma

(a) Let X, Y be weighted categories and $\varphi\colon f \to g\colon X \to Y$ a natural transformation of arbitrary functors. Then the Lipschitz weight of φ as a functor $X \otimes \mathrm{w}\mathbf{2} \to Y$ is

$$||\varphi|| = \max(||f||, ||g||, |\varphi|), \qquad (6.41)$$

where $|\varphi|$ is the reduced weight $|\varphi|$ defined above, in (6.40).

(b) The interval $\mathrm{w}\mathbf{2}$ is not exponentiable in $\mathrm{w}_\infty\mathbf{Cat}$.

Proof (a) The following computations give the thesis, where $a\colon x \to x'$ is in X, $u = (0 \to 1)$ is the non-trivial arrow of $\mathrm{w}\mathbf{2}$ and $b = \varphi(a, u) =$

$g(a) \circ \varphi(x)$

$$\varphi(a, \mathrm{id}(0)) = f(a), \qquad w(f(a)) \leqslant ||f||.w(a),$$

$$\varphi(a, \mathrm{id}(1)) = g(a), \qquad w(g(a)) \leqslant ||g||.w(a),$$

$$w(b) \leqslant w(f(a)) + w(\varphi(x')) \leqslant ||f||.w(a) + |\varphi|$$
$$\leqslant (||f|| \vee |\varphi|).(w(a) + 1) \leqslant (||f|| \vee |\varphi|).w(a, u).$$

(b) Suppose for a contradiction that w2 is exponentiable. It is easy to show that $Y^{\mathrm{w}2}$ must have as underlying category Y^2. Now, if X is discrete, a natural transformation $\varphi \colon f \to g \colon X \to Y$ is a Lipschitz functor $\varphi \colon X \otimes \mathrm{w}2 \to Y$ if and only if $|\varphi|$ is finite; but, viewed as a functor $\varphi \colon X \to Y^2$, it only reaches identity maps, and every weight of Y^2 makes it into a Lipschitz functor. $\qquad\square$

6.3.6 The fundamental weighted category

Let us come back to a δ-metric space X, and construct its fundamental weighted category.

Let us recall that an extended path in X is a δ_∞-map $a \colon \delta\mathbf{I} \to X$.

An extended *double path* is a δ_∞-map $A \colon \delta\mathbf{I}^2 \to X$. A *2-path* is a double path whose faces ∂_1^α are degenerate, and a 2-homotopy $A \colon a \prec_2 b \colon x \to x'$ between its faces ∂_2^α, which have the same endpoints. A 2-homotopy class of paths $[a]$ is a class of the equivalence relation $\simeq_2$ spanned by the preorder $\prec_2$.

Since $\delta_\infty \mathbf{Mtr}$ is a dI4-category, made concrete by the standard point $\{*\}$, we already have the *fundamental category* $\uparrow\!\Pi_1(X)$ of the δ-metric space X (Section 4.5.7). An object is a point of X; an arrow $[a] \colon x \to x'$ is a 2-homotopy class of paths from x to x'; composition is induced by concatenation of consecutive paths, and identities come from degenerate paths

$$[a] + [b] = [a + b], \qquad 0_x = [e(x)] = [0_x]. \tag{6.42}$$

We make $\uparrow\!\Pi_1(X)$ into the *fundamental weighted category* $\mathrm{w}\Pi_1(X)$, by enriching it with a weight. This is defined on an arrow $\xi \colon x \to x'$, by evaluating the length $L(a)$ of the extended paths which belong to this 2-homotopy class (6.1.8), and taking their greatest lower bound

$$w(\xi) = \inf_{a \in \xi} L(a). \tag{6.43}$$

The axioms of (sub)additive weights follow immediately from the properties of L (Definition 6.1.8)

$$w(0_x) = 0, \qquad w([a] + [b]) \leqslant w[a] + w[b].$$

On a δ_∞-map $f \colon X \to Y$ of δ-metric spaces, we get a w_∞-functor

$$f_* = w\Pi_1(f) \colon w\Pi_1(X) \to w\Pi_1(Y),$$
$$w\Pi_1(f)(x) = f(x), \qquad w\Pi_1(f)[a] = f_*[a] = [fa], \qquad (6.44)$$
$$w(f_*[a]) \leqslant ||f||.w[a], \qquad ||f_*|| \leqslant ||f||.$$

All this forms a functor $w\Pi_1 \colon \delta_\infty \mathbf{Mtr} \to w_\infty \mathbf{Cat}$, with values in the category of additively-weighted small categories and Lipschitz functors.

This functor restricts to $\delta\mathbf{Mtr} \to w\mathbf{Cat}$, because of the last inequality above; but, of course, $w\Pi_1(X)$ *is still based on extended paths*. In particular, $w\Pi_1$ preserves Lipschitz isomorphisms and isometric isomorphisms.

Finally, a δ_∞-homotopy $\varphi \colon f \to g \colon X \to Y$ yields a Lipschitz natural transformation

$$\varphi_* \colon f_* \to g_* \colon w\Pi_1(X) \to w\Pi_1(Y),$$
$$w(\varphi_*(x)) = w[\varphi(x)] \leqslant ||\varphi||, \qquad ||\varphi_*|| \leqslant ||\varphi||, \qquad (6.45)$$

so that $w\Pi_1 \colon \delta_\infty \mathbf{Mtr} \to w_\infty \mathbf{Cat}$ is a morphism of dh1-categories, as well as its restriction $\delta\mathbf{Mtr} \to w\mathbf{Cat}$ to the 1-Lipschitz case.

Here also, the fundamental weighted category of X is related to the fundamental groupoid of the underlying space UX, by an obvious *comparison* functor

$$w\Pi_1(X) \to \Pi_1(UX), \qquad\qquad x \mapsto x, \qquad [a] \mapsto [a]. \qquad (6.46)$$

6.3.7 Geodesics

In a δ-metric space X, we say that an extended path $a \colon x \to x'$ is a *homotopic geodesic* if it realises the weight of its class, $L(a) = w[a]$, which amounts to saying that $L(a) \leqslant L(a')$ for all extended paths $a' \simeq_2 a$.

We say that X is *geodetically simple* if every arrow $\xi \colon x \to x'$ of its fundamental weighted category $w\Pi_1(X)$ has *some* representative a which realises its weight: $L(a) = w(\xi)$; the path a is then a homotopic geodesic.

We say that X is *1-simple* if its fundamental category $w\Pi_1(X)$ is a preorder: all hom-sets have at most one arrow.

The δ-metric spaces $\delta\mathbf{R}^n$, $\delta\mathbf{R}^{\otimes n}$ are geodetically simple and 1-simple; all their convex subspaces are also (cf. Section 3.2.7(a)). The pierced plane $(!\delta\mathbf{R})^2 \setminus \{0\}$ is not geodetically simple, nor 1-simple. The δ-metric sphere $\delta\mathbf{S}^1$ is geodetically simple and not 1-simple.

Being geodetically simple is 'somehow' related to completeness of the δ-metric, as it appears from these examples. Notice that a non-complete space, like $\delta]0,1[\subset \delta\mathbf{R}$, can be geodetically simple, but it is also true that all its extended paths $x \to x'$ (between two given points) stay in the compact subspace $\delta[x, x']$.

6.3.8 Pasting theorem
('Seifert–van Kampen' for fundamental weighted categories)

Let X be a δ-metric space; let X_1, X_2 be two subspaces and $X_0 = X_1 \cap X_2$.

If $X = \text{int}(X_1) \cup \text{int}(X_2)$, the following diagram of weighted categories and contracting functors (induced by inclusions) is a pushout in w**Cat**

$$
\begin{array}{ccc}
\text{w}\Pi_1 X_0 & \xrightarrow{u_1} & \text{w}\Pi_1 X_1 \\
{\scriptstyle u_2}\downarrow & & \downarrow{\scriptstyle v_1} \\
\text{w}\Pi_1 X_2 & \xrightarrow[v_2]{} & \text{w}\Pi_1 X
\end{array}
\tag{6.47}
$$

Proof As in Theorem 3.2.6. $\qquad\square$

6.3.9 Homotopy monoids

The *fundamental weighted monoid* $\text{w}\pi_1(X, x)$ of the δ-metric space X at the point x is the (additively) weighted monoid of endo-arrows $x \to x$ in $\text{w}\Pi_1(X)$. It forms a functor from the (obvious) category $\delta\mathbf{Mtr.}$ of *pointed δ-metric spaces*, to the category of weighted monoids (Section 6.3.1)

$$
\text{w}\pi_1 \colon \delta\mathbf{Mtr.} \to \text{w}\mathbf{Mon}, \qquad \text{w}\pi_1(X, x) = \text{w}\Pi_1(X)(x, x). \tag{6.48}
$$

This functor is *strictly* homotopy invariant: a *pointed homotopy* $\varphi\colon f \to g\colon (X, x) \to (Y, y)$ has, by definition, a trivial path at the base-point ($\varphi(x) = 0_y$), whence the naturality square of every endomap $a\colon x \to x$ of X gives $f_*[a] = g_*[a]$ (as for d-spaces, see Section 3.2.5).

6.4 Minimal models

This is a brief exposition of how the minimal models developed in Chapter 3 for the fundamental category of a d-space can be enriched, in the present weighted setting.

6.4.1 The fundamental weighted category of a square annulus

Let us begin with an elementary example, enriching with a δ-metric the 'square annulus' analysed in Section 3.1.1, as an ordered space.

We start now from the δ-metric space $\delta\mathbf{I}^{\otimes 2}$, with

$$\delta(\mathbf{x}, \mathbf{y}) = \begin{cases} (y_1 - x_1) + (y_2 - x_2), & \text{if } x_1 \leqslant y_1, \; x_2 \leqslant y_2, \\ \infty, & \text{otherwise.} \end{cases} \qquad (6.49)$$

Its underlying ordered topological space (cf. (6.19)) is the ordered topological square $\uparrow\mathbf{I}^2$ (with euclidean topology and product order).

Taking out the *open* square $]1/3, 2/3[^2$ (marked with a cross), we get the square annulus $X \subset \delta\mathbf{I}^{\otimes 2}$, with the induced δ-metric

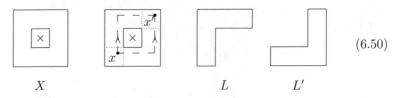

$$(6.50)$$

$$X \qquad\qquad\qquad\qquad L \qquad L'$$

Its extended paths are the Lipschitz *order-preserving* maps $\delta[0, 1] \to X$ defined on the standard δ-interval; they move 'rightward and upward' (in the weak sense). Extended homotopies of such paths are Lipschitz order-preserving maps $\uparrow[0, 1]^2 \to X$.

As a consequence of the 'van Kampen' theorem recalled above (using the subspaces L, L'), the fundamental weighted category $C = \mathrm{w}\Pi_1(X)$ is the category described (in Section 3.1.1) for the underlying ordered space, equipped with the appropriate weight. Moreover, the weight of an arrow can always be realised as the length of some representative: X is geodetically simple (Section 6.3.7).

Thus, the weighted category C is 'essentially represented' by the full weighted subcategory E on four vertices $0, p, q, 1$ (the central cell does not commute), *where each of the four generating arrows has weight* $2/3$, and the weight of E is linear (i.e. strictly additive on composition)

$$0 \to p \rightrightarrows q \to 1 \qquad (6.51)$$

$$E$$

The situation can be analysed as in Section 3.1.1, adding the information due to the weight:

- the action begins at 0, from where we move to the point p, *with weight 2/3*;
- p is an (effective) future branching point, where we have to choose between two paths, *each of them of weight 2/3*, which join at q, an (effective) past branching point;
- from where we can only move to 1, again *with weight 2/3*, where the process ends.

In order to make precise how E can 'model' the category C, we have proved in Chapter 3 that E is both future equivalent and past equivalent to C, and actually is the 'join' of a minimal future model with a minimal past model of the latter. All this can now be enriched with weights.

6.4.2 Future equivalence of weighted categories

The notion of future equivalence can be easily transferred from **Cat** (Section 3.3) to the 2-category $\mathrm{w}_\infty\mathbf{Cat}$, since it makes sense and works well in any 2-category.

Thus, a *future equivalence* $(f, g; \varphi, \psi)$ between the weighted categories C, D consists of a pair of Lipschitz functors and a pair of Lipschitz natural transformations, the *units*, satisfying two coherence conditions:

$$f\colon X \rightleftarrows Y \colon g \qquad \varphi\colon 1_X \to gf, \qquad \psi\colon 1_Y \to fg,$$
$$f\varphi = \psi f\colon f \to fgf, \qquad \varphi g = g\psi\colon g \to gfg \qquad (coherence) \tag{6.52}$$

and is said to be *elementary*, or 1-Lipschitz, if both functors and both natural transformations are.

Future equivalences compose (Section 3.3.3), and yield an equivalence relation of weighted categories; the elementary ones do not. Dually, *past equivalences* have *counits*, in the opposite direction.

In particular, an *elementary future retract* $i\colon C_0 \subset C$ will be a full weighted subcategory having a reflector $p \dashv i$ which is 1-Lipschitz, has a 1-Lipschitz unit $\eta\colon 1_C \to ip$ and a trivial counit $pi = 1$. The coherence conditions of the adjunction $(\eta i = 1_i, \, p\eta = 1_p)$ show that the four-tuple $(i, p; 1, \eta)$ is an elementary future equivalence.

A (weighted) *pf-presentation* of the weighted category C (extending Section 3.5.2) will be a diagram consisting of an elementary past retract P and an elementary future retract F of C (which are thus a full coreflective and a full reflective weighted subcategory, respectively) with elementary adjunctions $i^- \dashv p^-$ and $p^+ \dashv i^+$

$$P \underset{p^-}{\overset{i^-}{\rightleftarrows}} C \underset{i^+}{\overset{p^+}{\rightleftarrows}} F \qquad (6.53)$$

$$\varepsilon\colon i^- p^- \to 1_C \qquad (p^- i^- = 1, \ p^- \varepsilon = 1, \ \varepsilon i^- = 1),$$
$$\eta\colon 1_C \to i^+ p^+ \qquad (p^+ i^+ = 1, \ p^+ \eta = 1, \ \eta i^+ = 1).$$

6.4.3 Spectra

Coming back to the square annulus X (Section 6.4.1), the weighted category $C = \mathrm{w}\Pi_1(X)$ has a least *full reflective* weighted subcategory F, which is future equivalent to C and minimal as such. Its objects form the *future spectrum* $\mathrm{sp}^+(C) = \{p, 1\}$ (Section 3.8.1); the full weighted subcategory $F = \mathrm{Sp}^+(C)$ on these objects is also called a *future* (weighted) *spectrum* of C

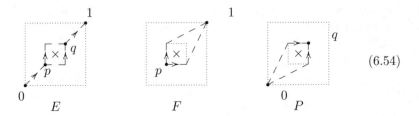

$$(6.54)$$

Dually, we have the least *full coreflective* weighted subcategory $P = \mathrm{Sp}^-(C)$, on the *past spectrum* $\mathrm{sp}^-(C) = \{0, q\}$.

Together, they form a (weighted) pf-presentation of C (cf. (6.53)), called the *spectral* pf-presentation. Moreover the (weighted) *pf-spectrum* $E = \mathrm{Sp}(C)$ is the full weighted subcategory of C on the set of objects $\mathrm{sp}(C) = \mathrm{sp}^-(C) \cup \mathrm{sp}^+(C)$ (Section 3.8.5). E is a *strongly minimal injective model* of the weighted category C (Theorem 3.8.8).

6.5 Spaces with weighted paths

We introduce a second framework for weighted algebraic topology, which is more complicated than δ-metric spaces but has finer quotients, as we shall see in Section 6.7. The relationship between the two notions is dealt with in Section 6.6.

6.5.1 *Main definitions*

A *w-space X*, or *space with weighted paths*, will be a topological space together with a *weight function* $w \colon X^{\mathbf{I}} \to [0, \infty]$, or *cost function* (also written as w_X) defined on the *set* of its (continuous) paths, which satisfies three axioms concerning the constant paths 0_x, the path-concatenation $a + b$ of consecutive paths and *strictly* increasing reparametrisation:

(wsp.0) $w(0_x) = 0$, for all points x of X;

(wsp.1) $w(a + b) \leqslant w(a) + w(b)$, for all consecutive paths a, b;

(wsp.2) $w(a\rho) \leqslant w(a)$, for all paths a and all *strictly increasing continuous maps* $\rho \colon \mathbf{I} \to \mathbf{I}$.

It is easy to see that the last condition, in the presence of the others, is equivalent to asking that $w(a\rho) \leqslant w(a)$, for all paths a and all *increasing continuous maps* $\rho \colon \mathbf{I} \to \mathbf{I}$ which are constant on a *finite* number of subintervals. It is also equivalent to the conjunction of the following two conditions:

(wsp.1') $\max(w(a), w(b)) \leqslant w(a + b)$, for all consecutive paths a, b;

(wsp.2') $w(a\rho) = w(a)$, for all paths a and all *increasing homeomorphisms* $\rho \colon \mathbf{I} \to \mathbf{I}$.

We shall say that a path is *free*, *feasible* or *unfeasible* when, respectively, its cost is 0, finite or ∞.

A w-space will be said to be *linear*, or *strictly additive*, if $w(a + b) = w(a) + w(b)$; see Section 6.5.5 for examples. We shall see in Section 6.6 that linear w-spaces form a coreflective subcategory. Note that we are not asking that the weight function be continuous with respect to the compact-open topology of $X^{\mathbf{I}}$; in most examples, this will only be true if we restrict w to the feasible paths (or, equivalently, if we topologise $[0, \infty]$ letting ∞ be everywhere dense, which might be interesting).

If X, Y are w-spaces, a *w-map* $f \colon X \to Y$, or *map of w-spaces*, or *1-Lipschitz map* will be a continuous mapping which decreases costs: $w(f \circ a) \leqslant w(a)$, for all (continuous) paths a of X. More generally, a *Lipschitz map*, or w_∞-*map* $f \colon X \to Y$ is a continuous mapping which has a finite Lipschitz constant $\lambda \in [0, \infty[$, in the sense that $w(f \circ a) \leqslant \lambda.w(a)$, for all continuous paths a in X.

We thus have the category w**Top** of w-spaces and w-maps, embedded in the category w_∞**Top** of w-spaces and w_∞-maps. Again, we distinguish between *isometric isomorphisms* (of w**Top**) and *Lipschitz isomorphisms* (of w_∞**Top**).

The forgetful functor $U \colon \mathrm{w}\mathbf{Top} \to \mathbf{Top}$ has left and right adjoints $D \dashv U \dashv D'$, where the discrete weight of DX is the highest possible one, with $w(a) = 0$ on the constant paths and ∞ on all the others, while the *natural*, indiscrete weight of $D'X$ is the lowest possible one, where all paths have a null cost. Except if otherwise stated, *when viewing a topological space as a weighted one we will use the embedding* $D' \colon \mathbf{Top} \to \mathrm{w}\mathbf{Top}$, where all paths are free.

Note now that the weight function of the w-space X acts on the continuous mappings $a \colon \mathbf{I} \to UX$, *with values in the underlying topological space*; we shall go on writing down, pedantically, such occurrences of U. (Viewing these paths as w-maps $D\mathbf{I} \to X$ would also be correct, but confusing.)

Here also, reversing paths by the involution $r \colon \mathbf{I} \to \mathbf{I}$, $r(t) = 1-t$, gives the *opposite* w-space and forms a (covariant) involutive endofunctor, called a *reversor*

$$R \colon \mathrm{w}\mathbf{Top} \to \mathrm{w}\mathbf{Top}, \qquad R(X) = X^{\mathrm{op}}, \qquad w^{\mathrm{op}}(a) = w(ar). \quad (6.55)$$

A w-space will be said to be *reversible* if it is invariant under the reversor (in accord with our general terminology for symmetries, after the clashes of the metric case, in Section 6.1.2). It is *reversive* if it is isometrically isomorphic to its opposite space. The notation $X \leqslant X'$ will mean that these w-spaces have the same underlying topological space and $w_X \leqslant w_{X'}$; equivalently, the identity of the underlying space is a w-map $X' \to X$.

6.5.2 The weight of a map

A continuous mapping $f \colon UX \to UY$ between w-spaces takes paths of UX into paths of UY, and inherits two weights from the category of weighted sets.

Letting the path a vary in $\mathbf{Top}(\mathbf{I}, UX)$ and $\lambda \in [0, \infty]$, we have the *additive weight*

$$\begin{aligned} |f|_0 &= \sup_a \left(w(f \circ a) - w(a)\right) \\ &= \min\{\lambda \mid \forall a, \ w(f \circ a) \leqslant \lambda + w(a)\}, \end{aligned} \quad (6.56)$$

and the *multiplicative weight*, or *Lipschitz weight*:

$$\begin{aligned} \|f\| &= |f|_1 = \sup_a \left(w(f \circ a)/w(a)\right) \\ &= \min\{\lambda \mid \forall a, \ w(f \circ a) \leqslant \lambda . w(a)\}. \end{aligned} \quad (6.57)$$

The first distinguishes w-maps with the condition $|f|_0 = 0$. But we shall only use the second, written as $||f||$, which distinguishes w-maps with the condition $||f|| \leqslant 1$, and w_∞-maps with the condition $||f|| < \infty$. With this weight, w**Top** and w_∞**Top** are multiplicatively weighted categories: all identities have weight $||1_X|| \leqslant 1$ and composition gives $||gf|| \leqslant ||f|| \cdot ||g||$.

If X is a w-space and $\lambda \in [0, \infty[$, we write λX for the same topological space equipped with the weight $\lambda.w_X$. A w_∞-map $f \colon X \to Y$ with $||f|| \leqslant \lambda$ is the same as a w-map $\lambda X \to Y$.

6.5.3 Limits

The category w**Top** has all limits and colimits, computed as in **Top** and equipped with the adequate w-structure.

Thus, for a product ΠX_i, a path $a \colon \mathbf{I} \to U(\Pi X_i)$ of components $a_i \colon \mathbf{I} \to U X_i$ has weight $w(a) = \sup w(a_i)$. For a sum ΣX_i, a path $a \colon \mathbf{I} \to U(\Sigma X_i)$ lives in *one* component $U X_i$ and inherits the weight from the latter.

Given a pair of parallel w-maps $f, g \colon X \to Y$, the equaliser is the topological one, with the restricted weight function. The coequaliser is the topological coequaliser Y/R, with the induced weight characterised in the theorem below.

Linear w-spaces are not closed under (even binary) products, as we see below. But they are closed under subspaces (obviously), all colimits (by adjointness) and tensor product (see Section 6.5.6).

The category w_∞**Top** has finite limits and colimits, which can be constructed as above. Such objects are only determined up to *Lipschitz* isomorphism, but we shall keep the previous constructions as privileged ones. Thus, when we write $X \times Y$ in w_∞**Top**, we still mean that its weight is the l_∞-weight, with $w(a, b) = \max(w(a), w(b))$; isomorphic constructions will have different names (cf. Section 6.5.6). It is easy to verify that $||f \times g|| \leqslant \max(||f||, ||g||)$.

We say that the group G (in additive notation) *acts* on the w-space X if it acts on the underlying topological space, and moreover all the homeomorphisms $x \mapsto x + g$ are w-maps. The last condition is equivalent to saying that, for every path $a \colon \mathbf{I} \to U X$ and every $g \in G$, $w(a) = w(a + g)$; i.e. that the weight function of X is invariant under the action of G.

The *orbit* w-space X/G is the generalised coequaliser of all the maps $X \to X$, $x \mapsto x + g$ (for $g \in G$). This structure is characterised below, in a simpler way than for a general quotient of w-spaces. (We have seen similar behaviour for d-spaces, in Section 1.4.2.)

6.5.4 Theorem (Quotients of w-spaces)

(a) Given a pair of parallel w-maps $f, g \colon X \to Y$, the coequaliser is the topological coequaliser Y/R, with the induced weight

$$w(b) = \inf \left(\textstyle\sum_i w_Y(a_i) \right) \qquad (b \colon \mathbf{I} \to (UY)/R), \qquad (6.58)$$

the inf being taken on all finite families $(a_1, ..., a_n)$ of paths in UY such that their projections on the quotient, $pa_i \colon \mathbf{I} \to (UY)/R$, are consecutive, and give $b = ((pa_1) + \cdots + (pa_n))\rho$ (by n-ary concatenation and reparametrisation along an increasing homeomorphism $\mathbf{I} \to \mathbf{I}$).

Of course, if there are no such families, $w(b) = \inf(\emptyset) = \infty$.

(b) If the group G acts on the w-space X, the orbit w-space X/G is the usual topological space, equipped with the weight

$$w(b) = \inf \left(w_X(a) \right) \qquad (b \colon \mathbf{I} \to (UX)/G), \qquad (6.59)$$

the inf being taken on all paths $a \colon \mathbf{I} \to UY$ such that $pa = b$.

Proof (a) The only non-trivial point is verifying that this weight-function satisfies the axioms (wsp.1, 1') of Section 6.5.1. Take $b = b' + b'' \colon \mathbf{I} \to Y/R$.

First, every pair of decompositions of b', b''

$$b' = ((pa'_1) + \cdots + (pa'_m))\rho', \qquad b'' = ((pa''_1) + \cdots + (pa''_n))\rho'',$$

gives a decomposition $b = ((pa'_1) + \cdots + (pa''_n))\rho$; therefore

$$(w_Y(a'_1) + \cdots + w_Y(a'_m)) + (w_Y(a''_1) + \cdots + w_Y(a''_n)) \geqslant w(b),$$

and $w(b') + w(b'') \geqslant w(b)$.

Second, given a decomposition $b = ((pa_1) + \cdots + (pa_n))\rho$, we can always assume that $n = 2k$ (otherwise, we insert a constant path, without modifying $\sum_i w_Y(a_i)$). Then, for suitable reparametrisations ρ', ρ'', we have

$$((pa_1) + \cdots + (pa_k)) + ((pa_{k+1}) + \cdots + (pa_{2k})) = b\rho^{-1} = b'\rho' + b''\rho'',$$
$$b'\rho' = (pa_1) + \cdots + (pa_k),$$
$$w(b') \leqslant w_Y(a_1) + \cdots + w_Y(a_k) \leqslant w_Y(a_1) + \cdots + w_Y(a_{2k}),$$

It follows that $w(b') \leqslant w(b)$, and similarly $w(b'') \leqslant w(b)$.

(b) Again, we only have to verify the axioms (wsp.1, $1'$) of Section 6.5.1. Take $b = b' + b'' : \mathbf{I} \to X/G$.

Every path $a : \mathbf{I} \to X$ which projects to b can be (uniquely) decomposed as $a = a' + a''$, into its two halves, and these project to b', b''.

Conversely, given any pair of paths $a', a'' : \mathbf{I} \to X$ which project to b', b'', we can always assume that they are consecutive in X (up to replacing a'' with a suitable path *of the same weight*, $a'' + g$, as in Section 1.4.2). Therefore, letting a, a', a'' vary as specified above, we have

$$w(b) = \inf (w_X(a)) \leqslant \inf (w_X(a') + w_X(a''))$$
$$= \inf (w_X(a')) + \inf (w_X(a'')) = w(b') + w(b'').$$
$$w(b') = \inf (w_X(a')) \leqslant \inf (w_X(a' + a'')) = w(b).$$

$\square$

6.5.5 Standard models

The *standard weighted real line*, or *w-line* w$\mathbf{R}$, will be the euclidean line with the following weight on all paths $a : \mathbf{I} \to \mathbf{R}$, equivalently defined by its span or length in $\delta\mathbf{R}$ (Definition 6.1.8)

$$w(a) = \mathrm{spn}(a) = L(a). \tag{6.60}$$

Thus, $w(a)$ is finite if and only if a is a (weakly) *increasing* path, and then $w(a) = a(1) - a(0)$. (General relations between δ-metric spaces and w-spaces will be studied in the next section.)

The *n-dimensional real w-space* w$\mathbf{R}^n$, a cartesian power in w**Top**, has $w(a) = \sup_i (a_i(1) - a_i(0))$ for all increasing paths $a : \mathbf{I} \to \mathbf{R}^n$ (with respect to the product order of $\mathbf{R}^n$, $x \leqslant x'$ if and only if $x_i \leqslant x'_i$ for all i). Plainly, w$\mathbf{R}$ is linear while every higher dimensional w$\mathbf{R}^n$ is not.

The *standard w-interval* w$\mathbf{I}$ has the subspace structure of the w-line; the *standard w-cube* w$\mathbf{I}^n$ is its n-th power, and a subspace of w$\mathbf{R}^n$. These w-spaces are not reversible (for $n > 0$), but reversive; in particular, the canonical reflecting isomorphism

$$r : \text{w}\mathbf{I} \to (\text{w}\mathbf{I})^{\mathrm{op}}, \qquad r(t) = 1 - t, \tag{6.61}$$

will be used to *reflect* paths and homotopies.

The *standard weighted circle* w$\mathbf{S}^1$ will be the coequaliser in w**Top** of each of the following two pairs of maps (equivalently)

$$\partial^-, \partial^+ : \{*\} \rightrightarrows \text{w}\mathbf{I}, \qquad \partial^-(*) = 0, \qquad \partial^+(*) = 1, \tag{6.62}$$

$$\mathrm{id}, f : \text{w}\mathbf{R} \rightrightarrows \text{w}\mathbf{R}, \qquad f(x) = x + 1. \tag{6.63}$$

The 'standard realisation' of the first coequaliser above is the quotient $(\mathbf{wI})/\partial\mathbf{I}$, which identifies the endpoints; a feasible path turns around the circle in a given direction, and its weight measures the length of the path with respect to the length of the circle: $w(a) = L(a)$ in $\delta\mathbf{S}^1$. The Lipschitz-isomorphic structure $2\pi.\mathbf{wS}^1$ is also of interest. Both are linear.

More generally, the *weighted n-dimensional sphere* will be the quotient of the weighted cube $\mathbf{wI}^n$ modulo its (ordinary) boundary $\partial\mathbf{I}^n$, while $\mathbf{wS}^0$ has the discrete topology and the unique w-structure

$$\mathbf{wS}^n = (\mathbf{wI}^n)/(\partial\mathbf{I}^n) \quad (n > 0), \qquad \mathbf{wS}^0 = \mathbf{S}^0 = \{-1, 1\}. \quad (6.64)$$

All weighted spheres are reversive. Again, $\mathbf{wS}^1$ is linear while the higher spheres are not.

6.5.6 Tensor product

The tensor product $X \otimes Y$ of two w-spaces (similar to the tensor product in $\mathbf{w}^+$) will be the cartesian product of the underlying topological spaces, with an l_1-weight (instead of the l_∞-weight, which is used in the cartesian product of w-spaces)

$$w_\otimes(a, b) = w_X(a) + w_Y(b). \quad (6.65)$$

Here $(a, b) \colon \mathbf{I} \to X \times Y$ denotes the path of components $a \colon \mathbf{I} \to X$, $b \colon \mathbf{I} \to Y$. This tensor product defines a symmetric monoidal structure on w**Top**, with identity the singleton space $\{*\}$.

Linear w-spaces are closed under tensor product. In particular, all tensor powers $(\mathbf{wR})^{\otimes n}$, $(\mathbf{wI})^{\otimes n}$ and $(\mathbf{wS}^1)^{\otimes n}$ are linear. The following theorem shows that all of them are exponentiable in w**Top**, with respect to the tensor product; in particular, this holds for the tensor power $(\mathbf{wI})^{\otimes n}$, which is what is relevant for homotopy.

This tensor product extends to $\mathbf{w}_\infty$**Top**, with $\|f \otimes g\| \leqslant \max(\|f\|, \|g\|)$. In this category the tensor product is isomorphic to the cartesian one, but we keep distinguishing these realisations.

6.5.7 Theorem (Exponentiable w-spaces)

*Let Y be a linear w-space with a locally compact Hausdorff topology. Then Y is exponentiable in w**Top**, with respect to the previous tensor product.*

For every w-space Z, the internal hom

$$Z^Y = \mathrm{w}\mathbf{Top}(Y,Z) \subset \mathbf{Top}(UY,UZ), \qquad (6.66)$$

is the set of w-maps, equipped with the compact-open topology (restricted from $(UZ)^{UY}$) and the w-structure where a path $c\colon \mathbf{I} \to U(ZY) \subset (UZ)UY$ has the following weight

$$W(c) = \sup_b \left(w_Z(\mathrm{ev}\circ(c,b)) - w_Y(b)\right) \qquad (b\colon \mathbf{I} \to UY). \qquad (6.67)$$

Here, $\lambda - \mu$ is the truncated difference in $\mathbf{w}^+$, while the evaluation mapping

$$\mathrm{ev}\colon Z^Y \otimes Y \to Z, \qquad (6.68)$$

is the restriction of the topological one. This mapping is a w-map, and yields the counit of the adjunction.

Proof (Note. The same argument, conveniently simplified, shows that $\delta\mathbf{Mtr}$ is monoidal closed.)

We defer to the end the technical part showing that (6.66) and (6.67) do define a w-structure.

First, the evaluation mapping (6.68) satisfies the inequality $w_\otimes(c,b) \geqslant w_Z(\mathrm{ev}\circ(c,b))$, because of the symmetric monoidal closed structure of $\mathbf{w}^+$:

$$W(c) \geqslant w_Z(\mathrm{ev}\circ(c,b)) - w_Y(b) \qquad (\forall b\colon \mathbf{I} \to UY),$$
$$w_\otimes(c,b) = W(c) + w_Y(b) \geqslant w_Z(\mathrm{ev}\circ(c,b)).$$

Second, the pair $(Z^Y, \mathrm{ev}\colon Z^Y \otimes Y \to Z)$ is a universal arrow from the functor $- \otimes Y$ to the object Z: given a w-space X and a w-map $f\colon X \otimes Y \to Z$, we have to prove that there is precisely one w-map $g\colon X \to Z^Y$ such that f factors as:

$$\mathrm{ev}\circ(g \otimes Y)\colon X \otimes Y \to Z^Y \otimes Y \to Z. \qquad (6.69)$$

Indeed, since Y is exponentiable in $\mathbf{Top}$, there exists precisely one continuous mapping $g\colon UX \to (UZ)^{UY}$ such that $f = \mathrm{ev}\circ(g\times UY)$, and it will be sufficient to prove the following two facts.

(a) $\mathrm{Im}(g) \subset Z^Y$. For $x \in X$, we must prove that $g(x)\colon Y \to Z$ is a w-map. And indeed, for every path $b\colon \mathbf{I} \to UY$

$$w(g(x)\circ b) = w(f\circ(0_x,b)) \leqslant w_\otimes(0_x,b) = w_X(0_x) + w_Y(b) = w_Y(b).$$

(b) The mapping g is a w-map $X \to Z^Y$. And indeed, for every path $a \colon \mathbf{I} \to UA$

$$W(ga) = \sup_b \left(w_Z \left(\mathrm{ev} \circ (ga, b) \right) - w_Y (b) \right) = \sup_b \left(w_Z \left(f(a, b) \right) - w_Y (b) \right)$$
$$\leqslant \sup_b \left(w_\otimes (a, b) - w_Y (b) \right) = w_X (a).$$

Finally, we verify the axioms for the weight W of the internal hom. First, the constant path $0_h \colon \mathbf{I} \to Z^Y$ at an arbitrary w-map $h \colon Y \to Z$ gives

$$W(0_h) = \sup_b \left(w_Z \left(\mathrm{ev} \circ (0_h, b) \right) - w_Y (b) \right) = \sup_b \left(w_Z (hb) - w_Y (b) \right) = 0.$$

Second, to prove (wsp.1), let $c = c' + c''$ be a concatenation of paths in $U(Z^Y)$. We can always rewrite a path $b \colon \mathbf{I} \to UY$ as the concatenation $b = b' + b''$ of its two halves, so that, using the assumption that Y *is linear*:

$$W(c' + c'') = \sup_b \left[\left(w_Z \left(\mathrm{ev} \circ (c' + c'', b) \right) \right) - w_Y (b) \right]$$
$$= \sup_{b'b''} \left[w_Z \left(\mathrm{ev} \circ (c' + c'', b' + b'') \right) - w_Y (b' + b'') \right]$$
$$\text{(for all consecutive paths } b', b'' \text{ in } UY),$$
$$= \sup_{b'b''} \left[w_Z \left(\left(\mathrm{ev} \circ (c', b') \right) + \left(\mathrm{ev} \circ (c'', b'') \right) \right) - w_Y (b') - w_Y (b'') \right]$$
$$\leqslant \sup_{b'b''} \left[w_Z \left(\mathrm{ev} \circ (c', b') \right) - w_Y (b') + w_Z \left(\mathrm{ev} \circ (c'', b'') \right) - w_Y (b'') \right]$$
$$\leqslant W(c') + W(c'').$$

The last inequality comes from the fact that the last term amounts to the previous sup for arbitrary paths b', b'' in Y. (Note: for δ-metric spaces, one would use the fact that all 'paths' $(y, y') \colon \mathbf{2} \to Y$ *can* be rewritten as a trivial 'concatenation' $(y, y') + (y', y')$, with $d(y, y') = d(y, y') + d(y', y')$.)

Now, for (wsp.1'), we can make our least upper bound smaller by restriction to those paths $b \colon \mathbf{I} \to Y$ which are constant on $[1/2, 1]$, so that $b = b' + b''$ with an arbitrary b' and b'' constant at the terminal of b':

$$W(c' + c'') \geqslant \sup_{b'} \left(w_Z \left(\left(\mathrm{ev} \circ (c', b') \right) + \left(\mathrm{ev} \circ (c'', b'') \right) \right) - w_Y (b' + b'') \right)$$
$$\geqslant \sup_{b'} \left(w_Z \left(\mathrm{ev} \circ (c', b') \right) - w_Y (b') \right) = W(c'),$$

where, again, we have used the linear property of Y: $w(b) = w(b') + w(b'') = w(b')$.

Last, for (wsp.2′), given an increasing homeomorphism $\rho\colon \mathbf{I} \to \mathbf{I}$, every path b' in UY can be rewritten as $b\rho$, with $b = b'\rho^{-1}$, so that:

$$W(c\rho) = \sup_b \left(w_Z(\mathrm{ev}\circ(c\rho, b\rho)) - w_Y(b\rho)\right)$$
$$= \sup_b \left(w_Z(\mathrm{ev}\circ(c, b)\circ\rho) - w_Y(b\rho)\right) = W(c).$$

$\square$

6.5.8 Elementary and extended paths

Let X be a w-space. An *elementary path* (resp. an *extended path*, or *Lipschitz path*) in X will be a 1-Lipschitz (resp. a Lipschitz) map $a\colon \mathrm{w}\mathbf{I} \to X$.

Thus, a continuous mapping $a\colon \mathbf{I} \to X$ is an elementary path if and only if $||a|| \leqslant 1$, for the Lipschitz weight (6.57) recalled below, and is an extended path if and only if $||a|| < \infty$

$$||a|| = \min\{\lambda \in [0, \infty] \mid \forall\rho\colon \uparrow\!\mathbf{I} \to \uparrow\!\mathbf{I}, \ w(a\rho) \leqslant \lambda.(\rho(1) - \rho(0))\}. \quad (6.70)$$

(Notice that ρ varies in the set of increasing maps $\mathbf{I} \to \mathbf{I}$.) Again, elementary paths are not closed under concatenation.

Thus, $w(a) \leqslant ||a||$. A path of finite weight $w(a)$ need not be Lipschitz, as one readily sees considering the square-root function $a\colon \mathbf{I} \to \mathbf{R}$, which in w$\mathbf{R}$ has $w(a) = 1$, but $||a|| = \infty$.

The *reflected* (elementary or extended) path is obtained in the usual way

$$a^{\mathrm{op}} = ar\colon \mathrm{w}\mathbf{I} \to X^{\mathrm{op}}, \qquad r(t) = 1 - t. \quad (6.71)$$

A *reversible* extended path is a mapping $a\colon \mathbf{I} \to X$ such that both a and a^{op} are extended paths $\mathrm{w}\mathbf{I} \to X$.

In the category w$\mathbf{Top}$, the pasting of two copies of the standard weighted interval, one after the other, can be realised as $\mathrm{w}[0, 2] \subset \mathrm{w}\mathbf{R}$ (or as $2.\mathrm{w}\mathbf{I}$, cf. Section 6.5.2), which is of no help to concatenate paths parametrised on w$\mathbf{I}$, in w$\mathbf{Top}$. But in $\mathrm{w}_\infty\mathbf{Top}$ this pasting can be realised as w$\mathbf{I}$ (which is Lipschitz-isomorphic to $\mathrm{w}[0, 2]$), by the *standard concatenation pushout*

$$
\begin{array}{ccc}
\{*\} & \xrightarrow{\ \partial^+\ } & \mathrm{w}\mathbf{I} \\
{\scriptstyle\partial^-}\big\downarrow & \overset{}{\dashrightarrow} \big\downarrow{\scriptstyle c^-} & \\
\mathrm{w}\mathbf{I} & \xrightarrow[\ c^+\ ]{} & \mathrm{w}\mathbf{I}
\end{array}
\qquad
\begin{array}{l}
c^-(t) = t/2, \\[1.5em]
c^+(t) = (t+1)/2.
\end{array}
\qquad (6.72)
$$

(Again, the diagram above lives in w**Top**, but is a pushout only in w_∞**Top**.) Now, given two consecutive Lipschitz paths $a, b \colon w\mathbf{I} \to X$, with $a(1) = b(0)$, we get a concatenated path

$$a + b \colon w\mathbf{I} \to X, \qquad \|a + b\| \leqslant 2.(\|a\| + \|b\|), \qquad (6.73)$$

as follows from the following proposition (or using the pushout $2.w\mathbf{I}$, in w**Top**).

We can now treat homotopies as in the case of δ-metric spaces, in Section 6.2. We define the fundamental weighted category and the fundamental weighted monoids of a w-space as in Section 6.3. In particular, concatenation is based on the following result.

6.5.9 Proposition

For every w-space X, the functor $X \times - \colon w_\infty\mathbf{Top} \to w_\infty\mathbf{Top}$ preserves the standard concatenation pushout (6.72).

Moreover, if a map $f \colon X \times w\mathbf{I} \to Y$ comes from the pasting of two 'consecutive' maps $f_0, f_1 \colon X \times w\mathbf{I} \to Y$, we have the following upper bound for its Lipschitz weight

$$\|f\| \leqslant 2.(\|f_0\| + \|f_1\|) \qquad (f_0 = f \circ (X \times c^-), \ f_1 = f \circ (X \times c^+)). \ (6.74)$$

Equivalently, one can use the Lipschitz-isomorphic functor $X \otimes -$.

Proof In **Top**, the preservation holds because the subspaces $UX \times [0, 1/2]$ and $UX \times [1/2, 1]$ form a finite closed covering of $UX \times \mathbf{I}$, so that each mapping defined on the latter and continuous on such closed parts is continuous (as already remarked in Section 1.1.2).

Consider then a (topological) map $f \colon UX \times \mathbf{I} \to UY$ coming from the pasting of two maps f_0, f_1 on the topological pushout $UX \times \mathbf{I}$

$$f(x, t) = \begin{cases} f_0(x, 2t), & \text{for } 0 \leqslant t \leqslant 1/2, \\ f_1(x, 2t - 1), & \text{for } 1/2 \leqslant t \leqslant 1. \end{cases}$$

Let now $(a, \rho) \colon \mathbf{I} \to UX \times \mathbf{I}$ be any feasible path; in particular, $\rho \colon \mathbf{I} \to \mathbf{I}$ is an increasing map. If the image of ρ is contained in the first half of $\mathbf{I}$, then $f \circ (a, \rho) = f_0(a, 2\rho)$ and

$$w(f \circ (a, \rho)) \leqslant \|f_0\|.(w(a) \vee 2w(\rho)) \leqslant 2.\|f_0\|.w(a, \rho).$$

A similar argument holds for the second half. Otherwise, since ρ is increasing, we have $\rho(t_1) = 1/2$ at some interior point $t_1 \in \,]0, 1[$; we can *assume* that $t_1 = 1/2$ (up to pre-composing with an *increasing homeomorphism* $\sigma \colon \mathbf{I} \to \mathbf{I}$, which does not modify the weight of paths, by (wsp.2′)).

Now, the path $f \circ (a, \rho) \colon \mathbf{I} \to UY$ is the concatenation of two paths $c_i \colon \mathbf{wI} \to UY$ which factor through the Lipschitz maps f_i

$$c_0(t) = f \circ (a(t/2), \rho(t/2)) = f_0 \circ (a(t/2), 2\rho(t/2)),$$
$$c_1(t) = f \circ (a((t+1)/2), \rho((t+1)/2))$$
$$= f_1 \circ (a((t+1)/2), 2\rho((t+1)/2) - 1).$$

and finally we can conclude that f is Lipschitz, with the upper bound (6.74)

$$w(f \circ (a, \rho)) \leqslant w(c_0) + w(c_1) \leqslant (\|f_0\| + \|f_1\|).(w(a) \vee 2w(\rho))$$
$$\leqslant 2.(\|f_0\| + \|f_1\|).w(a, \rho).$$

$\square$

6.6 Linear and metrisable w-spaces

The *span* and *length* function of a δ-metric space X, defined in Definition 6.1.8, allow us to construct the w-spaces $\mathbf{spn}X$ (Section 6.6.2) and $\mathbf{L}X$ (Section 6.6.3); the latter is linear.

6.6.1 Linear w-spaces

First, we want to observe that linear w-spaces form a full subcategory $\mathrm{Lw}_\infty\mathbf{Top}$ of $\mathrm{w}_\infty\mathbf{Top}$, which has a *coreflector* L, right adjoint to the embedding U

$$U \colon \mathrm{Lw}_\infty\mathbf{Top} \rightleftarrows \mathrm{w}_\infty\mathbf{Top} \colon L \qquad (U \dashv L). \qquad (6.75)$$

In fact, for a w-space X, there is a *linearised* w-space $L(X)$ on the same underlying topological space, endowed with the least linear weight $L \geqslant w$

$$L(a) = \sup_t \Sigma_j w(a(t_{j-1}, t_j)) \qquad (0 = t_0 < t_1 < \ldots < t_p = 1),$$
$$a(t_{j-1}, t_j)(t) = a((1-t).t_{j-1} + t.t_j) \qquad (0 \leqslant t \leqslant 1).$$
$$(6.76)$$

Note that we have written $a(t_{j-1}, t_j) \colon \mathbf{I} \to X$ the restriction of the path a to the interval $[t_{j-1}, t_j]$, *reparametrised* on the standard interval.

Thus $L(X) \geqslant X$, and $L(X) = X$ if and only if the w-space X is linear. These relations give, respectively, the counit $UL \to 1$ and the unit $LU = 1$ of the adjunction.

All this restricts to contractions, yielding the full coreflective subcategory $Lw\mathbf{Top} \subset w\mathbf{Top}$ of linear w-spaces. Therefore, the latter are closed under colimits in $w\mathbf{Top}$.

6.6.2 Span-metrisable w-spaces

Now, let us construct an adjunction

$$\boldsymbol{\delta} \colon w_\infty \mathbf{Top} \rightleftarrows \delta_\infty \mathbf{Mtr} \colon spn, \qquad \boldsymbol{\delta} \dashv spn. \qquad (6.77)$$

First, the functor

$$\boldsymbol{\delta} \colon w_\infty \mathbf{Top} \to \delta_\infty \mathbf{Mtr}, \quad ||\boldsymbol{\delta}f|| \leqslant ||f||, \qquad (\boldsymbol{\delta} \colon w\mathbf{Top} \to \delta\mathbf{Mtr}), \quad (6.78)$$

sends a w-space X to the δ-metric space $\boldsymbol{\delta}X$, consisting of the same set with the *geodetic* δ-metric associated to the weight

$$\delta(x, x') = \inf_a w(a), \qquad (6.79)$$

where $a \colon d\mathbf{I} \to X$ varies in the set of extended paths in X, from x to x'.

The δ-metric spaces obtained in this way, from w-spaces, will be said to be *geodetic*. Plainly, if $f \colon X \to X'$ is a w_∞-map, $\boldsymbol{\delta}f = f \colon \boldsymbol{\delta}X \to \boldsymbol{\delta}X'$ is continuous and satisfies the inequality of (6.78), whence it is a δ_∞-map (and 1-Lipschitz if f is).

Second, the functor

$$\mathbf{spn} \colon \delta_\infty \mathbf{Mtr} \to w_\infty \mathbf{Top}, \quad ||\mathbf{spn}(f)|| \leqslant ||f||,$$
$$(\mathbf{spn} \colon \delta\mathbf{Mtr} \to w\mathbf{Top}), \qquad\qquad\qquad (6.80)$$

has essentially been constructed in Section 6.1. For a δ-metric space Y, we let $\mathbf{spn}Y$ be the same set equipped with the symmetric topology (Section 6.1.6) and the weight-function spn (see (6.18))

$$spn(a) = \sup_t \delta(a(t_0), a(t_1)) \qquad (0 \leqslant t_0 < t_1 \leqslant 1), \qquad (6.81)$$

which we have already proved to satisfy the axioms of w-spaces (Definition 6.1.8).

The w-spaces obtained in this way will be said to be *span-metrisable*. On maps, we take again the same underlying mapping.

These two functors form an idempotent adjunction $\boldsymbol{\delta} \dashv \mathbf{spn}$, which restricts to a (covariant) Galois connection whenever we fix the underlying set. In fact, both functors do not change the underlying set; unit and counit reduce to the following inequalities

$$X \geqslant \mathbf{spn}(\boldsymbol{\delta}X), \qquad \boldsymbol{\delta}(\mathbf{spn}Y) \geqslant Y, \qquad (6.82)$$

where X is a w-space and Y a δ-metric space. (For idempotent adjunctions, see [AT], Section 6 and [LS], Lemma 4.3.)

This adjunction gives an equivalence between the full subcategories of:

(a) *span-metrisable w-spaces*, characterised by the condition $X = \mathbf{spn}(\delta X)$, or equivalently by the condition $X = \mathbf{spn}(Y)$ for a suitable δ-metric structure Y (on the same set);

(b) *geodetic δ-metric spaces*, characterised by the condition $Y = \delta(\mathbf{spn}Y)$, or equivalently by the condition $Y = \delta(X)$ for some weighted structure X on the associated topological space.

Restricting to 1-Lipschitz maps, span-metrisable w-spaces form a reflective subcategory of w**Top**, closed under limits, while geodetic δ-metric spaces form a coreflective subcategory of δ**Mtr**, closed under colimits.

Within the examples of Section 6.5.5, the standard w-line is span-metrisable, $w\mathbf{R} = \mathbf{spn}(\delta\mathbf{R})$, and the standard δ-line is geodetic, $\delta(w\mathbf{R}) = \delta\mathbf{R}$. Similarly, in higher dimension, $w\mathbf{R}^n = \mathbf{spn}(\delta\mathbf{R}^n)$ and $\delta(w\mathbf{R}^n) = \delta\mathbf{R}^n$; this also holds for the standard interval and its powers.

The standard δ-circle $\delta\mathbf{S}^1 = \delta(w\mathbf{S}^1)$ is geodetic, while the circle $\mathbf{S}^1$ with the euclidean metric of $\mathbf{R}^2$ is not, since $\delta(\mathbf{spn}(\mathbf{S}^1))$ has the obvious geodetic distance, which is bigger. The standard w-circle $w\mathbf{S}^1$ is not span-metrisable, since the weight (i.e. length) of its feasible paths has no finite upper bound, while the δ-metric of $\delta\mathbf{S}^1 = \delta(w\mathbf{S}^1)$ cannot exceed 1.

6.6.3 The length adjunction

The span-adjunction (6.77) and the adjunction of linear w-spaces (6.75) give a composed adjunction, which is again idempotent

$$\delta: Lw_\infty\mathbf{Top} \rightleftarrows \delta_\infty\mathbf{Mtr} : \mathbf{L} \qquad (\delta \dashv \mathbf{L}). \qquad (6.83)$$

Here, δ is the restriction of the functor (6.78), and equips a linear w-space X with the geodetic δ-metric $\delta(x, x') = \inf_a w(a)$. On the other hand, $\mathbf{L} = L\circ\mathbf{spn}$ takes a δ-metric space Y to the same set equipped with the symmetric topology (Section 6.1.6) and with the (linear) weight-function L which we have already defined in Definition 6.1.8

$$L(a) = \sup_t \Sigma_j \, \delta(a(t_{j-1}), a(t_j)) \qquad (0 = t_0 < t_1 < ... < t_p = 1).$$
$$(6.84)$$

Here also, maps are left 'unchanged' and $||\boldsymbol{\delta}f|| \leqslant ||f||$, $||\mathbf{L}f|| \leqslant ||f||$, so that the adjunction restricts to contractions.

6.6.4 Length-metrisable w-spaces

The length adjunction (6.83) also becomes a (covariant) Galois connection when we fix the underlying set: unit and counit reduce to inequalities

$$X \geqslant \mathbf{L}(\boldsymbol{\delta}X), \qquad \boldsymbol{\delta}(\mathbf{L}Y) \geqslant Y, \qquad (6.85)$$

where X is a linear w-space and Y a δ-metric space.

The adjunction thus gives an equivalence between the full subcategories of:

(a) *length-metrisable w-spaces*, characterised by the condition $X = \mathbf{L}(\boldsymbol{\delta}X)$, or equivalently by the condition $X = \mathbf{L}Y$ for some δ-metric structure Y on the same set (all such w-spaces are linear);

(b) *linearly geodetic δ-metric spaces*, characterised by the condition $Y = \boldsymbol{\delta}(\mathbf{L}Y)$, or $Y = \boldsymbol{\delta}X$ for some *linear* weight X on the associated topological space.

Thus, a linearly geodetic δ-metric space is geodetic; the converse need not be true. For instance, the δ-metric subspace $Y \subset \boldsymbol{\delta}\mathbf{R}^2$ consisting of the union of the two axes is geodetic, but not linearly geodetic: the points $y = (-1, 0)$ and $y' = (0, 1)$ have $\delta(y, y') = 1$ but all feasible paths a in Y, from y to y', have length $L(a) = 2$.

The two notions of 'metrisability' of w-spaces are not comparable. Indeed, the w-line $\mathbf{wR}$ is metrisable in both senses. The w-plane $\mathbf{wR}^2$ is span-metrisable and not linear, hence not length-metrisable. The standard w-circle $\mathbf{wS}^1$ (Section 6.5.5) is only length-metrisable. Finally, in Theorem 6.7.2, we will show that the irrational rotation w-space W_ϑ has a trivial δ-metric on $\boldsymbol{\delta}W_\vartheta$ (always zero), whence it is neither span- nor length-metrisable.

6.6.5 Directed spaces

Finally, we have a forgetful functor

$$\mathbf{d} \colon w_\infty \mathbf{Top} \to \mathrm{d}\mathbf{Top}, \qquad (6.86)$$

which sends a w-space to the same topological space, equipped with the distinguished paths obtained from the feasible ones, *by reparametrisation along weakly increasing maps* $\mathbf{I} \to \mathbf{I}$.

Composing $\mathbf{L}\colon \delta_\infty \mathbf{Mtr} \to \mathrm{Lw}_\infty \mathbf{Top}$ with the latter, we get the forgetful functor $\delta_\infty \mathbf{Mtr} \to \mathrm{d}\mathbf{Top}$ already considered in Section 6.1.9, which distinguishes the L-feasible paths of a δ-metric space – already closed under increasing reparametrisation.

There is an obvious comparison (of categories, of course)

$$\mathrm{w}\Pi_1(X) \to \uparrow\Pi_1(\mathrm{d}X), \qquad x \mapsto x, \quad [a] \mapsto [a], \qquad (6.87)$$

and one can prove that it is an isomorphism of categories for every w-space X whose feasible paths are already closed under reparametrisation along weakly increasing maps $\mathbf{I} \to \mathbf{I}$.

6.7 Weighted non-commutative tori and their classification

Throughout this section ϑ is an irrational number. We now introduce the *irrational rotation w-spaces* W_ϑ (Section 6.7.1), which have a classification similar to the irrational rotation C*-algebras A_ϑ, including the metric aspects which cannot be obtained with the cubical sets C_ϑ (Chapter 2) or the d-spaces D_ϑ (classified here, in Section 6.7.6).

Analogous results have been obtained in [G10] for 'normed' cubical sets and their 'normed' homology – an earlier approach to weighted algebraic topology.

6.7.1 Irrational rotation w-spaces

The irrational rotation C*-algebras A_ϑ and their classifications – up to isomorphism or up to strong Morita equivalence – have been reviewed in Section 2.5.1. We have also introduced a family of cubical sets C_ϑ, whose classification up to isomorphism is the same as the classification of A_ϑ up to strong Morita equivalence.

We define now the *irrational rotation w-space*

$$W_\vartheta = (\mathrm{w}\mathbf{R})/G_\vartheta, \qquad (6.88)$$

whose feasible paths are the *projection of the feasible paths* of $\mathrm{w}\mathbf{R}$, as we prove below. On the additive group $\mathbf{R}$ and its subgroup G_ϑ we use the standard weight

$$w(x) = \delta(0, x), \qquad (6.89)$$

i.e. $w(x) = x$ when $x \geqslant 0$, and $w(x) = \infty$ otherwise. We also have the (restricted) standard weight $w(x) = x$ on the additive monoids $\mathbf{R}^+$ and $G_\vartheta^+ = G_\vartheta \cap \mathbf{R}^+$, formed by the elements of finite weight.

6.7.2 Theorem

(a) The fundamental weighted monoid of W_ϑ at each point $\overline{x} \in \mathbf{R}/G_\vartheta$ is isometrically isomorphic to the additive weighted monoid G_ϑ^+, via the weight function

$$w \colon \mathrm{w}\pi_1(W_\vartheta, \overline{x}) \to [0, \infty[, \qquad \mathrm{Im}(w) = G_\vartheta^+. \tag{6.90}$$

(b) Let us choose a representative $x \in \mathbf{R}$ of $\overline{x}$; for every feasible *path $\overline{a} \colon \mathrm{w}\mathbf{I} \to W_\vartheta$ starting at $\overline{x}$ there is precisely one increasing path $a \colon \mathrm{w}\mathbf{I} \to \mathbf{R}$ which lifts it and starts at x. Moreover, the weight of $\overline{a}$ in W_ϑ coincides with the weight of a, $w(a) = a(1) - a(0) = a(1) - x$.*

(c) The w-space W_ϑ is linear; the associated metric space δW_ϑ (cf. (6.78)) is indiscrete, with $\delta(\overline{x}, \overline{y})$ always zero, so that W_ϑ is neither span- nor length-metrisable.

Proof We begin by proving (b), applying Theorem 6.5.4(b) which characterises the weight of the orbit w-space W_ϑ.

Take a feasible path $\overline{a} \colon \mathrm{w}\mathbf{I} \to W_\vartheta$ starting at $\overline{x}$, and choose a representative $x \in \mathbf{R}$ of the latter. Since $w(\overline{a}) < \infty$ is the greatest lower bound of the weights of the paths in $\mathbf{R}$ which lift it, there exists some feasible (i.e. increasing) path $a \colon \mathrm{w}\mathbf{I} \to \mathrm{w}\mathbf{R}$ which lifts a.

Now, up to G_ϑ-translations, we may assume that a starts at x (without changing its weight). But there is only one path which satisfies these conditions. Indeed, if b also does, the image of the continuous mapping $a - b \colon \mathbf{I} \to \mathbf{R}$ must be contained in G_ϑ, which is totally disconnected; thus $a - b$ is constant, and $a(0) = x = b(0)$ gives $a = b$. It follows that $w(\overline{a}) = w(a)$, where a is the unique path in $\mathbf{R}$ which starts at x and lifts $\overline{a}$.

For (c), the fact that W_ϑ is linear follows from (a) and the linearity of the weight in $\mathrm{w}\mathbf{R}$. The other assertions are obvious, taking into account the characterisations of span- and length-metrisable w-spaces, in Sections 6.6.2 and 6.6.4.

For (a), let us consider the weight function (6.90). First, we show that its image is G_ϑ^+. For a loop $\overline{a}$, we have $w(\overline{a}) = w(a)$ where the (increasing) lifting a starts at x and ends at some $x' \geqslant x$, which also projects to $\overline{x}$; thus $w(\overline{a}) = x' - x \in G_\vartheta^+$. On the other hand, if $g \in G_\vartheta^+$, any increasing path $a \colon x \to x + g$ projects to a loop at $\overline{x}$, whose weight is g.

Finally, we must prove that the weight function is injective. Let $\overline{a}, \overline{b}$ be two loops at $\overline{x}$ with the same weight $g \in G_\vartheta^+$, and let a, b be their liftings which start at x; they have again the same weight g, which means that they end at the same point $x' = x + g$. Then, the increasing path $c = a \vee b \colon \mathbf{I} \to \mathbf{R}$ also goes from x to x'; since $a \leqslant c$, the affine interpolation from a to c is an extended 2-homotopy $a \prec_2 c$ (cf. Section 6.3.6); similarly, $b \prec_2 c$ and $a \simeq_2 b$, whence $[\bar{a}] = [\bar{b}]$. $\qquad\square$

6.7.3 Theorem (Isometric classification)

Let ϑ, ϑ' be irrationals. The w-spaces $w\mathbf{R}/G_\vartheta$ and $w\mathbf{R}/G_{\vartheta'}$ are isometrically isomorphic if and only if $G_\vartheta^+ = G_{\vartheta'}^+$ (as subsets of $\mathbf{R}$), if and only if $G_\vartheta = G_{\vartheta'}$ (in the same sense), if and only if $\vartheta' \in \mathbf{Z} \pm \vartheta$.

Proof If our w-spaces are isometrically isomorphic, their fundamental weighted monoids (independently of the base point) are also: $G_\vartheta^+ \cong G_{\vartheta'}^+$ (isometrically). Since the values of the weight $w \colon G_\vartheta^+ \to \mathbf{R}$ form the set G_ϑ^+, it follows that $G_\vartheta^+ = G_{\vartheta'}^+$, which implies that G_ϑ (the additive subgroup of $\mathbf{R}$ generated by G_ϑ^+) coincides with $G_{\vartheta'}$. If this is the case, then $\vartheta = a + b\vartheta'$ and $\vartheta' = c + d\vartheta$ for suitable integers a, b, c, d; whence $\vartheta = a + bc + bd\vartheta$ and $d = \pm 1$, so that $\vartheta' = c \pm \vartheta$. Finally, if $\vartheta' \in \mathbf{Z} \pm \vartheta$, then $G_\vartheta = G_{\vartheta'}$ and $w\mathbf{R}/G_\vartheta = w\mathbf{R}/G_{\vartheta'}$. $\qquad\square$

6.7.4 Theorem (Lipschitz isomorphic classification)

Let ϑ, ϑ' be irrationals. The w-spaces $w\mathbf{R}/G_\vartheta$ and $w\mathbf{R}/G_{\vartheta'}$ are Lipschitz isomorphic if and only if the equivalent conditions of the following lemma (6.7.5) hold.

Proof One implication follows from Theorem 6.7.2: if our w-spaces are Lipschitz isomorphic, their fundamental weighted monoids G_ϑ^+ and $G_{\vartheta'}^+$ are also, by the functorial properties of $w\Pi_1$ (Section 6.3.6).

For the converse, let ϑ' belong to the closure $\{\vartheta\}_{RT}$; it suffices to consider the cases $\vartheta' \in \vartheta + \mathbf{Z}$ and $\vartheta' = \vartheta^{-1}$. In the first case, G_ϑ and $G_{\vartheta'}$ coincide, as well as their action on $w\mathbf{R}$; in the second, the Lipschitz isomorphism of weighted spaces

$$f \colon w\mathbf{R} \to w\mathbf{R}, \qquad f(t) = |\vartheta|.t, \tag{6.91}$$

restricts to a group-isomorphism $f' \colon G_\vartheta \to G_{\vartheta'}$, consistent with the actions: $f(t + g) = f(t) + f'(g)$; therefore (6.91) induces a Lipschitz isomorphism $w\mathbf{R}/G_\vartheta \to w\mathbf{R}/G_{\vartheta'}$. $\qquad\square$

6.7.5 Lemma

Let ϑ, ϑ' be irrationals. The following conditions are equivalent:

(a) the weighted groups G_ϑ and $G_{\vartheta'}$ are Lipschitz isomorphic;

(b) the weighted monoids G_ϑ^+ and $G_{\vartheta'}^+$ are Lipschitz isomorphic;

(c) G_ϑ and $G_{\vartheta'}$ are isomorphic as ordered groups (with respect to the total orders induced by $\mathbf{R}$);

(d) ϑ and ϑ' are conjugate under the action of $\mathrm{GL}(2, \mathbf{Z})$ (Section 2.5.1);

(e) ϑ' belongs to the closure $\{\vartheta\}_{RT}$ of $\{\vartheta\}$ under the mappings $R(t) = t^{-1}$ and $T^{\pm 1}(t) = t \pm 1$.

Proof The equivalence of the last three conditions has been proved in Lemma 2.5.7.

Further, (a) implies (b), because G_ϑ^+ is the monoid of elements of G_ϑ having a finite weight. And (b) implies (c), because G_ϑ is the group canonically associated to the cancellative monoid G_ϑ^+, ordered with the latter as a positive cone.

Finally, to prove that (e) implies (a), let ϑ' belong to the closure $\{\vartheta\}_{RT}$; as in the proof of the previous theorem, it suffices to consider the cases $\vartheta' \in \vartheta + \mathbf{Z}$ and $\vartheta' = \vartheta^{-1}$. In the first, $G_\vartheta = G_{\vartheta'}$; in the second, the Lipschitz isomorphism of weighted spaces $f \colon \mathrm{w}\mathbf{R} \to \mathrm{w}\mathbf{R}$ considered above (in (6.91)) restricts to a Lipschitz isomorphism of weighted abelian groups $G_\vartheta \to G_{\vartheta'}$.   □

6.7.6 Classifying the irrational rotation d-spaces

Consider now the irrational-rotation d-space $D_\vartheta = {\uparrow}\mathbf{R}/G_\vartheta$, defined in (2.78), viewed now as $\mathbf{d}W_\vartheta$.

By (6.87), it follows that its fundamental category is isomorphic to the category which underlies $\mathrm{w}\Pi_1(W_\vartheta)$, forgetting the weight of the latter. Therefore, at any point $\bar{x}$, the fundamental monoid ${\uparrow}\pi_1(D_\vartheta, \bar{x})$ is isomorphic to the monoid G_ϑ^+.

By the same argument as in the proof of Theorem 6.7.4, the d-spaces D_ϑ and $D_{\vartheta'}$ are isomorphic if and only if the equivalent conditions of Lemma 6.7.5 hold.

6.8 Tentative formal settings for weighted algebraic topology

We only sketch a few ideas, based on the previous structures for δ-metric spaces, w-spaces, weighted categories, together with another structure which we mention here: weighted cubical sets.

6.8.1 A tentative definition

Let us say that a *concrete symmetric wI4-category* is a symmetric dI4-homotopical category (Section 4.2.6)

$$\mathbf{A}_\infty = (\mathbf{A}_\infty, R, I, \partial^\alpha, e, r, g^\alpha, s, J, c, z),$$

which is made concrete by a standard point E (Section 4.5.7) and equipped with a *weight* function, defined on each set of paths

$$w = w_X : \mathbf{A}_\infty(\mathbf{I}, X) \to [0, \infty] \qquad (\mathbf{I} = I(E)). \tag{6.92}$$

The following axioms on the family (w_X) (X varying in $\mathrm{Ob}\mathbf{A}_\infty$) are assumed (for every point $x\colon E \to X$, every pair of consecutive paths $a, b\colon \mathbf{I} \to X$, every isomorphism $\rho\colon \mathbf{I} \to \mathbf{I}$, and every map $f\colon X \to Y$)

$$w_X(0_x) = 0,$$

$$w_X(a) \vee w_X(b) \leqslant w_X(a + b) \leqslant w_X(a) + w_X(b), \tag{6.93}$$

$$w_X(a\rho) = w_X(a), \qquad w_{X^{\mathrm{op}}}(a^{\mathrm{op}}) = w_X(a), \qquad ||f|| < \infty.$$

In the last condition we are using the *multiplicative weight*, or *Lipschitz weight* of a map $f\colon X \to Y$, which is defined as follows (as for w-spaces, in Section 6.5.2):

$$||f|| = \min\{\lambda \in [0, \infty] \mid \forall a \in \mathbf{A}_\infty(\mathbf{I}, X), \ w_Y(f \circ a) \leqslant \lambda.w_X(a)\}. \tag{6.94}$$

With this weight, $\mathbf{A}_\infty$ is a multiplicatively weighted category: all identities have $||\mathrm{id}X|| \leqslant 1$ and composition gives $||gf|| \leqslant ||f||.||g||$. The same is true of its wide subcategory $\mathbf{A}_1$ formed of all objects and *1-Lipschitz maps*, with $||f|| \leqslant 1$.

6.8.2 Examples

In all the following cases

(a) $\mathbf{A}_\infty = \delta_\infty \mathbf{Mtr} \supset \delta\mathbf{Mtr} = \mathbf{A}_1$;
(b) $\mathbf{A}_\infty = \mathrm{w}_\infty \mathbf{Top} \supset \mathrm{w}\mathbf{Top} = \mathbf{A}_1$;
(c) $\mathbf{A}_\infty = \mathrm{w}_\infty \mathbf{Cat} \supset \mathrm{w}\mathbf{Cat} = \mathbf{A}_1$;

the hypotheses above are satisfied. Moreover, the wide subcategory $\mathbf{A}_1$ is a concrete symmetric dI2-homotopical subcategory of $\mathbf{A}_\infty$, and – in its own right – a concrete symmetric dIP2-homotopical category.

Such hypotheses should be sufficient to work as in the relative setting of Section 5.8, and to enrich the fundamental category with an additive weight induced by the family (w_X).

Notice also that, in all these examples, we have a family of functors

$$\lambda_{\mathbf{A}} : \mathbf{A}_\infty \to \mathbf{A}_\infty \qquad (\lambda \in [0, \infty[), \tag{6.95}$$

which, perhaps, should be taken into account in a formal setting.

6.8.3 A defective case

Weighted cubical sets (introduced in [G10] as normed cubical sets) form a 'defective' case:

$$\mathrm{w}_\infty \mathbf{Cub} \supset \mathrm{w}\mathbf{Cub}. \tag{6.96}$$

A *weighted cubical set* is a cubical set X equipped with a sequence of weights which annihilate on degenerate elements

$$w : X_n \to [0, +\infty], \qquad w(e_i(a)) = 0 \qquad (a \in X_n). \tag{6.97}$$

We do *not* require any coherence condition for faces, nor any restriction on the weight of a point; for instance, a degenerate edge must have weight zero, but its vertices can have any weight.

The category $\mathrm{w}_\infty \mathbf{Cub}$ contains all morphisms of cubical sets $f : X \to Y$, with a finite weight:

$$||f|| = \min\{\lambda \in [0, \infty] \mid \forall x \in X_n, \ w(f_n(x)) \leqslant \lambda . w(x)\}, \tag{6.98}$$

while its wide subcategory $\mathrm{w}\mathbf{Cub}$ only contains the *weak contractions*, with $w(f_n(x)) \leqslant w(x)$, for all $x \in X_n$.

Here $\mathrm{w}_\infty \mathbf{Cub}$ and $\mathrm{w}\mathbf{Cub}$ are *only* dI1-categories. Therefore, to study the homotopy of weighted cubical sets, one should likely use a *relative* framework, like $\mathbf{Cub} \to \mathrm{d}\mathbf{Top}$, with a weighted geometric realisation as a forgetful functor (between pairs of categories)

$$w\mathcal{R} : (\mathrm{w}_\infty \mathbf{Cub}, \mathrm{w}\mathbf{Cub}) \to (\mathrm{w}_\infty \mathbf{Top}, \mathrm{w}\mathbf{Top}). \tag{6.99}$$

Appendix A
Some points of category theory

In this book, category theory is used extensively, if at an elementary level. The notions of category, functor and natural transformation are used throughout, together with standard tools like limits, colimits and adjoint functors.

The brief review of this appendix is also meant to fix the notation used here. Proofs can be found in the texts mentioned in Section A1.1, except for some non-standard points at the end of this chapter.

A1 Basic notions

A1.1 Smallness

Something must be said about set-theoretical aspects, to make precise the meaning of the category of 'all' sets', or 'all' topological spaces, and so on.

We work within the NBG (von Neumann–Bernays–Gödel) theory, where there are *sets* and *classes*, and the class of all sets or all spaces makes sense. In a category **A** the objects form a class Ob**A** and the morphisms form a class Mor**A**; but, for every pair X, Y of objects, we assume that the morphisms $X \to Y$, also called maps or arrows, form a *set* $\mathbf{A}(X, Y)$. The category is said to be *small* if the class Ob**A** is a *set*, and large otherwise.

This approach is followed in Mitchell's book [Mi] and, essentially, also in Adàmek, Herrlich and Strecker [AHS]; a brief exposition of NBG can be found in the Appendix of [Ke]. Thus, **Set** is the (large) category of sets and mappings, **Top** is the (large) category of topological spaces and continuous mappings, etc. Small categories and their functors also form a (large) category, **Cat**.

397

One can easily translate everything in the other set-theoretical setting widely used in category theory, which is based on *universes*: see the texts of Mac Lane [M3] and Borceux [Bo]. Then, a basic universe is chosen, and a *small* set is any element of the latter; **Set** is defined as the category of small sets, **Top** as the category of small topological spaces (i.e. having a small underlying set), and so on. This is slightly more complicated, but has the advantage of allowing one to consider categories of large categories, making use of a hierarchy of universes.

A1.2 Basic terminology

We assume that the reader is familiar with the very basic concepts and notation of category theory, like:

- category; the identity morphism id$X$ (or 1_X) of an object X in a category; isomorphism (or iso), monomorphism (or mono) and epimorphism (or epi); retract, split monomorphism (or section) and split epimorphism (or retraction), in a category;
- functor; the identity functor id$\mathbf{C}$ (or $1_\mathbf{C}$) of a category $\mathbf{C}$; faithful and full functor; forgetful functor between categories of structured sets;
- subcategory and its inclusion functor, full subcategory; cartesian product of categories and its projection functors.

(In a small category we may use an additive notation, for composition and identities, see Section 8 of the Introduction.)

Let us recall something about the two-dimensional structure of categories, which will be further analysed below (Sections A5.1 and A5.2). A natural transformation $\varphi\colon F \;\to\; G\colon \mathbf{C} \;\to\; \mathbf{D}$, between functors $F, G\colon \mathbf{C} \to \mathbf{D}$, consists of the following data:

- for each object X of $\mathbf{C}$, a morphism $\varphi X\colon FX \to GX$ in $\mathbf{D}$ (called the *component* of φ on X, and also written as φ_X, or φ^X, or φ),

so that, for every arrow $f\colon X \to X'$ in $\mathbf{C}$, we have a commutative square in $\mathbf{D}$:

$$
\begin{array}{ccc}
FX & \xrightarrow{\;\varphi X\;} & GX \\
{\scriptstyle Ff}\big\downarrow & & \big\downarrow{\scriptstyle Gf} \\
FX' & \xrightarrow[\;\varphi X'\;]{} & GX'
\end{array}
\qquad\qquad
\begin{array}{l}
\varphi X'.F(f) = G(f).\varphi X \\[2mm]
(\textit{naturality condition}).
\end{array}
\qquad \text{(A.1)}
$$

In particular, *the identity of a functor* $F \colon \mathbf{C} \to \mathbf{D}$ is the natural transformation $\mathrm{id}F \colon F \to F$, of components $(\mathrm{id}F)_X = \mathrm{id}(FX)$.

Natural transformations have a *vertical composition*

$$
\mathbf{C} \xrightarrow[\quad H \quad]{\substack{F \\ \downarrow \varphi \\ \downarrow \psi}} \mathbf{D}
\qquad
\begin{aligned}
&\psi\varphi \colon F \to H, \\[2mm]
&(\psi\varphi)(X) = \psi X.\varphi X \colon FX \to HX,
\end{aligned}
\qquad (A.2)
$$

and a *whisker composition*, or *reduced horizontal composition*, with functors

$$
\mathbf{C}' \xrightarrow{\ H\ } \mathbf{C} \xrightarrow[\quad G \quad]{\substack{F \\ \downarrow \varphi}} \mathbf{D} \xrightarrow{\ K\ } \mathbf{D}' \qquad (A.3)
$$

$$
K\varphi H \colon KFH \to KGH \colon \mathbf{C}' \to \mathbf{D}', \quad (K\varphi H)(X') = K(\varphi(HX')).
$$

An *isomorphism of functors* is a natural transformation $\varphi \colon F \to G$ which is invertible (with respect to vertical composition).

A1.3 Universal properties, products and equalisers

Many definitions in category theory are based on a universal property.

For instance, in a category $\mathbf{C}$, the *product* of a family $(X_i)_{i \in I}$ of objects (indexed on a *set* I) is defined as an object X equipped with a family of morphisms $p_i \colon X \to X_i$ $(i \in I)$, called *projections*, which satisfy the following universal property:

(i) for every object Y and every family of morphisms $f_i \colon Y \to X_i$, there exists a unique morphism $f \colon Y \to X$ such that, for all $i \in I$, $p_i f = f_i$.

The solution need not exist. But it is determined up to a unique *coherent* isomorphism, in the sense that if also Y is a product of the family $(X_i)_{i \in I}$ with projections $q_i \colon Y \to X_i$, then the morphism $f \colon X \to Y$ which commutes with all projections (i.e. $q_i f = p_i$, for all indices i) is an isomorphism. Therefore, one speaks of *the* product of the family (X_i), denoted as $\prod_i X_i$.

We say that a category $\mathbf{C}$ *has products* (resp. *finite products*) if every family of objects indexed *on a set* (resp. on a finite set) has a product in $\mathbf{C}$.

In particular, the product of the empty family of objects $\emptyset \to \mathrm{Ob}\mathbf{C}$ means an object X (equipped with no projections) such that for every object Y (equipped with no maps) there is a unique morphism $f \colon Y \to X$ (satisfying no conditions). The solution is called the *terminal* object of $\mathbf{C}$;

again, it need not exist, but is determined up to a unique isomorphism. It can be written as $\top$.

In **Set** and **Top**, all products exist, and are the usual cartesian ones. It is easy to prove that a category has finite products if and only if it has binary products $X_1 \times X_2$ and a terminal object.

Products are a basic instance of a much more general concept recalled below, the limit of a functor (Section A2.1). Another basic instance is the *equaliser* of a pair $f, g\colon X \to Y$ of 'parallel' maps of **C**; this is (an object E with) a map $m\colon E \to X$ such that $fm = gm$ and the following universal property holds:

(ii) every map $h\colon Z \to X$ such that $fh = gh$ factors uniquely through m (i.e. there exists a unique map $w\colon Z \to E$ such that $mw = h$).

The equaliser morphism m is necessarily a monomorhism, and, by definition, a *regular mono*. In **Set** (resp. **Top**), the equaliser of two parallel maps $f, g\colon X \to Y$ is the embedding in X of the maximal subset (resp. subspace) of X on which they coincide. Therefore, regular monomorphisms coincide with monomorphisms (or injective mappings) in **Set**, but 'amount' to inclusion of subspaces in **Top**.

It is easy to prove that a split monomorphism is always a regular mono.

A1.4 Duality, sums and coequalisers

If **C** is a category, the *opposite* (or *dual*) category, written $\mathbf{C}^{\mathrm{op}}$ or $\mathbf{C}^*$, has the same objects as **C** and 'reversed' arrows,

$$\mathbf{C}^{\mathrm{op}}(X, Y) = \mathbf{C}(Y, X), \tag{A.4}$$

with 'reversed composition' $g * f = f.g$ and the same identities.

Every notion of category theory has a dual notion, which comes from the opposite category (or categories): thus, monomorphism and epimorphism are dual to each other, while isomorphism is a self-dual notion. Dual notions are often distinguished by the prefix 'co-'.

The *sum*, or *coproduct*, of a family $(X_i)_{i \in I}$ of objects of **C** is dual to their product. Explicitly, it is an object X equipped with a family of morphisms $u_i\colon X_i \to X$ ($i \in I$), called *injections*, which satisfy the following universal property:

(i*) for every object Y and every family of morphisms $f_i\colon X_i \to Y$, there exists a unique morphism $f\colon X \to Y$ such that, for all $i \in I$, $fu_i = f_i$.

Again, if the solution exists, it is determined up to a unique coherent isomorphism. The sum of the family (X_i) is denoted as $\sum_i X_i$, or $X_1 +$

$\cdots + X_n$ in a finite case. The sum of the empty family is the *initial* object $\bot$: this means that, for every object X, there is precisely one map $\bot \to X$.

The *coequaliser* of a pair $f, g \colon X \to Y$ of parallel maps of **C** is a map $p \colon Y \to C$ such that $pf = pg$ and:

(ii*) every map $h \colon Y \to Z$ such that $hf = hg$ factors uniquely through p (i.e. there exists a unique map $w \colon C \to Z$ such that $wp = h$).

A reader not familiar with these notions should begin by performing these constructions in **Set** and **Top**. In **Top**, a *regular epimorphism* (i.e. a coequaliser map) amounts to a projection on a quotient space. Sums and coequalisers are particular instances of the *co*limit of a functor (Section A2.1).

A1.5 Isomorphism and equivalence of categories

(a) An *isomorphism of categories* is a functor $F \colon \mathbf{C} \to \mathbf{D}$ which is invertible. This means that F admits an *inverse*, i.e. a functor $G \colon \mathbf{D} \to \mathbf{C}$ such that $GF = \mathrm{id}\mathbf{C}$ and $FG = \mathrm{id}\mathbf{D}$.

For instance, the category **Ab** of abelian groups is (clearly) isomorphic to the category of **Z**-modules (and **Z**-homomorphisms). Being isomorphic categories is written as $\mathbf{C} \cong \mathbf{D}$.

(b) More generally, an *equivalence of categories* is a functor $F \colon \mathbf{C} \to \mathbf{D}$ which is invertible up to isomorphism of functors (Section A1.2), i.e. there exists a functor $G \colon \mathbf{D} \to \mathbf{C}$ such that $GF \cong \mathrm{id}\mathbf{C}$ and $FG \cong \mathrm{id}\mathbf{D}$.

An *adjoint equivalence of categories* is a coherent version of this notion, namely a four-tuple $(F, G, \eta, \varepsilon)$ where:

- $F \colon \mathbf{C} \to \mathbf{D}$ and $G \colon \mathbf{D} \to \mathbf{C}$ are functors;
- $\eta \colon \mathrm{id}\mathbf{C} \to GF$ and $\varepsilon \colon FG \to \mathrm{id}\mathbf{D}$ are isomorphisms of functors;
- $F\eta = (\varepsilon F)^{-1} \colon F \to FGF$, $\eta G = (G\varepsilon)^{-1} \colon G \to GFG$ (*coherence conditions*).

The following conditions on a functor $F \colon \mathbf{C} \to \mathbf{D}$ are equivalent, forming a very useful *characterisation of equivalences*:

(i) F is an equivalence of categories;

(ii) F can be completed to an adjoint equivalence of categories

$$(F, G, \eta, \varepsilon);$$

(iii) F is faithful, full and *essentially surjective on objects*.

The last condition means that: for every object Y of $\mathbf{D}$ there exists some object X in $\mathbf{C}$ such that $F(X)$ is isomorphic to Y in $\mathbf{D}$. The proof of the equivalence of these three conditions is rather long and requires the axiom of choice, for classes.

One says that two categories $\mathbf{C}, \mathbf{D}$ are *equivalent*, written $\mathbf{C} \simeq \mathbf{D}$, if there exists an equivalence of categories, as above. This is indeed an equivalence relation, as follows easily from the previous characterisation.

For instance, the category of finite sets (and mappings between them) is equivalent to its full subcategory of finite cardinals, which is small (and therefore cannot be isomorphic to the former).

A1.6 A digression on mathematical structures and categories

When studying a mathematical structure with the help of category theory, it is crucial to choose the 'right' kind of structure and the 'right' kind of morphisms, so that the result is sufficiently general and 'natural' to have good properties (with respect to the goals of our study), even if we are interested in more particular situations.

For instance, the category $\mathbf{Top}$ of topological spaces and continuous mappings is a natural framework for studying topology. Among its good properties there is the fact that all (co)products and (co)equalisers exist, and are computed *as in* $\mathbf{Set}$, then equipped with a suitable topology. (More generally, this is true of all limits and colimits, and is a consequence of the fact that the forgetful functor $\mathbf{Top} \to \mathbf{Set}$ has a left *and* a right adjoint, see below.) Hausdorff spaces are certainly important, but it is often better to view them in $\mathbf{Top}$, as their category is less well behaved: coequalisers exist, but are not computed as in $\mathbf{Set}$, i.e. preserved by the forgetful functor to $\mathbf{Set}$.

Many category theorists would agree with [M3], saying that even $\mathbf{Top}$ is not sufficiently good, because it is not a cartesian closed category, and prefer, for instance, the category of compactly generated spaces; however – since homotopy theory is our goal – we are essentially satisfied with the fact that the standard interval is exponentiable in $\mathbf{Top}$ (with all its cartesian powers, see Section A4.3).

Similarly, if we are interested in ordered sets, it is generally better to view them in the category of *pre*ordered sets and (weakly) increasing mappings, where (co)products and (co)equalisers not only exist, but again are computed *as in* $\mathbf{Set}$, with a suitable preorder. On the other hand, the category of *totally* ordered sets does not have (even binary) products, and – generally speaking – is of little interest; nevertheless, one should not forget that the category $\mathbf{\Delta}$ of finite positive ordinals (and increasing maps) is important as a basis of presheaves (see Section A1.8).

Another point to be kept in mind is that the isomorphisms of the category (i.e. its invertible arrows) should indeed 'preserve' the structure we are interested in, or we risk studying something different from our purpose. As a trivial example, the category T of topological spaces and *all* mappings between them has practically nothing to do with topology: an isomorphism of T is any bijection between topological spaces. Indeed, T is *equivalent to the category of sets* (according to the previous definition, in Section A1.5), and is a 'deformed' way of looking at the latter.

Less trivially, the category M of metric spaces and continuous mappings misses crucial properties of metric spaces, since its invertible morphisms do not preserve completeness. In fact, M is equivalent to the category of *metrisable topological spaces* and continuous mappings, and should be viewed in this way. A 'reasonable' category of metric spaces should be based on *Lipschitz* maps, or – more particularly – on weak contractions (see Section 6.1).

A1.7 Categories of functors

Let $\mathbf{S}$ be a small category and $S = \mathrm{Ob}\mathbf{S}$ its set of objects. For any category $\mathbf{C}$, one writes $\mathbf{C}^{\mathbf{S}}$ for the category whose objects are the functors $F\colon \mathbf{S} \to \mathbf{C}$ and whose morphisms are the natural transformations $\varphi\colon F \to G\colon \mathbf{S} \to \mathbf{C}$, with vertical composition.

Notice that the natural transformations between two given functors $F, G\colon \mathbf{S} \to \mathbf{C}$ do form a *set*

$$\mathbf{C}^{\mathbf{S}}(F, G) = \mathrm{Nat}(F, G), \tag{A.5}$$

since this class can be embedded in a product of sets indexed on a set: $\prod_{i \in S} \mathbf{C}(F(i), G(i))$. Moreover, if $\mathbf{C}$ is also small, $\mathbf{C}^{\mathbf{S}}$ is too.

In particular, the ordinal category $\mathbf{2}$ (with two objects $0, 1$ and one non-identity arrow, $0 \to 1$), gives $\mathbf{C}^{\mathbf{2}}$, the *category of morphisms* of $\mathbf{C}$, where a map $(u_0, u_1)\colon f \to g$ is a commutative square of $\mathbf{C}$; these are composed as below, on the right

$$
\begin{array}{ccc}
A_0 & \xrightarrow{u_0} & B_0 \\
f\downarrow & & \downarrow g \\
A_1 & \xrightarrow[u_1]{} & B_1
\end{array}
\qquad
\begin{array}{ccccc}
A_0 & \xrightarrow{u_0} & B_0 & \xrightarrow{v_0} & C_0 \\
f\downarrow & & \downarrow g & & \downarrow h \\
A_1 & \xrightarrow[u_1]{} & B_1 & \xrightarrow[v_1]{} & C_1
\end{array}
\tag{A.6}
$$

A natural transformation $\varphi\colon F \to G\colon \mathbf{A} \to \mathbf{B}$ can be viewed as a functor $\mathbf{A} \times \mathbf{2} \to \mathbf{B}$ or, equivalently, as a functor $\mathbf{A} \to \mathbf{B}^{\mathbf{2}}$.

A functor $F\colon \mathbf{C} \to \mathbf{Set}$ is said to be *representable* if it is isomorphic to a functor $\mathbf{C}(C, -)\colon \mathbf{C} \to \mathbf{Set}$, for some object C in $\mathbf{C}$ (which is determined by F, up to isomorphism). Then, the Yoneda lemma describes the natural transformations $F \to G$, for every functor $G\colon \mathbf{C} \to \mathbf{Set}$ [M3].

A1.8 Categories of presheaves

A functor $\mathbf{S}^{\mathrm{op}} \to \mathbf{C}$, defined on the opposite category $\mathbf{S}^{\mathrm{op}}$, is also called a *presheaf* of $\mathbf{C}$ on the (small) category $\mathbf{S}$. They form a category $\mathrm{Psh}(\mathbf{S}, \mathbf{C}) = \mathbf{C}^{\mathbf{S}^{\mathrm{op}}}$, whose arrows are the natural transformations between such functors.

The small category $\mathbf{S}$ is canonically embedded in its presheaf category $\mathbf{Set}^{\mathbf{S}^{\mathrm{op}}}$, by the *Yoneda embedding*

$$y\colon \mathbf{S} \to \mathbf{C}^{\mathbf{S}^{\mathrm{op}}}, \qquad y(i) = \mathbf{S}(-, i)\colon \mathbf{S}^{\mathrm{op}} \to \mathbf{Set}, \qquad (\mathrm{A}.7)$$

which sends every object i to the corresponding *representable* presheaf (Section A1.7).

Taking as $\mathbf{S}$ the category $\mathbf{\Delta}$ of finite positive ordinals (and increasing maps), one gets the category $\mathrm{Smp}\mathbf{C} = \mathbf{C}^{\mathbf{\Delta}^{\mathrm{op}}}$ of simplicial objects in $\mathbf{C}$, and – in particular – the category of simplicial sets $\mathbf{Smp} = \mathrm{Smp}\mathbf{Set} = \mathbf{Set}^{\mathbf{\Delta}^{\mathrm{op}}}$. The Yoneda embedding sends the ordinal n to the simplicial set Δ^n, freely generated by one simplex of dimension n.

Cubical objects also form a presheaf category $\mathrm{Cub}\mathbf{C} = \mathbf{C}^{\mathbb{I}^{\mathrm{op}}}$, where $\mathbb{I}$ is the subcategory of $\mathbf{Set}$ consisting of the *elementary cubes* $2^n = \{0, 1\}^n$, together with the maps $2^m \to 2^n$ which delete some coordinates and insert some 0's and 1's, without modifying the order of the remaining coordinates. For cubical objects with connections and/or symmetries, viewed as presheaves, see [GM]. (Cubical sets are studied in Section 1.6. Sheaves on a site $(\mathbf{S}, J)$ are recalled in Section 5.1.3.)

A1.9 Universal arrows

We end this section by recalling a general way of formalising universal properties, based on a functor $U\colon \mathbf{A} \to \mathbf{C}$ and an object X of $\mathbf{C}$.

A *universal arrow from the object X to the functor U* is a pair $(A, \eta\colon X \to UA)$ consisting of an object A of $\mathbf{A}$ and arrow η of $\mathbf{C}$ which is universal, in the sense that every similar pair $(B, f\colon X \to UB)$ factors uniquely through (A, η): in other words, there exists a unique $g\colon A \to B$ in **A** such that the following triangle commutes in **C**

$$X \xrightarrow{\ \eta\ } UA$$
$$\begin{array}{c} \qquad\qquad \downarrow{\scriptstyle Ug} \qquad\qquad Ug.\eta = f. \qquad\qquad \text{(A.8)} \\ f\searrow \quad \\ UB \end{array}$$

Dually, a *universal arrow from the functor U to the object X* is a pair $(A, \varepsilon\colon UA \to X)$ consisting of an object A of **A** and arrow ε of **C** such that every similar pair $(B, f\colon UB \to X)$ factors uniquely through (A, ε), i.e. there exists a unique $g\colon B \to A$ in **A** such that the following triangle commutes in **C**

$$UA \xrightarrow{\ \varepsilon\ } X$$
$$\begin{array}{c} {\scriptstyle Ug}\uparrow \quad \nearrow \qquad\qquad \varepsilon.Ug = f. \qquad\qquad \text{(A.9)} \\ \quad\ f \\ UB \end{array}$$

A reader who is not familiar with these notions might begin by constructing the universal arrow from a set X to the forgetful functor **Ab** $\to$ **Set**, or from a group G to the inclusion functor **Ab** $\to$ **Gp**. Then, one can describe (co)products and (co)equalisers in a category **C** as universal arrows for suitable functors (Section A2.4 may be of help).

A2 Limits and colimits

A2.1 Main definition

The categorical notion of the *limit* of a functor contains, as particular cases, cartesian products, equalisers (Section A1.3), pullbacks and the classical projective limits.

Let **S** be a small category and $X\colon \mathbf{S} \to \mathbf{C}$ a functor, written in 'index notation' (for $i \in S = \mathrm{Ob}\mathbf{S}$ and $a\colon i \to j$ in **S**):

$$X\colon \mathbf{S} \to \mathbf{C}, \qquad i \mapsto X_i, \qquad a \mapsto (X_a\colon X_i \to X_j), \qquad \text{(A.10)}$$

as we think of X as a diagram of shape **S** in **C**.

A *cone* for X is an object A of **C** equipped with a family of maps $(f_i\colon A \to X_i)_{i \in S}$ in **C** such that the following triangles commute

$$A \xrightarrow{\ f_i\ } X_i$$
$$\begin{array}{c} \qquad\qquad \downarrow{\scriptstyle X_a} \qquad\qquad (a\colon i \to j \text{ in } \mathbf{S}). \qquad\qquad \text{(A.11)} \\ f_j\searrow \quad \\ X_j \end{array}$$

The *limit* of $X : \mathbf{S} \to \mathbf{C}$ is a *universal cone* $(L, (u_i : L \to X_i)_{i \in S})$. This means a cone of X such that every cone $(A, (f_i : A \to X_i)_{i \in S}$ factors uniquely through the former; in other words, there is a unique map $f : A \to L$ such that, for all $i \in S$, $u_i f = f_i$.

The solution need not exist. When it does, it is determined up to a unique coherent isomorphism, and the object L is denoted as $\mathrm{Lim}(X)$.

The definition of *colimit* is dual: a universal *cocone*.

A2.2 Particular cases

The product ΠX_i of a family $(X_i)_{i \in S}$ of objects of $\mathbf{C}$ is the limit of the corresponding functor $X : \mathbf{S} \to \mathbf{C}$, defined on the *discrete* category whose objects are the elements of the index set S (and whose morphisms only consist of the formal identities of such objects).

The equaliser in $\mathbf{C}$ of a pair of parallel morphisms $f, g : X_0 \to X_1$ is the limit of the obvious functor defined on the category $0 \rightrightarrows 1$.

The pullback of a pair of morphisms with the same codomain $f_i : X_i \to X_0$ $(i = 1, 2)$ is the limit of the obvious functor defined on the category $1 \to 0 \leftarrow 2$. This amounts to the usual definition: an object A equipped with two maps $u_i : A \to X_i$ which form a commutative square with f_1 and f_2, in a universal way:

$$\begin{array}{ccc} A & \xrightarrow{u_1} & X_1 \\ {\scriptstyle u_2}\downarrow & & \downarrow{\scriptstyle f_1} \\ X_2 & \xrightarrow[f_2]{} & X_0 \end{array} \qquad (\mathrm{A.12})$$

that is, $f_1 u_1 = f_2 u_2$, and for every triple (B, v_1, v_2) such that $f_1 v_1 = f_2 v_2$, there exists a unique map $w : B \to A$ such that $u_1 w = v_1$, $u_2 w = v_2$.

It is easy to show that A can be constructed as the equaliser of the two maps $f_i p_i : X_1 \times X_2 \to X_0$, when such limits exist in our category.

Sums and coequalisers are dual to products and equalisers. The colimit of a pair of morphisms $f_i : X_0 \to X_i$ (with the same domain) is called a pushout. A category with binary sums $X_1 + X_2$ and coequalisers has all pushouts.

A2.3 Complete categories and the preservation of limits

A category **C** is said to be *complete* (resp. *finitely complete*) if it has a limit for every functor **S** → **C** defined over a *small* category (resp. a *finite* category).

One says that a functor $F: \mathbf{C} \to \mathbf{D}$ *preserves the limit*

$$(L, (u_i: L \to X_i)_{i \in S})$$

of a functor $X: \mathbf{S} \to \mathbf{C}$ if the cone $(FL, (Fu_i: FL \to FX_i)_{i \in S})$ is the limit of the composed functor $FX: \mathbf{S} \to \mathbf{D}$. One says that F *preserves limits* if it preserves those limits *which exist* in **C**. Analogously for the preservation of products, equalisers, etc.

One proves, by a constructive argument, that a category is complete (resp. finitely complete) if and only if it has equalisers and products (resp. finite products). Moreover, if **C** is complete, a functor $F: \mathbf{C} \to \mathbf{D}$ preserves all limits (resp. all finite limits) if and only if it preserves equalisers and products (resp. finite products).

Dually, a category is said to be *cocomplete* if it has all colimits; and all colimits can be constructed from sums and coequalisers.

A2.4 Limits and colimits as universal arrows

Consider the category $\mathbf{C}^{\mathbf{S}}$ of functors **S** → **C** and their natural transformations (Section A1.7). The *diagonal functor*

$$D: \mathbf{C} \to \mathbf{C}^{\mathbf{S}}, \quad (DA)_i = A, \quad (DA)_a = \mathrm{id}A \quad (i \in S, \ a \text{ in } \mathbf{S}), \quad (A.13)$$

sends an object A to the constant functor at A, defined above, and a morphism $f: A \to B$ to the natural transformation $Df: DA \to DB: \mathbf{S} \to \mathbf{C}$ whose components are constant at f.

Then, the limit of a functor $X: \mathbf{S} \to \mathbf{C}$ in **C** is the same as a universal arrow $(L, \varepsilon: DL \to X)$ from the functor D to the object X of $\mathbf{C}^{\mathbf{S}}$.

Dually, the colimit of X in **C** is the same as a universal arrow $(L, \eta: X \to DL)$ from the object X of $\mathbf{C}^{\mathbf{S}}$ to the functor D.

A3 Adjoint functors

A3.1 Main definitions

An *adjunction* $F \dashv G$, making a functor $F: \mathbf{C} \to \mathbf{D}$ *left adjoint* to a functor $G: \mathbf{D} \to \mathbf{C}$, can be equivalently presented in four main forms.

(An elegant, concise proof of the equivalence can be seen in [M3]; again, one needs the axiom of choice.)

(i) We assign two functors $F\colon \mathbf{C} \to \mathbf{D}$ and $G\colon \mathbf{D} \to \mathbf{C}$ together with a family of bijections

$$\varphi_{XY}\colon \mathbf{D}(FX, Y) \to \mathbf{C}(X, GY) \qquad (X \text{ in } \mathbf{C}, Y \text{ in } \mathbf{D}),$$

which is natural in X, Y. More formally, the family (φ_{XY}) is an invertible natural transformation between functors in two variables

$$\varphi\colon \mathbf{D}(F(-), .) \to \mathbf{C}(-, G(.))\colon \mathbf{C}^{\mathrm{op}} \times \mathbf{D} \to \mathbf{Set}.$$

(ii) We assign a functor $G\colon \mathbf{D} \to \mathbf{C}$ and, for every object X in $\mathbf{C}$, a universal arrow from the object X to the functor G

$$(F_0 X, \eta X\colon X \to GF_0 X).$$

(ii*) We assign a functor $F\colon \mathbf{C} \to \mathbf{D}$ and, for every object Y in $\mathbf{D}$, a universal arrow from the functor F to the object Y

$$(G_0 Y, \varepsilon Y\colon FG_0 Y \to Y).$$

(iii) We assign two functors $F\colon \mathbf{C} \to \mathbf{D}$ and $G\colon \mathbf{D} \to \mathbf{C}$, together with two natural transformations

$$\eta\colon \mathrm{id}\mathbf{C} \to GF \qquad (\text{the } \textit{unit}), \qquad\qquad \varepsilon\colon FG \to \mathrm{id}\mathbf{D} \qquad (\text{the } \textit{counit}),$$

which satisfy the *triangular identities*: $\varepsilon F.F\eta = \mathrm{id}F$, $G\varepsilon.\eta G = \mathrm{id}G$

$$\begin{array}{ccc}
F \xrightarrow{\ F\eta\ } FGF & \qquad & G \xrightarrow{\ \eta G\ } GFG \\
\underset{\mathrm{id}F}{\searrow} \quad \downarrow{\scriptstyle \varepsilon F} & & \underset{\mathrm{id}G}{\searrow} \quad \downarrow{\scriptstyle G\varepsilon} \\
F & & G
\end{array} \qquad (\mathrm{A}.14)$$

The term 'unit' is motivated by the monad associated to the adjunction (Section A4.4).

A3.2 Remarks

The previous forms have different features. Form (i) is the classical definition of an adjunction, and is at the origin of the name (compare with adjoint maps of Hilbert spaces). Form (ii) is used when one starts from an 'easily defined' functor and wants to construct its *left* adjoint. Form (ii*) is dual to the previous one, and used in a dual way, to construct right adjoints. Form (iii) is adequate to the formal theory of adjunctions (and makes sense in an abstract 2-category).

Duality of categories interchanges left and right adjoint.

An adjoint equivalence (Section A1.5) amounts to an adjunction where the unit and counit are both invertible.

A3.3 Main properties of adjunctions

(a) *Uniqueness.* Given a functor, its left adjoint (if it exists) is uniquely determined up to isomorphism.

(b) *Composing adjoint functors.* Given two consecutive adjunctions

$$F\colon \mathbf{C} \rightleftarrows \mathbf{D} \colon G, \qquad \eta\colon 1 \to GF, \quad \varepsilon\colon FG \to 1,$$
$$H\colon \mathbf{D} \rightleftarrows \mathbf{E} \colon K, \qquad \rho\colon 1 \to KH, \quad \sigma\colon HK \to 1, \tag{A.15}$$

there is a composed adjunction from the first to the third category:

$$HF\colon \mathbf{C} \rightleftarrows \mathbf{E} \colon GK,$$
$$G\rho F.\eta\colon 1 \to GK.HF, \qquad \sigma.H\varepsilon K\colon HF.GK \to 1. \tag{A.16}$$

(c) *Adjoints and limits.* A left adjoint preserves (the existing) colimits, a right adjoint preserves (the existing) limits.

(d) *Faithful and full adjoints.* Suppose we have an adjunction $F \dashv G$, with counit $\varepsilon\colon FG \to 1$. Then

  (i) G is faithful if and only if all the components εY of the counit are epimorphisms;
 (ii) G is full if and only if all the components εY of the counit are split monomorphisms;
(iii) G is full and faithful if and only if the counit is invertible.

A3.4 Reflective and coreflective subcategories

A subcategory $\mathbf{C}' \subset \mathbf{C}$ is said to be *reflective* (notice: not 'reflexive') if the inclusion functor $U\colon \mathbf{C}' \to \mathbf{C}$ has a left adjoint, and *coreflective* if U has a right adjoint.

For instance, $\mathbf{Ab}$ is reflective in $\mathbf{Gp}$, while the full subcategory of $\mathbf{Ab}$ formed by torsion abelian groups is coreflective in $\mathbf{Ab}$.

A3.5 The adjoint functor theorem (P. Freyd)

Let $G\colon \mathbf{D} \to \mathbf{C}$ be a functor defined on a complete category. Then G has a left adjoint if and only if it preserves all limits and:

(Solution set condition) *for every X in $\mathbf{C}$ there exists a solution set, i.e. a set of objects $S(X)$ in $\mathbf{D}$ such that every morphism $f\colon X \to GY$ (with Y in $\mathbf{D}$) factors as*

$$X \xrightarrow{\ f_0\ } GY_0 \qquad\qquad Gg.f_0 = f, \tag{A.17}$$

with f and Gg mapping into GY

for some $Y_0 \in S(X)$, f_0 in $\mathbf{C}$ and g in $\mathbf{D}$.

A4 Monoidal categories, monads, additive categories

A4.1 Monoidal categories

A *monoidal category* $(\mathbf{C}, \otimes, E)$ is a category equipped with a *tensor product*, which is a functor in two variables

$$\mathbf{C} \times \mathbf{C} \to \mathbf{C}, \qquad (A, B) \mapsto A \otimes B. \tag{A.18}$$

Without entering into details, this operation is assumed to be associative up to a natural isomorphism $(A \otimes B) \otimes C \cong A \otimes (B \otimes C)$, and the object E is assumed to be an identity, up to natural isomorphisms $E \otimes A \cong A \cong A \otimes E$. All these isomorphisms form a 'coherent' system, which allows one to forget them and write $(A \otimes B) \otimes C = A \otimes (B \otimes C)$, $E \otimes A = A = A \otimes E$. See [M2, Ke1, EK, Ke2].

A *symmetric* monoidal category is further equipped with a symmetry isomorphism, coherent with the other ones:

$$s(X, Y)\colon X \otimes Y \to Y \otimes X. \tag{A.19}$$

The latter cannot be omitted: notice that $s(X, X)\colon X \otimes X \to X \otimes X$ is not the identity, generally.

A4.2 Exponentiable objects and internal homs

In a *symmetric* monoidal category $\mathbf{C}$, an object A is said to be *exponentiable* if the functor $- \otimes A\colon \mathbf{C} \to \mathbf{C}$ has a right adjoint, often written as $(-)^A\colon \mathbf{C} \to \mathbf{C}$ or $\mathrm{Hom}(A, -)$, and called an *internal hom*.

Since adjunctions compose, it follows easily that all its powers A^n are also exponentiable, with

$$\mathrm{Hom}(A^n, -) = (\mathrm{Hom}(A, -))^n. \tag{A.20}$$

A symmetric monoidal category is said to be *closed* if all its objects are exponentiable. The category **Ab** of abelian groups is symmetric monoidal closed, with respect to the usual tensor product and Hom functor.

In the non-symmetric case, one should consider a left and a right Hom functor, as it happens for cubical sets (see (1.156)).

A4.3 Cartesian closed categories

A category **C** with finite products has a symmetric monoidal structure given by the categorical product. This structure is called *cartesian*.

Then, **C** is said to be *cartesian closed* if all objects are exponentiable for this structure. **Set** is cartesian closed. **Cat** is cartesian closed, with the internal hom $\mathbf{Cat}(\mathbf{S}, \mathbf{C}) = \mathbf{C}^{\mathbf{S}}$ described in Section A1.7. Every category of presheaves of sets is cartesian closed.

Ab is not cartesian closed: for every abelian group $A \neq 0$, the product $- \times A$ does not preserves sums, and cannot have a right adjoint.

Top is not cartesian closed: for a fixed Hausdorff space A, the product $- \times A$ preserves quotients (if and) only if A is locally compact ([Mi], Theorem 2.1 and footnote (5)). But, as a crucial fact for homotopy, the standard interval **I** is exponentiable, with all its powers; more generally, each locally compact Hausdorff space is exponentiable (as recalled in Section 1.1.2).

A4.4 Monads and adjunctions

A *monad* in the category **C** is a triple (T, η, μ) where $T \colon \mathbf{C} \to \mathbf{C}$ is an endofunctor, $\eta \colon 1 \to T$ and $\mu \colon T^2 \to T$ are natural transformations (called the *unit* and *multiplication* of the monad), and the following diagrams commute:

$$
\begin{array}{ccc}
T \xrightarrow{\;\eta T\;} T^2 \xleftarrow{\;T\eta\;} T & \qquad & T^3 \xrightarrow{\;T\mu\;} T^2 \\
\searrow\ \downarrow{\scriptstyle\mu}\ \swarrow & & {\scriptstyle\mu T}\downarrow \qquad \downarrow{\scriptstyle\mu} \\
T & & T^2 \xrightarrow{\;\mu\;} T
\end{array}
\tag{A.21}
$$

It is easy to verify that an adjunction

$$F\colon \mathbf{C} \rightleftarrows \mathbf{A} \colon U, \qquad\qquad \eta\colon 1 \to UF, \qquad \varepsilon\colon FU \to 1 \qquad (A.22)$$

yields a monad (T, η, μ) on $\mathbf{C}$, where $T = UF\colon \mathbf{C} \to \mathbf{C}$, η is the unit of the adjunction and $\mu = U\varepsilon F\colon UF.UF \to UF$.

A4.5 Algebras for a monad

Given an arbitrary monad, as above, one defines the category $\mathbf{C}^T$ of *T-algebras*, or *Eilenberg–Moore algebras* for T.

An object is a pair $(X, a\colon TX \to X)$ consisting of an object X of $\mathbf{C}$ and a map a (the *algebraic structure*) satisfying two coherence axioms: the following diagrams commute

$$
\begin{array}{ccc}
X \xrightarrow{\ \eta X\ } TX & & T^2X \xrightarrow{\ Ta\ } TX \\
\searrow \ \downarrow a & & \mu X \downarrow \downarrow a \\
X & & TX \xrightarrow[a]{} X
\end{array}
\qquad (A.23)
$$

A morphism of T-algebras $f\colon (X, a) \to (Y, b)$ is a morphism $f\colon X \to Y$ of $\mathbf{C}$ which preserves the algebraic structures, in the sense that $fa = b.Tf$.

There is an adjunction

$$F^T \colon \mathbf{C} \rightleftarrows \mathbf{C}^T \colon U^T,$$
$$\eta^T = \eta\colon 1 \to U^T F^T, \qquad \varepsilon^T \colon F^T U^T \to 1, \qquad (A.24)$$

whose associated monad coincides with the given one.

A functor $U\colon \mathbf{A} \to \mathbf{C}$ is said to be *monadic*, or to make $\mathbf{A}$ *monadic over* $\mathbf{C}$, if it has a left adjoint $F\colon \mathbf{C} \to \mathbf{A}$ and moreover the following *comparison functor* from $\mathbf{A}$ to the category of algebras $\mathbf{C}^T$ of the monad associated to the adjunction

$$K\colon \mathbf{A} \to \mathbf{C}^T, \qquad K(A) = (UA, U\varepsilon A\colon UFUA \to UA), \qquad (A.25)$$

is an equivalence of categories. One also says that $\mathbf{A}$ is *algebraic* over $\mathbf{C}$ (via U).

For instance, the category $\mathbf{Ab}$ of abelian groups is algebraic over $\mathbf{Set}$ (via the usual forgetful functor); the same holds for all categories of 'equationally defined algebras'. Less obviously, the category of compact Hausdorff spaces is algebraic over $\mathbf{Set}$ [M3].

A4.6 Additive categories

Let us recall that a *preadditive category* **C** is a category *enriched on the symmetric monoidal category* **Ab**. Explicitly, this means that every hom-set $\mathbf{C}(X, Y)$ is equipped with a structure of abelian group and that composition is bilinear. The zero element of $\mathbf{C}(X, Y)$ is written $0_{XY} : X \to Y$.

We also need the more general notion of a *category enriched on abelian monoids* (e.g. for directed chain complexes), which is defined in a similar way.

Let **C** be enriched on abelian monoids. The following conditions on the object Z are equivalent:

(a) Z is terminal;
(b) Z is initial;
(c) $\mathbf{C}(X, X)$ is the null group;
(d) $\mathrm{id} Z = 0_{ZZ}$.

In this case Z is the zero object, often written as 0.

In the same situation, given two objects X_1, X_2, their *biproduct* $X = X_1 \oplus X_2$ comes with *injections* $u_i : X_i \to X$ and *projections* $p_i : X \to X_i$ satisfying the following equivalent properties:

(i) (X, p_1, p_2) is the product of X_1, X_2 and the injections have components $u_1 = (\mathrm{id} X_1, 0)$, $u_2 = (0, \mathrm{id} X_2)$;
(ii) (X, u_1, u_2) is the sum of X_1, X_2 and the projections have 'co-components' $p_1 = [\mathrm{id} X_1, 0]$, $p_2 = [0, \mathrm{id} X_2]$;
(iii) the following relations hold:

$$X_1 \quad \xrightarrow{u_1} \quad X_2 \qquad\qquad p_i u_i = \mathrm{id} X_i$$

$$X \qquad\qquad (i = 1, 2), \qquad\qquad (A.26)$$

$$X_1 \quad \xleftarrow{p_1} \quad X_2 \qquad\qquad u_1 p_1 + u_2 p_2 = \mathrm{id} X.$$

Therefore, in a category enriched on abelian monoids, the existence of binary products is equivalent to the existence of binary sums, which are called biproducts and written $X_1 \oplus X_2$.

An *additive category* is a preadditive category with finite biproducts. A preadditive category is finitely complete if and only if it is additive and has kernels.

A5 Two-dimensional categories and mates

We end this review with some less standard subjects of category theory, which are of interest here.

A5.1 Sesquicategories

Let us begin with the notion of an 'h-category' [G2], which was introduced by Kamps under the name of 'generalised homotopy system' [Km2]. An *h-category* $\mathbf{C}$ is a category equipped with:

(a) for each pair of parallel morphisms $f, g \colon X \to Y$, a set of *2-cells*, or *homotopies*, $\mathbf{C}_2(f, g)$ whose elements are written as $\varphi \colon f \to g \colon X \to Y$ (or $\varphi \colon f \to g$), so that each map f has a *trivial* (or *degenerate*, or *identity*) *endocell* $\mathrm{id}f \colon f \to f$;

(b) a *whisker composition*, or *reduced horizontal composition*, for homotopies and maps

$$X' \xrightarrow{\ h\ } X \xRightarrow[g]{f}{\downarrow\varphi} Y \xrightarrow{\ k\ } Y' \qquad (A.27)$$

$$k \circ \varphi \circ h \colon kfh \to kgh \colon X' \to Y',$$

also written as $k\varphi h$. These data must satisfy the following axioms:

$$
\begin{aligned}
k' \circ (k \circ \varphi \circ h) \circ h' &= (k'k) \circ \varphi \circ (hh') & &\textit{(associativity)}, \\
1_Y \circ \varphi \circ 1_X &= \varphi, \qquad k \circ \mathrm{id}f \circ h = \mathrm{id}(kfh) & &\textit{(identities)}.
\end{aligned}
\qquad (A.28)
$$

This structure can be viewed as a category enriched over the category **Gph** of (small) reflexive graphs, equipped with the symmetric monoidal closed structure described in Section 4.3.3.

A *sesquicategory* [St] is further equipped with a *concatenation*, or *vertical composition* of 2-cells $\psi.\varphi$, which is associative, has for identities the trivial 2-cells and is consistent with whisker composition:

$$X \xRightarrow[h]{f}{\substack{\downarrow\varphi \\ \downarrow\psi}} Y \qquad \psi.\varphi \colon f \to h \colon X \to Y, \qquad (A.29)$$

$$
\begin{aligned}
\chi.(\psi.\varphi) &= (\chi.\psi).\varphi, \qquad \varphi.\mathrm{id}f = \varphi = \mathrm{id}g.\varphi, \\
k \circ (\psi.\varphi) \circ h &= (k \circ \psi \circ h).(k \circ \varphi \circ h).
\end{aligned}
\qquad (A.30)
$$

A5.2 Two-categories

A *2-category* is a sesquicategory which satisfies the following *reduced interchange property*:

$$X \quad \overset{f}{\underset{g}{\xrightarrow{\downarrow\varphi}}} \quad Y \quad \overset{h}{\underset{k}{\xrightarrow{\downarrow\psi}}} \quad Z \qquad (\psi\circ g).(h\circ\varphi) = (k\circ\varphi).(\psi\circ f). \qquad (A.31)$$

To recover the usual definition [Be, KS], one defines the *horizontal composition* of 2-cells φ, ψ which are *horizontally consecutive*, as in diagram (A.31)

$$\psi\circ\varphi = (\psi\circ g).(h\circ\varphi) = (k\circ\varphi).(\psi\circ f) \colon hf \to kg \colon X \to Z. \qquad (A.32)$$

Then, one proves that the horizontal composition of 2-cells is associative, has identities (any identity 2-cell of an identity arrow) and satisfies the *middle-four interchange property* with vertical composition (an extension of the previous reduced interchange property):

$$X \quad \overset{\downarrow\varphi}{\underset{\downarrow\psi}{\xrightarrow{\hspace{1cm}}}} \quad Y \quad \overset{\downarrow\sigma}{\underset{\downarrow\tau}{\xrightarrow{\hspace{1cm}}}} \quad Z \qquad (\tau.\sigma)\circ(\psi.\varphi) = (\tau\circ\psi).(\sigma\circ\varphi). \qquad (A.33)$$

The prime example of such a structure is the 2-category **Cat** of small categories, functors and natural transformations.

Notice: the usual definition of a 2-category [Be, KS] is based on the complete horizontal composition, rather than on the reduced one. But practically one generally works with the reduced horizontal composition; and there are important cases of sesquicategories where the reduced interchange property does not hold (and one does not define a complete horizontal composition): for instance, the sesquicategory $Ch_{\bullet}D$ of chain complexes, chain morphisms and homotopies.

A5.3 Natural transformations and mates

(a) Let us have two adjunctions (between ordinary categories)

$$\begin{aligned} F \colon \mathbf{X} \rightleftarrows \mathbf{Y} \colon U, \qquad & \eta \colon 1 \to UF, \qquad \varepsilon \colon FU \to 1, \\ F' \colon \mathbf{X}' \rightleftarrows \mathbf{Y}' \colon U', \qquad & \eta' \colon 1 \to U'F', \qquad \varepsilon' \colon F'U' \to 1, \end{aligned} \qquad (A.34)$$

and two functors $H \colon \mathbf{X} \to \mathbf{X}'$, $K \colon \mathbf{Y} \to \mathbf{Y}'$. Then there is a bijection between sets of natural transformations:

$$\mathrm{Nat}(HU, U'K) \to \mathrm{Nat}(F'H, KF), \qquad \lambda \mapsto \mu, \qquad (A.35)$$

$$\mu = \varepsilon' KF.F' \lambda F.F' H\eta \colon F'H \to F'HUF \to F'U'KF \to KF,$$
$$\lambda = U'K\varepsilon.U'\mu U.\eta' HU \colon HU \to U'F'HU \to U'KFU \to U'K,$$

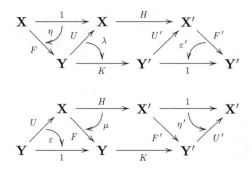

The natural transformations λ, μ are said to be *mates under the adjunctions* (A.34).

(More generally, as shown in [KS], 2.2, this holds true for *internal* adjunctions in a 2-category, and can be formalised as an isomorphism between two double categories.)

(b) In particular, if $\mathbf{X} = \mathbf{X}'$, $\mathbf{Y} = \mathbf{Y}'$, $H = \mathrm{id}\mathbf{X}$ and $K = \mathrm{id}\mathbf{Y}$, we have a bijection

$$\mathrm{Nat}(U, U') \to \mathrm{Nat}(F', F). \qquad (A.36)$$

(c) The case of interest for the *cylinder/cocylinder* adjunction of homotopy theory is even more particular: we have an adjunction $I \dashv P$ of endofunctors of the category $\mathbf{Y}$. Then, we have composed adjunctions $I^n \dashv P^n$ for their powers, and bijections

$$\mathrm{Nat}(I^n, I^k) \to \mathrm{Nat}(P^k, P^n) \qquad (n, k \geqslant 0). \qquad (A.37)$$

A5.4 Mates and limits

Under this correspondence of mates, colimits of natural transformations of left adjoints correspond to limits of natural transformations of right adjoints. The following instance is of interest here.

Let us assume we have four adjunctions (for $i = 0, ..., 3$)

$$F_i \colon \mathbf{X} \rightleftarrows \mathbf{Y} \colon U_i, \qquad \eta_i \colon 1 \to U_i F_i, \qquad \varepsilon_i \colon F_i U_i \to 1, \qquad (A.38)$$

with four natural transformations f, g, h, k and their mates f', h', g', h', as in the following diagrams

$$
\begin{array}{ccc}
F_0 X \xrightarrow{\ fX\ } F_1 X & \qquad & G_0 Y \xleftarrow{\ f'Y\ } G_1 Y \\
\end{array}
$$

$$ \tag{A.39} $$

Then, every left-hand square is a pushout in **Y** if and only if every right-hand square is a pullback in **X**.

Indeed, assuming that the left-hand squares are pushouts, take – for an arbitrary X in **X** – two maps $u_i : X \to G_i Y$ $(i = 1, 2)$ which commute with $f'Y$, $g'Y$. Then their 'adjoint' maps $v_i : F_i X \to Y$ commute with fX, gX, and yield a unique coherent morphism $v : F_3 X \to Y$. The adjoint map $u : X \to G_3 Y$ yields the unique morphism coherent with u_1, u_2.

Starting from a cylinder/cocylinder adjunction $I \dashv P$ of endofunctors of the category **A**, this result links the concatenation pushout J to the concatenation pullback Q: take $F_0 = \mathrm{id}\mathbf{A}$, $F_1 = F_2 = I$, $F_3 = J$ and $G_0 = \mathrm{id}\mathbf{A}$, $G_1 = G_2 = P$, $F_3 = Q$.

References

[AHS] Adámek, J., Herrlich, H. and Strecker G. (1990). *Abstract and Concrete Categories* (Wiley Interscience, New York).

[AT] H. Applegate, H. and Tierney, M. (1969). Categories with models, in *Seminar on Triples and Categorical Homology Theory*, Lecture Notes in Mathematics 80 (Springer, Berlin), pp. 156–244.

[AL] Aubry, M. and Lemaire, J.M. (1988). Homotopie d'algèbres de Lie et de leurs algèbres enveloppantes, in *Algebraic Topology – Rational Homotopy, Louvain-la-Neuve, 1986*, Lecture Notes in Math. Vol. 1318, pp. 26–30 (Springer-Verlag, Berlin).

[BBS] Bauer, A., Birkedal, L. and Scott, D.S. (2004). Equilogical spaces, *Theor. Comput. Sci.* **315**, 35–59.

[Ba] Baues, H.J. (1989). *Algebraic Homotopy* (Cambridge University Press, Cambridge).

[Be] Beck, J. (1969). Distributive laws, in *Seminar on triples and categorical homology theory*, Lecture Notes in Math. Vol. 80, pp. 119–140 (Springer-Verlag, Berlin).

[BBP] Bednarczyk, M.A., Borzyszkowski, A.M. and Pawlowski, W. (1999). Generalized congruences – epimorphisms in Cat, *Theory Appl. Categ.* **5**, No. 11, 266–280.

[Bl] Blackadar, B. (1986). *K-Theory for Operator Algebras* (Springer-Verlag, Berlin).

[Bo] Borceux, F. (1994). *Handbook of Categorical Algebra 1–3* (Cambridge University Press, Cambridge).

[BK] Borceux, F. and Korostenski, M. (1991). Open localizations, *J. Pure Appl. Algebra* **74**, 229–238.

[Bk] Bourbaki, N. (1961). *Topologie Générale, Ch. 10* (Hermann, Paris).

[Br] Brown, R. (1968). *Elements of Modern Topology* (McGraw-Hill, New York). Second edition: *Topology*, Ellis Horwood, Chichester 1988. Third edition: *Topology and Groupoids*, BookSurge, LLC, Charleston 2006.

[BH1] Brown, R. and Higgins, P.J. (1981). On the algebra of cubes, *J. Pure Appl. Algebra* **21**, 233–260.

[BH2] Brown, R. and Higgins, P.J. (1981). Colimit theorems for relative homotopy groups, *J. Pure Appl. Algebra* **22**, 11–41.

[BH3] Brown, R. and Higgins, P.J. (1987). Tensor products and homotopies for ω-groupoids and crossed complexes, *J. Pure Appl. Algebra* **47**, 1–33.

[BL] Brown, R. and Loday, J.L. (1987). Van Kampen theorems for diagrams of spaces, *Topology* **26**, 311–335.

[C1] Connes, A. (1980). C^*-algèbres et géométrie différentielle, *C.R. Acad. Sci. Paris* **A 290**, 599–604.

[C2] Connes, A. (1994). *Noncommutative Geometry* (Academic Press, San Diego CA).

[C3] Connes, A. (2000). A short survey of noncommutative geometry, *J. Math. Physics* **41**, 3832–3866.

[CP] Cordier, J.M. and Porter, T. (1996). Categorical aspects of equivariant homotopy, *Appl. Categ. Structures* **4**, 195–212.

[Cu] Curtis, E.B. (1971). Simplicial homotopy theory, *Adv. Math.* **6**, 107–209.

[Cr] Crans, S. (1995). Quillen closed model structures for sheaves, *J. Pure Appl. Algebra* **101**, 35–57.

[Dr] Dror Farjoun, E. (1987). Homotopy and homology of diagrams of spaces, in *Algebraic Topology, Proceedings, Seattle 1985*, Lecture Notes in Math. Vol. 1286, pp. 93–134 (Springer-Verlag, Berlin).

[Eh] Ehresmann, A.C. (2002). Localization of universal problems. Local colimits, *Appl. Categ. Structures* **10**, 157–172.

[EK] Eilenberg, S. and Kelly, G.M. (1966). Closed categories, in *Proc. Conf. Categorical Algebra, La Jolla 1965*, pp. 421–562 (Springer-Verlag, Berlin).

[FhR] Fahrenberg, U. and Raussen, M. (2007). Reparametrizations of continuous paths, *J. Homotopy Relat. Struct.* **2**, 93–117.

[FjR] Fajstrup, F. and Rosický, J. (2008). A convenient category for directed homotopy, *Theory Appl. Categ.* **21**, No. 1, 7–20.

[FGR1] Fajstrup, F., Goubault E. and Raussen M. (1998). Detecting deadlocks in concurrent systems, in *CONCUR'98*, Nice, Lecture Notes in Comput. Sci., Vol. 1466, pp. 332–347 (Springer-Verlag, Berlin).

[FGR2] Fajstrup, F., Goubault E. and Raussen M. (2006). Algebraic topology and concurrency, *Theor. Comput. Sci.* **357**, 241–178. (Revised version of a preprint at Aalborg, 1999.)

[FRGH] Fajstrup, F., Raussen, M., Goubault, E. and Haucourt, E. (2004). Components of the fundamental category, *Appl. Categ. Structures* **12**, 81–108.

[Ga1] Gaucher, P. (2000). Homotopy invariants of higher dimensional categories and concurrency in computer science, *Math. Struct. in Comp. Science* **10**, 481–524.

[Ga2] Gaucher, P. (2003). A model category for the homotopy theory of concurrency, *Homology Homotopy Appl.* **5**, no. 1, 549–599.

[GG] Gaucher, P. and Goubault, E. (2003). Topological deformation of higher dimensional automata, *Homology, Homotopy Appl.* **5**, 39–82.

[Ge1] Gersten, S.M. (1971). Homotopy theory of rings and algebraic K-theory, *Bull. Amer. Math. Soc.* **77**, 117–119.

[Ge2] Gersten, S.M. (1971). Homotopy theory of rings, *J. Algebra* **19**, 396–415.

[Go] Goubault, E. (2000). Geometry and concurrency: a user's guide, in *Geometry and Concurrency*, Math. Structures Comput. Sci. **10**, no. 4, 411–425.

[GH] Goubault, E. and Haucourt, E. (2007). Components of the fundamental category. II, *Appl. Categ. Structures* **15**, no. 4, 387–414.

[G1] Grandis, M. (1993). Cubical monads and their symmetries, in *Proceedings of the Eleventh International Conference on Topology, Trieste 1993*, Rend. Ist. Mat. Univ. Trieste, **25**, 223–262.

[G2] Grandis, M. (1994). Homotopical algebra in homotopical categories, *Appl. Categ. Structures* **2**, 351–406.

[G3] Grandis, M. (1996). Cubical homotopical algebra and cochain algebras, *Ann. Mat. Pura Appl.* **170**, 147–186.

[G4] Grandis, M. (1997). Categorically algebraic foundations for homotopical algebra, *Appl. Categ. Structures* **5**, 363–413.

[G5] Grandis, M. (1999). On the homotopy structure of strongly homotopy associative algebras, *J. Pure Appl. Algebra* **134**, 15–81.

[G6] Grandis, M. (2002). An intrinsic homotopy theory for simplicial complexes, with applications to image analysis, *Appl. Categ. Structures* **10**, 99–155.

[G7] Grandis, M. (2002). Directed homotopy theory, II. Homotopy constructs, *Theory Appl. Categ.* **10**, No. 14, 369–391.

[G8] Grandis, M. (2003). Directed homotopy theory, I. The fundamental category, *Cah. Topol. Géom. Différ. Catég.* **44**, 281–316.

[G9] Grandis, M. (2003). Ordinary and directed combinatorial homotopy, applied to image analysis and concurrency, *Homology Homotopy Appl.* **5**, No. 2, 211–231.

[G10] Grandis, M. (2004). Normed combinatorial homology and noncommutative tori, *Theory Appl. Categ.* **13**, No. 7, 114–128.

[G11] Grandis, M. (2004). Inequilogical spaces, directed homology and noncommutative geometry, *Homology Homotopy Appl.* **6**, 413–437.

[G12] Grandis, M. (2005). Directed combinatorial homology and noncommutative tori (The breaking of symmetries in algebraic topology), *Math. Proc. Cambridge Philos. Soc.* **138**, 233–262.

[G13] Grandis, M. (2005). Equilogical spaces, homology and noncommutative geometry, *Cah. Topol. Géom. Différ. Catég.* **46**, 53–80.

[G14] Grandis, M. (2005/06). The shape of a category up to directed homotopy, *Theory Appl. Categ.* **15**, No. 4, 95–146.

[G15] Grandis, M. (2006). Modelling fundamental 2-categories for directed homotopy, *Homology Homotopy Appl.* **8**, 31–70.

[G16] Grandis, M. (2006). Lax 2-categories and directed homotopy, *Cah. Topol. Géom. Différ. Catég.* **47**, 107–128.

[G17] Grandis, M. (2006). Absolute lax 2-categories, *Appl. Categ. Structures* **14**, 191–214.

[G18] Grandis, M. (2006). Quotient models of a category up to directed homotopy, *Theory Appl. Categ.* **16**, No. 26, 709–735.

[G19] Grandis, M. (2007). Categories, norms and weights, *J. Homotopy Relat. Struct.* **2**, No. 9, 171–186.

[G20] Grandis, M. (2007). The fundamental weighted category of a weighted space (From directed to weighted algebraic topology), *Homology Homotopy Appl.* **9**, 221–256.

[G21] Grandis, M. (2007). Directed algebraic topology, categories and higher categories, *Appl. Categ. Structures* **15**, 341–353.

[GP] Grandis, M. and Paré, R. (2004). Adjoints for double categories *Cah. Topol. Géom. Différ. Catég.* **45**, 193–240.

[GMc] Grandis, M. and MacDonald, J. (1999). Homotopy structures for algebras over a monad, *Appl. Categ. Structures* **7**, 227–260.

[GM] Grandis, M. and Mauri, L. (2003). Cubical sets and their site, *Theory Appl. Categ.* **11**, No. 8, 185–211.

[GT] Grandis, M. and Tholen, W. (2006). Natural weak factorisation systems, *Archivum Mathematicum (Brno)* **42**, 397–408.

[Gy1] Gray, J.W. (1965). Sheaves with values in a category, *Topology* **3**, 1–18.

[Gy2] Gray, J.W. (1974). *Formal Category Theory: Adjointness for 2-Categories*, Lecture Notes in Mathematics, Vol. 391 (Springer-Verlag, Berlin).

[HK1] Hardie, K.A. and Kamps, K.H. (1987). Homotopy over B and under A, *Cah. Topol. Géom. Différ. Catég.* **28**, 183–195.

[HK2] Hardie, K.A. and Kamps, K.H. (1989). Track homotopy over a fixed space, *Glasnik Mat.* **24**, 161–179.

[HK3] Hardie, K.A. and Kamps, K.H. (1992). Variations on a theme of Dold, *Can. Math. Soc. Conf. Proc.* **13**, 201–209.

[HKP] Hardie, K.A., Kamps, K.H. and Porter T. (1991). The coherent homotopy category over a fixed space is a category of fractions, *Topology Appl.* **40**, 265–274.

[HW] Hilton, P.J. and Wylie, S. (1962). *Homology Theory* (Cambridge University Press, Cambridge).

[J1] James I.M. (1971). Ex homotopy theory I, *Ill. J. of Math.* **15**, 329–345.

[J2] James I.M. (1995). Introduction to fibrewise homotopy, in *Handbook of Algebraic Topology*, pp. 169–194 (North-Holland, Amsterdam).

[JT] Janelidze, G. and Tholen, W. (1999). Functorial factorization, well-pointedness and separability, *J. Pure Appl. Algebra* **142**, 99–130.

[Jo] Johnstone, P.T. (1975). Adjoint lifting theorems for categories of algebras, *Bull. London Mat. Soc.* **7**, 294–297.

[KV] Karoubi, M. and Villamayor, O. (1971). K-théorie algébrique et K-théorie topologique, I, *Math. Scand.* **28**, 265–307.

[Km1] Kamps, K.H. (1968). Faserungen und Cofaserungen in Kategorien mit Homotopiesystem, Dissertation, Saarbrücken.

[Km2] Kamps, K.H. (1970). Über einige formale Eigenschaften von Faserungen und h-Faserungen, *Manuscripta Math.* **3**, 237–255.

[KP] Kamps, K.H. and Porter, T. (1997). *Abstract Homotopy and Simple Homotopy Theory* (World Scientific Publishing Co., River Edge NJ).

[Ka1] Kan, D.M. (1955). Abstract homotopy I, *Proc. Nat. Acad. Sci. U.S.A.* **41**, 1092–1096.

[Ka2] Kan, D.M. (1956). Abstract homotopy II, *Proc. Nat. Acad. Sci. U.S.A.* **42**, 255–258.

[Ke] Kelley, J.L. (1955). *General Topology* (Van Nostrand, Princeton).

[Ke1] Kelly, G.M. (1964). On Mac Lane's conditions for coherence of natural associativities, commutativities, etc. *J. Algebra* **1**, 397–402.

[Ke2] Kelly, G.M. (1982). *Basic Concepts of Enriched Category Theory* (Cambridge University Press, Cambridge).

[Ke3] Kelly, G.M. (1987). On the ordered set of reflective subcategories, *Bull. Austral. Math. Soc.* **36**, 137–152.

[KL] Kelly, G.M. and Lawvere, F.W. (1989). On the complete lattice of essential localizations, in *Actes du Colloque en l'Honneur du Soixantième Anniversaire de R. Lavendhomme, Bull. Soc. Math. Belg.* **A41**, 289–319.

[KS] Kelly, G.M. and R. Street, R. (1974). Review of the elements of 2-categories, in *Category Seminar, Sydney 1972/73* (Lecture Notes in Math. 420, Springer, Berlin), pp. 75–103.

[Ky] Kelly, J.C. (1963). Bitopological spaces, *Proc. London Math. Soc.* **13**, 71–89.

[Kr] Krishnan S. (to appear). A convenient category of locally preordred spaces, *Appl. Categ. Structures.* DOI 10.1007/s10485-008-9140-9

[LS] Lambek, J. and Scott, P.J. (1986). *Introduction to Higher Order Categorical Logic* (Cambridge University Press, Cambridge).

[Lw1] Lawvere, F.W. (1974). Metric spaces, generalized logic and closed categories, *Rend. Sem. Mat. Fis. Univ. Milano* **43**, 135–166.

[Lw2] Lawvere, F.W. (1996). Unity and identity of opposites in calculus and physics, in *The European Colloquium of Category Theory (Tours, 1994)*, Appl. Categ. Structures **4**, 167–174.

[M1] Mac Lane, S. (1963). *Homology* (Springer, Berlin).

[M2] Mac Lane, S. (1963). Natural associativity and commutativity, *Rice Univ. Studies* **49**, 28–46.

[M3] Mac Lane, S. (1971). *Categories for the Working Mathematician* (Springer, Berlin).

[MM] Mac Lane, S. and Moerdijk, I. (1992). *Sheaves in Geometry and Logic* (Springer, Berlin).

[MS] MacDonald, J. and Stone, A. (1989). The natural number bialgebra, *Cahiers Top. Géom. Diff.* **30**, 349–363.

[Ms] Massey, W. (1980). *Singular Homology Theory* (Springer, Berlin).

[My] May, J.P. (1967). *Simplicial Objects in Algebraic Topology* (Van Nostrand, Princeton).

[Mi] Michael, E. (1968). Local compactness and Cartesian products of quotient maps and k-spaces, *Ann. Inst. Fourier (Grenoble)* **18**, 281–286.

[Mt] Mitchell, B. (1971). *Theory of Categories* (Academic Press, New York.)

[MoS] Moerdijk, I. and Svensson, J.A. (1993). Algebraic classification of equivariant homotopy 2-types, I, *J. Pure Appl. Algebra* **89**, 187–216.

[Mn] Munkholm, H.J. (1978). DGA algebras as a Quillen model category – Relations to shm maps, *J. Pure Appl. Algebra* **13**, 221–232.

[Mu] Munkres, J.R. (1984). *Elements of Algebraic Topology* (Perseus, Cambridge MA).

[Mr] Murphy, G.J. (1990). *C*-Algebras and Operator Theory* (Academic Press, Boston MA).

[N1] Niefield, S. (1982). Cartesianness: topological spaces, uniform spaces, and affine schemes, *J. Pure Appl. Algebra* **23**, 147–167.

[N2] Niefield, S. (1986). Adjoints to tensor for graded algebras and coalgebras, *J. Pure Appl. Algebra* **4**, 155–161.

[P1] Patchkoria, A. (2000). Homology and cohomology monoids of presimplicial semimodules, *Bull. Georgian Acad. Sci.* **162**, 9–12.

[P2] Patchkoria, A. (2000). Chain complexes of cancellative semimodules, *Bull. Georgian Acad. Sci.* **162**, 206–208.

[PV] Pimsner, M. and Voiculescu, D. (1980). Imbedding the irrational rotation C*-algebra into an AF-algebra, *J. Operator Th.* **4**, 93–118.

[Pr] Pratt, V. (2000). Higher dimensional automata revisited, *Math. Struct. Comp. Science* **10**, 525–548.

[Pu] Puppe, D. (1958). Homotopiemengen und ihre induzierten Abbildungen, I *Math. Z.* **69**, 299-344.

[Qn] Quillen, D.G. (1967). *Homotopical Algebra*, Lecture Notes in Mathematics 43 (Springer, Berlin).

[Ra1] Raussen, M. (2003). State spaces and dipaths up to dihomotopy, *Homotopy Homology Appl.* **5**, 257–280.

[Ra2] Raussen, M. (2007). Invariants of directed spaces, *Appl. Categ. Structures* **15**, no. 4, 355–386.

[Ri1] Rieffel, M.A. (1981). C*-algebras associated with irrational rotations, *Pacific J. Math.* **93**, 415–429.

[Ri2] Rieffel, M.A. (1988). Projective modules over higher-dimensional noncommutative tori, *Canad. J. Math.* **40**, 257–338.

[Ro] Rosolini, G. (2000). Equilogical spaces and filter spaces, in *Categorical studies in Italy (Perugia, 1997)*. Rend. Circ. Mat. Palermo (2) Suppl. No. **64**, 157–175.

[Sc] Scott, D. (1996). A new category? Domains, spaces and equivalence relations, Unpublished manuscript. http://www.cs.cmu.edu/Groups/LTC/

[Se] Serre, J.P. (1951). Homologie singulière des espaces fibrés. Applications, *Ann. of Math.* **54**, 425–505.

[Sp] Specker, E. (1950). Additive Gruppen von Folgen ganzer Zahlen, *Portugaliae Math.* **9**, 131–140.

[Sf] Stasheff, J.D. (1963). Homotopy associativity of H-spaces. II, *Trans. Amer. Math. Soc.* **108**, 293–312.

[St] Street, R. (1996). Categorical structures, in *Handbook of Algebra*, Vol. 1, pp. 529–577 (North Holland, Amsterdam).

[Ta] Tapia, J. (1987). Sur la pente du feuilletage de Kronecker et la cohomologie étale de l'espace des feuilles, *C. R. Acad. Sci. Paris Sér. I Math.* **305**, 427–429.

[Wh] Whitehead, G.W. (1978). *Elements of Homotopy Theory* (Springer, Berlin).

Glossary of symbols

Index